3-

Success in Mathematics

Other Success Studybooks

Biology

Book-keeping for the Small Business

Chemistry

Commerce

Geography: Human and Regional

Geography: Physical and Mapwork

Economics

Investment

Nutrition

Principles of Accounting

Principles of Accounting: Answer Book

Twentieth Century World Affairs

Success in
MATHEMATICS

Walter Van Stigt, Ph.D.

Dean of Students and Principal Lecturer in Mathematics,
Whitelands College,
Roehampton Institute of Higher Education,
University of London.

CONSULTANT EDITOR

Nigel Warwick, B.Sc., F.I.M.A.

University of London School Examinations Department

John Murray

First published 1974
Reprinted 1976

Printed in Great Britain by
Fletcher & Son Ltd
Norwich

Cased 0 7195 2924 7
Paperback 0 7195 2923 9

Foreword

Many people go through school and beyond with only the flimsiest idea of what mathematics is all about. When confronted by a situation where mathematical calculation is necessary they will flounder through mechanical processes which they never properly understood in the first place. Weakness in maths is frequently the stumbling-block to a promising career, and it is a sad fact that countless people never manage to make up for a bad mathematical start.

There are many reasons for this and one of the most important is: lack of suitable material from which to work. Textbooks for class use rarely give detailed explanations, and more 'general' books, designed for adults, are often too superficial or too specialized. What is needed is a book containing *all* the basic material, presented in such a way that a reader can come to terms with it in his own way, in his own time, in whole or in part, and in whatever order he wishes. *Success in Mathematics* has been written for precisely that purpose.

Everyone needs mathematics, but each person needs a different kind of mathematics, depending on the job he does or the type of person he is. Yet there is a solid core of knowledge that must be acquired as a basis for future progress or specialization. We can say, roughly, that O-level mathematics covers this 'core', and any student who works towards O level or something similar will have a strong foundation from which to build and enlarge.

Success in Mathematics does two things which have never before been attempted in the same book. First, it offers a comprehensive course which can be used by anyone for self-instruction, revision or reference. For this reason it includes topics which are important in their everyday application to practical situations, and which contribute to a wider understanding of present-day mathematics (for example, *the slide rule*, *the language of sets and the use of Venn diagrams* and *linear programming*). Secondly, *Success in Mathematics* covers the main topics required by all British GCE examining boards in their 'traditional' O-level syllabuses, combining the modern approach with traditional material.

The book is divided into Units of study, each following a logical pattern of progression, and each fully cross-referenced so that you can easily look back, or ahead, to sections which relate to the topic under discussion. You should find this especially helpful if you are working only on certain parts of the course, or following it in a different order from that given.

The text is designed primarily to give you insight and understanding of basic mathematical concepts and procedures. Explanations are fully reinforced by worked examples and illustrations. Exercises are given in appropriate places throughout each Unit and, as they form an integral part of the process of learning, you will find greatest benefit by working them at the points where they occur. They are not examination questions so if you are working for an examination, get some of the past papers of your examining board, and study the topics covered and the kind of questions asked.

Mathematics is often called a language, a mode of expression and thought that can be used with increasing fluency and creativity once the grammar and vocabulary have been learned. What mathematics has to offer you, therefore, is something beyond the immediate practical concerns of examinations or professional life. It can offer one of the most exciting and rewarding of intellectual activities, a means of entering into a new dimension of thought and experience.

W. VS.

Acknowledgments

I have great pleasure in thanking all who, in many ways, have contributed to the production of this book. I am particularly grateful to Nigel Warwick for his invaluable advice and suggestions at all stages of planning and writing, and to Irene Slade who originated the 'Success Studybook' series and guided us, with encouragement and expertise, from the beginning. Thanks are also due to David Petchey, who criticized the text and made many useful suggestions, to Leslie Basford, who read and advised on the final version, and to my wife Judith who typed the manuscript.

Especially warm thanks must go to Dr Jean Macqueen for her generous and knowledgeable contribution in editing the book and preparing it so meticulously for press.

W. VS.

Contents

Part Three: Algebra and Graphs

Part One

Number Systems and Their Operations

Unit One

Number Systems

1.1 Introduction

The history of human civilization is the story of man's power to create new tools to enable him to survive the changing hazards of his environment. A leading chapter in this story is his creation of numbers of various kinds and complexity as his needs demanded.

Man alone has the ability to link things together in his mind, to group them and count them. In the Stone Age he did not need to invent large numbers; even today, the languages of some primitive tribes have no words for numbers beyond 2 or 3. On the other hand, the sophistication of life among the early Babylonians can be gauged from their language which, more than thirty centuries before Christ, had 59 different words for the first 59 counting numbers.

Number symbols, *numerals*, in their crudest forms of notches on sticks or rows of pebbles were used in even earlier cultures to record the number of possessions such as sheep and cattle, each notch representing one item. This principle of a direct one-to-one correspondence between a number of marks and items to be counted also applied in the oldest and simplest systems. Look, for example, at the symbols representing the first few counting numbers in Egyptian, Babylonian, Mayan and Roman writings (Fig. 1.1).

Numeration Systems

Modern	1	2	3	4	5	6	7	8	9	10	20	30	40	50	60
Babylonian	Y	YY	YYY	YYY Y	YYY YY	YYY YYY	YYY YYY Y	YYY YYY YY	YYY YYY YYY	<	<<	<<<	<<<<	<<<<<	Y
Egyptian hieroglyphic	|	||	|||	||||	||| ||	||| |||	|||| |||	|||| ||||	|||| |||||	∩	∩∩	∩∩∩	∩∩∩∩	∩∩∩∩∩	∩∩∩ ∩∩∩
Mayan	•	••	•••	••••	—	• —	•• —	••• —	•••• —	=	• ⊖	• =	•• ⊖	•• =	••• ⊖
Roman	I	II	III	IV	V	VI	VII	VIII	IX	X	XX	XXX	XL	L	LX
Arabic	١	٢	٣	٤	٥	٦	٧	٨	٩	١٠	٢٠	٣٠	٤٠	٥٠	٦٠

Fig. 1.1

The growing complexity of civilized society required increasingly larger numbers, which in turn required simplification for easy management and recognition. By placing dots and strokes in a definite pattern and grouping them, the number could more readily be recognized and larger numbers recorded. Such an arrangement is cumbersome for large numbers, and the possibility of replacing a certain number of dots by a single, different symbol was discovered (look at the Mayan and Roman symbols for 5, and the Egyptian and Babylonian symbols for 10). From this it was a natural step to count all numbers using this compound unit—the symbol which represents in itself a number of individuals—as the fundamental grouping, or *base*. In some culture groups the base was 5 or 10 (the number of fingers on one or two hands), but other numbers were also used for the purpose, including 2, 4, 20 and even 60.

A still greater advance in number notation was made when the same symbols were used to represent both units and compound units and the different values were indicated by the *position* of the symbol, a principle which had already been applied in the *abacus*, one of the early instruments man invented to help him as his calculations became increasingly more demanding.

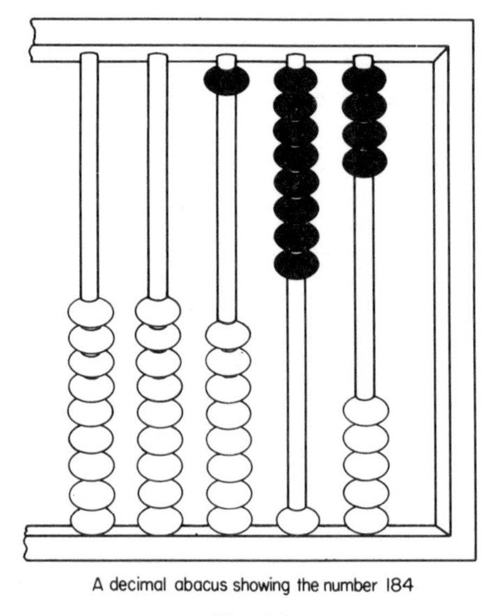

A decimal abacus showing the number 184

Fig. 1.2

The need to measure, the consequent development of geometry and the demands of advancing commerce and science prompted the creation of other kinds of numbers, different from the counting numbers discussed so far. They included *positive* and *negative* numbers, *rational* numbers ('fractions') and *real* numbers. These are the numbers we shall look at in detail later in this Unit. Human facility in handling this enlarged range of numbers evolved as the numbers themselves evolved. Even ancient civilizations had developed skills of multiplication and division from the elementary operations of addition and subtraction. Man's intelligence responded to the need for complicated calculations produced by the increased complexity of daily life by creating the necessary tools.

In the 1970s civilization has reached the point where almost all human problems are translated into questions that can be posed in numerical terms and solved by vast machines. These can calculate at speeds almost incomprehensible to the human mind. Without computers most modern achievements in architecture, engineering and science would be impossible. Indeed the well-being of large sections of the world population depends on them. They are the most recent illustration of the power of the human intellect to create tools as and when they are needed.

However, much of the treasure of man's mathematical knowledge remains unknown to most of us because of its size and complexity. Our remarkable adaptability makes us accept our present power through numbers, and the sophistication of our number systems, as a matter of course—just as we have come to accept man's journeys to the moon and back. The true nature of the various kinds of number, so clearly shown in the historical evolution of numbers, remains obscure to us; we identify them too easily with their written symbols, and perform the various arithmetical operations mechanically. Yet if you are to come to a real understanding of number, and use it efficiently, you must first understand the nature of the different number systems and the operations of arithmetic. This will now be explained and you should follow each concept carefully if you are to appreciate fully the reasoning we shall arrive at in later sections of the book. You probably use instinctively the ideas and processes we shall discuss, having been familiar with them from childhood. But don't be tempted to skip the step-by-step explanation that follows. Your success in the study of mathematics depends on it.

1.2 The Natural Numbers

The simplest numbers are the *natural numbers*. They are the counting numbers 1, 2, 3, 4, etc., which originally only related to *sets* of 'concrete' objects, such as 3 foxes, 4 sons, 2 daughters, and so on. It is an interesting fact that when numbers were first introduced into language, they were *adjectives* and always qualified a noun.

Later, it became possible to consider numbers without any relationship to concrete objects. The word 'three', for example, became a noun and represented the very abstract notion of 'that which the set of 3 apples has in common with the set of 3 men, the set of 3 houses, etc.'

Our present system of notation includes ten basic symbols: the Arabic numerals 1, 2, 3, 4, 5, 6, 7, 8, 9, a symbol 0 for zero and a highly ingenious system of *base* and *place value*. It is a *decimal* system, that is, the fundamental grouping, or the base, is ten (from the Latin word 'decem', ten). Counting in this system, which is called a system in the base ten, can be thought of as an arrangement of objects in rows of ten, squares of ten by ten, cubes of ten by ten by ten, etc.

A number of dots, for example:

can be rearranged into:

The number of dots could then be recorded as:

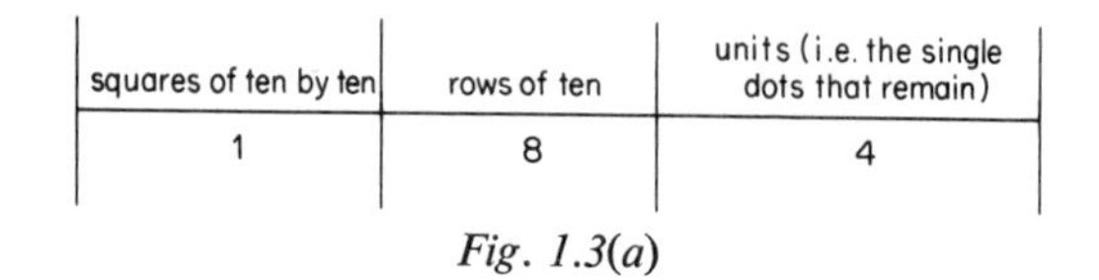

squares of ten by ten	rows of ten	units (i.e. the single dots that remain)
1	8	4

Fig. 1.3(a)

or as: 1 square of 10 (i.e. 10 rows of 10 dots each),
8 rows of 10,
4 single dots.

This is usually written as 1 8 4 (i.e. $100 + 80 + 4$). (Compare this with the illustration of the abacus, Fig. 1.2.)

The choice of 10 as the base is completely arbitrary; any other number can serve as a base. Indeed, modern computing science uses the system in the base 2, which is called the *binary system.* A number of dots, such as:

can be rearranged into:

This is then recorded as:

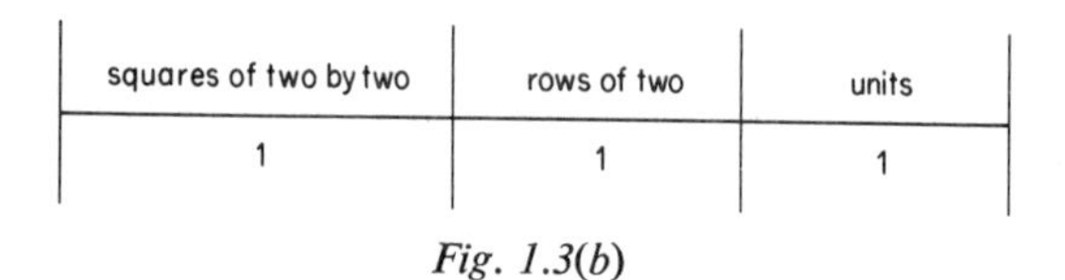

squares of two by two	rows of two	units
1	1	1

Fig. 1.3(b)

or as: 1 square of 2 by 2 (i.e. 2 rows of 2 dots each),
1 row of 2 dots,
1 single dot,

and written simply as 1 1 1 (equivalent to 7 in the decimal system, because $1 \times 4 + 1 \times 2 + 1 = 7$). This is sometimes written as $1\,1\,1_2$ to indicate that the base is 2.

1.3 Positive and Negative Numbers

You are already familiar with some everyday number systems in which there are two distinct and opposite kinds of values. For example, a bank account takes note of money paid in and money paid out; book-keeping balances the gains and losses made by a business; a contour map records heights both above and below sea-level; temperature measurements similarly are quoted in terms of degrees above or below an arbitrary zero. In each of these cases, there is a zero point which is common to both kinds of number: for example, when there is no money in the account and no money is owed to the bank; when there is neither a profit nor a loss and the business 'breaks even'; when height observations are made exactly at sea-level. The relations 'more' or 'less', 'smaller than' or 'greater than', can be represented on either side of the zero point.

In mathematics we speak of *positive* and *negative* values. A system including both positive and negative values can be represented by a scale divided into two parts by a zero point. Such a scale can be drawn vertically, for example, for measuring height or temperature, as in Fig. 1.4(*a*).

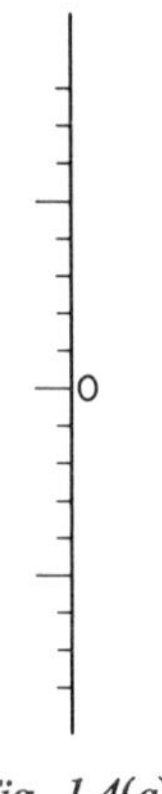

Fig. 1.4(a)

The more usual *mathematical* model of a number system including positive and negative values is a horizontal number line. One point on the line is chosen to represent zero (0); points on the line to the right of 0 represent positive numbers, and points on the line to the left of 0 represent negative numbers.

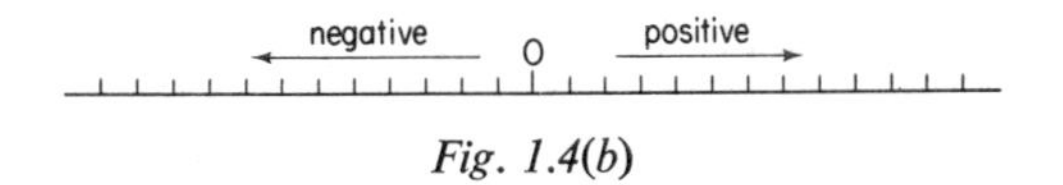

Fig. 1.4(b)

In this model the relations of 'smaller than' and 'greater than'—referred to in mathematics as *order relations*—are represented by *direction*. 'To the right of' represents the relation 'greater than' or 'more than'. 'To the left of' represents the relation 'smaller than' or 'less than'. Thus on the number line the number 3 lies to the left of 4 since 3 is less than 4, which in mathematical notation is written $3 < 4$. Similarly, 4 lies to the right of 3 since 4 is more than 3, or in mathematical notation $4 > 3$. Numbers of a system which includes both positive and negative values are therefore sometimes called *directed numbers*.

In banking practice, the distinction between positive and negative used to be made by printing the numbers in black ink for positive and red ink for negative. In mathematics, the sign + or − is attached to numbers, for example $^{+}3$ and $^{-}4$. Because of this sign, the positive and negative numbers are sometimes called *signed numbers*.

1.4 The System of the Integers

The natural numbers, when given both positive and negative values, constitute the system of the *integers* (from the Latin word *integer*, meaning 'whole'). The integers are the set of all positive and negative whole numbers together with zero.

A visual model of the system of the integers is the *number line*.

Any convenient arbitrary length is first chosen as a unit of measurement to represent the number 1. The number 2 is then represented by a *line segment* (see Section 14.4) of length 2 units, the number 3 by a line segment 3 units long, and so on.

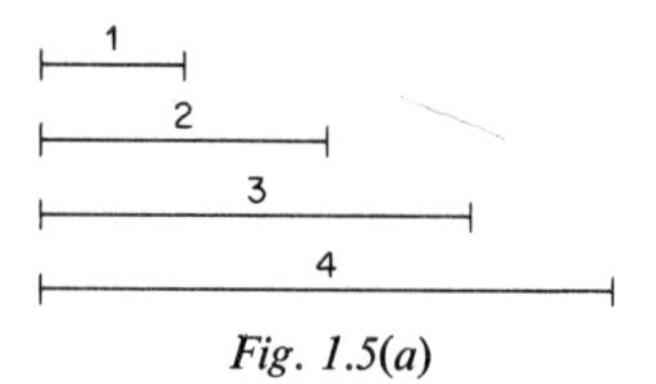

Fig. 1.5(a)

These segments, when measured to the right of a fixed point on a line—the zero point, 0—represent the *positive integers*. When measured to the left of the zero point, they represent the *negative integers*.

$^{-}4$ $^{-}3$ $^{-}2$ $^{-}1$ 0 $^{+}1$ $^{+}2$ $^{+}3$ $^{+}4$

Fig. 1.5(b)

If we associate each number with the end-points of a line segment, the model of the set of integers is not itself a line, but a set of points spaced at equal intervals in a linear arrangement.

$^{-}4$ $^{-}3$ $^{-}2$ $^{-}1$ 0 $^{+}1$ $^{+}2$ $^{+}3$ $^{+}4$

Fig. 1.5(c)

1.5 Ratios

When comparing two sets, such as a set of 8 apples and a set of 2 apples, we can express the relation between these two sets by saying 'There are 4 times as many apples in the first set as there are in the second' or 'There are 4 apples in the first set for every apple in the second'. Such a numerical relation, or *ratio*, between two sets is expressed in mathematics as 4:1 or 4/1. We can describe it as a four-to-one correspondence, every 4 apples in the first set being matched with 1 apple in the second. More precisely, the relation is 4:1 or 4/1, if we consider the set of 8 apples first; if we consider the set of 2 apples first, it is 1:4 or 1/4.

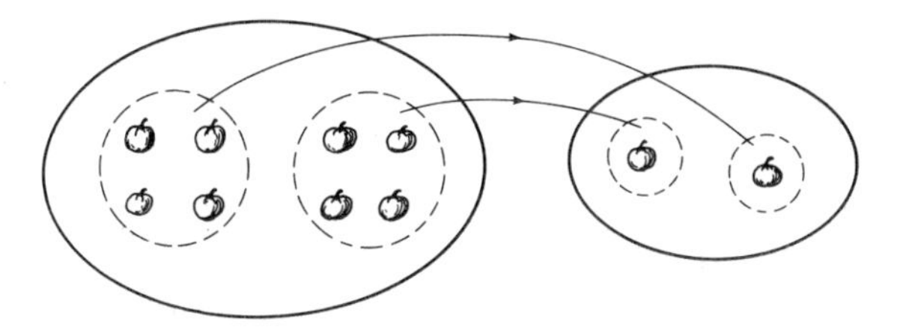

Fig. 1.6

1.6 Equivalent Ratios

A given numerical relation can be expressed equally validly in various ways; for example, the ratio between a set of 12 girls and a set of 18 boys can be expressed as

12:18, or
6:9 (to every 6 girls there are 9 boys), or
4:6 (to every 4 girls there are 6 boys), or
2:3 (to every 2 girls there are 3 boys).

This is possible because both the set of 12 girls and the set of 18 boys can be partitioned either into 2 equal groups, or *subsets*, of 6 girls and 9 boys respectively; or into 3 subsets of 4 girls and 6 boys respectively; or into 6 subsets of 2 girls and 3 boys respectively (Fig. 1.7).

The ratios 12:18, 6:9, 4:6 and 2:3 all express the numerical relation between the same two sets. They are *equivalent ratios*. The fourth ratio, 2:3, illustrated in the last diagram, is the *simplest form* of the ratios 12:18, 6:9 and 4:6.

1.7 Exercises

1. Find some ratios equivalent to the following:

   8:12, 12:36 and 3:4.

2. Simplify the ratios

   3/9, 14/21 and 6/15.

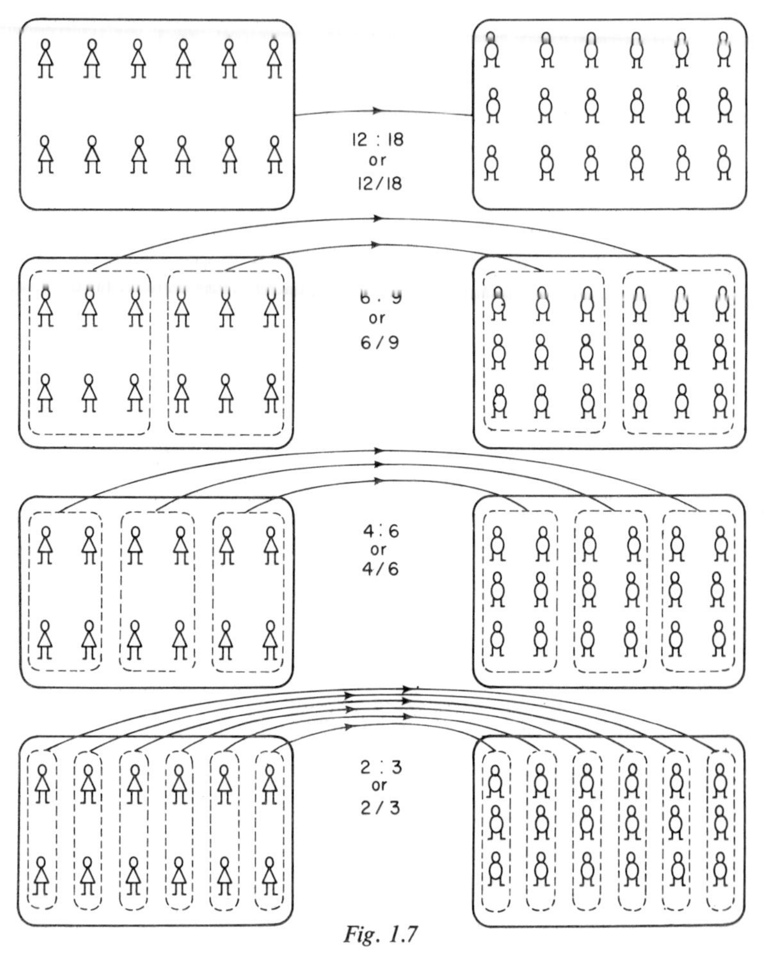

Fig. 1.7

3. Formulate a rule that can be used to simplify ratios. [Think about how you did Exercise 2, and try to write down instructions which would explain your method to someone who had not yet tackled that exercise.]

1.8 Rational Numbers

If we need to measure quantities such as length, area and volume, we choose a *unit* of measurement. Measuring such quantities is then simply finding and expressing the numerical relation, the ratio, between the size of the given quantity and the size of the chosen unit of measurement. Thus if we say that a line segment measures 17 centimetres long, we mean that there is a ratio of 17:1, or 17/1, between the length of the line segment and that of the centimetre.

The number line, which is based on length and on a unit of measurement, could therefore also be shown as:

Fig. 1.8(a)

Since units of measurements are quite arbitrarily chosen, and usually for reasons of convenience only, then clearly quantities exist which are smaller than the chosen unit. For example, there are obviously lengths smaller than one centimetre. Their relation to the unit can also be expressed by means of a ratio. If, for instance, the length of a line segment is such that the total length of 4 such segments is the same as the unit length, the ratio of the length of this segment to the unit length is 1:4 or 1/4, that is:

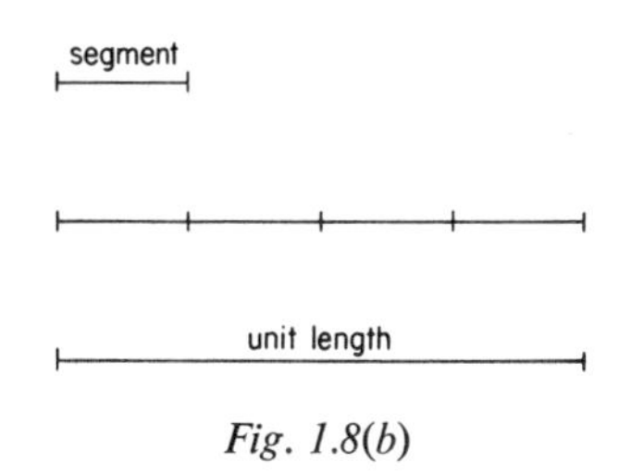

Fig. 1.8(b)

Here the length of the line segment is a kind of 'sub-unit' of measurement, i.e. the line segment, smaller than the unit, has been used to measure the unit length.

Ratios as numbers

Ratios can themselves be treated as numbers; they can be represented by points on the number line, and they have a definite order. For example the ratio, or the *rational number*, 1/4 is associated with the line segment whose length is in the ratio 1:4 to the unit length. Its corresponding point on the number line is found by measuring the length of this segment from the zero point.

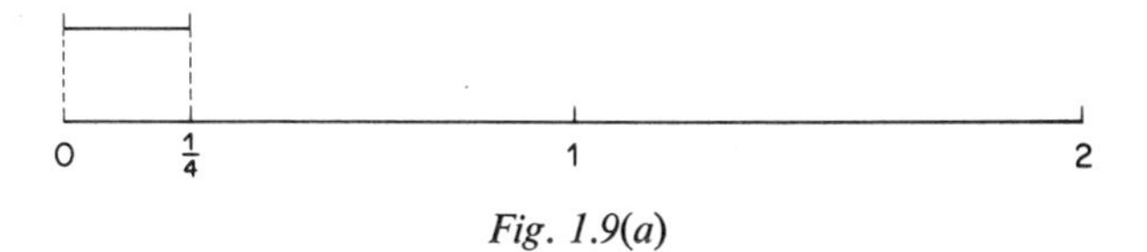

Fig. 1.9(a)

We can now *order* ratios, or rational numbers. That is, we can arrange them in order of size, since we can relate each one to an appropriate length; for example, $1/4 < 1/3$ because on the number line 1/4 lies to the left of 1/3.

The number line can be further graduated, if the length of a sub-unit is used as a unit of measurement. For example, we can use the line segment representing 1/4 as a unit of measurement like this:

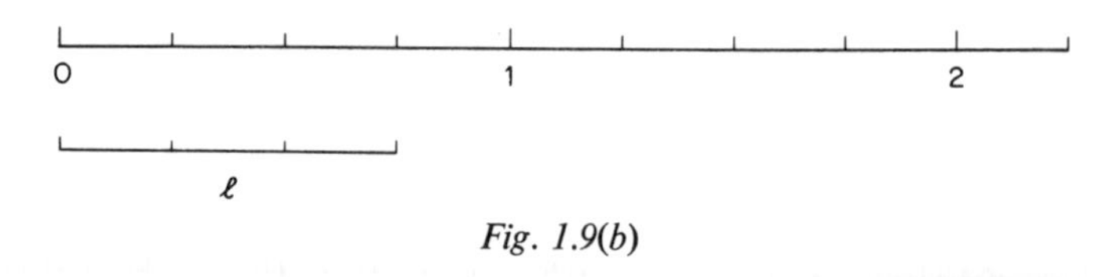

Fig. 1.9(b)

The line segment, *l*, has the ratio of 3:4 or 3/4, when compared with the unit length. The number line can then be shown as:

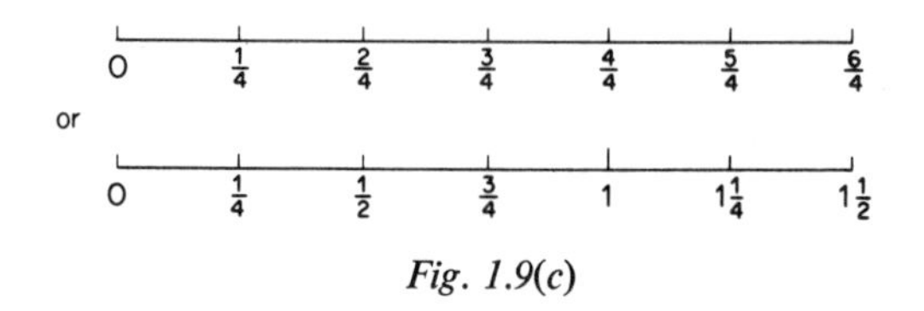

Fig. 1.9(c)

All ratios, or rational numbers, can be represented in this way by points on the number line.

1.9 Irrational Numbers

(You can leave this section until a second reading if you prefer.)

An interesting fact to consider here is that early Greek mathematicians discovered that there are lengths whose relation to a given unit length cannot be expressed exactly by any ratio. When they measured the lengths of the sides of a triangle like that in Fig. 1.10,

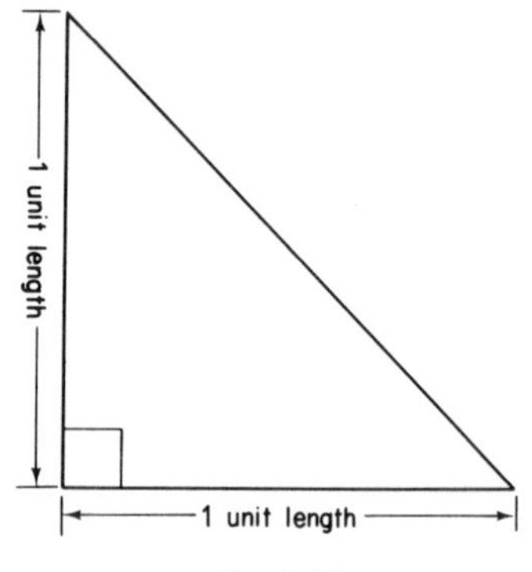

Fig. 1.10

they found that the length of the third side was a number of units which, when multiplied by itself (i.e. *squared*) is equal to 2. But no such rational number can be found. Indeed, besides the innumerable rational points on the number line, there are innumerable points associated with lengths which cannot be expressed in terms of any ratio (or rational number). These points are irrational points and the numbers they represent are called *irrational numbers*.

1.10 Exercises

1. Using an arbitrary unit length (not too small), draw a triangle as illustrated above. Use the same unit length for the construction of a number line. Using compasses, mark the length l of the third side of the triangle along the number line, measuring from the zero point.

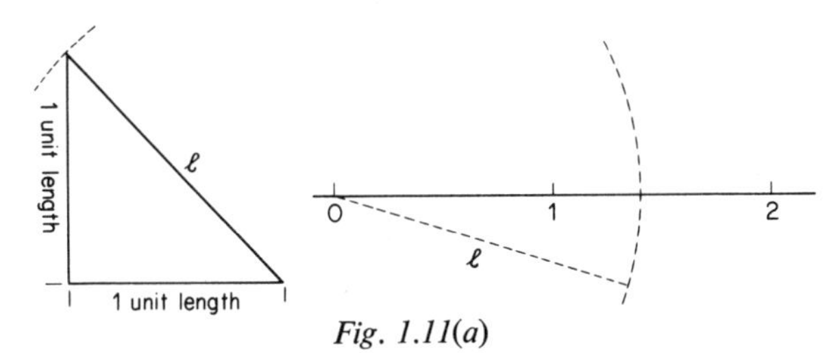

Fig. 1.11(a)

   The line segment l represents the number whose square is 2. (It can indeed be proved that this number, written $\sqrt{2}$, cannot be expressed exactly as a ratio.)

2. A famous irrational number has been given the name π, the Greek letter *pi*. It can be estimated approximately in the following way:

   Draw a circle, and use the diameter (see Section 17.6) of this circle (the longest line segment in the circle, that is, a line segment passing through the centre and with its end-points on the circle) as the unit length of a number line. Now measure the length of the circumference (see Section 17.6) of the circle with string or a paper strip, and mark this length along the number line from the zero point. The length of this line segment represents the number π which is seen to have a value a little greater than 3.

   Alternatively, you may find it easier and also more accurate to use a circular tin, as in Fig. 1.11(*b*), instead of a circle drawn with compasses.

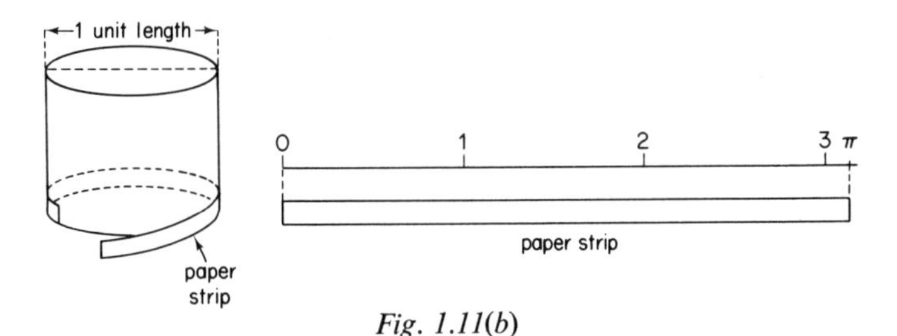

Fig. 1.11(b)

1.11 Real Numbers

The system of all numbers which can be represented by line segments or points on the number line is called *the system of real numbers*. It includes the natural numbers, the integers, the rational numbers and the irrational numbers, positive as well as negative.

Unit Two

Operations on Numbers: Addition and Subtraction of Integers

2.1 Addition and Subtraction of Natural Numbers

Addition, subtraction, multiplication and division are *operations* on numbers. In each case we start with a given number and the mathematical operation transforms this number: usually it is either increased or decreased. These arithmetical operations can all be represented by movements along the number line.

The operation of *adding a natural number* is represented on the number line by a *shift to the right*. For example, the operation of adding 2 to any chosen number, written $+ 2$, is represented by a shift of 2 unit lengths to the right of the starting-point.

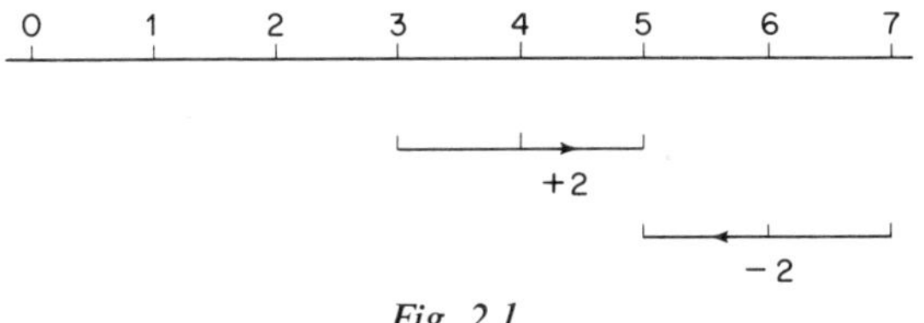

Fig. 2.1

Example

$3 + 2$ (in words, 2 added to 3):
start at 3 and move 2 unit lengths to the right; the resulting number is 5.

Similarly, the operation of *subtracting a natural number* is represented by a *shift to the left* in our model of the number line. For example, subtracting 2 from a chosen number, written $- 2$, is a shift of two unit lengths to the left.

Example

$7 - 2$ (in words, 2 subtracted from 7):
start at 7 and move 2 unit lengths to the left; the resulting number is 5.

2.2 Addition of Natural Numbers with the Simple Slide Rule

You can make a simple slide rule which is based on the number line, and which uses shifts along the line to represent arithmetical operations. It illustrates both addition and subtraction, and we can use it to perform calculations. (This simple slide rule is not the same as the slide rule which is commercially available. We shall encounter the commercial slide rule, and discover its usefulness as a tool in multiplication and division, in Unit 6 of this book.)

Cut two strips of paper or card, and mark the natural numbers at equal intervals along the strips. (If squared paper is used, the markings are already there.)

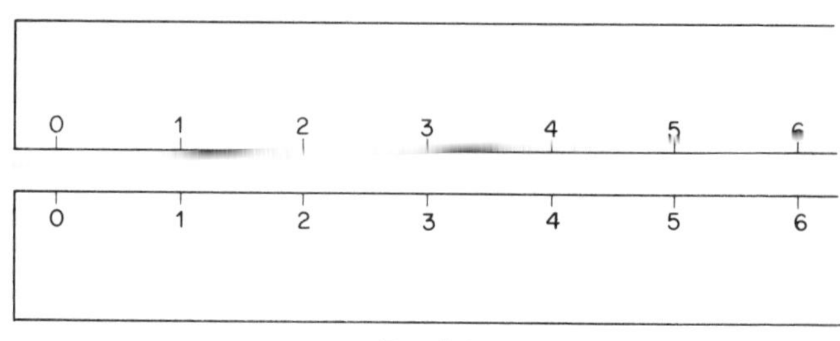

Fig. 2.2

Place the two strips beside each other, as in Fig. 2.2. Move the upper strip 2 unit lengths to the right. The zero point on the upper strip now faces the point 2 on the lower strip.

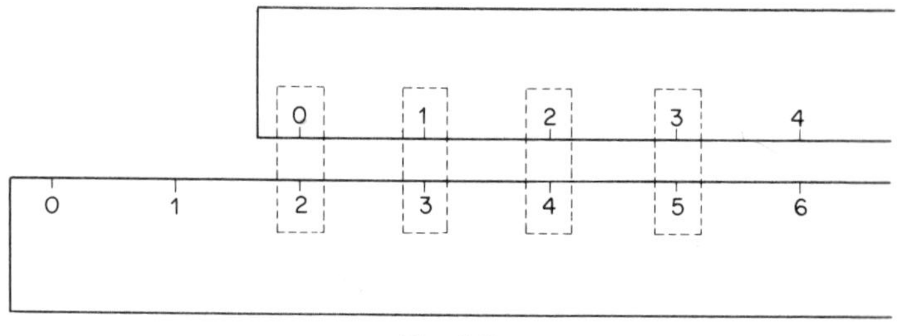

Fig. 2.3

With the strips placed like this, you can read off the result of adding 2 to any number represented on the number line, by observing the pairs of numbers which now lie opposite each other. It is clear that:

$$1 + 2 = 3,$$
$$2 + 2 = 4,$$
$$3 + 2 = 5.$$

2.3 Exercises

1. Use your home-made slide rule to practise some simple additions.
2. Using the slide rule, calculate

$$3 + 2 \quad \text{and}$$
$$2 + 3.$$

Similarly, calculate

$$1 + 4 \quad \text{and}$$
$$4 + 1.$$

What do you notice?

You are demonstrating the *commutative law* for addition, that is, the result of addition is not affected by interchanging, or changing the order of, the numbers to be added.

3. Similarly, first calculate and then comment on:

$$(3 + 1) + 2 \text{ [first add 1 to 3, and then add 2 to the result],}$$
$$\text{and } 3 + (1 + 2) \text{ [first add 2 to 1, and then add the result to 3].}$$

You are demonstrating the *associative law* for addition, that is, the result of an addition involving 3 numbers is not affected by the order of the operations.

Check that these laws still hold for additions involving more than 3 numbers.

2.4 Addition of Signed Numbers

Within a system that includes both negative and positive values, the positive whole numbers (positive integers) are identified with the natural numbers. *Adding a positive whole number* to any number, negative or positive, is therefore again represented by a shift to the right.

For example, the addition of $^+2$, written $+ \ ^+2$, is a shift of 2 unit lengths to the right. (Look at Fig. 2.3 again.) This relation holds equally well if the starting-point is a negative number; for example,

$$^-1 + {}^+2 = {}^+1,$$

as shown in Fig. 2.4.

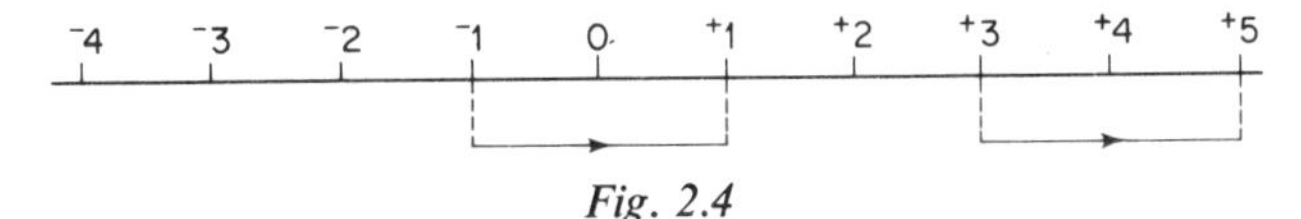

Fig. 2.4

This at once becomes clear if we extend the simple slide rule to include the negative integers; we can then use it for a wider range of calculations. As before, we add $^+2$ by moving the upper strip 2 unit lengths to the right, so that the zero point of the upper strip faces the number $^+2$ on the lower strip. Again, you can read off the result of the addition of $^+2$ directly (Fig. 2.5).

For example,

$$^-3 + {}^+2 = {}^-1,$$
$$^-2 + {}^+2 = 0.$$

As an exercise, practise some additions on your simple slide rule, adding positive numbers to negative numbers.

If we need examples to clarify the ideas involved in addition (and subtraction) of positive and negative numbers, we must look for them in areas in everyday life where positive and negative values exist. Compare the addition

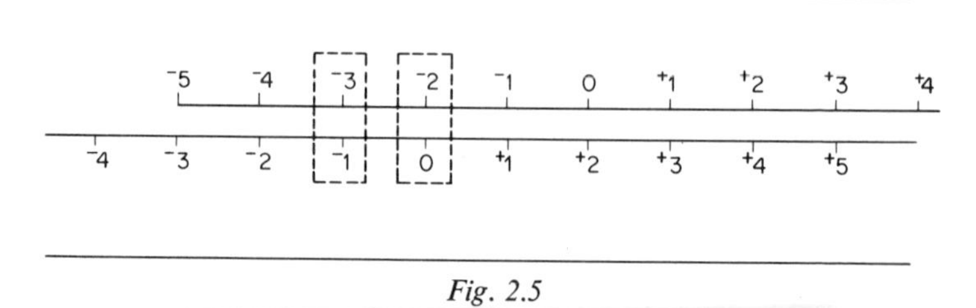

Fig. 2.5

of $^{+}5$ with the effect of a payment of £5 into a bank account. If there is already £60 in the account the payment results in an increased balance of £65. But if the account is overdrawn by £30, this same payment brings the balance to £25 'in the red'.

We can write this in mathematical language:

$$^{+}60 + {}^{+}5 = {}^{+}65,$$

and

$$^{-}30 + {}^{+}5 = {}^{-}25.$$

2.5 Adding a Negative Number

'Addition' in everyday language normally implies an increase.

In an appropriate everyday situation, the addition of a negative number brings about an actual *decrease*. In accounting, for example, adding a loss of £15 will *reduce* the total balance of the account by that amount.

We can examine the effect of adding a negative number by studying a series of additions, in which we gradually reduce the number to be added, first using $^{+}4$, then $^{+}3$, $^{+}2$, and so forth. Each time the number to be added is one less, and the result is also one less.

For example,

$$^{+}2 + {}^{+}4 = {}^{+}6;$$
$$^{+}2 + {}^{+}3 = {}^{+}5;$$
$$^{+}2 + {}^{+}2 = {}^{+}4;$$
$$^{+}2 + {}^{+}1 = {}^{+}3.$$

This process can be continued past the zero point if we remember that $^{-}1$ is one less than zero, $^{-}2$ one less than $^{-}1$, and so on; so that

$$^{+}2 + 0 = {}^{+}2;$$
$$^{+}2 + {}^{-}1 = {}^{+}1;$$
$$^{+}2 + {}^{-}2 = 0;$$
$$^{+}2 + {}^{-}3 = {}^{-}1, \text{etc.}$$

Use your simple slide rule to perform these operations; you will notice that at each calculation the upper strip is moved one unit length to the *left*. Continuing this process past the zero point (when the zeros are opposite each other for 'addition of 0'), it is clear that we can add $^{-}1$ by moving the upper strip one unit length to the *left*.

Addition of a negative number is therefore represented on the number line by a *shift to the left*.

We can add negative numbers using the simple slide rule by the same procedure that we use to add positive numbers. First, put the zero point of the upper strip opposite the negative number to be added on the lower strip. To add $^{-}3$ (written $+\,^{-}3$), for instance, place the zero point of the upper strip opposite the point $^{-}3$ on the lower strip:

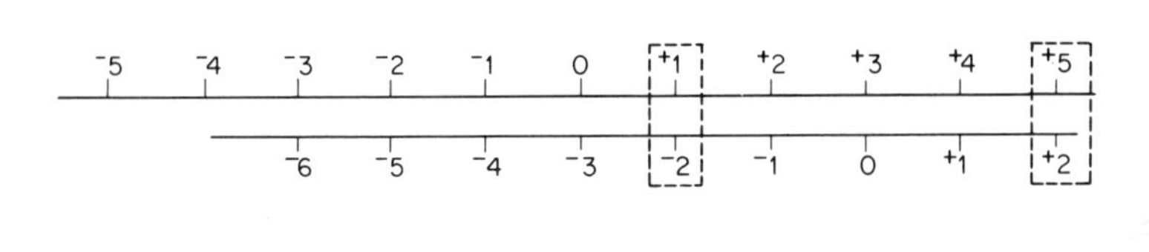

Fig. 2.6

For example,

$$^{+}5 + {}^{-}3 = {}^{+}2;$$
$$^{+}1 + {}^{-}3 = {}^{-}2.$$

2.6 Exercises

1. Using the simple slide rule, calculate

$$^{+}3 + {}^{-}2;$$
$$^{+}4 + {}^{+}4;$$
$$^{-}1 + {}^{-}5.$$

2. Make up your own examples of additions of negative numbers to both positive and negative numbers.

2.7 Subtraction

Addition is always intuitively seen as a form of joining one number on to another, but the mathematical operation of subtraction reflects a wider range of activities, and can be thought of in the following various ways:

Finding the difference. We can think of the numerical relation between two sets, or the relation between two numbers, in terms of their *difference*. For example, the difference between a set of 5 apples and a set of 3 apples is 2 apples; the age difference between a man of 42 and a boy of 12 is 30 years.

On the number line, the difference between two numbers is represented by the distance between the points which represent them. For example, the difference between 3 and 7 is 4; the difference between $^{+}2$ and $^{-}6$ is 8. The difference is understood better if we also know which number of the two is the larger. Thus we know that 7 is 4 more than 3 (or 4 is 3 less than 7); $^{+}2$ is 8 more than $^{-}6$, since $^{+}2$ lies 8 unit lengths to the right of $^{-}6$.

Taking away. Subtraction, in particular subtraction of natural numbers, is usually considered as a process of 'taking away', that is, removing a number of elements from a set of things. This *subset* is separated and removed, and we then consider the size of the remaining subset.

Inverse addition. While the operation of adding a natural number to a number increases that number and is therefore represented by a shift to the right on the number line, the operation of subtracting a natural number from a number decreases the number and is represented by a shift to the left on the number line. This reversal of direction means that subtraction can be thought of as the opposite of, or 'inverse', addition, and as such we can extend its use to both positive and negative numbers.

In any of these aspects, subtraction as a mathematical operation is represented on the number line by a shift in the direction opposite to that representing addition.

2.8 Subtraction of Signed Numbers

We have already seen (see Section 2.4) that, in the system consisting of the positive and negative integers, the positive integers are identified with the natural numbers. Subtraction of positive numbers is therefore represented by a shift to the left on the extended number line.

On the simple slide rule, we add $^{+}5$ by moving the upper strip 5 unit lengths to the *right*. We subtract $^{+}5$, written $- {}^{+}5$, by moving the upper strip 5 unit lengths to the *left*, that is, starting with both zeros together, we move the upper strip to the left so that its 0 is opposite the $^{-}5$ of the lower scale.

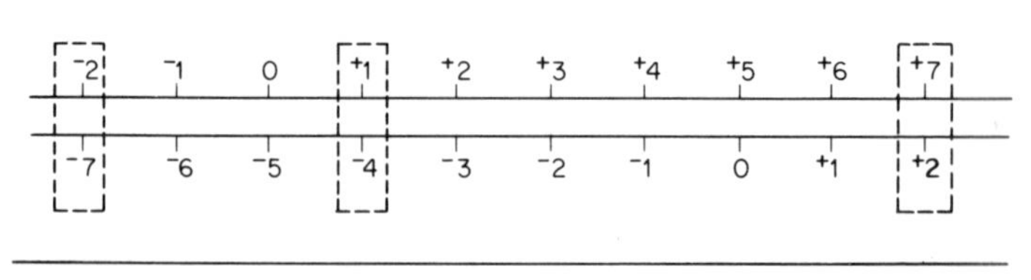

Fig. 2.7

Now we can read off the result of the operation $- {}^{+}5$ directly along the lower strip; for example,

$$^{+}7 - {}^{+}5 = {}^{+}2;$$
$$^{+}1 - {}^{+}5 = {}^{-}4;$$
$$^{-}2 - {}^{+}5 = {}^{-}7.$$

2.9 Exercises

1. Practise subtraction of some positive numbers, using your slide rule.
2. Compare the movements of the slide rule when calculating $^{+}6 + {}^{-}2$ and $^{+}6 - {}^{+}2$, and other similar examples. What do you conclude about the effect of $+ {}^{-}2$ and $- {}^{+}2$? You will see that the effect is the same in both cases.

Moving the upper strip of the slide rule to the right has the same effect as moving the lower strip to the left; the relative positions of the two strips are the same when the movement is completed. We made it a rule to move the upper strip to the right when adding natural numbers; but as the two parts of the slide rule are identical, it would have been equally correct to move the lower strip to the left, the results then being recorded on the upper strip. We can also add and subtract by the same movement of the slide rule, reading off the results on either the upper or the lower strip. If we wish to examine, for example, the effect of both $+ {}^{+}3$ and $- {}^{+}3$ on a number, we place the zero point of the upper strip opposite the point $^{+}3$ on the lower strip. Then $^{+}5 + {}^{+}3$ is read off on the lower strip in the usual way, i.e. the number $^{+}8$ which is opposite the number $^{+}5$ on the upper strip.

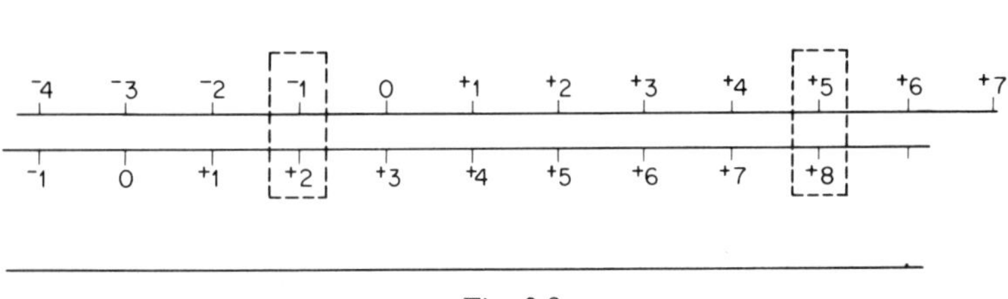

Fig. 2.8

Without moving the slide rule, we can read off the effect of $- {}^{+}3$ on the upper strip: $^{+}2 - {}^{+}3$ is found by reading off the number on the upper strip which is opposite the number $^{+}2$ on the lower strip, that is, $^{-}1$. However, the simple slide rule (like the commercial slide rule which you will meet later) is a *tool*; to use it efficiently you should develop a *fixed procedure*, practise it often and use it whenever you can.

2.10 Subtracting a Negative Number

Now let us find out the effect of *subtracting a negative number*. If we think of subtraction as 'inverse addition', that is, a shift in the direction opposite to that for addition, subtracting a negative number is clearly represented by a shift to the *right* on the number line. Thus, if the effect of $+ {}^{-}3$ was found by a shift of 3 unit lengths to the left, then the effect of $- {}^{-}3$ must be found by a shift of 3 unit lengths to the right.

2.11 Exercises

1. Follow your general procedure for subtraction using the simple slide rule, and calculate the effect of $- {}^-2$; for example, calculate ${}^+4 - {}^-2$, ${}^-3 - {}^-2$ and other examples of your own. (Use the procedure described in Section 2.5: subtract a number by placing that number on the upper strip opposite the zero point of the lower strip, and read off the answers on the lower strip.)
2. Now complete the following table, which shows the appropriate direction of the shifts for addition and subtraction of the signed number 2:

$+ {}^+2$: to the right	$- {}^+2$: to the left
$+ {}^-2$: to the left	$- {}^-2$: ?

2.12 Subtraction as 'Finding the Difference'

We can think of the subtraction of a negative number in terms of the *difference* aspect of subtraction. A subtraction which involves the numbers ${}^+3$ and ${}^-4$, for example, expresses the difference between these two numbers on the number line (Fig. 2.9).

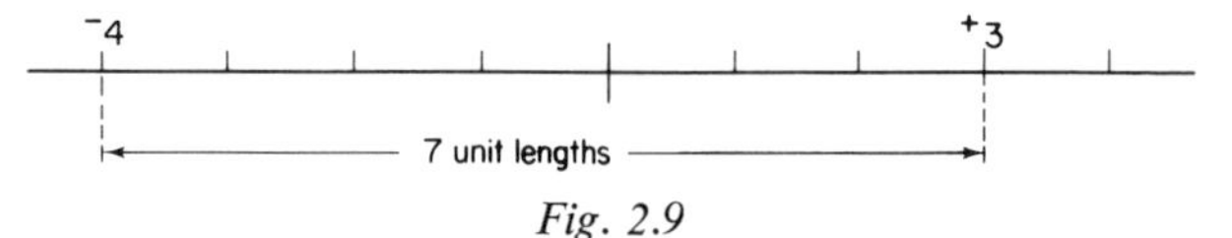

Fig. 2.9

This number can be either ${}^+7$ or ${}^-7$, depending on whether the subtraction produces a 'surplus' or a 'shortage'. There is a surplus whenever we subtract a smaller number from a larger number (a smaller number lies to the left of a larger number on the number line), and a shortage when we subtract a larger number from a smaller number.

Therefore, $${}^+3 - {}^-4 = {}^+7,$$

and $${}^-4 - {}^+3 = {}^-7.$$

The notion of subtraction as *taking away* can only be applied to negative numbers if we think of situations in everyday life where negative values occur, and where a meaning can be given to 'taking away', for example, the effect of the 'removal' of a loss on the overall balance of a firm's accounts. Can you think of another example?

2.13 Exercises

1. Use the simple slide rule and the difference method to calculate

$${}^+4 - {}^-2;$$
$${}^-3 - {}^-3;$$
$${}^-6 - {}^-2;$$
$${}^-3 - {}^-5.$$

2. Make up some more examples for further practice.

2.14 Some Practical Hints

(*a*) When no confusion can arise, the positive number sign may be omitted altogether; for instance, $^{+}3$ is simply written as 3, and $+\,^{+}3$ as $+3$.

(*b*) Addition of a negative number and subtraction of a positive number are both represented by a shift on the number line to the left; for example, the effect of $+\,^{-}2$ and $-\,^{+}2$ is the same. They can therefore be interchanged, and are usually written simply as -2.

(*c*) Sometimes $^{-}2$ is written as (-2) and $^{+}2$ as $(+2)$. In such cases, both $+(-2)$ and $-(+2)$ can be written as -2.

(*d*) Subtracting a negative number was found to be equivalent to adding the corresponding positive value. Therefore, for example, $-\,^{-}4$ can be written as $+\,^{+}4$ or simply as $+4$; and, as above, $-(-4)$ and $+(+4)$ can both be written as $+4$.

(*e*) The two preceding paragraphs can be summarized as:

two like signs are equivalent to $+$ *and two unlike signs are equivalent to* $-$

(*f*) When we need to simplify a long expression involving addition and subtraction of positive and negative numbers, it is often easiest to combine the positive numbers and the negative numbers first, just as an accountant totals profits and losses before calculating the final balance. This can be done in various ways.

Example

Calculate

$$^{+}7 - {}^{+}2 + {}^{-}3 + {}^{+}6 - {}^{-}3 + {}^{-}6.$$

Replace the subtractions by the equivalent additions:

$$^{+}7 + {}^{-}2 + {}^{-}3 + {}^{+}6 + {}^{+}3 + {}^{-}6.$$

Combine positive and negative numbers and simplify:

$$^{+}16 + {}^{-}11 = {}^{+}5.$$

Alternatively, the original expression

$$^{+}7 - {}^{+}2 + {}^{-}3 + {}^{+}6 + {}^{+}3 + {}^{-}6$$

can first be simplified to

$$7 - 2 - 3 + 6 + 3 - 6;$$

and then, combining the numbers preceded by the addition sign and the numbers preceded by the subtraction sign,

$$\boxed{7 + 6 + 3} - \boxed{2 + 3 + 6} = 16 - 11 = 5.$$

2.15 Exercise

Make up some long expressions involving addition and subtraction of positive and negative numbers, then simplify and calculate the results.

Checking your answer

Subtraction is the inverse operation of addition, and vice versa; these facts provide us with a simple means of checking the results of our calculations. For example, if the calculation $^{+}3 + {}^{-}2 = {}^{+}1$ is correct, then $^{+}1 - {}^{-}2$ should be $^{+}3$. You may find this easier to grasp if you say it in words: 'If $^{-}2$ added to $^{+}3$ gives $^{+}1$, then $^{-}2$ taken from $^{+}1$ should give $^{+}3$'. Similarly, if the subtraction $^{+}3 - {}^{-}2 = {}^{+}5$ is correct, then $^{+}5 + {}^{-}2 = {}^{+}3$.

Notation. The phrase 'if . . . is correct, then . . .' is often abbreviated in mathematics, and expressed by the symbol $\Rightarrow$. The examples in the last paragraph could then be written more concisely:

$$^{+}3 + {}^{-}2 = {}^{+}1 \Rightarrow {}^{+}1 - {}^{-}2 = {}^{+}3,$$

and

$$^{+}3 - {}^{-}2 = {}^{+}5 \Rightarrow {}^{+}5 + {}^{-}2 = {}^{+}3.$$

2.16 Another View of the Positive and Negative Numbers

In our system of numbers, there is an interesting relation between $^{+}2$ and $^{-}2$, between $^{+}4$ and $^{-}4$, etc.; they are pairs of numbers at the same distance from the zero point on the number line but on different sides. Indeed, you can easily show that the number line is symmetrical about the zero point. If you place a double-sided mirror upright on the number line on the zero point (try it), the negative numbers appear as mirror images of the positive numbers; vice versa, the positive numbers appear as mirror images of the negative numbers.

The negative sign $^{-}$ can itself be seen as a symbol indicating the relation 'is the mirror image of . . .', 'is the reflection of . . .', or more precisely 'is the *negative* of . . .'. Thus $^{-}3$, or $^{-}({}^{+}3)$ is the mirror image of $^{+}3$, the reflection of $^{+}3$, the inverse of $^{+}3$ or simply the negative of $^{+}3$.

Put the mirror again on the zero point; you can see clearly that $^{+}3$ is the mirror image of $^{-}3$, or the negative of $^{-}3$, written $^{-}({}^{-}3)$. This means that we can treat the positive sign, $^{+}$, as a symbol indicating a double reflection, a 'there and back', or just an instruction to 'stay where you are', without changing the value or the sign of the number. For example, $^{+}({}^{+}6)$ is the positive of $^{+}6$, or simply $^{+}6$.

2.17 Exercises

1. Simplify the following:

$$^{+}(^{-}5);\ ^{-}(^{+}3);\ ^{-}(^{-}2);\ ^{-}(^{-}(^{-}4));\ ^{-}(^{-}(^{-}(^{-}7))).$$

2. Try to write down the rules which describe the effect of an even or odd number of positive and negative signs acting on a number. Make up some more examples to test your rules.

Unit Three

Operations on Numbers: Multiplication and Division

3.1 Multiplication of Natural Numbers

Multiplication of natural numbers is a straightforward process of repeated addition. For example, 5×3 is simply a shorthand way of writing

$$3 + 3 + 3 + 3 + 3.$$

If we represent each unit by a dot we can always illustrate this kind of multiplication by a rectangular arrangement of dots. For example, 5×3 can be represented by

```
•   •   •   •   •

•   •   •   •   •

•   •   •   •   •
```

Fig. 3.1

This rectangular arrangement of dots obviously also represents the multiplication 3×5, or $5 + 5 + 5$; thus, like addition, multiplication is commutative, that is $5 \times 3 = 3 \times 5$, and so on.

Primary schoolchildren and primitive tribes calculate *products* (that is, the results of multiplication operations) of smaller numbers by actually adding repeatedly, and the results are then committed to memory (the familiar 'multiplication tables'). Many modern computers and desk calculators still perform multiplication by repeated addition.

Long multiplication

There are various *algorithms* (that is, procedures) for performing long multiplication. You will have been taught one particular method at your primary school, and there is no need to change your habits. Probably, however, you have come to perform this procedure purely mechanically, and this is the right time to examine the method you have adopted.

The details of individual procedures and notations differ, but the underlying mathematical laws are the same. We will examine, for example, the multiplication 4×237:

$$4 \times 237 = 237 + 237 + 237 + 237.$$

We know that 237 is simply $200 + 30 + 7$, or 2 hundreds + 3 tens + 7 units, which we can abbreviate to $2h + 3t + 7u$.

Using this abbreviation,

$$4 \times 237 = (2h + 3t + 7u) + (2h + 3t + 7u) + (2h + 3t + 7u) + (2h + 3t + 7u).$$

The commutative and associative laws (see Section 2.3) for addition allow us to rearrange this and write:

$$4 \times 237 = (2h + 2h + 2h + 2h) + (3t + 3t + 3t + 3t) + (7u + 7u + 7u + 7u),$$

or $\quad 4 \times 237 = 4 \times 2h + 4 \times 3t + 4 \times 7u.$

We have demonstrated here the *distributive* law of multiplication, that is,

$$\begin{aligned} 4 \times 237 = 4(2h + 3t + 7u) &= 4 \times 2h + 4 \times 3t + 4 \times 7u \\ &= 8h + 12t + 28u. \end{aligned}$$

Some adjustment is needed now, because of our decimal system (i.e. number system in the base ten) in which 10 units = 1 ten, 10 tens = 1 hundred, 10 hundreds = 1 thousand, and so on, so that we can write $12t = 1h + 2t$, and $28u = 2t + 8u$. (This last process is known as *carrying forward*.)
Therefore,

$$\begin{aligned} 4 \times 237 &= 8h + 12t + 28u \\ &= 8h + 1h + 2t + 2t + 8u \\ &= 9h + 4t + 8u \\ &= 948. \end{aligned}$$

If we need to multiply two larger numbers, such as 34×237, we can apply the distributive law to *both* numbers. Thus,

$$\begin{aligned} 34 \times 237 &= (30 + 4) \times (200 + 30 + 7) \\ &= 30 \times (200 + 30 + 7) + 4 \times (200 + 30 + 7) \\ &= 10 \times 3 \times (200 + 30 + 7) + 4 \times (200 + 30 + 7) \\ &= 10 \times (6h + 9t + 21u) + 8h + 12t + 28u \\ &= 10 \times (6h + 9t + 2t + 1u) + 9h + 4t + 8u \\ &= 10 \times (7h + 1t + 1u) + 9h + 4t + 8u \\ &= 79h + 14t + 18u \\ &= 80h + 5t + 8u = 8058. \end{aligned}$$

Perhaps you were taught to write down this calculation thus:

$$\begin{array}{r} 237 \\ \times\ 34 \\ \hline 948 \\ 7110 \\ \hline 8058 \\ \hline \end{array}$$

If so, compare this process (usually called 'long multiplication') with the foregoing step-by-step working. You will see that the reasoning is the same in both cases.

3.2 Exercises

1. Carry out a multiplication, such as 3×268, in your usual way, and at every step try to account for what you have done.
2. Try other numbers; then extend your examination to multiplications involving two large numbers.

3.3 Multiplication of Signed Numbers

(*a*) Multiplying by a natural number

Multiplications such as $4 \times {}^{-}3$, or $2 \times {}^{+}5$, where the first number, the *multiplier*, is a natural number, are easily interpreted as repeated additions of a negative or positive number. For example,

$$4 \times {}^{-}3 \text{ is simply } {}^{-}3 + {}^{-}3 + {}^{-}3 + {}^{-}3 = {}^{-}12,$$
$$2 \times {}^{+}5 = {}^{+}5 + {}^{+}5 = {}^{+}10.$$

You can find examples in ordinary life of such multiplications; for example, selling 3 articles, each at a loss of £10, means a total loss of £30 to the shop.

(*b*) Multiplying by a signed number

We cannot easily interpret multiplications where the multiplier is a signed number, such as ${}^{+}3 \times {}^{+}4$ or ${}^{-}5 \times {}^{+}6$, in terms of examples of everyday life where negative and positive values occur. Such multiplications can have meaning only if the positive and negative numbers are taken in their abstract mathematical sense and if the signs, taken in the sense of 'the positive of' or 'the negative of', are applied to the whole of the multiplication.
For example, ${}^{+}4 \times {}^{-}2$ should be read as: 'the positive of' $4 \times {}^{-}2$, or ${}^{+}(4 \times {}^{-}2) = {}^{+}({}^{-}2 + {}^{-}2 + {}^{-}2 + {}^{-}2) = {}^{+}({}^{-}8) = {}^{-}8$.
Also, ${}^{-}3 \times {}^{+}6$ should be read as: 'the negative of' $3 \times {}^{-}6$, or ${}^{-}(3 \times {}^{+}6) = {}^{-}({}^{+}6 + {}^{+}6 + {}^{+}6) = {}^{-}({}^{+}18) = {}^{-}18$.
Similarly:
${}^{-}2 \times {}^{-}7 = {}^{-}(2 \times {}^{-}7) = {}^{-}({}^{-}7 + {}^{-}7) = {}^{-}({}^{-}14) = {}^{+}14$.

3.4 Exercises

1. Draw up some simple rules concerning the sign of the product of two signed numbers. (You will see that these rules are very similar to hint (*d*) in Section 2.14.)

2. Use the simple slide rule or number line to calculate $^+5 \times {}^-2$ and $^-5 \times {}^+2$. In what way are these two operations different? Similarly, calculate $^-3 \times {}^-4$ and $^+3 \times {}^+4$, and note how the operations are different.
3. Compare such multiplications as $^+5 \times {}^+7$ and $^+7 \times {}^+5$, $^+2 \times {}^-3$ and $^-3 \times {}^+2$, $^-3 \times {}^-4$ and $^-4 \times {}^-3$. Is there a commutative law (compare Section 2.3) for multiplication of signed numbers?

Summarizing, we can state the following rules:

(*a*) The product of two positive numbers is a positive number.
(*b*) The product of a positive number and a negative number is a negative number.
(*c*) The product of two negative numbers is a positive number.

We can write down these general rules in mathematical shorthand:

For any two numbers N and M,

(*a*)	$^+N \times {}^+M = {}^+(NM)$
(*b*) (i)	$^-N \times {}^+M = {}^-(NM)$
(ii)	$^+N \times {}^-M = {}^-(NM)$
(*c*)	$^-N \times {}^-M = {}^+(NM)$

3.5 Multiplication of More than Two Numbers

Multiplication, like addition, involves only two numbers; but we can then use the resulting product as one of the two numbers of another multiplication. For example, we can multiply 2 by 3, and then use the product 6 to multiply 4,

or $$(3 \times 2) \times 4 = 6 \times 4 = 24.$$

As an exercise, calculate $3 \times (4 \times 5)$ working out the multiplication 4×5 first; now compare your result with $(3 \times 4) \times 5$, calculating 3×4 first. Try some similar examples. You are demonstrating the *associative* law for multiplication, that is, that when three numbers are combined by the operation of *multiplication*, the order in which they are combined does not affect the result. (We have assumed the associative law already, when we calculated 34×237 in Section 3.1.)

Three or more *signed numbers* can be multiplied in the same way.
For example, $^+3 \times {}^-2 \times {}^-4$ can be calculated in various ways:

$$({}^+3 \times {}^-2) \times {}^-4 = {}^-6 \times {}^-4 = {}^+24,$$

or $$^+3 \times ({}^-2 \times {}^-4) = {}^+3 \times {}^+8 = {}^+24,$$

or $$({}^+3 \times {}^-4) \times {}^-2 = {}^-12 \times {}^-2 = {}^+24.$$

3.6 Exercises

1. Make up and work out some multiplications involving three or more signed numbers.
2. Extend the rules governing the sign of the product of two signed numbers (Section 3.4) to multiplication of three signed numbers. Compare these extended rules with Exercise 2 of Section 2.17.
3. Write down a general rule for the sign of the product of a chain of signed numbers. [Hint: try a chain of positive numbers and see if the sign of the product is affected by the length of the chain; then try chains involving one negative number, two negative numbers, three negative numbers, and so on.]

3.7 Division of Natural Numbers

The meaning of the arithmetical operations of addition, subtraction and multiplication of natural numbers is very clearly related to the meanings of these words in everyday language. These operations can easily be visualized in terms of sets of 'concrete' things; their results are the sets which are left after the operation, which answer questions like 'What have I got if I add 2 apples to 3 apples?', or '. . . if I take 4 apples away from 7 apples?', or '. . . if I have 4 times 5 apples?'

Division, as we usually understand it, is less specific and its result is less clearly 'a set which is left after the operation'. Dividing a set is primarily a matter of splitting the set into a number of subsets, or *partitioning* a set into subsets; mathematical division presupposes that these subsets are all of equal size, i.e. that they contain the same number of things (or *elements*). A set of 12 apples, for example, can be partitioned into 3 subsets containing 4 apples each. The immediate result of such a division or partitioning is simply the set as partitioned, in this example, the set of 12 apples partitioned into 3 subsets of 4 apples each. If we partition a 'concrete' set, there are two possibilities:

(*a*) The size of each subset, i.e. the number of elements in each subset, is known. The set is then partitioned by taking away one subset after another. For example, to partition a set of 12 apples in subsets of 4 apples each, we can keep separating sets of 4 apples from the main set, until none are left.

(*b*) The required number of equal subsets is known; then we partition the set into equal subsets by progressively putting the elements of the set into each partition. For example, if our set of 12 apples is to be divided into 3 equal subsets, we can take one apple at a time from the set and place it in turn in one of 3 containers (A, B and C in Fig. 3.2).

There are thus two aspects of mathematical division. The *quotient*, the result of mathematical division, provides the answer *either* to the question

(*a*) How many equal subsets are there?

or to the question

(*b*) How many elements does each of these subsets contain?

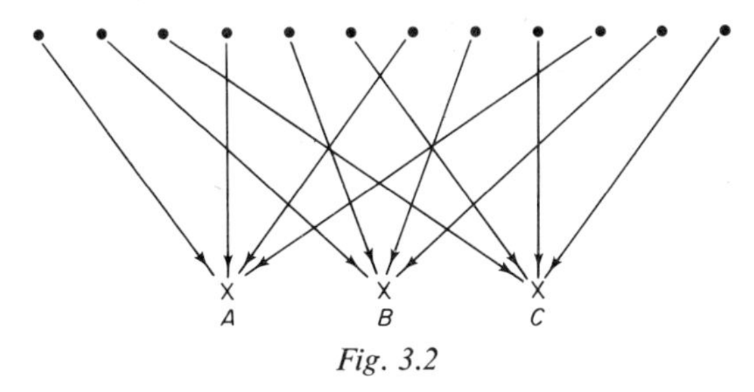

Fig. 3.2

Practical problems involve both kinds of division.

A simple example of division of type (*a*) is: How many ice-creams can be bought for 50 pence, if ice-creams cost 5 pence each?

Notice that the answer to such a division is just a *number*, in this case 10, the *number* of ice-creams which can be bought.

Examples of division of type (*b*) all involve some form of *equal distribution.* They are found in problems concerning sums of money to be equally shared between a number of people, in problems concerning averages, and many others. The answer to a question of this kind is not just a number; it includes a reference to the original quantity, or set, that has been divided. For example, if a line segment of length 28 cm is divided into 4 equal parts, the length of each segment is 7 cm (28 cm ÷ 4 = 7 cm); if the sum of £120 is equally shared between 3 people, each share is £40 (£120 ÷ 3 = £40). It is meaningless to say that the length of a line segment is 7, or that a sum of money is 40; the units must always form a part of the result of a division of this kind.

The distinction between these two aspects of division is not always vitally important, but sometimes it can clarify the various aspects of mathematical division and their methods of computation.

3.8 Division as Repeated Subtraction

Division is clearly seen as repeated subtraction when we calculate the number of equal subsets in a given set. The simplest way of finding how many times a subset of 3 apples is contained in a set of 15 apples is to actually remove sets of 3 apples repeatedly until there are no apples left, while we count the number of times such a set is taken away.

$15a - 3a = 12a$ (one set taken away);
$12a - 3a = 9a$ (two sets taken away);
$9a - 3a = 6a$ (three sets taken away);
$6a - 3a = 3a$ (four sets taken away);
$3a - 3a = 0$ (five sets taken away; no apples remain).

The situation is a little different if, for example, we need to share 12 apples equally between 4 children; but again, the 12 apples can be equally distributed by giving each child one apple at a time, so effectively removing a subset of 4 apples at each round, and each child's share is then 3 apples.

The method of repeated subtraction is employed by computers or desk calculators when they are used for calculations involving division. It is also the method used in long division, as a close examination of your own procedure for long division will show.

3.9 Exercise

Examine your method of calculating $2808 \div 34$, and notice how 34, or multiples of 34, are subtracted again and again. If you can get hold of a desk calculator or an abacus, examine the procedure for division.

3.10 Division as Inverse Multiplication

Division—regarded as a process of partitioning into equal subsets—is the inverse of multiplication, which joins a number of equal subsets into a total; we can see this clearly in the question posed by the operation of division: 'how many such equal subsets are contained in a given set?' or 'how many *times* is a certain subset contained in this set?' In a division such as $12 \div 3$ we are asked to find a number which, when multiplied by 3, produces 12, or $\boxed{?} \times 3 = 12$. Consider, also, multiplication as repeated addition; division as repeated subtraction is clearly the inverse of this process.

Checking your answer

Multiplication and division can thus be used as mutual checks. For example, to check the correctness of the division $595 \div 35 = 17$, calculate 17×35; the answer should be 595.

3.11 Division and Ratio

Mathematicians emphasize the close association between division and ratio by using the same symbol—the stroke, /—for both; for example, both $12 \div 3$ and $12:3$ can be written $12/3$. Indeed, continental mathematicians dispense with the special division symbol altogether, and use the colon symbol, : , for both ratio and division.

We described ratio earlier (Section 1.5) as a numerical relation or correspondence between two sets or numbers. We can think of division as establishing this ratio, and reducing it to its simplest form. In our example of the set of 12 apples to be divided into sets of 3 apples, the ratio is between the set of 12 apples and the set of 3 apples, that is $12:3$, which simplifies to $4:1$, or $4/1$, or simply 4.

3.12 Divisibility and Factors

Attempts to partition a set may prove impossible if we are restricted to using whole numbers only. A set of 13 people, for example, cannot be divided into 4 equal subsets, and the question 'how many subsets of 4 apples are contained in a set of 17 apples?' cannot be answered by a whole number. If we repeatedly subtract 4 apples, we will eventually be left with 1 apple, the *remainder*. If such a partition can be made exactly, if in other words there is no remainder, the number is said to be *divisible*. For example, the number 12 is divisible by 2; it is also divisible by 3, by 4 and by 6. The numbers 2, 3, 4 and 6 are called the *factors* of 12, which means that they divide 12 exactly (leaving a zero remainder).

Every number is divisible by itself and also by the number 1. Numbers that have no other factors are called *prime numbers*, for example 5, 11, 37.

3.13 Exercises

1. Write down the first 20 prime numbers.
2. Show that division by 0 is impossible and has no meaning. [Take division here in each of its meanings in turn: repeated subtraction, inverse multiplication and ratio.]

3.14 Divisibility Tests

If we wish to find whether one number is divisible by another, we can just perform the division and check whether there is a remainder or not. However, there are simple tests for divisibility by some of the smaller numbers, and they can be further extended on the basis of the following principles:

Principle 1

If two numbers are divisible by a certain number, then their sum is also divisible by that number; for example, since both 6 and 18 are divisible by 3, then 24 ($= 6 + 18$) is also divisible by 3. It follows that if a number is divisible by a certain number, any multiple of it is also divisible by that number; for example, 63 is divisible by 7, so 126 ($= 2 \times 63$, or $63 + 63$) is also divisible by 7.

Principle 2

If a number is divisible by two different prime numbers it is also divisible by the product of these two numbers; for example, since 42 is divisible by 2 and also by 3, then it is divisible by 6 ($= 2 \times 3$).

The truth of these principles can be established as an exercise using set-diagrams, as in Section 1.5.

Divisibility by 2. Numbers divisible by 2 are called *even numbers*; other whole numbers are called *odd numbers*. Knowing that 2, 4, 6, 8 and 10 are divisible by 2 and applying Principle 1, we recognize all even numbers by the digit in the unit position; if this is 0, 2, 4, 6 or 8 the number is even. For example, the number 34 is recognized as an even number by the last digit, which is 4; this is because

$$\begin{aligned} 34 &= 30 + 4 \\ &= 3 \times 10 + 4. \end{aligned}$$

When we are testing a number to see whether it is even, we can ignore all digits on the left of the unit position, as all multiples of 10 are divisible by 2.

Divisibility by 4, 8, etc. A similar divisibility test can be designed for 4 and 8. Since the number 100 (and therefore all multiples of 100) are divisible by 4, we can ignore all but the last two digits of a number to be tested for divisibility by 4. For example, to test 345 684 for divisibility by 4, we need only test 84 for divisibility by 4.

As an exercise, devise similar tests for divisibility by 8, 16, and so on.

Divisibility by 3 and 9. It is interesting that, as inspection of the '3 times' or '9 times' tables shows, the digits of any multiple of 3 add up to a number which is also divisible by 3 (at least in our decimal number system). The number 452 678 526 is divisible by 3 because $4 + 5 + 2 + 6 + 7 + 8 + 5 + 2 + 6$ $(= 45)$ is divisible by 3. The number 45 is divisible by 9 as well, and this shows that our original number is also divisible by 9. The validity of these tests can be proved, but the proof has been omitted here for simplicity.

Divisibility by 5. We know that 5 and 10 are divisible by 5, and if we apply Principle 1, we can establish a test for divisibility by 5 on the same lines as that for divisibility by 2: any multiple of 5 will have either a 0 or a 5 in the unit position.

Divisibility by 6. Application of Principle 2 gives a divisibility test for 6: a number is divisible by 6 if it is divisible by 2 and by 3. For example, 32 454 is divisible by 6 because it is divisible by 2 (the last digit is an even number) and it is *also* divisible by 3 ($3 + 2 + 4 + 5 + 4$ is divisible by 3).

Principle 2 in fact not only applies to products of prime numbers but also to products of numbers which are *relatively prime*, that is, numbers which have no factors in common or no *common factors*. For example, 3 and 4 are relatively prime. Therefore, if a number is divisible by 3 and also by 4, it is also divisible by 12 ($= 3 \times 4$).

3.15 Exercises

1. Devise a test for divisibility by 15. [Hint: $15 = 5 \times 3$.]
2. Devise some divisibility tests for 12, 20, etc.
3. Satisfy yourself that Principle 2 does not apply to products of numbers which have some common factor, for example, 4 and 10, or 6 and 14.

3.16 Factorizing Numbers

The process of expressing a number as the product of its factors is called *factorization.* Sometimes a number can be factorized in various ways, for example $36 = 2 \times 18$, or $36 = 3 \times 12$, or $36 = 6 \times 6$. Also $36 = 2 \times 2 \times 9$, or $36 = 2 \times 2 \times 3 \times 3$. The most useful form is factorization into prime factors, and there is only one way in which a number can be factorized into prime factors. The order of the factors is unimportant, but it is sometimes convenient to write down the factors in order of size, for example:

$$480 = 2 \times 2 \times 2 \times 2 \times 2 \times 3 \times 5.$$

Notation. The multiplication sign, $\times$, is often replaced by a simple dot, especially in cases where there is a chain of numbers to be multiplied. The factorization of 480 is then written $480 = 2.2.2.2.2.3.5$ or, in mathematical shorthand, as $480 = 2^5.3.5$. (2^5 is read as '2 to the power 5', '2 to the fifth power' or just '2 to the fifth', and simply means $2 \times 2 \times 2 \times 2 \times 2$, or 32.)

3.17 Exercises

1. Factorize into prime factors the following numbers: 24; 210; 396.
2. Find *all* the prime factors that these three numbers have in common, and so calculate their *highest common factor* (HCF), that is, the largest whole number that divides all these numbers exactly.
3. Find *all* the prime factors of 10, 12 and 18. By collecting together *just* enough of these factors, calculate their *lowest common multiple* (LCM), that is, the smallest whole number which is a multiple of each of these numbers.

3.18 Division of Signed Numbers

The rules governing division of signed numbers are easily established if we consider division as inverse multiplication.

Starting with some simple numerical examples such as $^{+}12 \div {}^{+}4 = \ldots$ or $^{+}18 \div {}^{-}3 = \ldots$, ask yourself the question 'which number multiplied by $^{+}4$ produces $^{+}12$?', that is $\boxed{?} \times {}^{+}4 = {}^{+}12$, or 'which number multiplied by $^{-}3$ produces $^{+}18$?', that is $\boxed{?} \times {}^{-}3 = {}^{+}18$.

3.19 Exercise

Make up some simple divisions of various kinds involving signed numbers, such as $^{+}24 \div {}^{+}3$, $^{-}36 \div {}^{+}9$, and so on; continue until you are able to deduce some general rules governing the sign of the quotient.

3.20 General Rules

From the general rules governing the multiplication of signed numbers we deduce the general rules for division of signed numbers:

For any two numbers N and M:

$$(a)\ {}^{+}N \times {}^{+}M = {}^{+}(NM) \Rightarrow {}^{+}(NM) \div {}^{+}M = {}^{+}N$$
$$(b)\ {}^{-}N \times {}^{-}M = {}^{+}(NM) \Rightarrow {}^{+}(NM) \div {}^{-}M = {}^{-}N$$
$$(c)\ {}^{+}N \times {}^{-}M = {}^{-}(NM) \Rightarrow {}^{-}(NM) \div {}^{-}M = {}^{+}N$$
$$(d)\ {}^{-}N \times {}^{+}M = {}^{-}(NM) \Rightarrow {}^{-}(NM) \div {}^{+}M = {}^{-}N$$

Now we can summarize the rules for both multiplication and division of two signed numbers:

If both numbers have the same sign (i.e. both are positive or both are negative), *their product and their quotient are positive. If one number is positive and the other negative, their product and their quotient are negative.*

Unit Four

Operations on Rational Numbers and Point Notation

4.1 Addition and Subtraction

The addition and subtraction of rational numbers (Section 1.8) can be considered in terms of shifts of appropriate length along the number line, in the same way that we studied the addition and subtraction of integers.

4.2 Exercise

Make a simple slide rule like the one used in Unit 2, but with a larger scale, covering the range from $^{-}2$ to $^{+}2$. Mark along it intervals of length $\frac{1}{8}$ (that is, one-eighth of a unit length) and indicate the rational numbers represented by the corresponding points:

$$^{-}2,\ ^{-}1\tfrac{7}{8},\ ^{-}1\tfrac{3}{4},\ ^{-}1\tfrac{5}{8},\ \ldots,\ 0,\ ^{+}\tfrac{1}{8},\ ^{+}\tfrac{1}{4},\ ^{+}\tfrac{3}{8},\ ^{+}\tfrac{1}{2},\ \ldots,\ ^{+}1\tfrac{7}{8},\ ^{+}2.$$

Use this simple slide rule and the shift method to calculate:

$$\begin{array}{l}\frac{1}{2} + \frac{1}{4};\\ \frac{1}{4} + \frac{3}{8};\\ ^{+}1 + {}^{-}\frac{1}{4};\\ ^{-}\frac{1}{2} + {}^{+}\frac{1}{4};\\ ^{-}\frac{3}{8} + {}^{-}\frac{1}{4}.\end{array}$$

We can use this slide rule to carry out addition and subtraction involving larger numbers and fractions, such as, for example, $15\frac{3}{4} + 3\frac{1}{8}$, if we bear in mind that $15\frac{3}{4} = 15 + \frac{3}{4}$ and $3\frac{1}{8} = 3 + \frac{1}{8}$,

so that
$$\begin{aligned}15\tfrac{3}{4} + 3\tfrac{1}{8} &= 15 + \tfrac{3}{4} + 3 + \tfrac{1}{8}\\ &= 15 + 3 + \tfrac{3}{4} + \tfrac{1}{8}\\ &= 18\tfrac{7}{8}.\end{aligned}$$

4.3 The Arithmetical Processes of Addition and Subtraction of Rational Numbers

When we extended the use of the number line model to rational numbers less than the number 1, the unit, we adopted a unit of measure smaller than the unit length. We called it a *sub-unit*, and represented it symbolically by its ratio with the unit, for example $\frac{1}{2}$ or $\frac{1}{3}$ or $\frac{1}{4}$. When we write down a rational number, we indicate the sub-unit we are using by the number below the line, the *denominator*; the number above the line, the *numerator*, denotes the number of sub-units concerned. For example, the rational number $\frac{3}{4}$ has numerator 3 and denominator 4; it describes 3 sub-units each of value one-fourth of one whole unit.

If we use $\frac{1}{4}$ as the sub-unit, and add $\frac{1}{4}+\frac{1}{4}+\frac{1}{4}$ we obtain a number represented by a length which has the ratio 3:4 or $\frac{3}{4}$ with the unit length; in other words, $\frac{3}{4}$ is the same as $\frac{1}{4}+\frac{1}{4}+\frac{1}{4}$. Additions and subtractions such as $\frac{3}{4}+\frac{2}{4}$, $\frac{4}{10}-\frac{3}{10}$, then, are carried out as follows:

$$\frac{3}{4}+\frac{2}{4}=(\frac{1}{4}+\frac{1}{4}+\frac{1}{4})+(\frac{1}{4}+\frac{1}{4})=\frac{5}{4};$$

$$\frac{4}{10}-\frac{3}{10}=(\frac{1}{10}+\frac{1}{10}+\frac{1}{10}+\frac{1}{10})-(\frac{1}{10}+\frac{1}{10}+\frac{1}{10})=\frac{1}{10}.$$

These operations are straightforward if the rational numbers concern the same basic sub-unit—quarters, thirds, tenths, for example. Two rational numbers can be added if the denominators are the same; as the denominator indicates the sub-unit, it will also be the denominator in the sum. We find the numerator of the sum by adding the numerators of the numbers to be added, for example:

$$\frac{3}{10}+\frac{4}{10}=\frac{3+4}{10}=\frac{7}{10}.$$

The subtraction of rational numbers of this kind is equally simple:

$$\frac{5}{10}-\frac{3}{10}=\frac{5-3}{10}=\frac{2}{10}.$$

We can write this in general terms as

$$\frac{N}{L}+\frac{M}{L}=\frac{N+M}{L}.$$

If we need to add or subtract rational numbers whose denominators are not the same, we must replace one or both numbers by equivalents in such a way that both have the same denominator.

Examples

1. $\frac{1}{2}+\frac{1}{8}$.

   Replacing $\frac{1}{2}$ by its equivalent $\frac{4}{8}$,

$$\frac{1}{2}+\frac{1}{8}=\frac{4}{8}+\frac{1}{8}=\frac{5}{8}.$$

2. $\frac{1}{6}+\frac{3}{4}$.

   Replacing $\frac{1}{6}$ by $\frac{2}{12}$, and $\frac{3}{4}$ by $\frac{9}{12}$,

$$\frac{1}{6}+\frac{3}{4}=\frac{2}{12}+\frac{9}{12}=\frac{2+9}{12}=\frac{11}{12}.$$

3. $\frac{1}{5} + \frac{1}{3}$.
   Replacing $\frac{1}{5}$ by $\frac{3}{15}$, and $\frac{1}{3}$ by $\frac{5}{15}$,

$$\frac{1}{5} + \frac{1}{3} = \frac{3}{15} + \frac{5}{15} = \frac{3+5}{15} = \frac{8}{15}.$$

Choosing the appropriate equivalent depends on the denominators of the numbers involved. If, as in Example 1, one denominator is a factor of the other, it is easy to replace this denominator so that it is the same as the other (in this example, replace $\frac{1}{2}$ by $\frac{4}{8}$). If the two denominators have a common factor, as in Example 2, use any common multiple, preferably their lowest common multiple (see Exercise 3 of Section 3.17), as the denominator in both equivalents. If the denominators have no common factors their lowest common multiple is just their product (Example 3).

Some practical hints

(*a*) A rational number such as $\frac{4}{6}$ or $\frac{2}{4}$, where numerator and denominator have a factor in common, is a perfectly 'respectable' number. However, in the final answer to a problem it is usual and preferable to replace it by the simplest equivalent; for example, write $\frac{4}{6}$ as $\frac{2}{3}$ and $\frac{2}{4}$ as $\frac{1}{2}$.

(*b*) The procedure for replacing rational numbers by their simplest equivalent (reducing them to their *lowest terms*) was discovered in Exercise 1.7, that is, by dividing both numerator and denominator by their common factor(s). This can be done gradually by repeated division by the common factors, for example,

$$\frac{72}{336} = \frac{72 \div 4}{336 \div 4} = \frac{18}{84} = \frac{18 \div 6}{84 \div 6} = \frac{3}{14}.$$

Alternatively, if this is easier, first find the highest common factor of the numerator and the denominator and then carry out a single division:

$$\frac{72}{336} = \frac{72 \div 24}{336 \div 24} = \frac{3}{14} \text{ (since the HCF of 72 and 336 is 24).}$$

4.4 Multiplication of Rational Numbers

Multiplication of a rational number by a whole number is simply repeated addition. For example,

$$5 \times \frac{1}{4} = \frac{1}{4} + \frac{1}{4} + \frac{1}{4} + \frac{1}{4} + \frac{1}{4} = \frac{1+1+1+1+1}{4} = \frac{5 \times 1}{4} = \frac{5}{4}.$$

$$4 \times \frac{2}{3} = \frac{2}{3} + \frac{2}{3} + \frac{2}{3} + \frac{2}{3} = \frac{8}{3}.$$

Similarly, you can see from these examples that in multiplying a rational number by an integer, the numerator is multiplied by that integer; for example:

$$5 \times \frac{1}{4} = \frac{5 \times 1}{4} = \frac{5}{4}.$$

Multiplication by a rational number

The process of multiplication by a natural number is a process of *magnification* or *enlargement*. The result of multiplying by the number 3, for example, is three times as large as the original number; the ratio between the product (the magnification) and the original is 3:1, or $\frac{3}{1}$, and the multiplier (or the multiplying factor or the enlargement factor) is precisely that ratio, $\frac{3}{1}$. Similarly, in a multiplication by 2 the magnifying factor is $\frac{2}{1}$; in a multiplication by 4 the multiplying factor is $\frac{4}{1}$, and so forth.

Anyone interested in photography will realize that as well as magnification to twice, three times, four times the original size, there are innumerable intermediate magnifications.

Think of the blowing up of a balloon as a process of magnification. Blowing up a balloon to twice its original size, for example, is a gradual, continuous process, passing through many intermediate stages. At each stage, the balloon is larger than it was at first, but not yet twice its original size.

Returning to multiplication of numbers as represented on the number line, consider enlarging a length of 2 units to a length of 3 units. The ratio between the magnification and the original is then 3:2, or $\frac{3}{2}$. This process of magnification is expressed mathematically as $\frac{3}{2} \times 2 = 3$.

If we multiply a given length by, in turn, the multiplying ratios 4 $(=\frac{4}{1})$, 3 $(=\frac{3}{1})$, 2 $(=\frac{2}{1})$, $\frac{3}{2}$, etc., the magnification becomes gradually smaller, and more and more like the original. When we use the multiplying ratio $\frac{1}{1}$, the 'magnification' is of the same size as the original. If we then use a multiplying ratio even smaller than 1 (but greater than 0), the original is again changed or transformed, but this time the transformation is not a magnification but a contraction; the original is made smaller. Application of a multiplying ratio of 1:2 or $\frac{1}{2}$ produces a result which has the ratio 1:2 with the original, in other words reducing the original to half its size.

If N is any number, $\frac{1}{2} \times N$ is what we mean in everyday language by 'half of N', or $N \div 2$, or $\frac{N}{2}$. If N is a whole number, the result can be found as follows:

$$\frac{1}{2} \times 3 = 3 \div 2 = \frac{3}{2};$$
$$\frac{1}{2} \times 4 = 4 \div 2 = \frac{4}{2} = 2;$$
$$\frac{1}{2} \times 5 = 5 \div 2 = \frac{5}{2}.$$

If N is itself a rational number, $\frac{N}{2}$ can be found in a similar way. To calculate, for example, $\frac{1}{2} \times \frac{1}{3}$ or 'half of' $\frac{1}{3}$, we express $\frac{1}{3}$ by its equivalent $\frac{2}{6}$—that is, $\frac{1}{6} + \frac{1}{6}$; and we can then obtain half of $\frac{1}{3}$, or half of $\frac{2}{6}$, or:

$$\frac{1}{2} \times \frac{1}{3} = \frac{1}{2} \times \frac{2}{6} = \frac{1}{6}.$$

In this example, both the numerator and the denominator of $\frac{1}{3}$ were multiplied by 2; then the numerator was divided by 2. Now multiplication by 2 followed by division by 2 leaves a number unchanged; so we may write immediately:

$$\frac{1}{2} \times \frac{1}{3} = \frac{1 \times 1}{2 \times 3} = \frac{1}{6}.$$

If the numerator of a multiplying ratio is not the number 1, the same process can still be used; we can write a rational number such as $\frac{2}{3}$ just as well as $2 \times \frac{1}{3}$, so that, for example:

$$\tfrac{2}{3} \times \tfrac{1}{4} = 2 \times \tfrac{1}{3} \times \tfrac{1}{4} = 2 \times \frac{1}{3 \times 4} = \frac{2 \times 1}{3 \times 4} = \frac{2}{12} = \tfrac{1}{6}.$$

This means that, *to multiply two rational numbers*, we simply have *to multiply the two numerators and multiply the two denominators*. For example,

$$\tfrac{3}{4} \times \tfrac{5}{7} = \frac{3 \times 5}{4 \times 7} = \frac{15}{28}.$$

In general terms, using the letters a, b, c, d for whole numbers,

$$\boxed{\frac{a}{b} \times \frac{c}{d} = \frac{a \times c}{b \times d}}$$

We can simplify the result, if the numerator and denominator have a factor in common. The common factor is *cancelled*, by dividing both the numerator and denominator by this common factor: for example,

$$\tfrac{3}{4} \times \tfrac{1}{3} = \frac{3 \times 1}{4 \times 3} = \frac{1 \times 3 \div 3}{4 \times 3 \div 3} = \tfrac{1}{4}.$$

This 'cancelling' is usually done by crossing out common factors:

$$\tfrac{3}{4} \times \tfrac{1}{3} = \frac{\cancel{3} \times 1}{4 \times \cancel{3}} = \tfrac{1}{4} \text{ or simply } \frac{\cancel{3}}{4} \times \frac{1}{\cancel{3}} = \tfrac{1}{4};$$

$$\tfrac{2}{5} \times \tfrac{3}{8} = \frac{\cancel{2} \times 3}{5 \times \cancel{2} \times 4} = \frac{3}{5 \times 4} = \tfrac{3}{20}.$$

Note. Calculations with *mixed numbers*, that is, combinations or sums of whole numbers and rational numbers such as $4\frac{3}{4}$, are usually easier if we express the mixed numbers as proper ratios.

For example, $2\frac{1}{4} = 2 + \frac{1}{4} = \frac{8}{4} + \frac{1}{4} = \frac{9}{4}$

and $2\frac{1}{4} \times 1\frac{1}{3} = \frac{9}{4} \times \frac{4}{3} = \dfrac{3 \times \cancel{3}}{\cancel{4}} \times \frac{\cancel{4}}{\cancel{3}} = 3.$

4.5 Exercises

1. Calculate:

$$\tfrac{1}{3} \times \tfrac{2}{7};$$
$$\tfrac{2}{3} \times \tfrac{3}{5};$$
$$\tfrac{3}{4} \times \tfrac{4}{9};$$
$$\tfrac{2}{5} \times \tfrac{10}{3} \times \tfrac{3}{2};$$
$$3\tfrac{2}{7} \times 1\tfrac{2}{5}.$$

2. Extend these by further examples of your own, if you feel you need more practice.

4.6 Multiplication of Signed Rational Numbers

The general rules of multiplication and division of signed numbers also apply to *multiplication of signed rational numbers.*

For example, $$^{-}(\tfrac{1}{2}) \times {}^{-}(\tfrac{1}{3}) = {}^{+}(\tfrac{1}{6}).$$

If you need to multiply a chain of rational numbers, first decide on the sign of the product.

$$^{-}(\tfrac{1}{3}) \times {}^{+}(\tfrac{1}{4}) \times {}^{-}(\tfrac{3}{7}) \times {}^{-}(\tfrac{4}{5}) = {}^{-}(\tfrac{1}{3} \times \tfrac{1}{4} \times \tfrac{3}{7} \times \tfrac{4}{5}) = {}^{-}(\tfrac{1}{35}).$$

Sometimes positive or negative signs appear in the numerator or denominator of a rational number; the sign of the number itself is then found by using the rules governing the sign of quotients (see Section 3.20). For example,

$$\frac{^{-}2}{^{+}3} = {}^{-}\left(\frac{2}{3}\right);$$

and
$$\frac{^{-}3}{^{-}4} = {}^{+}\left(\frac{3}{4}\right).$$

4.7 Exercise

Calculate the following:

$$^{-}\left(\frac{5}{7}\right) \times {}^{-}\left(\frac{3}{5}\right) \times {}^{+}\left(\frac{1}{4}\right) \times {}^{+}\left(\frac{7}{3}\right);$$

$$^{-}2\tfrac{1}{3} \times \frac{^{+}4}{^{-}7} \times \frac{^{-}5}{^{+}8}.$$

4.8 Division of Rational Numbers

We can divide a rational number into equal parts by multiplying the denominator, as we have already seen.

In a division such as $\frac{2}{3} \div 5$, we can replace the rational number $\frac{2}{3}$ by its equivalent $\dfrac{5 \times 2}{5 \times 3}$ or $\dfrac{2+2+2+2+2}{15} = \dfrac{2}{15} + \dfrac{2}{15} + \dfrac{2}{15} + \dfrac{2}{15} + \dfrac{2}{15}$. We have now expressed the rational number $\frac{2}{3}$ as the sum of 5 equal rational numbers, each $\dfrac{2}{15}$. Therefore $\dfrac{2}{3} \div 5 = \dfrac{2}{15}$.

In this example, the numerator has been both multiplied and divided by 5 and therefore remains the same, and the denominator is only multiplied by 5. Clearly, therefore, dividing by 5 is the same as multiplying by $\frac{1}{5}$; for example,

$$\tfrac{2}{3} \div 5 = \tfrac{1}{5} \times \tfrac{2}{3} = \tfrac{2}{15}.$$

Division by a rational number

There are some cases of division of rational numbers, for example $\frac{12}{15} \div \frac{1}{15}$, that can be solved by asking the question 'how many times is $\frac{1}{15}$ contained in $\frac{12}{15}$?', that is $\frac{12}{15} \div \frac{1}{15} = 12$. Division in a case like this takes place in the sense of repeated subtraction.

In *all* cases of division of rational numbers, we can take division in the sense of inverse multiplication. A question such as $\frac{2}{3} \div \frac{5}{7} = \boxed{?}$ then becomes 'which rational number multiplied by $\frac{5}{7}$ produces $\frac{2}{3}$?', or $\frac{?}{?} \times \frac{5}{7} = \frac{2}{3}$.

Now we can replace $\frac{2}{3}$ by its equivalent $\frac{2 \times 5 \times 7}{3 \times 5 \times 7}$; the numerator is then expressed as a multiple of 5, and the denominator as a multiple of 7, so that $\frac{?}{?} \times \frac{5}{7} = \frac{2}{3} = \frac{5 \times (2 \times 7)}{7 \times (3 \times 5)}$.

The required rational number is then $\frac{2 \times 7}{3 \times 5} = \frac{14}{15}$ and we can see that

$$\frac{2}{3} \div \frac{5}{7} = \frac{2}{3} \times \frac{7}{5} = \frac{14}{15}.$$

In general terms, using the letters a, b, c and d for any four integers,

$$\boxed{\frac{c}{d} \div \frac{a}{b} = \frac{c}{d} \times \frac{b}{a} = \frac{c \times b}{d \times a}}$$

The notion of division as inverse multiplication now takes on a special literal meaning: division by a rational number $\frac{a}{b}$ is multiplication by the inverse of the ratio, $\frac{b}{a}$, that is, $\frac{a}{b}$ upside down!

4.9 Exercise

Calculate:

$\frac{3}{22} \div \frac{6}{11}$;
$2\frac{2}{3} \div \frac{4}{5}$ (see the Note at the end of Section 4.4);
$\dfrac{3\frac{2}{3} - 2\frac{1}{6}}{1\frac{2}{5}}$.

4.10 Rational Numbers in Point Notation

We have already discovered that addition and subtraction of rational numbers with the same (common) denominator is a simple matter of adding or subtracting numerators.

Calculation with rational numbers would therefore be much easier if we could handle them all as their equivalents with the same denominator.

Since 10 is the base of the number system in common use, the decimal system, the most convenient denominators are 10, 100, 1000, etc., that is powers (see Section 3.16) of 10, 10^1, 10^2, 10^3, etc. We say that rational numbers are changed to their *decimal equivalents* if the denominator of the equivalent is a power of 10.

To *change ordinary rational numbers* (that is, 'common' or 'vulgar' fractions) *to their decimal equivalents* we use the general procedure for finding equivalents (see Section 1.6). For example, to find the decimal equivalent of $\frac{1}{5}$, we must complete the statement

$$\tfrac{1}{5} = \frac{?}{\text{a power of 10}}.$$

Inspection shows that multiplication of both denominator and numerator by 2 will produce a rational number whose denominator is a power of 10, in this case, 10^1, or 10 itself:

$$\tfrac{1}{5} = \frac{2 \times 1}{2 \times 5} = \tfrac{2}{10}.$$

We can find the decimal equivalent of $\frac{3}{4}$ in a similar way: trying first 10 for the denominator, we find that 10 is not a multiple of 4. Next we try 10^2, or 100; $100 = 25 \times 4$, so,

$$\tfrac{3}{4} = \frac{25 \times 3}{25 \times 4} = \tfrac{75}{100}.$$

Only those rational numbers whose denominator is a factor of some power of 10 can be replaced by an exact decimal equivalent. We shall discuss other rational numbers in Section 4.20.

4.11 Exercise

Using the method of inspection, replace the following rational numbers by their decimal equivalents:

$$\tfrac{1}{4};\ \tfrac{1}{8};\ \tfrac{1}{20};\ \tfrac{3}{5}.$$

A rational number can be expressed in an infinite number of different equivalent forms; it can also be expressed as the sum of other rational numbers. For example,

$$\tfrac{1}{2} = \tfrac{2}{4} = \tfrac{1}{4} + \tfrac{1}{4};$$
$$\tfrac{1}{2} = \tfrac{3}{6} = \tfrac{1}{6} + \tfrac{1}{6} + \tfrac{1}{6};$$
$$\tfrac{1}{2} = \tfrac{4}{8} = \tfrac{2}{8} + \tfrac{2}{8} = \tfrac{2}{8} + \tfrac{1}{4}, \text{ etc.}$$

In the same way, a rational number whose denominator is a factor of a power of 10 can be broken up into an equivalent sum of rational numbers whose denominators are powers of 10. For example,

$$\frac{3}{25}=\frac{12}{100}=\frac{10+2}{100}=\frac{10}{100}+\frac{2}{100}=\frac{1}{10}+\frac{2}{100}=\frac{1}{10}+\frac{2}{10^2},$$

or

$$\frac{1}{8}=\frac{125}{1000}=\frac{100+20+5}{1000}=\frac{100}{1000}+\frac{20}{1000}+\frac{5}{1000}=\frac{1}{10}+\frac{2}{10^2}+\frac{5}{10^3}.$$

4.12 Positional Notation

In the decimal system, every number is in fact expressed as a sum of multiples of powers of 10, for example:

$3245 = (3\times1000) + (2\times100) + (4\times10) + 5 = 3(10^3) + 2(10^2) + 4(10) + 5.$

The power is not specifically mentioned, and the addition sign is not used. By convention only, the numbers are placed in order from left to right, the number on the extreme right indicating the number of units, the number next to it indicating the number of tens, the next number indicating the number of hundreds and so on. If we think of these numbers placed in columns, this could be shown as:

10^4	10^3	10^2	10	units
	3	2	4	5

Fig. 4.1

Each item in any of the columns has ten times the value of the same item in the column immediately on its right (and, of course, one-tenth of the same item in the column immediately on its left). For example, the number 333 could be shown as

10^2	10	units
3	3	3

Fig. 4.2

The 3 in the extreme left-hand column represents $3 \times 10^2 = 300$; the 3 in the middle column represents $3 \times 10 = 30 = \frac{1}{10} \times 300$, or 10×3, and the 3 in the extreme right-hand column represents 3 units.

This system of *positional notation*, where the value of the symbol depends on the place it occupies, can be extended in the same way to the right, beyond the unit column.

The column immediately to the right of the unit column is then the column of the tenths ($\frac{1}{10}$), the second column to the right of the unit column is the column of the hundredths $\left(\frac{1}{100} \text{ or } \frac{1}{10^2}\right)$, and so on.

...	10^2	10	units	$\frac{1}{10}$	$\frac{1}{10^2}$	$\frac{1}{10^3}$	...

Fig. 4.3

For example, consider the number $245\frac{3}{8}$; if the fractional part $\frac{3}{8}$ is replaced by the decimal equivalent $\frac{3}{10} + \frac{7}{100} + \frac{5}{1000}$, the number can be expressed as

$$(2 \times 100) + (4 \times 10) + (5 \times 1) + \left(3 \times \frac{1}{10}\right) + \left(7 \times \frac{1}{100}\right) + \left(5 \times \frac{1}{1000}\right),$$

or $$2(10^2) + 4(10) + 5(1) + 3\left(\frac{1}{10}\right) + 7\left(\frac{1}{10^2}\right) + 5\left(\frac{1}{10^3}\right).$$

We can show this in column form as:

10^2	10	units	$\frac{1}{10}$	$\frac{1}{10^2}$	$\frac{1}{10^3}$
2	4	5	3	7	5

Fig. 4.4

4.13 Unit Point

It is not always convenient to use columns and column headings, but we must indicate somehow which numbers indicate the units, which the tens, and so on. In Great Britain a *point* is used after the number which indicates the units; the number in the example above is written 245·375.

A rational number expressed in this *point notation* is often loosely called a decimal fraction, and the point a decimal point. But remember that any common or vulgar fraction, such as $\frac{5}{12}$, is written in the base 10, and is therefore decimal, while the point as an indicator of the unit position can be used—and is used—in any base.

For example in base 4,

$$321{\cdot}23_4 = (3 \times 4^2) + (2 \times 4) + 1 + \left(2 \times \frac{1}{4}\right) + \left(3 \times \frac{1}{4^2}\right);$$

and in base 2 (binary system),

$$101{\cdot}11_2 = (1 \times 2^2) + (1 \times 1) + \left(1 \times \frac{1}{2}\right) + \left(1 \times \frac{1}{2^2}\right).$$

There is nothing 'decimal' (i.e. to do with 10) about the point; it simply indicates the position of the units, and we will therefore call it the *unit point.* Neither is the part of the number written on the right of the point necessarily decimal (compare our examples of numbers in base 4 or base 2). In general terms, therefore, we speak of *point fractions*, although in the decimal system such fractions could of course be called *decimal fractions*, which are simply fractions in the base 10.

Note. The Metrication Board has recommended that the use of the comma to separate groups of 3 digits should be discontinued; instead, a space should be used to separate these groups, as for example in

$$3\ 645\ 131{\cdot}214$$

This is to avoid confusion with the continental practice of using the comma, instead of the dot, as the unit point.

4.14 Exercises

1. Put the following rational numbers into columns, as in Section 4.12:

$$\tfrac{12}{100};\ \tfrac{125}{1000};\ 3\tfrac{75}{100}.$$

   Now express these rational numbers as point fractions, using the unit point.
2. Find the decimal equivalents of the following rational numbers and then express them as point fractions:

$$\tfrac{2}{5};\ \tfrac{3}{8};\ \tfrac{3}{4};\ \tfrac{5}{16}.$$

4.15 Addition and Subtraction of Decimal Fractions

Operations on point fractions are all carried out in the same way as the operations on whole numbers.

When we add or subtract whole numbers in positional notation, we need to make sure that units are added to units, tens to tens, and so on; similarly, care must be taken when adding point fractions that tenths are added to tenths, hundredths to hundredths, and so on. Numbers are 'carried forward' if in any column the sum exceeds 9, as in the addition of whole numbers. For example,

$$\begin{aligned} 2{\cdot}342 + 1{\cdot}231 &= 2 + \tfrac{3}{10} + \tfrac{4}{100} + \tfrac{2}{1000} + 1 + \tfrac{2}{10} + \tfrac{3}{100} + \tfrac{1}{1000} \\ &= 2 + 1 + (\tfrac{3}{10} + \tfrac{2}{10}) + (\tfrac{4}{100} + \tfrac{3}{100}) + (\tfrac{2}{1000} + \tfrac{1}{1000}) \\ &= 3{\cdot}573. \end{aligned}$$

The simplest way is to add or subtract by vertical arrangement, making sure that the unit points are placed one above the other; for example:

$$\begin{array}{rr} & 321{\cdot}346 \\ + & 23{\cdot}712 \\ \hline & 345{\cdot}058 \end{array}$$

or

$$\begin{array}{rr} & 1{\cdot}800 \\ - & 0{\cdot}375 \\ \hline & 1{\cdot}425 \end{array}$$

The second of these examples illustrates the simplicity of adding and subtracting fractions in point notation, as compared with handling the same fractions in their 'common' form. Contrast, for example,

$$\begin{array}{rr} & 1{\cdot}800 \\ - & 0{\cdot}375 \\ \hline & 1{\cdot}425 \end{array}$$

with $$1\tfrac{4}{5} - \tfrac{3}{8} = \tfrac{9}{5} - \tfrac{3}{8} = \frac{72 - 15}{40} = \tfrac{57}{40} = 1\tfrac{17}{40}.$$

4.16 Multiplication of Decimal Fractions

We can multiply decimal fractions using the normal procedure for multiplying rational numbers. For example,

$$3{\cdot}7 \times 2{\cdot}45 = 3\tfrac{7}{10} \times 2\tfrac{45}{100} = \tfrac{37}{10} \times \tfrac{245}{100} = \frac{37 \times 245}{10 \times 100}$$
$$\frac{37 \times 245}{1000} = \tfrac{9065}{1000} = 9\tfrac{65}{1000} = 9{\cdot}065.$$

This process can be simplified if we consider the effect of multiplication by powers of 10 on the position of the unit point. Multiplying a whole number by 10 is equivalent to writing an additional zero at the end of the number; if the number contains a unit point, multiplication by 10 shifts the unit point one place to the right.

For example,
$$\begin{aligned} 10 \times 31 &= 10 \times (30 + 1) \\ &= 10 \times 30 + 10 \times 1 \\ &= 3 \times 100 + 1 \times 10 = 310. \\ 10 \times 4{\cdot}2 &= 10 \times (4 + \tfrac{2}{10}) \\ &= 10 \times 4 + \frac{10 \times 2}{10} \\ &= 4 \times 10 + 2 = 42. \end{aligned}$$

Similarly, multiplication by 100 produces a shift of the unit point 2 places to the right; for instance, $100 \times 3{\cdot}255 = 325{\cdot}5$; multiplication by 1000 moves the unit point 3 places to the right, and so on.

In the same way, multiplication of a number by 0·1 (or $\frac{1}{10}$, which is equivalent to division by 10) has the effect of moving the unit point one place to the left; for example:

$$\begin{aligned} 0{\cdot}1 \times 32{\cdot}4 &= \tfrac{1}{10} \times (3 \times 10 + 2 + \tfrac{4}{10}) \\ &= \frac{3 \times 10}{10} + \frac{2}{10} + \frac{4}{10 \times 10} \\ &= 3 + \tfrac{2}{10} + \tfrac{4}{100} = 3{\cdot}24. \end{aligned}$$

Multiplication by 0·01 (or $\frac{1}{100}$) moves the unit point 2 places to the left, and so on.

Conversely we can say that shifting the unit point one place to the left is equivalent to multiplying by 0·1, etc.; for example, the number 3·24 can be written as $0{\cdot}01 \times 324$. Using this idea, we can express a multiplication like $0{\cdot}37 \times 2{\cdot}4$ as

$$0{\cdot}01 \times 37 \times 0{\cdot}1 \times 24 = (0{\cdot}01 \times 0{\cdot}1) \times 37 \times 24 = 0{\cdot}001 \times 37 \times 24.$$

The multiplication $0{\cdot}37 \times 2{\cdot}4$ can therefore be carried out as the multiplication $37 \times 24 = 888$, followed by multiplying this product by 0·001 (the last operation implying a shift of the unit point 3 places to the left), and the result is 0·888. Notice that the number of digits after the unit point in the product is the same as the sum of the number of digits after the unit point in the numbers multiplied. We can use this principle as a rule in multiplying decimal fractions. For example,

0·001	×	0·01	=	0·000 01,
(3 digits after the point)		(2 digits after the point)		(5 digits after the point)

3·12	×	6·3	=	19·656.
(2 digits)		(1 digit)		(3 digits)

Be careful, however, when you apply this rule: you must include in the count any zeros that result from multiplication. For example,

0·2	×	0·5	=	0·10	= 0·1.
(1 digit)		(1 digit)		(2 digits)	

Note. You should always check whether the answer to a problem is reasonable, especially when working with decimal fractions. A multiplication which appears as $1{\cdot}25 \times 30{\cdot}54 = 381{\cdot}75$ is obviously wrong; a very rough approximation would show that the answer is a number more than 1×30 but less than 2×31, or in mathematical notation,

$$1 \times 30 < 1{\cdot}25 \times 30{\cdot}45 < 2 \times 31,$$

where $<$ means 'is less than'.

If the numerical calculation is correct, the unit point should be placed so as to satisfy these conditions, and

$$1{\cdot}25 \times 30{\cdot}54 = 38{\cdot}175.$$

4.17 Exercises

1. Calculate the following:

$$0{\cdot}01 \times 357{\cdot}4;$$
$$43{\cdot}56 \times 2{\cdot}35.$$

2. The rule for the number of digits to the right of the point in the product of decimal fractions also applies to 3 or more numbers. Show this by means of examples such as

$$0{\cdot}01 \times 0{\cdot}001 \times 0{\cdot}001;$$
$$0{\cdot}2 \times 3{\cdot}25 \times 5{\cdot}5.$$

4.18 Division of Decimal Fractions

In division, especially division of decimal fractions, always make sure that the number of units in the *quotient* (i.e. the result of division) is written immediately above the number of units in the *dividend* (i.e. the number to be divided), the number of tens in the quotient above the number of tens in the dividend, and so on. The unit point in the quotient is placed immediately above the unit point in the dividend; and division of decimal fractions is then as easy as long division of whole numbers. For example,

```
       14·7
65 ) 955·5
     65
     ---
     305
     260
     ----
      45 5
      45 5
      ----
         0
```

or, $$955{\cdot}5 \div 65 = 14{\cdot}7.$$

We use the same procedure if the dividend is a number which is smaller than the *divisor* (i.e. the number with which you are dividing), for example in the division $0{\cdot}0429 \div 13$,

```
       0·0033
13 ) 0·0429
         39
         ——
          39
          39
          ——
           0
```

or, $$0{\cdot}0429 \div 13 = 0{\cdot}0033.$$

If the divisor contains digits after the unit point, it is simplest to replace the calculation by an equivalent division where the divisor is a whole number, for example,

$$34 \div 1{\cdot}7 \left(\text{or } \frac{34}{1{\cdot}7}\right) = \frac{10 \times 34}{10 \times 1{\cdot}7} = \frac{340}{17} = 20,$$

or $$\frac{71{\cdot}3}{0{\cdot}31} = \frac{100 \times 71{\cdot}3}{100 \times 0{\cdot}31} = \frac{7130}{31}.$$

We can do this immediately by moving the unit point the same number of places to the right in both divisor and dividend; the number of places moved depends on the number of digits after the unit point in the divisor. For example,

$$8{\cdot}602 \div 3{\cdot}74 = 860{\cdot}2 \div 374.$$

The long division is then carried out in the usual way:

```
        2·3
374 ) 860·2
      748
      ———
      112 2
      112 2
      —————
          0
```

This rule of moving the unit point an equal number of places in both divisor and dividend can be used in more complicated situations when divisor and dividend are themselves products of numbers containing unit points. For example,

$$\frac{8{\cdot}45 \times 6{\cdot}5}{2{\cdot}75}$$

can be replaced by an equivalent division whose divisor is a whole number (275) by multiplying both dividend and divisor by 100:

$$\frac{100 \times 8{\cdot}45 \times 6{\cdot}5}{100 \times 2{\cdot}75} = \frac{845 \times 6{\cdot}5}{275},$$

or $$\frac{100 \times 8{\cdot}45 \times 6{\cdot}5}{100 \times 2{\cdot}75} = \frac{10 \times 8{\cdot}45 \times 10 \times 6{\cdot}5}{100 \times 2{\cdot}75} = \frac{84{\cdot}5 \times 65}{275}.$$

This illustrates the procedure to use when the divisor or dividend, or both, are products of decimal fractions. The unit point is moved by a number of places equal to the *sum* of the numbers of digits after the unit point in the terms which make up the divisor. This can be done in different ways; compare the following examples:

(*a*) $$\frac{3{\cdot}1 \times 4{\cdot}3}{4{\cdot}25} = \frac{31 \times 43}{425} \text{ or } \frac{310 \times 4{\cdot}3}{425} \text{ or } \frac{3{\cdot}1 \times 430}{425}$$

(moving the unit point(s) a total number of 2 places in both dividend and divisor);

(*b*) $$\frac{5{\cdot}27 \times 3{\cdot}45}{6{\cdot}4 \times 8{\cdot}95} = \frac{527 \times 34{\cdot}5}{64 \times 895} \text{ or } \frac{52{\cdot}7 \times 345}{64 \times 895}$$

(moving the unit point a total number of 3 places in both dividend and divisor).

4.19 Exercises

1. Replace $$\frac{7{\cdot}45 \times 4{\cdot}55}{3{\cdot}25 \times 5{\cdot}6}$$

   by an equivalent division where the divisor is a whole number, multiplying both dividend and divisor by the same power of 10.
2. Calculate:

$$308{\cdot}76 \div 83;$$
$$46{\cdot}99 \div 12{\cdot}7;$$
$$3{\cdot}648 \div 11{\cdot}4;$$
$$\frac{7{\cdot}2 \times 0{\cdot}5}{0{\cdot}24};$$
$$\frac{82{\cdot}5 \times 3{\cdot}9}{0{\cdot}52 \times 0{\cdot}11}.$$

4.20 Terminating and Non-terminating Decimal Fractions

We noted in Section 3.12 that many divisions did not have an exact integral answer; for example, 13 ÷ 4 = 3 *remainder* 1. We said that 13 is not divisible by 4, or that 4 is not a factor of 13. But if we extend the system of numbers to include the rational numbers, we have a precise answer: 13 ÷ 4 = $3\frac{1}{4}$. Using point fractions, we can carry out the division by long division, continued beyond the unit point:

$$\begin{array}{r} 3{\cdot}25 \\ 4\,\overline{)\,13{\cdot}00} \\ 12\phantom{{\cdot}00} \\ \hline 1\,0 \\ 8 \\ \hline 20 \\ 20 \\ \hline 0 \end{array}$$

Therefore, $13 \div 4 = 3{\cdot}25 = 3\frac{1}{4}$.

This is also the most convenient method of converting a rational number into its equivalent point fraction. For example,

$\frac{5}{8}$ or $5 \div 8$:

$$\begin{array}{r} 0{\cdot}625 \\ 8\,\overline{)\,5{\cdot}000} \\ 4\,8 \\ \hline 20 \\ 16 \\ \hline 40 \\ 40 \\ \hline 0 \end{array}$$

Therefore $\frac{5}{8} = 0{\cdot}625$.

Whenever the divisor (denominator) is a factor of 10, 100, or some other power of 10, the division will *terminate*, that is, at some point in the long division process the remainder will be zero.

But if the divisor is not a factor of some power of 10 (for example, if it is 3 or 7 or 9), the possibility arises that the division may not terminate; a digit, or a sequence of digits, may begin to recur in the quotient, repeating itself over and over again.

For example,

$\frac{8}{6}$ or $8 \div 6$:

$$\begin{array}{l} 1{\cdot}333\,3\ldots \\ 6\,\overline{)\,8{\cdot}000\,00} \\ 6 \\ \hline 2\,0 \\ 1\,8 \\ \hline 20 \\ 18 \\ \hline 20 \\ 18 \\ \hline \ldots \end{array}$$

Similarly, $\frac{4}{11}$ or $4 \div 11$.

```
      0·363 6363 . . .
 11 ) 4·000 0
      3 3
       70
       66
        40
        33
         70
         66
          40
          33
          . . .
```

Such non-terminating decimal fractions are therefore called repeating decimal fractions, or *recurring* decimal fractions. We can write the two examples above as $\frac{8}{6}$ or $8 \div 6 = 1{\cdot}\dot{3}$ and $\frac{4}{11}$ or $4 \div 11 = 0{\cdot}3\dot{6}\dot{3}$, respectively.

4.21 Numbers in Standard Form

There are some practical applications of the technique used in Section 4.16 for the multiplication of decimal fractions.

For example, we saw that the number 0·34 is equivalent to the product $0{\cdot}01 \times 34$ or $\frac{1}{100} \times 34$; also $5600 = 56 \times 100$, and so on. This equivalence means that we can express very large or very small numbers in a more convenient form.

If the distances of stars are given in kilometres or even in light-years, the numbers to be handled become truly astronomical. In ordinary notation, the distance between the Earth and a particular star could be written down as the unwieldy expression 345 000 000 000 000 000 000 000 000 000 000 km. If we wanted to have some idea of the size of such a number we would probably first count the number of zeros. It would be much more convenient to write 345×10^{30} or $34{\cdot}5 \times 10^{31}$ or $3{\cdot}45 \times 10^{32}$.

We can treat very small numbers in a similar way. For example, the radius of an atom of hydrogen is known to be about 0·000 000 000 03 m; this is more readily appreciated if we write it as $3 \times \dfrac{1}{10^{11}}$ m.

Multiplication by 100, or 10^2, moves the unit point 2 places to the right; multiplication by 10^3 moves the unit point 3 places to the right. In general, the *index* of the 10 (i.e. the little number indicating the power of 10) indicates the number of places the unit point is moved to the right, and vice versa.

Multiplication by $\frac{1}{10^2}$ similarly moves the unit point 2 places to the *left*, and so on. In Unit 9 we will see that we can write $\frac{1}{10^2}$ as 10^{-2}, or $\frac{1}{10^3}$ as 10^{-3}; the direction 'to the left' is then indicated by the negative sign.

Examples

$$375 \times 10^{12} = 375\,000\,000\,000\,000,$$

$$264 \times \frac{1}{10^4} \text{ (or } 264 \times 10^{-4}) = 0{\cdot}0264,$$

and vice versa,

$$\begin{aligned} 2\,453\,000\,000\,000 &= 245{\cdot}3 \times 10^{10} \\ &= 24{\cdot}53 \times 10^{11} \\ &= 2{\cdot}453 \times 10^{12}. \end{aligned}$$

We choose, as the usual form of presentation of a number, a form such as $2{\cdot}453 \times 10^{12}$, adjusting the value of the power of 10 so that the first number is more than 1 but less than 10, that is $1 < 2{\cdot}453 < 10$.

In general terms, we can express *any* number in the form

$$\boxed{A \times 10^n}$$

where A is a number between 1 and 10 ($1 < A < 10$), and n indicates the appropriate power of 10. A number written like this is said to be given in *standard form*.

4.22 Exercise

Express the following numbers in standard form:

823 500 000 000;
724·5;
29·37;
0·006 84;
0·136.

Part Two

Further Arithmetic and Applications of Arithmetic

Unit Five

Measurement and Metrication

5.1 Quantities

There is one obvious way of finding the size of a crowd of people: we can count the individuals. The size of a set can usually be found by counting the number of elements in it, although this number could be very large—like the number of grains of rice in a pan—and counting might take a very long time. The elements of a set of this kind are whole things, each one complete in itself, and in some sense indivisible. The merging of all the elements into one set produces a new kind of unit, for example a *crowd* of people, and this new unit is of a totally different nature from the elements themselves; for example, mankind is not a man, nor is a forest itself a tree.

Quantities such as length, area or volume, and materials like water or butter, are of a completely different nature; they are continuous quantities, which we can divide again and again into smaller and smaller parts. We can, for example, divide a length into smaller lengths, and then divide these again. But we shall never, at least within the limits set by ordinary methods of measurement, find the ultimate indivisible length which can be used as the natural unit for counting the 'size' of the length. Also, of course, the sum of any collection of different lengths will be a length itself, of the same kind as its constituent parts.

5.2 Measuring

Two quantities of the same kind, such as two lengths or two areas, can be *compared*, and we can then express the relation between their sizes. If we put two strips of paper side by side, we can decide whether they are the same length or not, and if they are not the same length, we can tell which is the longer. In the same way, if a flat object of a certain area can cover another completely and vice versa, we can deduce that their areas are the same. If, on the other hand, one area covers the other completely, but is not itself completely covered by the other, we can say that it is the larger in size of the two.

There is no *absolute* precision in measurement, and any statement concerning 'equality' in measurement must be an approximation. Two lengths may look the same to the human eye, but the use of a magnifying glass might indicate that they are not quite the same. We could make adjustments so that both lengths look the same under a magnifying glass, and yet still find that, seen through a microscope, they are not exactly the same, and so on.

5.3 Units of Measurement

Suppose that you wished to compare two lengths, such as the length of your foot and the length of a corridor. You might find that not only is the corridor longer than your foot, but that you could express this relation more exactly by a number. If you went 'pigeon stepping' down the corridor, you might perhaps find that the length of the corridor is 35 times the length of your foot. You have been using the length of your foot as a tool to find and express the size of another length. *A particular length has been arbitrarily chosen to serve as a unit of length.* Any length could serve as a unit for measuring other lengths, any area as a unit for measuring other areas, any volume as a unit for measuring other volumes, and so on. A circular area, such as

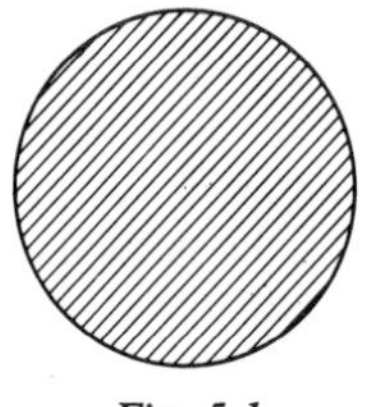

Fig. 5.1

or an irregularly shaped area, such as

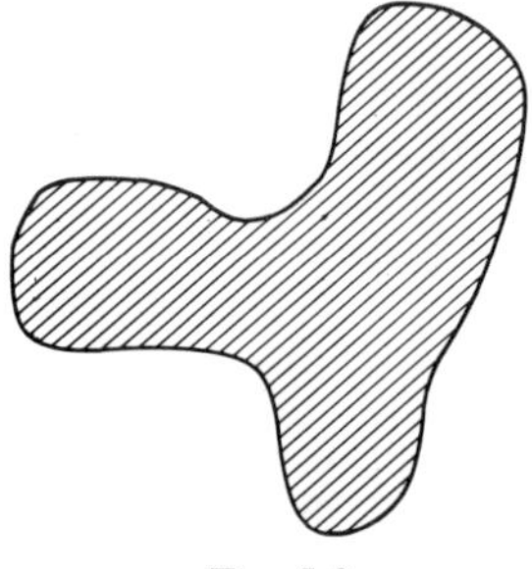

Fig. 5.2

could be used as a unit of area. But these would, however, be rather impractical, because it is impossible to cover a surface completely, without overlap, with shapes of this kind. It is more convenient to use units for measuring areas which have shapes that *tessellate*, that is, that can completely cover a flat surface without overlap when fitted together. For example, hexagons, triangles, parallelograms and squares tessellate (Fig. 5.3). The most convenient unit for measuring area is the square.

Similarly, tessellating shapes such as cuboids (i.e. brick-shaped objects) could serve as units for measuring volume; but for this purpose the most convenient unit is the cube.

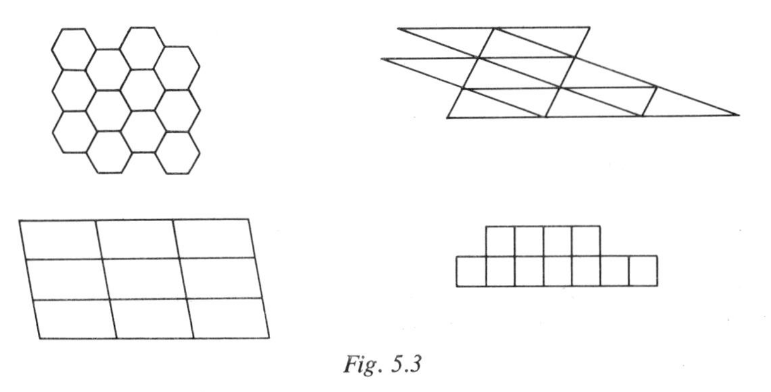

Fig. 5.3

Summarizing, we can say:

Measuring is comparing 'like' quantities—such as length with length, area with area—and expressing the relative size by means of a number. This comparison is usually with an agreed unit of measurement.

5.4 Uniform or Standard Measure

Quite obviously there is a need for world-wide uniformity of measures, especially as international trade increases and multi-national companies develop in number and importance. An internationally agreed system of units is now gradually being introduced throughout the world; it is a modernized form of the metric system, known as the *Système Internationale d'Unités*, or simply SI.

5.5 The International System

The metric system is the most widely used system of units; in its original form it was introduced on the whole of the European continent by Napoleon. In Great Britain the process of metrication, both in industry and in everyday life, is approaching completion; the USA have set a target date (1981) for complete metrication.

The International System, SI, differs from the traditional metric system in that it recommends only some of the metric units and their multiples and sub-multiples for use.

The main advantage of the metric system is its simplicity. Like our number system it is *decimal*, i.e. its base is ten. The larger units, or multiple units, are decimal multiples—that is, multiples of ten or powers of ten—of the basic unit of measurement, and the smaller units, referred to earlier as sub-units (see Section 1.8) are decimal fractions of the basic unit. For example, consider the measurement of length, taking the metre as the basic unit:

$$1 \text{ kilometre (km)} = 1000 \text{ metres (m), or } 10^3 \text{ m},$$
$$1 \text{ decimetre (dm)} = \tfrac{1}{10} \text{ m, or } 0{\cdot}1 \text{ m},$$
$$1 \text{ centimetre (cm)} = \tfrac{1}{100} \text{ m, or } 0{\cdot}01 \text{ m, or } \frac{1}{10^2} \text{ m},$$
$$1 \text{ millimetre (mm)} = \tfrac{1}{1000} \text{ m, or } 0{\cdot}001 \text{ m, or } \frac{1}{10^3} \text{ m}.$$

(The units hectometre (hm) and dekametre (dam) are seldom used.)

If we use an arrangement of columns, like the positional notation of our decimal number system, we can read measurements off in various units or sub-units, according to choice. For example, 112 cm:

km	hm	dam	m	dm	cm	mm
0	0	0	1	1	2	0

Fig. 5.4

This can be read as 1·12 m, or 11·2 dm, or 112 cm, or 1120 mm.

The *unit* point now clearly has a further significance—it indicates which *unit* of measurement we are using.

5.6 Quantities

A list of quantities, their basic units, the recommended multiples and sub-multiples and the recommended symbols for these, appears in Section 5.8.

We only need at present to concern ourselves with the quantities *length, area, volume, mass* and *temperature.*

By international agreement the word *mass* is used in preference to *weight.* Physicists use the term *weight* to describe a force, the kind of pulling force that the Earth exerts on every object (gravitation). This force depends on the position of the object. The nearer an object is to the centre of the Earth the greater the gravitational force upon it; far out in space, away from any other body, an object becomes weightless.

Mass is an invariant property of an object depending only on the internal structure of the object and its size. There is a close connexion between mass and weight: the gravitational force—weight—acting on an object depends on the mass of the object, and the masses of two objects are in the same ratio to each other as the ratio of their weights measured at the same point in space. The mass of an object is measured in terms of the gravitational force exerted on it, i.e. by weighing. In everyday use—and in this book—the distinction between mass and weight is not often made; the word 'weight' is unlikely to disappear from our vocabulary.

Volume and capacity

SI makes no distinction between volume and capacity, which both relate to three-dimensional space. The *volume* of an object is the size of the space it occupies; the *capacity*—usually of a container—is the size of the space determined by the internal dimensions of the container, or to put it simply, the volume that the container can hold.

5.7 Prefixes

Some of the prefixes and their symbols, recommended for use with SI units, are:

multiple		*prefix*	*symbol*	*example*
1 000 000	10^6	mega	M	megagram (Mg)
1 000	10^3	kilo	k	kilometre (km)
100	10^2	hecto	h	hectolitre (hl)
10	10	deka	da	dekametre (dam)
0·1	$\frac{1}{10}$	deci	d	decimetre (dm)
0·01	$\frac{1}{10^2}$	centi	c	centimetre (cm)
0·001	$\frac{1}{10^3}$	milli	m	milligram (mg)
0·000 001	$\frac{1}{10^6}$	micro	μ	microgram (μg)
0·000 000 001	$\frac{1}{10^9}$	nano	n	nanometre (nm)

In official, technical work—including examinations—we must adhere to the accepted conventions and carefully observe the rules which apply to the use of symbols and prefixes.

Note.

(*a*) All the prefixes are written in small (lower-case) type, as are the symbols abbreviating them, except M (mega).

(*b*) In official work, the underlined prefixes are preferred: mega, kilo and milli.

(*c*) No full stops are used after the symbol, nor is the plural 's' used; thus we write 20 cm and not 20 cms.

5.8 Summary

If we include the multiple units and sub-units which you are likely to encounter (although they are not all recommended for official use) we can summarize SI as follows:

Table 5.1 Some SI units and their symbols

quantity	*basic SI unit*	*symbol*	*recommended multiple units and sub-units*	*other units in common use*
length	metre	m	km dm cm mm μm	international nautical mile (n mile) 1 n mile = 1852 m
area	square metre	m^2	km^2 dm^2 cm^2 mm^2	hectare (ha) 1 ha = 10^4 m^2, are (a) 1 a = 100 m^2
volume	cubic metre	m^3	dm^3 cm^3 mm^3	hectolitre (hl) 1 hl = 100 l litre (l) 1 l = 1 dm^3 centilitre (cl) $1\text{ cl} = \frac{1}{10^2}\text{ l}$ millilitre (ml) $1\text{ ml} = \frac{1}{10^3}\text{ l}$
angle	radian	rad	mrad μrad	degree (°) $1° = \frac{\pi}{180}\text{ rad}$ minute (′) $1' = \frac{1°}{60}$ second (″) $1'' = \frac{1'}{60}$
time	second	s	ks ms μs ns	day (d) 1 d = 24 h hour (h) 1 h = 60 min minute (min) 1 min = 60 s
velocity	metre per second	m/s		kilometre per hour knot (kn) 1 kn = 1 nautical mile per hour = 0·514 444 m/s
mass	kilogram	kg	Mg g mg μg	tonne (t) 1 t = 10^3 kg
temperature	degree Celsius or kelvin	°C K		(degree Celsius is often called degree Centigrade)

5.9 Area and Volume

The conversion of one metric unit of area or volume into another needs some care. When all the prefixes (milli-, centi-, deci-, deka-, hecto-, kilo-) are used, every multiple unit or sub-unit of length is 10 times its predecessor; for example,

$$1 \text{ cm} = 10 \text{ mm},$$
$$1 \text{ dm} = 10 \text{ cm},$$
$$1 \text{ m} = 10 \text{ dm},$$

that is, the ratio of any two consecutive multiple units (or sub-units) of *length* is 1:10. We shall see that this does not apply to units of area or units of volume.

Area

1 square decimetre (1 dm^2) is the area of a square whose sides all have length 1 dm.

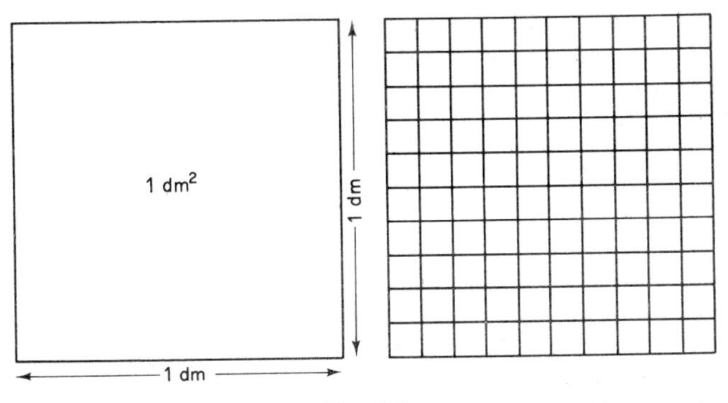

Fig. 5.5

The number of square centimetres (cm^2) in a square decimetre is 100; there are 10 rows of 10 square centimetres each. Similarly,

$$1 \text{ m}^2 = 10^2 \text{ dm}^2 \text{ or } 1 \text{ m}^2 = 100 \text{ dm}^2$$
and
$$1 \text{ cm}^2 = 10^2 \text{ mm}^2 \text{ or } 1 \text{ cm}^2 = 100 \text{ mm}^2.$$

The ratio of any two consecutive multiple units (or sub-units) of *area* is $1:10^2$, or 1:100.

Volume

A cubic decimetre (dm^3) is the volume of a cube whose edges are each 1 dm long (Fig. 5.6).

We could build up a cube of this volume using little cubes, each with edges of length 1 cm; we should need to pile up 10 layers of cubes, each layer made up

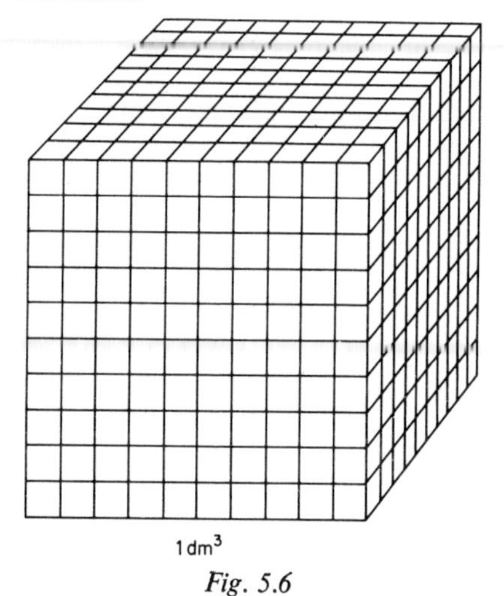

Fig. 5.6

of 100 cubes (laid out as 10 rows of 10 cubes each). We could then count all the cubes, and verify that 1 dm^3 = 1000 cm^3, or 1 dm^3 = 10^3 cm^3.

The ratio of any two consecutive multiple units (or sub-units) of *volume* is $1:10^3$ or $1:1000$.

5.10 Exercises

1. The British Imperial measures and the metric system are not directly related; and there is no advantage to be gained in working in British Imperial measures and then converting to metric units.

   If you have been brought up in the British Imperial system, try to forget it, and *think metric*.

   Take a metric tape-measure and find some familiar measures in metric units: your height, vital statistics, your span, the length of your stride, the size of your room, and so on. Try to get hold of some metric bathroom scales and weigh yourself. Find out how small a gram is (about 450 to the Imperial pound); find the temperature of your body (in °C) when healthy and sick; find what is the temperature on a cold day, a warm day, or in a comfortably heated room.

   In other words, make yourself thoroughly familiar with the metric system by making *direct associations* of everyday situations with metric measures.
2. Carry out some conversions of one metric unit into another; begin by completing the following:

   1·67 m = . . . dm = . . . cm;
   500 g = . . . kg;
   345 cm = . . . dm;
   0·45 dm = . . . cm.

3. How much water can be held in a tank which has the following internal dimensions: 200 cm long, 150 cm wide and 100 cm deep (or 200 cm by 150 cm by 100 cm)?

   If the empty tank weighs 1·3 t, what is the weight of the tank and its contents when it is full? (1 dm³ of water weighs 1 kg.)

5.11 Precision and Approximation

The arithmetical processes involving natural numbers, integers and rational numbers produce exact and precise results. There is absolute precision in a multiplication such as $3 \times 4 = 12$, and a statement such as '4 × 5 men = 20 men' needs no qualification.

Measurement, on the other hand, is by its very nature a matter of approximation (see Section 5.2); the precision of measurement always depends on the accuracy of the measuring instrument and the observing eye or recording device. There is always a *margin of error*, that is, there is a whole range of possible values on either side of the stated result. If, for example, we measure the length of a line as 119 cm using an instrument which is accurate to within 1 centimetre ('correct to the nearest centimetre') the true length of the line could be anywhere between 118·5 cm and 119·5 cm (Fig. 5.7).

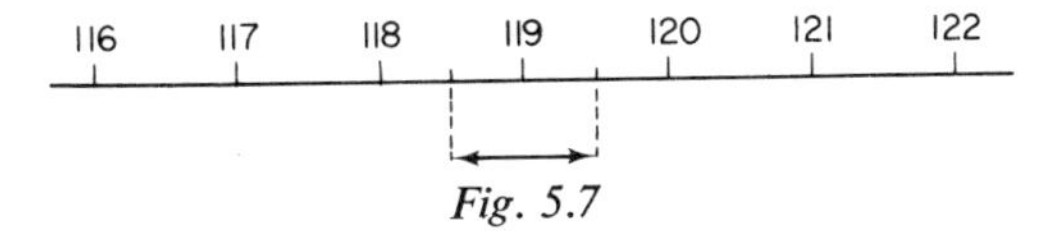

Fig. 5.7

The error on either side of 119 cm can be anything up to 0·5 cm, the lower limit being 118·5 cm and the upper limit 119·5 cm. If a length is given correct to the nearest metre, for example 1125 m, the greatest possible error is 0·5 m. The length is said to be '1125 metres plus or minus 0·5 m', which is written as 1125 m ± 0·5 m or (1125 ± 0·5) m.

We can choose our method of measurement according to the degree of accuracy required for our particular purpose. For example, much greater accuracy is required from the chemist who is making up a prescription than from the coal merchant weighing out his coal. Even if more accurate instruments are available and measurements can be given very precisely—for example, by counting every individual in a theatre audience—we may be able to ignore certain quantities as unimportant or irrelevant. We might know that the number of people admitted to a football match is 35 324 exactly (and box office accounts need this accurate knowledge); a journalist describing the match, however, might quote an attendance of 35 000, which gives a clear enough picture for his purpose of the size of the crowd: the number has been *rounded off* to the nearest thousand. The number 35 324 lies between 35 000 and 36 000, that is $35\,000 < 35\,324 < 36\,000$; it is 324 more than 35 000, and 676 less than 36 000. It is therefore nearer to 35 000 than to 36 000 and rounded off *downwards* to 35 000. Now look at the number 42 513; it is nearer to 43 000 than to 42 000, and can be rounded off *upwards* to 43 000.

The number 43 500 lies exactly half-way between 43 000 and 44 000. By convention we round off *upwards* whenever the first digit to be 'cast off' is 5 or more. Rounded off to the nearest thousand therefore, 43 500 becomes 44 000.

Consider now a sum of £85 537 682:

rounded off to the nearest million pounds it is £86 000 000;
rounded off to the nearest hundred thousand pounds it is £85 500 000;
rounded off to the nearest thousand pounds it is £85 538 000.

5.12 Significant Figures

When we round off certain numbers, we replace 'cast off' digits by zeros for positional reasons, that is to say, to make sure that the approximation has not changed the value of the digit which stands, for example, in the thousand or in the ten-thousand position. Take the number 94 784; rounded off to the nearest thousand, it becomes 95 000, the zeros ensuring that the values of the digits 9 and 5 are recognized, that is that there are 5 thousands and 9 ten-thousands. The digits remaining are called *significant figures*. In our example, there are 2 significant figures left after the rounding-off, 9 and 5, and we say that the number is given *correct to 2 significant figures*, sometimes abbreviated to 2 SF.

Rounded off to 3 significant figures, this number is 94 800.

To round off a whole number to 2 significant figures:

Replace the 2nd digit by the next higher digit if the 3rd digit is 5, 6, 7, 8 *or* 9—*otherwise leave the 2nd digit unchanged; then replace the 3rd and further digits by zeros.*

Similarly, when rounding off a number to 3 significant figures replace, if necessary, the 3rd digit by the next higher digit, then replace the 4th and further digits by zeros. For example, to round off 34 256 to 3 SF, we replace the 3rd digit (2) by 3 since the 4th digit is 5, and replace the 4th and 5th digits by zeros: 34 300. We then say that 34 256 = 34 300 correct to 3 significant figures (3 SF).

A number given correct to a certain number of significant figures represents not one exact number but a whole range of values—rather like a stated measurement (compare Section 5.11). For example, the number 654 000, given correct to 3 significant figures, represents all the numbers between 653 500 and 654 499 inclusive.

In any approximation the non-zero digits are obviously significant: for example, in the number 367 000 the first 3 digits are certainly significant. But zeros on the right of the last non-zero digit (in this case, to the right of the 7) are not always written there only for positional reasons; they may be significant too.

The number 367 000, rounded off to 4 SF, is 367 000, the 4 significant figures being 3670; in fact, the approximation 367 000 can be interpreted in different ways according to the accuracy that is stated.

367 000, given as correct to 3 SF, represents all numbers between 366 500 and 367 499 inclusive;
367 000, given as correct to 5 SF, represents all numbers between 366 995 and 367 004 inclusive.

5.13 Exercises

1. What is the margin of error in each of the following measurements?

   21 cm, correct to the nearest cm;
   35 mm, correct to the nearest mm;
   124 m, correct to the nearest m.

   State in each case the greatest possible error.
2. Round off the following:

   113 325 to the nearest thousand;
   29 642 to the nearest thousand;
   9964 to the nearest hundred.

3. Round off to 3 significant figures:

   37 425;
   890 214.

5.14 Significant Figures in Decimal (Point) Fractions

Zeros written to the left of the unit point indicate the position of thousands, hundreds and tens in numbers like 17 000, 300 and 40. Similarly, zeros are needed to the right of the unit point to indicate the position of tenths, hundredths and thousandths in numbers like 0·035 and 0·004. In all these examples, the zeros are written for *positional reasons only*, and are not significant figures. The first significant figure of 0·003 45 is 3, the first significant figure of 0·000 789 is 7, and so forth.

The rule for rounding off to a certain number of significant figures also applies to 'decimal fractions', except that any numbers cast off to the right of the unit point need not be replaced by zeros—they are *not* needed for positional reasons. For example, 43·3478 correct to 4 significant figures (in this case correct to the nearest $\frac{1}{100}$) is 43·35, not 43·3500; the two zeros are redundant. However, a zero (or zeros) should be written as the last digit(s) to the right of the unit point *if it is significant*, and not the result of casting off. For example, in rounding off 1·3014 correct to 3 SF, the digits 1 and 4 are cast off (and not replaced by zeros), but the third significant figure—0—is written, so that 1·3014 correct to 3 SF is 1·30.

Similarly 3·5972 correct to 3 SF is 3·60; and 5·003 correct to 3 SF is 5·00.

This means we can tell the number of significant figures in a decimal fraction from the form in which it is written; for example:

3·60 is given correct to 3 SF;
3·6 correct to 2 SF;
3·600 correct to 4 SF.

5.15 Exercise

Round off to 3 significant figures:

3·155;
296·4;
0·014 74;
5·7039.

5.16 Margins of Error in Calculations

Margins of error are cumulative in operations on approximate numbers—that is, errors introduced in approximations can accumulate, and quite large errors may be introduced. For example,

24 426 rounded off to the nearest thousand is	24 000
+13 247 rounded off to the nearest thousand is	+13 000
37 673 rounded off to the nearest thousand is 38 000, *not*	37 000

Again,

2948 rounded off to the nearest hundred is	2900
− 1356 rounded off to the nearest hundred is	− 1400
1592 rounded off to the nearest hundred is 1 600, *not*	1500

If the margin of error in each of two numbers is 100, the margin of error in their sum is 200; if we add three such numbers, the margin of error increases to 300, and so on.

Therefore, do not round off numbers until all your calculations are complete.

5.17 Accuracy to a Certain Number of Decimal Places

When we specify accuracy and 'rounding off' in terms of significant figures, we emphasize the accuracy of the first digits of a number, regardless of whether these digits represent millions, thousands, tenths or hundredths. Consider, for example, the numbers 1·2 and 35 000, both given as correct to 2 SF. The first two digits in each of these numbers are accurately given, but the margins of error are different; the margin of error in 1·2 is $\frac{1}{10}$ and the margin of error in 35 000 is 1000.

Sometimes we refer directly to the margin of error, as when a number is given as correct 'to the nearest thousand', 'to the nearest hundred', and so on. If, for example, numbers are quoted as correct to the nearest hundred, then in each case the margin of error is 100.

If the margin of error is less than 1, say $\frac{1}{10}$, a number is said to be correct 'to the nearest tenth', or, when we are using point notation, '*correct to* 1 *place of decimals*'.

We round off a number to a certain number of 'decimal places' in much the same way as we approximate to a given number of significant figures; but, instead of casting off the digits remaining after a certain number of figures have been considered, we cast off digits *after a certain position in the decimal fraction.* For example, in rounding off 1·453 27 to 3 places of decimals, the digits 2 and 7 are cast off (the digit representing the thousandths, 3, remains unchanged since the following digit is less than 5).

Also, 0·562 85 corrected to 3 places of decimals is 0·563; the number 0·834 02 corrected to 4 decimal places is 0·8340.

Both kinds of specification—to significant figures and to decimal places—have particular uses. In calculations involving multiplication and division, it is best to use approximations to significant figures, but in additions and subtractions, approximation by a common margin of error is more suitable. This becomes clear if we compare multiplication and addition of numbers in a specific example, using both kinds of approximation:

number	*correct to 2 SF*	*correct to the nearest hundred*
35 426	35 000	35 400
12	12	0
24	24	0

Multiplication of 35 426 by 24 obviously produces a very different result from multiplication by 12. The difference is reflected in the multiplications of their approximations to 2 SF:

$$12 \times 35\,000 = 420\,000,$$
$$24 \times 35\,000 = 840\,000$$

(these are valid approximations to the products $12 \times 35\,426$ and $24 \times 35\,426$). But if we approximate these numbers to the nearest hundred, the result is in each case multiplication by zero, and bears no relation to the original calculation.

On the other hand, the use of approximation to 2 SF in *adding* these numbers is meaningless: it amounts to an addition of 12 or 24 to a number with a margin of error of 1000. The relatively slight change made by adding 12 or 24 to a large number like 35 426 is better reflected if we add their approximations with a common margin of error; 35 000 + 0 = 35 000 is a good approximation of both 35 426 + 12 and 35 426 + 24.

5.18 Exercise

Round off the following numbers to 2 decimal places:

4·324 67;
0·058 61;
3·202 4.

5.19 Mathematical Tables

We can save ourselves a good deal of laborious manipulation of figures by making use of *mathematical tables*, like those which appear near the end of this book. You will need these tables for the work in the next few Units; it is worth a little practice to learn to use tables quickly and accurately.

Table 5.2 is a straightforward table of squares and cubes of whole numbers:

Table 5.2

number n	*its square* n^2	*its cube* n^3
1	1	1
2	4	8
3	9	27
4	16	64
5	25	125
6	36	216
7	49	343
8	64	512
9	81	729
10	100	1000
etc.		

The mathematical tables you will usually need are called *four-figure tables*; sometimes—as in the tables of squares and square roots—these 4 figures are the 4 significant figures, whilst in other tables results are given to 4 decimal places.

Most tables use a special arrangement which makes it possible to condense on to one page what would otherwise require far more space; they require *inspection* or other methods for a decision on the proper value. Such tables are the tables of *squares and square roots* and also the table of logarithms, which you will meet later.

5.20 The Arrangement of Four-figure Tables

Most four-figure tables use an arrangement of *rows* and *columns*. There are three main vertical divisions, which in the case of the square and square root tables refer to the 4 significant figures of the number whose square or square root is to be found. Look at the tables at the end of this book; you will see they are set out thus:

first two figures	*third figure* 0 1 2 3 4 5 6 7 8 9	*fourth figure* 1 2 3 4 5 6 7 8 9
10 11 12 13 . .		

The first two figures determine the row to be used, the third figure the column.

This means that if we need to find $(324{\cdot}5)^2$ in the table of squares, we first look down the first column for 32, which indicates the row to be used; in this row we select the column headed by the number 4—the third figure. We find there the number 1050.

	0	1	2	3	4	5	6	7	8	9	1	2	3	4	5	6	7	8	9
32	1024	1030	1037	1043	[1050]	1056	1063	1069	1076	1083	1	1	2	3	[3]	4	5	5	6

A small adjustment now has to be made depending on the fourth figure. These adjustments are given in the same row, in the columns on the extreme right of the page. Select the appropriate adjustment in the column headed by the number 5—the fourth figure of 324·5—in this case 3, and *add* this to the number 1050.

The figures 1053 that we have now found are the 4 significant figures of $(324{\cdot}5)^2$; but we do not yet know the value that these digits represent, or the position of the unit point.

5.21 Exercises

1. Using the table of squares find the 4 significant figures of each of the following:

$$(7{\cdot}256)^2;$$
$$(64{\cdot}15)^2;$$
$$(118{\cdot}5)^2.$$

2. Similarly find the 4 significant figures of the following:

$$(53{\cdot}629)^2;$$
$$(1{\cdot}034\,45)^2;$$
$$(678{\cdot}689)^2.$$

[Hint: when four-figure tables are used for calculations involving numbers of more than 4 significant figures these numbers must first be rounded off to 4 SF; for example, 53·629 to 53·63.]

5.22 The Table of Squares

We have already noticed the 'significance' of the first digits in the multiplication of two numbers. Consider also multiplications such as

$$23 \times 35,\ 230 \times 3500,\ 23\,000 \times 350\,000 \text{ or } 2{\cdot}3 \times 0{\cdot}35.$$

Apart from the zeros and the position of the unit point, the pairs of numbers contain the same digits, and we can show that their products contain the same digits too:

$$23 \times 35 = 805,$$

$$230 \times 3500 = 23 \times 10 \times 35 \times 10^2 = (23 \times 35) \times 10 \times 10^2 = 805 \times 10^3,$$

$$23\,000 \times 350\,000 = 23 \times 10^3 \times 35 \times 10^4 = (23 \times 35) \times 10^3 \times 10^4 = 805 \times 10^7,$$

$$2{\cdot}3 \times 0{\cdot}35 = 23 \times \frac{1}{10} \times 35 \times \frac{1}{10^2} = (23 \times 35) \times \frac{1}{10} \times \frac{1}{10^2} = 805 \times \frac{1}{10^3},$$

and so on.

We have the same situation when a number is squared, that is, multiplied by itself. For example,

$$25^2 = 25 \times 25 = 625,$$

$$250^2 = 250 \times 250 = 25 \times 10 \times 25 \times 10 = (25 \times 25) \times 10 \times 10 = 625 \times 10^2,$$

$$25\,000^2 = 25\,000 \times 25\,000 = 25 \times 10^3 \times 25 \times 10^3 = (25 \times 25) \times 10^3 \times 10^3 = 625 \times 10^6,$$

$$(2{\cdot}5)^2 = 2{\cdot}5 \times 2{\cdot}5 = 25 \times \frac{1}{10} \times 25 \times \frac{1}{10} = (25 \times 25) \times \frac{1}{10} \times \frac{1}{10} = 625 \times \frac{1}{10^2}$$

and so on.

The table of squares (and the table of square roots) provides information about the 4 significant figures of the number to be squared and of the square of that number.

For example, the table of squares can tell us that the square of any number whose 4 significant figures are 3824 will be a number with 1462 as its 4 significant figures. (Check this for yourself.) Whether these digits represent the number of millions, or thousands, tenths or hundredths—in other words the *position of the unit point*—has to be decided by inspection, and we can do this by a rough approximation, making use of a simple table of squares:

$1^2 = 1$	$10^2 = 100$	$100^2 = 10\,000$	$(0{\cdot}1)^2 = 0{\cdot}01$	$(0{\cdot}01)^2 = 0{\cdot}0001$
$2^2 = 4$	$20^2 = 400$	$200^2 = 40\,000$	$(0{\cdot}2)^2 = 0{\cdot}04$	$(0{\cdot}02)^2 = 0{\cdot}0004$
$3^2 = 9$	$30^2 = 900$	$300^2 = 90\,000$	$(0{\cdot}3)^2 = 0{\cdot}09$	$(0{\cdot}03)^2 = 0{\cdot}0009$
$4^2 = 16$	$40^2 = 1600$	$400^2 = 160\,000$	$(0{\cdot}4)^2 = 0{\cdot}16$	$(0{\cdot}04)^2 = 0{\cdot}0016$
$5^2 = 25$	$50^2 = 2500$	$500^2 = 250\,000$	$(0{\cdot}5)^2 = 0{\cdot}25$	$(0{\cdot}05)^2 = 0{\cdot}0025$
$6^2 = 36$	$60^2 = 3600$	$600^2 = 360\,000$	$(0{\cdot}6)^2 = 0{\cdot}36$	$(0{\cdot}06)^2 = 0{\cdot}0036$
$7^2 = 49$	$70^2 = 4900$	$700^2 = 490\,000$	$(0{\cdot}7)^2 = 0{\cdot}49$	$(0{\cdot}07)^2 = 0{\cdot}0049$
$8^2 = 64$	$80^2 = 6400$	$800^2 = 640\,000$	$(0{\cdot}8)^2 = 0{\cdot}64$	$(0{\cdot}08)^2 = 0{\cdot}0064$
$9^2 = 81$	$90^2 = 8100$	$900^2 = 810\,000$	$(0{\cdot}9)^2 = 0{\cdot}81$	$(0{\cdot}09)^2 = 0{\cdot}0081$

Approximating, for example, $(15{\cdot}65)^2$ on the basis of this table, we find that, since $10 < 15{\cdot}65 < 20$, then $10^2 < (15{\cdot}65)^2 < 20^2$ or $100 < (15{\cdot}65)^2 < 400$. Alternatively, we could write

$$10 < 15{\cdot}65 < 20 \Rightarrow 10^2 < (15{\cdot}65)^2 < 20^2 \Rightarrow 100 < (15{\cdot}65)^2 < 400.$$

The 4 significant figures of the required number are found in the table of squares: they are 2449. Now we can place the unit point unambiguously; $244{\cdot}9$ is the only possibility which satisfies the conditions $100 < (15{\cdot}65)^2 < 400$ or $100 < 244{\cdot}9 < 400$.

An easy way to find the position of the unit point is to express the number to be squared in the standard form—that is, in the form $A \times 10^n$ (see Section 4.21)—and approximate its square on the basis of the table of squares of whole numbers from 1 to 10. Remember that *both* parts of the number must be squared, including of course the power of 10. For example,

$$\begin{aligned} 235^2 &= (2{\cdot}35 \times 10^2)^2 \\ &= (2{\cdot}35 \times 10^2) \times (2{\cdot}35 \times 10^2) \\ &= 2{\cdot}35 \times 2{\cdot}35 \times 10^2 \times 10^2 \\ &= (2{\cdot}35)^2 \times 10^4. \end{aligned}$$

Since $2 < (2{\cdot}35) < 3$,

$$2^2 < (2{\cdot}35)^2 < 3^2 \text{ or } 4 < (2{\cdot}35)^2 < 9.$$

Therefore, using tables, $(2{\cdot}35)^2 = 5{\cdot}523$, to 4 SF, and

$$235^2 = 5{\cdot}523 \times 10^4 = 55\,230, \text{ to 4 SF.}$$

Similarly,

$$\begin{aligned} 3456^2 &= (3{\cdot}456 \times 10^3)^2 \\ &= (3{\cdot}456 \times 10^3) \times (3{\cdot}456 \times 10^3) \\ &= (3{\cdot}456)^2 \times 10^6. \end{aligned}$$

Since $3 < 3{\cdot}456 < 4$,

$$3^2 < (3{\cdot}456)^2 < 4^2 \text{ or } 9 < (3{\cdot}456)^2 < 16.$$

Therefore, using tables, $(3{\cdot}456)^2 = 11{\cdot}94$, to 4 SF

and $\qquad 3456^2 = 11{\cdot}94 \times 10^6 = 11\,940\,000$, to 4 SF.

5.23 Exercises

1. Using the simple table of squares in Section 5.22 make a rough approximation of the following:

$$12^2;\ 35^2;\ (3{\cdot}4)^2;\ (49{\cdot}35)^2;\ (0{\cdot}5345)^2.$$

Now, using the four-figure table of squares, find these squares correct to 4 significant figures.

2. Express the following numbers in standard form and find their squares correct to 4 significant figures:

$$34{\cdot}56;\ 1134;\ 784{\cdot}3;\ 5642;\ 0{\cdot}008\,423.$$

5.24 The Table of Square Roots

We have described the relation between pairs of numbers like 4 and 2, 9 and 3 or 16 and 4, as a 'square' relation; 4 is the square of 2, 9 is the square of 3, and so on.

If, however, we start with the second number in these number pairs, its relation to the first is described in mathematics by the words '*is the square root of*': 2 is the square root of 4, and 3 is the square root of 9. (Compare the relation between Mr and Mrs Jones described by the words: Mr Jones is the husband of Mrs Jones. There is a relation in the opposite sense too: Mrs Jones is the wife of Mr Jones.) Finding the square root of a number consists of finding another number which when squared equals that number. For example, if we are asked to find the square root of 25, we have to find the number which, when squared, equals 25. The symbol for this relation is $\sqrt{\ }$; $\sqrt{16} = 4$, $\sqrt{9} = 3$, etc.

Remember that a positive number can be the product of two positive *or* of two negative numbers, and that therefore the square root of a number always has two possible values, a positive and a negative one. For example, $\sqrt{25} = {}^{+}5$ but also $25 = ({}^{-}5)^2$, and therefore $\sqrt{25} = {}^{+}5$ or ${}^{-}5$, usually written $\sqrt{25} = {}^{\pm}5$. For convenience only, we will confine ourselves to the positive values for the present. (See Section 13.6.)

Further examination of the table of *squares* reveals an interesting and perhaps unexpected number relation.

$$2^2 = 4 \text{ so that } \sqrt{4} = 2;$$
$$20^2 = 400 \text{ so that } \sqrt{400} = 20;$$
$$200^2 = 40\,000 \text{ so that } \sqrt{40\,000} = 200;$$
etc.

In this sequence, in which we have squared consecutively 2, 20, 200, etc., the number of zeros is doubled each time: the sequence of squares is 4, 400, 40 000, and so on.

Now consider the other numbers which have 4 as their first digit and have an *odd* number of zeros, such as 40, 4000, 400 000, etc.

By approximation we find that $\sqrt{40}$ lies between 6 and 7, or $6 < \sqrt{40} < 7$ (because $6^2 < 40 < 7^2$); indeed, $\sqrt{40}$ can be calculated correct to 4 SF as 6·325. Similarly,

$$\sqrt{4000} = 63{\cdot}25;$$
$$\sqrt{400\,000} = 632{\cdot}5;$$
etc.

In fact, if the significant figures of any number are given, there are two alternatives for the significant figures of its square root. Four-figure tables of square roots will therefore always give two alternatives, and which of the two applies is found by inspection.

Example

Find $\sqrt{834{\cdot}2}$.

In four-figure tables the two alternatives given are

2888
9133

A rough approximation is: $400 < 834{\cdot}2 < 900$
or $20^2 < 834{\cdot}2 < 30^2$.

The answer we need therefore lies between 20 and 30; the only number between 20 and 30 that can be obtained by placing the unit point in either alternative is 28·88. Therefore, $\sqrt{834{\cdot}2} = 28{\cdot}88$ to 4 SF.

The 'doubling' aspect of squaring numbers—already mentioned—suggests a simple method for inspection and approximation when finding square roots:

Group the digits of a number in pairs, starting from the unit point, and choose the correct value of the square root from the two alternatives given in the table on the basis of the first number or pair of numbers on the left (ignoring zeros).

Example

Find $\sqrt{53\,240}$.

Pair off as follows: 5 | 32 | 40· |

$\sqrt{5}$ is approximately 2· . . .; so, of the two alternatives given in the square root table, $\frac{2308}{7297}$, choose 2308.

The position of the unit point can then be determined by inspection, or by counting the number of pairs of digits, or by placing the digits of the chosen alternative over these pairs (one digit over each pair):

$$\begin{array}{c|c|c|c} 2 & 3 & 0\cdot & 8 \\ 5 & 32 & 40\cdot & \end{array}$$

Therefore $\sqrt{53\,240} = 230{\cdot}8$.

This last method is particularly useful in finding square roots of numbers less than 1.

Example

Find $\sqrt{0{\cdot}007\,321}$.

$$0\cdot \mid 00 \mid 73 \mid 21$$

Approximating $\sqrt{73}$ as $8\cdot\ldots$ $(64 < 73 < 81$ or $8^2 < 73 < 9^2)$, we choose of the alternatives, $\frac{2706}{8557}$, the second namely 8557.

We place these digits, 8557, over the pairs as follows:

$$\begin{array}{c|c|c|c|c|c|} 0\cdot & 0 & 8 & 5 & 5 & 7 \\ 0\cdot & 00 & 73 & 21 & 00 & 00 \end{array}$$

Therefore $\sqrt{0{\cdot}007\,321} = 0{\cdot}085\,57$.

5.25 Exercise

Using the table of square roots, find the four significant figures of:

$\sqrt{376{\cdot}4}$;
$\sqrt{92{\cdot}47}$;
$\sqrt{0{\cdot}002\,512}$.

Select the appropriate alternative, and determine the position of the unit point by either of the methods in Section 5.24.

Unit Six

Scales and Progressions

6.1 Arithmetic Progressions and the Linear Scale

Employees' salaries are often calculated on the basis of a starting salary, plus fixed increments (increases) for each year of service. Suppose a man begins to work for a firm at a starting salary of £1000, and his annual increments are £75 each; then in successive years he will be paid salaries of £1000, £1075, £1150, £1225, £1300, and so on. Such a series of numbers is called an *arithmetic progression*. There is a fixed first number, the first *term* of the progression, and each further term differs from the preceding term by the same amount, the *common difference*.

Probably the simplest arithmetic progression (abbreviated to AP) is the series of natural numbers:

$$0,\ 1,\ 2,\ 3,\ 4,\ 5,\ 6,\ \ldots.$$

The multiplication tables are all familiar APs, in which the first term is always zero:

$$0,\ 2,\ 4,\ 6,\ 8,\ 10,\ \ldots$$
$$0,\ 3,\ 6,\ 9,\ 12,\ 15,\ \ldots$$
$$0,\ 4,\ 8,\ 12,\ 16,\ 20,\ \ldots$$

An arithmetic progression can start from any number, positive or negative, for example:

$$5,\ 8,\ 11,\ 14,\ 17,\ \ldots$$
$$2\tfrac{1}{4},\ 6\tfrac{1}{4},\ 10\tfrac{1}{4},\ 14\tfrac{1}{4},\ \ldots$$
$$^{-}7,\ ^{-}2,\ ^{+}3,\ ^{+}8,\ \ldots$$

Rational numbers can serve as the common difference, as in the AP

$$9,\ 11\tfrac{1}{2},\ 14,\ 16\tfrac{1}{2},\ 19,\ \ldots$$

and negative numbers, as in the AP

$$^{+}10,\ ^{+}7,\ ^{+}4,\ ^{+}1,\ ^{-}2,\ ^{-}5,\ \ldots$$

An arithmetic progression can be represented on the number line by a series of points placed the same distance apart; the progression 13, 17, 21, 25, ... then appears as in Fig. 6.1.

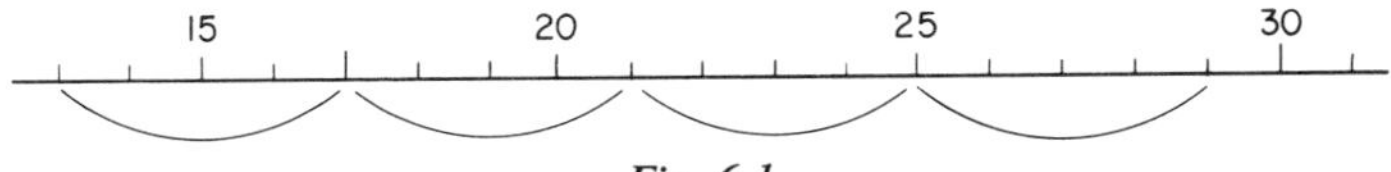

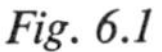

Fig. 6.1

Indeed, we could regard the scale of the number line itself as an arithmetic progression; we constructed it by repeatedly adding one unit (see Section 1.4). The scale of the number line, and any other scale where adding a fixed or constant number is represented by a fixed interval, is called a *linear scale*; the simple slide rule, introduced in Section 2.2, has this kind of scale.

6.2 Exercises

Clearly, if we want to build up an arithmetic progression, we need only know the first term and the common difference. (Use the first term on the number line as starting point and take 'steps' of the required length; compare the way you carried out addition using the simple slide rule.)

Write down the first 8 terms of the arithmetic progression:

1. which has first term 11 and common difference 7;
2. which has first term $^{-}2$ and common difference 8;
3. which has first term $1\frac{1}{3}$ and common difference $\frac{2}{3}$;
4. which has first term 12 and common difference $^{-}4$;
5. which has first term $^{-}5$ and common difference $^{-}2$.

6.3 The General Term of an Arithmetic Progression

If we need to find a particular term of an AP, for example the 12th term of an AP whose first term is 5 and common difference is 3, we could, of course, build up the AP as far as the required term. But this method would become rather laborious if we had to find, say, the 63rd term of the progression.

We can find an easier method if we look at the progression more closely.

Using the same example, the second term is found by adding the common difference, 3, to the first term, 5,: $5 + 3$;

the third term is found by adding 3 to the second term:

$$(5 + 3) + 3, \text{ that is, } 5 + (2 \times 3);$$

the fourth term is found by adding 3 to the third term:

$$(5 + 3 + 3) + 3 \text{ or } 5 + (3 \times 3).$$

Without completing all the step-by-step calculations, we can see straight away that the *twelfth* term can be found by adding 3 *eleven* times to the first term:

$$5 + (11 \times 3).$$

If we call the common difference d, and denote the first term by a, then for any AP we can say that:

the 2nd term is $a + d$;
the 3rd term is $a + 2d$;
the 4th term is $a + 3d$;
the 5th term is $a + 4d$;
etc.

This suggests a formula for any term of the progression:

$$\boxed{\ldots}\text{th term} = a + (\boxed{\ldots} - 1)\,d.$$

By writing, for example, the number 7 in each of the boxes we find the 7th term:

$$\boxed{7}\text{ th term} = a + (\boxed{7} - 1)\,d$$

or

$$7\text{th term} = a + 6d.$$

Drawing boxes is rather clumsy; we can use a letter instead, for which the appropriate number may be substituted. The formula is then:

$$\boxed{n\text{th term} = a + (n - 1)\,d}$$

This nth term is called the *general term* of the series. If the general term of an AP is given, all the terms can be found by successively substituting for n the numbers 1, 2, 3, and so on.

Sometimes we can simplify the formula for the general term; for example, if the first few terms of a series are given as 1, 3, 5, 7 . . .,

the first term is 1;
the common difference is 2;
and the general term, or nth term, is $1 + (n - 1)\,2$

$$= 1 + 2n - 2$$
$$= 2n - 1.$$

6.4 Exercises

1. Find the 15th term of the AP whose first term is 8 and common difference is 2.
2. Find the 21st term of the AP whose first term is $^{-}3$ and common difference is 5.
3. Write down the first 6 terms of an AP whose general term is given by the formula:

$$n\text{th term} = 2 + 3n.$$

4. Write down the formula for the general term, that is the nth term, of the APs:
   6, 10, 14, 18, . . .
   12, 9, 6, 3, . . .
   [Hint: decide first of all on the first term and the common difference; then use the formula in Section 6.3.]

5. The 4th term of an AP is 14 and the 12th term is 70; find the first term and the common difference.
[Hint: use the number line and consider the number of 'steps' between the two given terms.]

6.5 The Sum of the First *n* Terms of an AP

In an arithmetic progression of 3 terms only (or considering any 3 consecutive terms of an AP), the second term is the *middle term* in more than one way. It occupies the middle place in the series, and also lies half-way between the first and the third terms on the number line; for example, if 3 consecutive terms of an AP are 10, 16, 22, the middle term, 16, is 6 more than 10 and 6 less than 22 (Fig. 6.2).

10 16 22

Fig. 6.2

(The middle term is called the *arithmetic mean* of 10 and 22.)

If we add the 3 terms, we have:

$$10 + 16 + 22 \text{ or } (16 - 6) + 16 + (16 + 6) = 3 \times 16.$$

In all APs, the sum of any 3 consecutive terms is $3 \times$ middle term.

If m denotes the middle term, and d the common difference as before, the first of the 3 terms is $m - d$, the third term is $m + d$, and the sum of the 3 terms is:

$$m - d + m + m + d = 3m.$$

Note that the sum of the first and the third term, or of the first and the last term of the 3 terms, is twice the middle term.

Similarly, in a sequence of 5 terms, the third term is the middle term and the same kind of 'balance' exists; not only is the sum of the second and fourth terms twice the middle term, but also the sum of the first and the fifth terms.

Look, for example, at the arithmetic progression 4, 7, 10, 13, 16.

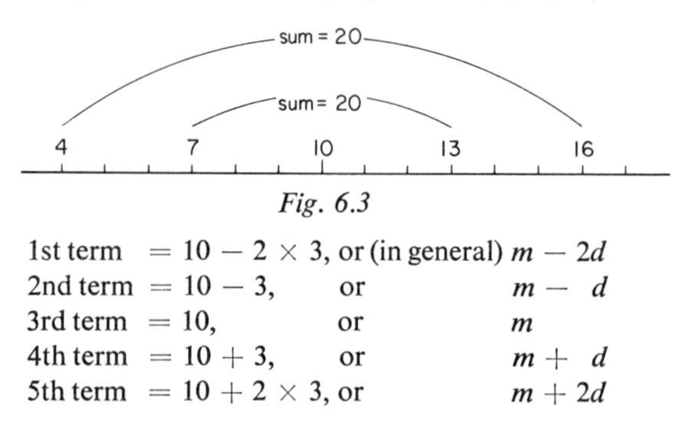

Fig. 6.3

1st term	$= 10 - 2 \times 3$,	or (in general)	$m - 2d$
2nd term	$= 10 - 3$,	or	$m - d$
3rd term	$= 10$,	or	m
4th term	$= 10 + 3$,	or	$m + d$
5th term	$= 10 + 2 \times 3$,	or	$m + 2d$

The sum of the first 5 terms $= 5 \times 10$, or (in general) $5m$.

Similar relations hold for all APs which have *one* middle term, i.e. all APs consisting of *an odd number of terms*:

sum of the first 7 terms (abbreviated to S_7) $= 7 \times$ middle term;
sum of the first 9 terms (abbreviated to S_9) $= 9 \times$ middle term;

and so on.

Since the middle term is half the sum of the first term, a, and the last term, l, or $m = \frac{1}{2}(a + l)$, a general formula for the sum of the first n terms is:

$$S_n = n \times \tfrac{1}{2}(a + l) \text{ or } S_n = \frac{n}{2}(a + l).$$

This formula also applies to APs with an even number of terms, for example, the AP 3, 7, 11, 15. There is a 'balancing number' half-way between the two middle terms, in this case the number 9; the sum of the two middle terms is twice this balancing number, and so also is the sum of the first and the last terms, that is, the balancing number is $\frac{1}{2}(a + l)$.

The general formula for the sum of the first n terms of an arithmetic progression in *all* cases is therefore:

$$\boxed{S_n = \frac{n}{2}(a + l)}$$

6.6 Exercises

1. Find the sum of the first 12 even numbers, 2, 4, . . ., 24.
   [Hint: $n = 12$, $a = 2$, $l = 24$.]
2. Using the general formula calculate $1 + 2 + 3 + 4 + 5 \ldots + 25$.
   [Hint: $n = 25$, $a = 1$, $l = 25$.]
3. Find a general formula for the sum of the first n positive integers.
4. The first 3 terms of an AP are 22, 26, 30; find the common difference, the 13th term and the sum of the first 15 terms.

6.7 Geometric Progressions

While an arithmetic progression is built up by repeatedly adding a constant amount to a given first number, repeated multiplication by a number builds up a different kind of series. If the starting prize in a game of 'Double Your Money' is £3, repeated doubling (multiplication by 2) would produce the series of possible prizes £3, £6, £12, £24, £48, £96, . . . There is a constant ratio between each two consecutive terms, in this case 2:1. To build up such a series, a *geometric progression*, we need to know only two numbers: the first term and the constant ratio, usually called the *common ratio*.

If, for example, the first term is 5 and the common ratio is 3, or 3/1, the progression begins: 5, 15, 45, 135, ...

The first term and the common ratio can be any numbers, whole numbers or rational numbers, positive or negative.

6.8 Exercises

Write down the first 6 terms of the geometric progressions (GPs) whose first term and common ratio are respectively:

1. 3 and 4;
2. $^{-}5$ and $^{+}2$;
3. 5 and $^{-}2$;
4. $\frac{1}{2}$ and 4;
5. 48 and $\frac{1}{2}$.

6.9 The General Term of a GP

If the first term in a GP is 7, and the common ratio is 3:

the 2nd term is 7×3,
the 3rd term is $7 \times 3 \times 3$, or 7×3^2,
the 4th term is $7 \times 3 \times 3 \times 3$, or 7×3^3,
the 5th term is $7 \times 3 \times 3 \times 3 \times 3$, or 7×3^4,

...

the 13th term is 7×3^{12},

...

the nth term is $7 \times 3^{n-1}$.

For any geometrical progression, if the first term is denoted by a and the common ratio by r, the general term is given by the formula

$$n\text{th term} = a \times r^{n-1}$$

6.10 Exercises

1. Use the general formula for the nth term of a GP to find the 7th term of the GPs whose first term and common ratio are given in Exercise 6.8; check, using your earlier answers.
2. If the first 3 terms of a GP are 4, 12, 36, what is the common ratio? What is the ratio between the 1st and 3rd terms? the 2nd and 4th terms? the 2nd and 5th terms?
3. If the 1st term in a GP is 2 and the 4th term is 54, what is the common ratio?
4. If the common ratio of a GP is $1\frac{1}{2}$ (or $\frac{3}{2}$) and the 3rd term is 36, what is the 1st term?

6.11 Geometric Mean

From the examples in Exercise 6.10 you will have discovered an interesting relation between 3 consecutive terms of a GP: there is again a kind of balance about the middle term. In the GP mentioned in Exercise 2 (4, 12, 36),
the third term (36) is 3 times the middle term (12),
the first term (4) is $\frac{1}{3}$ of the middle term or $12 \div 3$.
If the letter M denotes the middle term and r the common ratio, any 3 consecutive terms of a GP can be written as:

$$\frac{M}{r},\ M,\ r \times M.$$

The product of the first and the third term is then clearly the square of the middle term:

$$\frac{M}{r} \times rM = M^2,$$

or $\boxed{(\textit{first term} \times \textit{third term}) = (\textit{middle term})^2}$

Check this result, using examples of 3 consecutive terms in any GP:

for example, $4 \times 36 = 12^2$.

The middle term of 3 consecutive terms of a GP is called the *geometric mean* or *geometric average*. (We will see in the next section that on a 'geometric' or 'logarithmic scale' this geometric mean lies indeed half-way between the other two numbers.)

The formula (first term $\times$ third term) = (middle term)^2 can be used to find the middle term, or geometric mean, if the other two terms are known. For example, if the first of 3 terms of a GP is given as 5, and the third as 80:

$$(\text{middle term})^2 = 5 \times 80 = 400.$$

Therefore the middle term is $\sqrt{400} = 20$, and the 3 terms are 5, 20, 80; we can say that the *geometric mean* of the two numbers 5 and 80 is 20.

6.12 Exercise

Find the geometric mean of $\frac{1}{4}$ and 16, and of 3 and 75.

6.13 The Law of Growth and the Logarithmic Scale

The geometric progression, in which the value of the first term affects the difference between the terms, is a more usual phenomenon than the arithmetic progression. The increase in capital in the form of interest received from a bank depends on how much money one deposits in the bank in the first place, or, to take another example, populations do not increase by a fixed number of people in a given period, regardless of the original size of the population. Indeed the world population *doubles* about every 26 years. If this rate of growth is maintained we can estimate the world population, at intervals of about 26 years, as:

in 1935: 2 000 000 000;
in 1971: 4 000 000 000;
in 2007: 8 000 000 000;
in 2043: 16 000 000 000.

Similar examples of growth by multiplication are found in biology, such as the splitting of cells or the reproduction of bacteria and viruses.

If we want to illustrate the doubling of the world population at regular intervals of time, we could place the terms of the sequence at regular intervals along a line (Fig. 6.4).

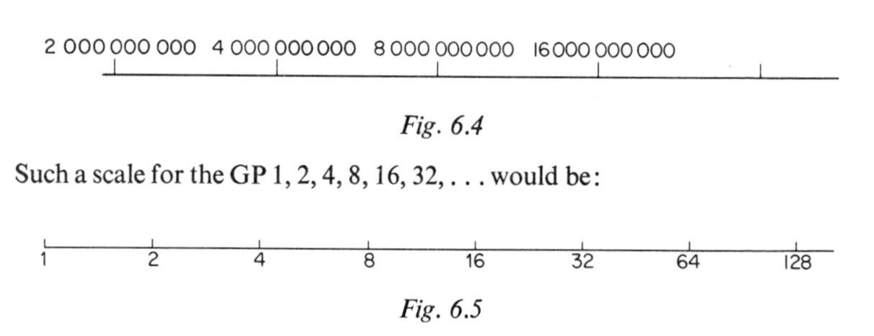

Fig. 6.4

Such a scale for the GP 1, 2, 4, 8, 16, 32, . . . would be:

Fig. 6.5

A similar scale for the GP 1, 3, 9, 27, . . . is:

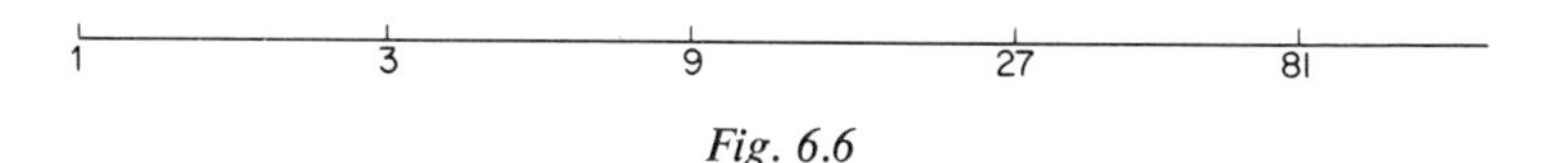

Fig. 6.6

We could incorporate these various scales into a single scale, usually called the *logarithmic scale* (Fig. 6.7).

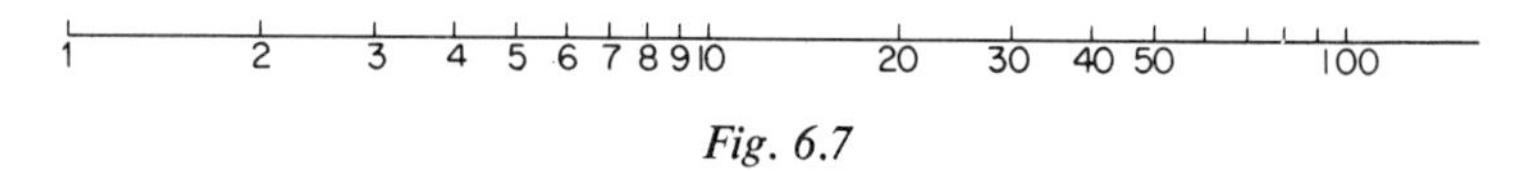

Fig. 6.7

Such a scale with all its intermediate points is difficult to construct, but fortunately graph paper with the logarithmic scale printed on it—logarithmic paper, or log paper—is commercially available. There are two kinds of log paper: *log–linear* or *semilog paper* (Fig. 6.8) on which one edge is marked with a logarithmic scale and the other with a linear scale, and *log–log paper* (Fig. 6.9), on which both edges are marked with a logarithmic scale.

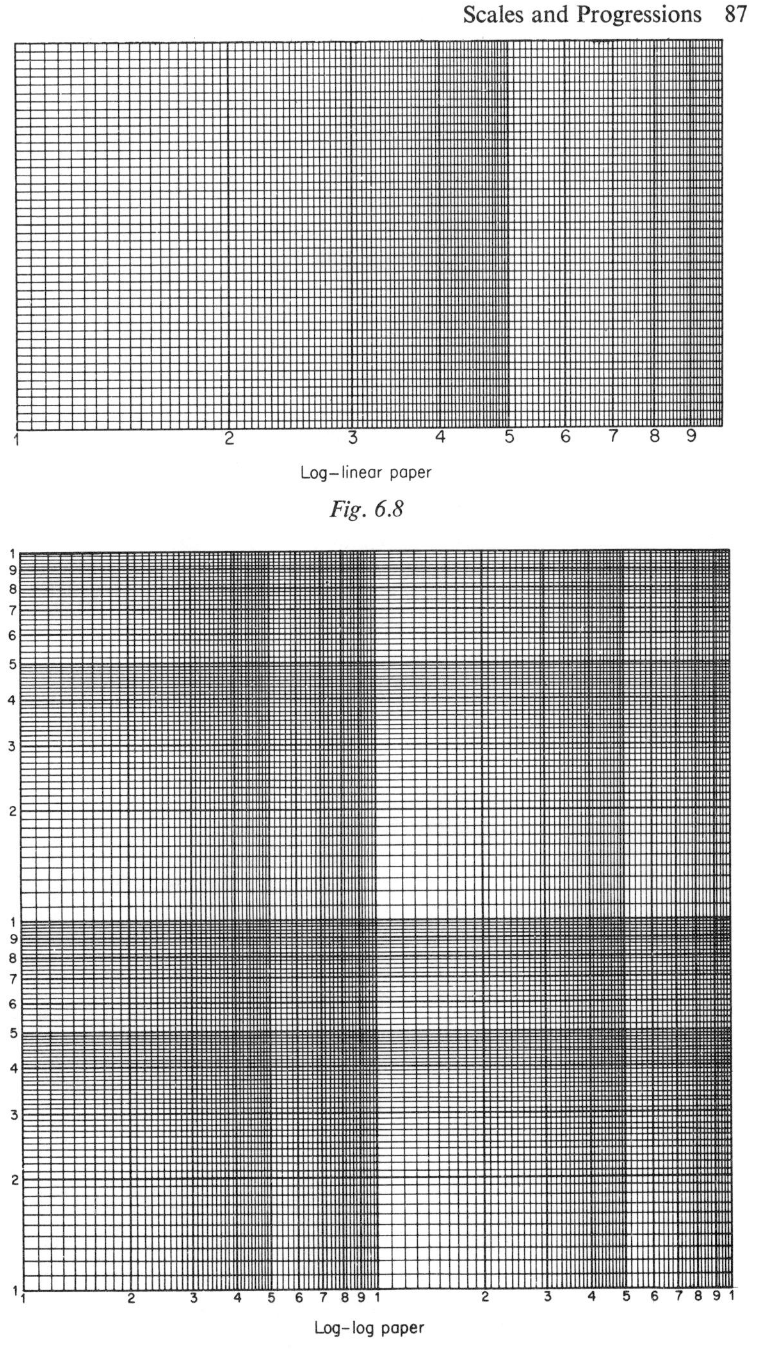
1 2 3 4 5 6 7 8 9
Log–linear paper
1 9 8 7 6 5 4 3 2 1 9 8 7 6 5 4 3 2 1
1 2 3 4 5 6 7 8 9 1 2 3 4 5 6 7 8 9 1
Log–log paper

Fig. 6.8

Fig. 6.9

6.14 Exercises

1. Draw a line on a large sheet of paper and carefully mark the points 1, 2, 3, 4, 5, 6, 7, 8, 9, 10, and 20 according to the logarithmic scale as given above. (Use log paper if possible.) Cut a strip of paper of the same length as the interval between 1 and 2. Place this strip along your logarithmic scale, and check whether your points 2, 4, 8, 16 are indeed the same distance apart; then, using this strip, continue along the line and mark the points 32, 64, 128, 256, etc.

   Each time we 'take a step' of the length of this strip to the right, we multiply by 2. Check this by 'taking steps' starting from 3 (Fig. 6.10).

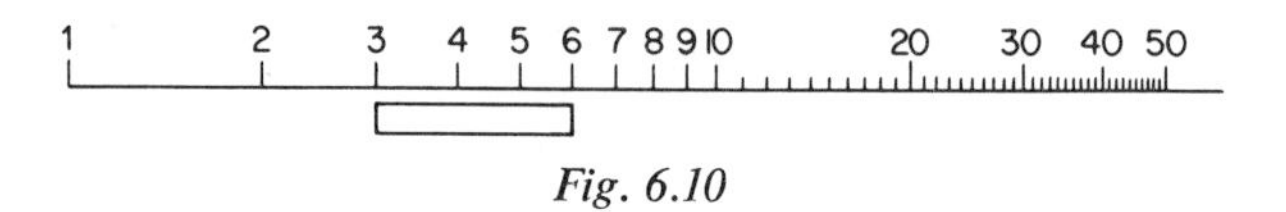

Fig. 6.10

   Continue this process, marking the points 24, 48, 96, 192, ...
2. Now cut off a strip of paper of the same length as the interval between 1 and 3; repeat the process using 'steps' of this length to multiply by 3.
3. Repeat this exercise using strips of length 5, 7 and 11.
4. Show on your logarithmic scale that division can be done by 'taking steps' to the left. (For example, Fig. 6.10 illustrates $6 \div 2 = 3$ as well as $2 \times 3 = 6$.)

6.15 The Slide Rule

Instead of cutting strips of different lengths to multiply by different numbers, we can use a single strip of paper on which all these lengths are marked. We then have a slide rule which is essentially the same as the commercial slide rule.

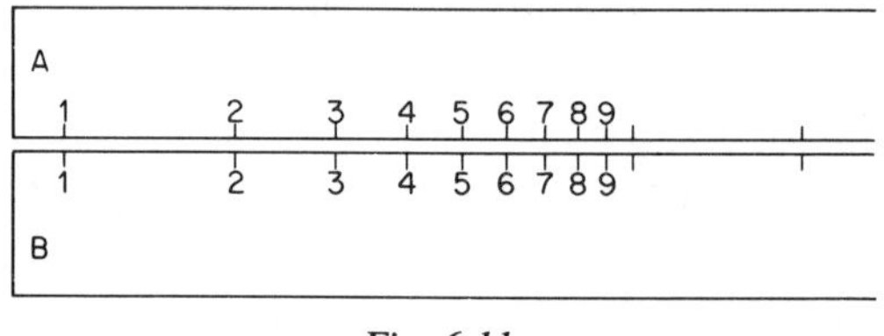

Fig. 6.11

If logarithmic paper is used, the intermediate markings will represent fractions.

Mark the two strips A and B, and use them to multiply just as in Exercise 6.14. For example, for *multiplication by* 2: move the B-scale to the right so that the number 1 on the B-scale faces the number 2 on the A-scale (Fig. 6.12).

The product is found on the A-scale, opposite the number to be multiplied on the B-scale. For example, opposite the number 3 on the B-scale is the

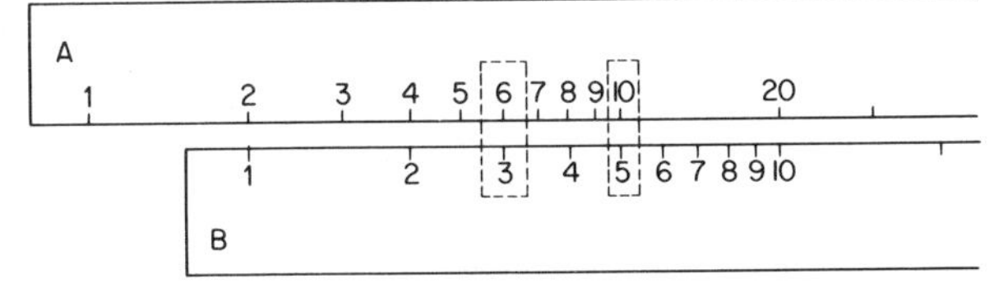

Fig. 6.12

number 6 on the A-scale, illustrating $2 \times 3 = 6$; opposite the number 5 on the B-scale is the number 10 on the A-scale.

Without adjusting the slide rule, we can also read off *division by* 2: opposite the number to be divided (the *dividend*) on the A-scale, we find the quotient on the B-scale. For example, opposite the number 8 on the A-scale, we find the number 4 on the B-scale ($8 \div 2 = 4$), and so on.

We can multiply and divide point fractions by using the intermediate markings on the logarithmic scale on the slide rule between the whole numbers; for example, if we place the number 1 on the B-scale opposite the number 2·4 on the A-scale, we can read off multiplication by 2·4 or division by 2·4.

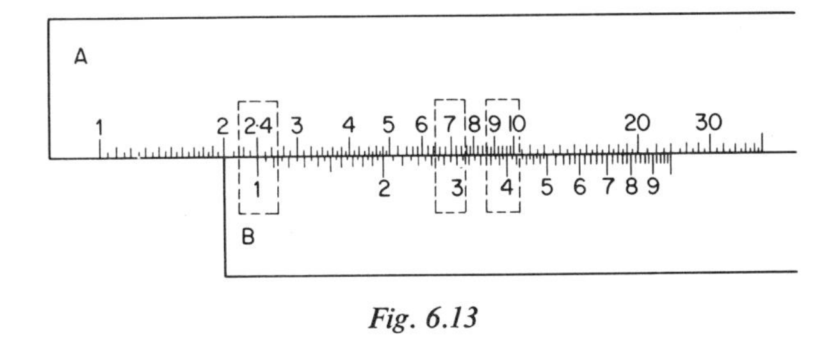

Fig. 6.13

For example,

$$2{\cdot}4 \times 3 = 7{\cdot}2,$$
$$9{\cdot}6 \div 2{\cdot}4 = 4.$$

6.16 Exercises

1. Use your slide rule to calculate:

3×4;	$3 \times 1{\cdot}5$;
$12 \div 3$;	$7{\cdot}5 \div 3$;
$2{\cdot}5 \times 4$;	$2{\cdot}5 \times 1{\cdot}2$;
$7{\cdot}5 \div 2{\cdot}5$;	$8{\cdot}5 \div 2{\cdot}5$.

2. Make up further examples of simple multiplications and divisions, and practise the procedures of multiplication and division on your slide rule.

6.17 The Commercial Slide Rule

At this stage you will find it helpful to use a proper commercial slide rule.

Most commercial slide rules are more than just slide rules; they display all sorts of information, useful to the various kinds of expert. They may provide data on cubes, logarithms and trigonometric functions, and there are special slide rules for chemists and engineers. You will probably find a simple, inexpensive, all-purpose slide rule the most useful; the more specialized, sophisticated variety can be confusing because of the amount of unwanted information.

The essential part of the slide rule consists of two sets of scales: the A- and B-scales with a logarithmic scale from 1 to 100, the C- and D-scales with a logarithmic scale from 1 to 10.

Note. Some slide rules use different letters to indicate the various scales; A, B, C and D are the most usual indications. Check that your own slide rule corresponds with the description given here; and if not, what indications it uses.

The B- and C-scales are given on the *slide*; a *rider* or *cursor* is provided for accuracy in taking readings.

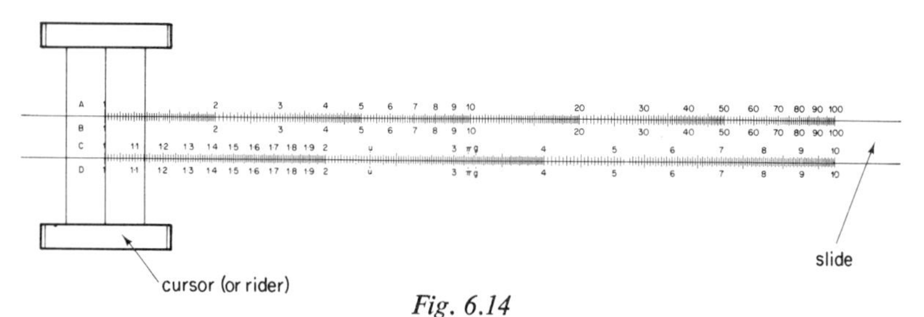

Fig. 6.14

In Unit 2 we used a 'simple slide rule' with a linear scale for addition and subtraction; from now on we will use the term 'slide rule' only for the slide rule with the logarithmic scale.

6.18 Procedures

We emphasized earlier that a *fixed procedure* is essential if you are to use the slide rule efficiently. You may adopt your own procedure, as long as it is correct and becomes a fixed habit. If you have not used a slide rule before, we suggest the following *procedure for multiplication* using the example $2{\cdot}4 \times 3{\cdot}5$:

(*a*) Place the inscribed line of the rider on the multiplier (2·4) on the A-scale.
(*b*) Move the slide so that the number 1 of the B-scale is exactly under the line of the rider.

(*c*) While holding the slide in this position, move the rider so that its line is over the number to be multiplied (3·5) on the B-scale.
(*d*) Read the answer (8·4) under the line of the rider on the A-scale.

You can use the same procedure with the C- and D-scales: in the above instructions just read C for A and D for B.

For *division* you can follow a reverse procedure. We will use the example 9 ÷ 3·6 as calculated on the C- and D-scales:

(*a*) Place the line of the rider over the dividend (9) on the D-scale.
(*b*) Move the slide so that the number 3·6 on the C-scale is exactly under the line of the rider, opposite the number 9 on the D-scale.
(*c*) Read the answer (2·5) on the D-scale opposite the number 1 on the C-scale.

6.19 Exercises

1. Practise the procedure for multiplication and division using the examples in Exercise 1 of Section 6.16.
2. Practice is the only way to ensure accuracy in the reading of intermediate markings and estimating the final figure in the result.
   Take special care to be as accurate as possible while calculating the following:

   1·45 × 6·2;
   2·15 × 4·8;
   6·4 × 3·5;
   16 ÷ 1·25;
   8·6 ÷ 1·65;
   7·25 ÷ 1·25.

6.20 Accuracy

The approximate nature of measurement limits the accuracy of the slide rule and of calculations performed with it.

The slide rule's accuracy depends first of all on its size; longer slide rules allow more, and more accurate, sub-divisions. For the same reason the C- and D-scales on all slide rules are more accurate than the A- and B-scales. Moreover, the left-hand end of the slide rule is more accurate than the right-hand end, because the logarithmic scale is a contracting number scale, so that for example the space between 1 and 2 is the same as the space between 5 and 10.

But most numbers can be read accurately on the slide rule to 3 significant figures; using the C- and D-scales numbers can even be read off to 4 significant figures, especially on the left-hand end.

6.21 Exercise

Using the rider, find the following numbers on the slide rule, first on the C- and D-scales, then on the A- and B-scales:

1·75; 4·55; 1·345;
3·45; 6·45; 1·026;
1·38; 9·85.

6.22 Numbers not on the Slide Rule

So far, we have only considered numbers between 1 and 10 or between 1 and 100; indeed, the numbers as given on the slide rule are all between 1 and 10 (C- and D-scales) or 1 and 100 (A- and B-scales). But the slide rule can be used for calculations involving *any* numbers, if we express them in standard form (Section 4.21), or if we use simple approximations; the figures on the slide rule then represent the significant figures only.

For example, to calculate 78×315 we can write $7{\cdot}8 \times 3{\cdot}15 \times 10^3$; we can then use the slide rule to find $7{\cdot}8 \times 3{\cdot}15$ and multiply the result by 10^3 or 1000.

Alternatively, we can find the 3 significant figures of 78×315 with the slide rule (236), and place the unit point by inspection, using a rough approximation:

$$70 \times 300 < 78 \times 315 < 80 \times 350$$
$$21\,000 < 78 \times 315 < 28\,000$$

Therefore, $$78 \times 315 = 23\,600 \text{ (to 3 SF)}.$$

For maximum accuracy, use the C- and D-scales whenever possible; these have their limitations, however, in that a number is more easily 'off the scale'.

6.23 Exercise

Calculate correct to 3 SF:

$12{\cdot}8 \times 4{\cdot}65$;
238×650;
$4\,300\,000 \times 3{\cdot}25$;
$4{\cdot}25 \times 0{\cdot}62$;
$0{\cdot}0265 \times 1{\cdot}75$;
$8{\cdot}65 \div 13{\cdot}8$;
$5{\cdot}25 \div 44{\cdot}5$.

6.24 Calculations Involving more than Two Numbers

When more than two numbers are to be multiplied, for example

$$2{\cdot}3 \times 1{\cdot}65 \times 3{\cdot}25$$

the rider (or cursor) can be left in position at the end of the multiplication $2{\cdot}3 \times 1{\cdot}65$, marking the product $(2{\cdot}3 \times 1{\cdot}65)$; the multiplication $(2{\cdot}3 \times 1{\cdot}65) \times 3{\cdot}25$ is then effected by moving the slide so that the number 1 of the slide is under the line of the rider. This sequence of steps is illustrated in Fig. 6.15.

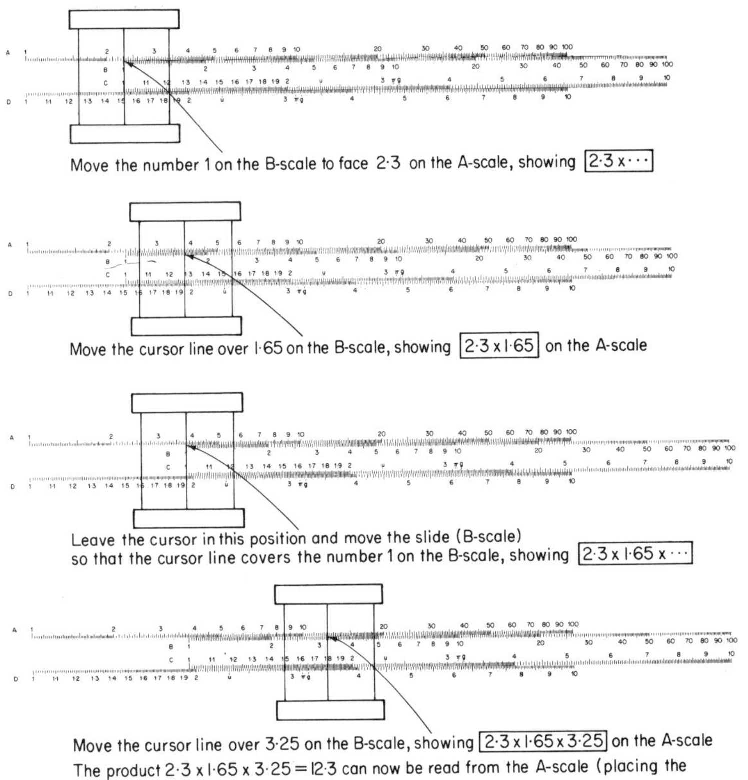

Fig. 6.15

6.25 Exercises

1. Using this last method, calculate:

$$3{\cdot}14 \times 1{\cdot}06 \times 2{\cdot}35.$$

2. In the same way calculate:

$$17{\cdot}6 \times 32{\cdot}5 \times 128.$$

3. Starting with a simple example such as $2 \times 3 \div 2{\cdot}5$, establish a similar procedure for expressions involving multiplication and division.

6.26 The A- and D-Scales as Tables of Squares and Square Roots

Since the D-scale is 'twice as long' as the A-scale, the two scales, if perfectly lined up, provide a table of squares and square roots.

To find the square of a number, for example $(2{\cdot}2)^2$, place the line of the rider exactly on that number on the D-scale and read off the answer on the A-scale. Fig. 6.16 illustrates $(2{\cdot}2)^2 = 4{\cdot}8$ correct to 2 significant figures.

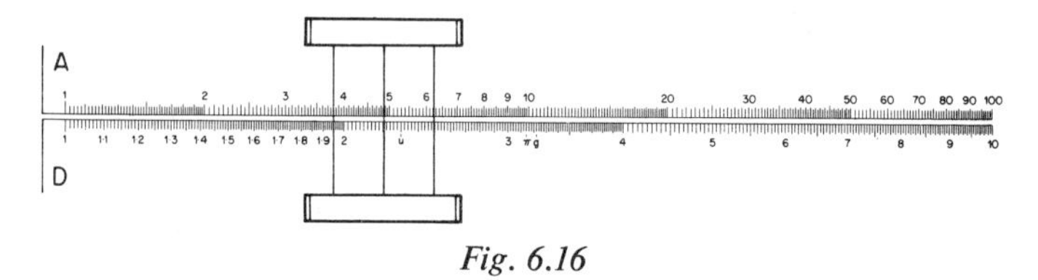

Fig. 6.16

Similarly, the square root of a number is found by placing the line of the rider over that number on the A-scale and reading off the answer on the D-scale.

6.27 Exercise

Use the examples of Exercises 5.21, 5.23 and 5.25 to practise finding squares and square roots of numbers by slide rule.

The slide rule is the most practical and efficient calculator easily available to you. You must be thoroughly familiar with it in order to use it really efficiently.

Use the slide rule as much as possible for calculations later in this book and in everyday life.

Unit Seven

Ratio and Rate in Practical Applications

7.1 Ratio

Ratio was introduced in Section 1.5 as the numerical relation or correspondence between the number of objects, called *elements*, in two sets or their subsets. We used the example of a set of 12 girls and a set of 18 boys; by partitioning both sets into an equal number of subsets, the numerical relation 12:18 or 12/18 was shown to be equivalent to 6:9 or 6/9 ('to every 6 girls there are 9 boys') and to 4:6 or 4/6 ('to every 4 girls there are 6 boys'), and to 2:3 or 2/3 ('to every 2 girls there are 3 boys').

We found, too, that ratios can be treated as numbers themselves—the rational numbers, discussed in Section 1.8.

Now we will consider two aspects of ratios with very useful practical applications: cross-multiplication and extensions of ratios.

7.2 Cross-multiplication of Ratios

If A is a set of 3 elements and B is a set of 5 elements, the ratio between the number of elements in A and the number of elements in B is 3:5. Using the abbreviation $n(A)$ for 'the number of elements in the set A', we can write

$$n(A):n(B) = 3:5.$$

Fig. 7.1 shows the sets A and B as 'enclosures' and the elements as dots.

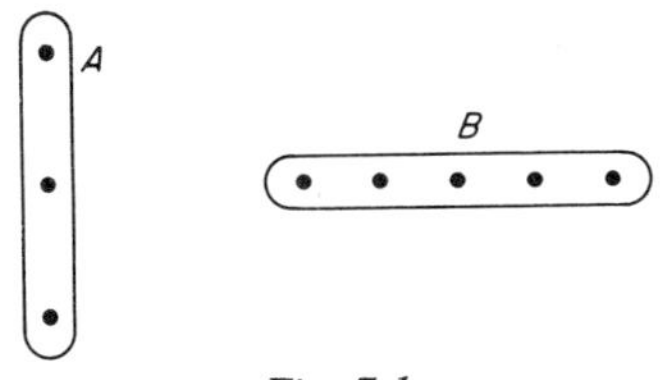

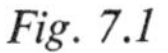

Fig. 7.1

A rectangular array of 15 dots

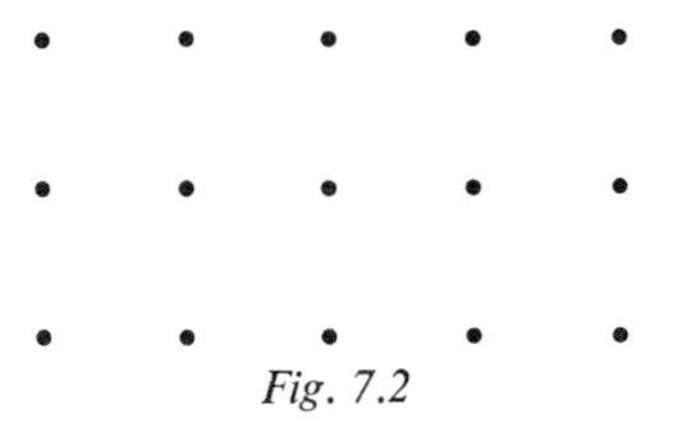

Fig. 7.2

was used in Section 3.1 (Fig. 3.1) to illustrate the commutative law for multiplication, that is:

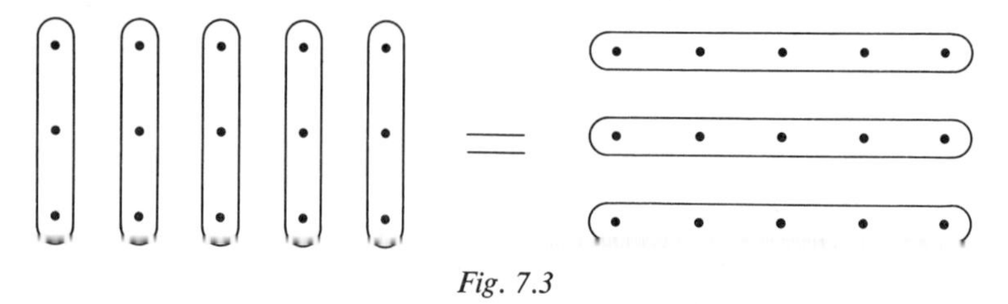

Fig. 7.3

$$5 \text{ columns of } 3 \text{ dots} = 3 \text{ rows of } 5 \text{ dots}$$

or

$$5 \times 3 = 3 \times 5;$$

or, in terms of the sets A and B and the number of elements in each set,

$$5 \times n(A) = 3 \times n(B).$$

This is known as *cross-multiplication.*

If $n(A):n(B) = 3:5$ is written in the form $\dfrac{n(A)}{n(B)} = \dfrac{3}{5}$, the terms of the ratios are 'cross-multiplied'.

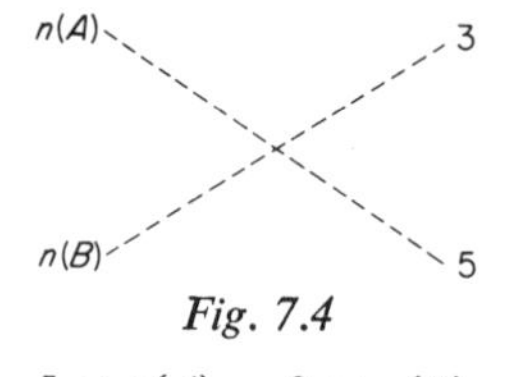

Fig. 7.4

$$5 \times n(A) = 3 \times n(B).$$

Any two equivalent ratios can be cross-multiplied. For example, since

$$\tfrac{2}{3} = \tfrac{8}{12}, \text{ then } 2 \times 12 = 8 \times 3,$$

or in symbolic notation,

$$\tfrac{2}{3} = \tfrac{8}{12} \Rightarrow 2 \times 12 = 8 \times 3.$$

Taking another example,

$$\tfrac{3}{5} = \tfrac{6}{10} \Rightarrow 3 \times 10 = 6 \times 5.$$

7.3 Problems

Mathematics is mainly concerned with solving *problems*, that is, analysing complicated situations in which we have just enough relevant information for a complete explanation. We try to use this information, by various methods, to find out about other aspects of the situation which are not directly known.

In the following example, cross-multiplication is chosen as the method of solving a problem concerning ratios.

Example

The ratio of boys to girls in a school is 5:6. There are 360 girls in the school. How many boys are there in the school?

Using B for the set of boys and G for the set of girls,

$$\frac{\text{number of boys in the school}}{\text{number of girls in the school}} = \frac{n(B)}{n(G)} = \frac{5}{6}.$$

Since $n(G) = 360$,

$$\frac{n(B)}{360} = \frac{5}{6}.$$

Cross-multiplying, we can write:

$$6 \times n(B) = 5 \times 360 = 1800.$$

Therefore since $6 \times 300 = 1800$, $n(B) = 300$. There are therefore 300 boys in the school.

Note. At the end of this example we used the 'balancing principle' of equations; we will discuss this more fully in Unit 11. According to this principle we may multiply, or divide, both sides of the equation (or equivalence) by the same number without changing the equation's validity; so that if $6 \times n(B) = 1800$, then, dividing both sides by 6, we have $n(B) = 300$.

7.4 Exercises

By cross-multiplication, find the missing number:

1. $\frac{\boxed{?}}{5} = \frac{8}{20}$.
2. $\frac{7}{\boxed{?}} = \frac{21}{6}$.

7.5 Ratios as Multiplying Factors

In problems concerning ratios, the ratio can be regarded simply as a *multiplying factor*. If, for example, the ratio between two sets of apples is said to be 4:1, then there are 4 times as many apples in the first set as there are in the second; conversely, the number of apples in the second set is a quarter of the number of apples in the first set.

Using the example discussed in Section 7.2,

$$\frac{n(A)}{n(B)} = \frac{3}{5} \Rightarrow n(A) = \frac{3}{5} \times n(B) \text{ and } n(B) = \frac{5}{3} \times n(A).$$

Example

If the ratio of two sums of money is given as 4:7 and the larger sum is £350, we can say immediately that the smaller sum is $\frac{4}{7} \times$ £350 = £200.

Scale

Any good map or plan of a district or building displays—usually in one corner—its *scale*, that is, the ratio between a length or distance on the map and the corresponding actual distance. Look for examples in any atlas, or on an architect's or engineer's drawings.

If, for example, the scale of the plan of a building is given as 1:200, the distance between two points on the plan is $\frac{1}{200}$ of the corresponding distance on the actual building; vice versa, any distance on the building is 200 times the corresponding distance on the plan. The scale of a plan, given in the form of a multiplying factor (for example $\frac{1}{200}$), is sometimes called the *representative fraction* (RF) of the plan.

Ratios are always *numerical* relations between sets of elements, or between the sizes of two quantities of the *same kind*, for example between distances, between sets of people, or between amounts of money. Whenever ratios are used for comparing quantities or sizes, these must be expressed in terms of the *same* unit of measurement.

For example, if we need to find the ratio between 70 cm and 3·5 m, we must first express both lengths in a common unit:

70 cm and 350 cm,

or 7 dm and 35 dm.

The required ratio is then 7:35, or 1:5, or $\frac{1}{5}$.

Similarly, the ratio between 75p and £3 is the ratio between 75p and 300p and therefore 75:300, or 1:4, or $\frac{1}{4}$.

Notice that these ratios are *numbers*: $\frac{1}{5}$, not $\frac{1}{5}$ cm; and $\frac{1}{4}$, not $\frac{1}{4}$p.

7.6 Exercises

1. The scale of a map of Italy is 1:500 000; what is the actual distance between Verona and Venice, if the distance between the two towns on this map is 29·8 cm?
2. Find the ratios between:

   £500 and £800;
   £2 and 50p;
   100 days and 1 year.

7.7 Ratios of Three (or more) Terms

If we have three sets A, B and C, containing 4, 8 and 6 apples respectively (Fig. 7.5)

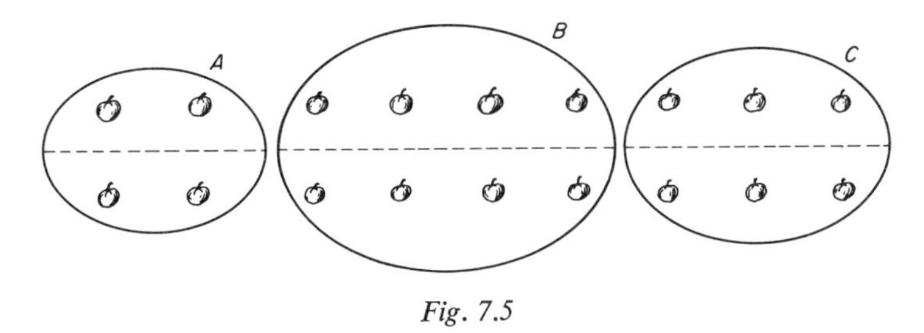

Fig. 7.5

we can compare the number of elements in A with those of B, or with those of C; we can also compare the number of elements of B with those of C. These comparisons are expressed as:

$$n(A):n(B) = 4:8 = 2:4 = 1:2,$$
$$n(A):n(C) = 4:6 = 2:3,$$
$$n(B):n(C) = 8:6 = 4:3.$$

More concisely, we may write:

$$n(A):n(B):n(C) = 4:8:6 = 2:4:3.$$

On the other hand, if such a ratio of 3 (or more) terms is given, the ratio between every pair of terms is immediately known. Suppose that the ratio of the ages of four girls, Natalie, Elizabeth, Tracy and Kathryn, is given as 2:3:4:5; then we can say that Natalie's age is $\frac{2}{3}$ of Elizabeth's, $\frac{2}{4}$ or $\frac{1}{2}$ of Tracy's and $\frac{2}{5}$ of Kathryn's; Elizabeth's age is $\frac{3}{2}$ of Natalie's, $\frac{3}{4}$ of Tracy's, and so on. If we know one girl's age, all the others can then be found. If, for example, Kathryn is 10 years old, Natalie's age is $\frac{2}{5} \times 10 = 4$, Elizabeth's age is $\frac{3}{5} \times 10 = 6$ and Tracy's age is $\frac{4}{5} \times 10 = 8$.

7.8 Construction of Further Ratios

If we have two given sets, we can construct other sets from them, and then establish ratios between these.

Suppose a class has 14 boys and 20 girls. We can construct from the set of boys (B) and the set of girls (G) the set of all children in the class, which will contain $14 + 20 = 34$ elements; or we can consider the set containing the difference of the number of elements of these two sets (6 elements). The ratio between the number of boys and the number of girls is

$$n(B):n(G) = 14:20 = 7:10.$$

If the set of boys, or the set of girls, is compared to the whole class (C), the ratios are

$$n(B):n(C) = 14:34 = 7:17;$$
$$n(G):n(C) = 20:34 = 10:17.$$

Combining these ratios,

$$n(B):n(G):n(C) = 7:10:17.$$

Indeed, if we are told any ratio between two sets, we can find at once the ratio of either set with their sum and with their difference. If the ratio between the numbers of elements of two sets A and B is given as 3:5, and we denote the 'sum set' by S and the 'difference set' by D, we can write:

$$n(A):n(B):n(S):n(D) = 3:5:8:2,$$

because $3 + 5 = 8$, and $5 - 3 = 2$.

The individual ratios between $n(A)$, $n(B)$, $n(S)$ and $n(D)$ can be read off immediately, for example:

$$n(A):n(S) = 3:8,$$
$$n(B):n(S) = 5:8,$$
$$n(A):n(D) = 3:2, \text{ and}$$
$$n(B):n(D) = 5:2.$$

Examples

1. A sum of £480 is to be divided into two amounts in the ratio 3:7.

   The ratio of the smaller amount to the sum of the two amounts is 3:10; the smaller amount is therefore $\frac{3}{10} \times £480 = £144$; the larger amount is $\frac{7}{10} \times £480 = £336$.

2. The ratio of the masses of nickel, manganese and copper in an alloy is 1:3:21. If the total mass of an article made of this alloy is 75 kg, the amounts of the various metals in this article are:

   nickel: $\frac{1}{25} \times 75 \text{ kg} = 3 \text{ kg};$
   manganese: $\frac{3}{25} \times 75 \text{ kg} = 9 \text{ kg};$
   copper: $\frac{21}{25} \times 75 \text{ kg} = 63 \text{ kg}.$

   Checking: $3:9:63 = 1:3:21.$

7.9 Exercises

1. If 3 people are 2 years, 16 years and 24 years old respectively, what is the ratio of their ages (in its simplest form)?

2. A television set is sold at £56. This price includes a Value Added Tax (VAT) of £16.
   What is the retail price without VAT?
   What is the ratio between VAT and the retail price with VAT?
   What is the ratio between VAT and the retail price without VAT?
3. Three people share an inheritance in the ratio 1:1:3. The share of the person who receives the most is £500 more than the share of each of the others.
   What is the value of the inheritance, and how much do the other two people each receive?
   [Hint: extend the given ratios by expressing the ratio between the share of each person, and the difference between the shares.]

7.10 Percentage

A particular ratio can be expressed in an unlimited number of ways. Consider, for example, the ratio 1:4; although we usually write it in this simple form (in which the terms of the ratio have no common factor), it is adequately expressed by any of its equivalents, 2:8, 3:12, 20:80, 25:100, etc., or by a phrase such as 'one in every four'.

An important equivalent ratio is the one whose second term is 100, especially if the comparison is with some kind of 'whole'. In our example this equivalent is 25:100 or $\frac{25}{100}$, usually written as 25 *per cent* (Latin, 'for every hundred'), or as 25%.

For example, instead of giving the ratio of men in a whole college population as 6:10, or 3:5, we could say that $\frac{60}{100}$ or 60% of the college population are men.

Ordinary ratios or fractions can be changed to this *percentage form* by the usual processes of finding equivalents, namely inspection or division (see Section 1.6), or by cross-multiplication.

For example, $\frac{2}{5} = \dfrac{20 \times 2}{20 \times 5} = \frac{40}{100} = 40\%,$

or $\frac{2}{5} = \dfrac{2{\cdot}00}{5} = 0{\cdot}40 = 40\%,$

or, considering $\frac{2}{5} = \dfrac{\boxed{?}}{100}$ and cross-multiplying,

$$2 \times 100 = 5 \times \boxed{?} \Rightarrow \boxed{?} = 40.$$

There is a close connexion between a point fraction and the percentage form of a ratio. Since the second digit after the unit point indicates the number of hundredths, percentages can be read off immediately from point fractions.

For example,

$$0{\cdot}04 \left(\text{or } \frac{4}{100}\right) = 4\%;$$
$$0{\cdot}12 \left(\text{or } \frac{12}{100}\right) = 12\%;$$
$$0{\cdot}125 \left(\text{or } \frac{12{\cdot}5}{100}\right) = 12{\cdot}5\%;$$
$$1{\cdot}5 \left(\text{or } \frac{150}{100}\right) = 150\%.$$

If the percentage form is required, it is often easier first to change a fraction or ratio to its decimal equivalent, and then to express it as a 'percentage'.

For example, 1:8, or $\frac{1}{8}$, $= 0{\cdot}125 = 12{\cdot}5\%$.

7.11 Exercises

1. Express the following ratios as percentages:

   3:5;
   1:16;
   2:25;
   $\frac{1}{20}$;
   $\frac{3}{40}$.

   Express also the other ratios of Exercise 2 of Section 7.6 in percentage form.
2. Express the ratio between £50 and £400 in percentage form.

7.12 Rate

If only the ratio between two sets or quantities is given, we know nothing about the actual size of each of these sets or quantities. Indeed, the same ratio exists between many pairs of sets; the ratio 1:2 exists between a set of 2 apples and a set of 4 apples, between a set of 3 apples and a set of 6 apples, between a set of 150 apples and a set of 300 apples, and so on.

Often, while the total size of a quantity may change, or *vary*, the ratio between the parts remains the same, or *constant*. If, for example, the ratio of zinc to copper in a particular kind of brass is 3:7 (or 30% zinc and 70% copper), then the ratio of zinc to copper in any article made of this brass is 3:7, no matter how big or small the article.

We only speak of *ratio* when sets or quantities of the *same kind* are compared, such as the sizes of groups of people, sums of money or lengths.

When sets or quantities of *different kinds* are related, we use the word *rate*; we speak for instance, of 'a rate of pay of 85 pence per hour' (money–time), the price of butter as 45 pence per kilogram (money–weight of butter) or of an average speed of 50 kilometres per hour (distance–time).

In all these cases, the rate is a *constant* relation between the sizes of the two quantities concerned. For example, if hiring a car is charged at a rate of 8 pence per kilometre,

a journey of 40 kilometres costs 40 × 8 pence, and
a journey of 100 kilometres costs 100 × 8 pence.

If we state a ratio, we need only give the appropriate rational number; but if we state a rate we must *always* give the two quantities concerned and the unit of measurement. For example, average speed is written as 100 kilometres per 2 hours, or 50 kilometres per (one) hour. Rates can be written in a form similar to that of ratios, for example,

$$100 \text{ km per 2 hours} = \frac{100 \text{ km}}{2 \text{ hours}} \text{ or } \frac{50 \text{ km}}{1 \text{ hour}},$$

or, more usually, 50 km/ h.

We may even speak of 'equivalent' rates. For instance, a laundry charge of 40 pence per 2 sheets is equivalent to a charge of 20 pence per sheet, or 80 pence per 4 sheets, a constant rate of 40 kilometres per hour is equivalent to 80 kilometres per 2 hours, or

$$\frac{40 \text{ km}}{1 \text{ hour}} = \frac{80 \text{ km}}{2 \text{ hours}} = \frac{120 \text{ km}}{3 \text{ hours}}$$

and so on.

Replacing a rate in this way by one of its equivalents only alters the numbers, not the unit of measurement.

If we need to change the units of measurement, it is advisable to substitute the appropriate equivalent units first, and then calculate the rate. For example, to state a speed of 50 kilometres per hour in terms of metres per second,

$$\frac{50 \text{ km}}{1 \text{ h}} = \frac{50 \times (1 \text{ km})}{1 \times (1 \text{ h})} = \frac{50 \times (1000 \text{ m})}{1 \times (3600 \text{ s})} = \frac{50\,000 \text{ m}}{3600 \text{ s}}$$

$$= \frac{500 \text{ m}}{36 \text{ s}} = \frac{13\frac{8}{9} \text{ m}}{1 \text{ s}} = 13\tfrac{8}{9} \text{ m/s}.$$

7.13 Exercises

1. If a car travels at a constant speed of 45 km per hour (45 km/h), how long will a journey of 270 km take?
2. What rate in metres per second is equivalent to a speed of 45 km per hour?
3. Local government services—welfare, police and fire services, for example—are financed by taxes raised within the local authority area; for this purpose the size and amenities of properties are assessed and a certain number (a sum in pounds) is attached to each property: its *rateable value*. Every householder pays local taxes on the basis of the rateable value of his home and an annual *rate* of '. . . pence in the pound'; so that, if the rateable value

of a house is £120 and the rate for this year is 75 pence in the pound, the amount the householder will pay this year is 120×75 pence $=$ £90.

The total rateable value of all the properties in a town is £350 000. Find the revenue obtained by the town council from a rate of 75 pence in the pound.

7.14 Interest

When commodities such as cars or houses are hired or let, a charge is made. Whenever *money* is borrowed, for example for house purchase (a mortgage), or to buy a large item such as a car or a television set (hire purchase), or in the form of a bank loan, the charge made is called *interest*; this, of course, is paid in addition to paying back the amount borrowed. If money has been saved and is deposited with a bank or building society, the customer, then the lender, *receives* interest.

The amount of the charge, the interest, depends on the amount of money borrowed and on the period of time for which it is borrowed. Interest is both a rate and a ratio. If the rate of interest is 5% per annum (Latin, 'for each year', abbreviated to p.a.), the amount of interest payable per year (rate) is in the ratio 5:100 to the amount of the loan.

Example

The interest on a sum of £4000 at a rate of 5% p.a. is

$$\tfrac{5}{100} \times £4000 = £200 \text{ each year.}$$

On a sum of £25, the interest at 7% p.a. is

$$\tfrac{7}{100} \times £25 = £1{\cdot}75 \text{ per year.}$$

Interest is usually calculated on the basis of a period of one year, but for short-term loans the period is sometimes one month; interest rates are then given as . . . % per month.

A capital sum which is invested or lent is called the *principal sum*, or simply *principal* (abbreviated to P). A general formula for the annual interest on a capital of £P at the rate of R% is then

$$\text{annual interest} = \frac{R}{100} \times P.$$

In the example above, the annual interest on £4000 at a rate of 5% is

$$\tfrac{5}{100} \times £4000 = £200.$$

7.15 Exercises

1. What is the annual interest payable on a bank loan of £500 if the rate of interest is $7\frac{1}{2}\%$ per annum, and no capital repayments are made (i.e. the capital sum borrowed remains the same, or constant)?
2. What is the interest to be paid on a sum of £250 at the end of one year if the rate of interest is $6\frac{1}{4}\%$ p.a.?

7.16 Simple Interest

If the amount of the loan remains the same, and the rate of interest remains the same, the interest to be paid will be the same every year; we call this *simple interest*.

For example, if a sum of £3000 is deposited with a bank at an interest rate of 6% p.a., it will produce an income of $\frac{6}{100} \times £3000 = £180$ at the end of one year. If this £180 is taken out of the bank, the principal of £3000 will again produce £180 in interest at the end of the second year. Over the two years the total income from interest on £3000 is therefore $2 \times £180$. If again this interest is not left in the bank, that is, it is not added to the principal, the interest (at the rate of 6%) on the £3000 for the third year will again be £180. The total amount of interest received over 3 years is

$$3 \times £180 = £540.$$

If the general formula for annual interest on a capital sum, £P, at a rate of R% is used, the simple interest over two years

$$= 2 \times \frac{R}{100} \times £P;$$

and over 3 years the simple interest is

$$3 \times \frac{R}{100} \times £P.$$

This suggests a formula for simple interest on a sum of £P at a rate of R% p.a. over n years:

$$\text{simple interest} = n \times \frac{R}{100} \times £P$$

or

$$\boxed{\text{simple interest} = £\frac{n \times R \times P}{100}}$$

Note that if n is a number of *years*, R must be a rate *per annum*.

7.17 Compound Interest

In the last example, the sum invested, or the amount of the loan, remained the same year after year, and therefore the interest remained the same. But if after one year the interest is not taken out of the bank but added to the main capital sum, the principal in the second year is then larger than during the first year, and the interest at the end of the second year will also be more than at the end of the first year. This growing interest is called *compound interest.*

The interest and the growth of the principal can be calculated year by year.

Example

Calculate the compound interest on £3000 at 5% p.a.

principal		=	£3000
interest after 1st year:	$\frac{5}{100} \times$ £3000	=	£ 150
principal after 1 year		=	£3150
interest after 2nd year,	$\frac{5}{100} \times$ £3150	=	£ 157·50
principal after 2 years		=	£3307·50
interest after 3rd year	$\frac{5}{100} \times$ £3307·50	=	£ 165·375
principal after 3 years		=	£3472·875
etc.			

The total compound interest over 3 years = £472·875. You will have recognized this growing increase through compound interest as an example of a geometric progression (Section 6.7). Section 7.21 deals more fully with this aspect.

7.18 Exercises

1. Using the method given above, calculate the compound interest on £500 for 3 years at 4% p.a.
2. Find the difference between the total simple interest and the total compound interest on £5000 for 2 years at 6% p.a.

   Is there a difference between the simple interest and the compound interest at the end of the first year?

7.19 Increase or Decrease as a Multiplying Factor

We can express an increase or a decrease in a capital sum, or in any other quantity, as a *percentage increase*; for example, a gain of £50 on a capital investment of £200 is a gain of 25% or a quarter of the original investment, because $\frac{50}{200} = \frac{1}{4} = \frac{25}{100} = 25\%$. Similarly, a loss in weight of 12·5 kg on a weight of 62·5 kg is a loss of 20%, or one-fifth, because

$$\frac{12\cdot5}{62\cdot5} = \frac{125}{625} = \frac{5}{25} = \frac{1}{5} = \frac{20}{100} = 20\%.$$

Such an increase or decrease can be described as an addition or subtraction, and alternatively, equally validly, as a *multiplication*. For example, losing a fifth of one's weight (W) is the same as multiplying one's weight by $\frac{4}{5}$:

$$1W - \tfrac{1}{5}W = \tfrac{4}{5} \times W.$$

Gaining a quarter of one's capital (P) is the same as multiplying it by $\frac{5}{4}$:

$$1P + \tfrac{1}{4}P = \tfrac{5}{4} \times P.$$

The fractions changing W and P, $\frac{4}{5}$ and $\frac{5}{4}$ in these cases, are the *multiplying factors* for making the appropriate decrease or increase.

In terms of 'percentage', an increase of 5%, say, is equivalent to a multiplication by $\frac{105}{100}$ or 1·05:

$$1 + \tfrac{5}{100} = \tfrac{100}{100} + \tfrac{5}{100} = \tfrac{105}{100} \text{ or } 1\cdot05.$$

A decrease of 8% is the same as a multiplication by $\frac{92}{100}$ or 0·92:

$$1 - \tfrac{8}{100} = \tfrac{100}{100} - \tfrac{8}{100} = \tfrac{92}{100} \text{ or } 0\cdot92.$$

7.20 Exercises

1. What percentage increase or decrease is:

   a gain of £45 on £1500?
   a loss of £20 on £500?

2. Write down the multiplication factors to bring about:

   an increase of 12%;
   an increase of $4\frac{1}{2}$%;
   a decrease of 6%;
   a decrease of $7\frac{1}{2}$%.

7.21 A Formula for Finding Compound Interest

Receiving an interest of 5% on capital at the end of a year and adding it to the capital is equivalent to multiplying this capital by $\frac{105}{100}$ or by 1·05.

In general terms, receiving an interest of $R\%$ on a capital sum of £P and adding this to the principal is equivalent to multiplying the principal by

$$\left(1 + \frac{R}{100}\right) \text{ or } \left(\frac{100 + R}{100}\right).$$

After one year the principal P has then grown to $\dfrac{100 + R}{100} \times P$.

Example

A principal of £3000 invested at a rate of 5% p.a. has grown at the end of one year to

$$1{\cdot}05 \times £3000 = £3150.$$

If the investment is left untouched the sum will again grow by compound interest at the rate of 5%, and be multiplied by 1·05 at the end of the second year, that is,

$$\begin{aligned} & 1{\cdot}05 \times £3150 \\ \text{or} \quad & 1{\cdot}05 \times (1{\cdot}05 \times £3000) \\ = \; & 1{\cdot}05 \times 1{\cdot}05 \times £3000 \\ = \; & (1{\cdot}05)^2 \times £3000 = £3307{\cdot}50. \end{aligned}$$

At the end of the third year the amount is:

$$\begin{aligned} 1{\cdot}05 \times £3307{\cdot}50 \text{ or } 1{\cdot}05 \times 1{\cdot}05 \times 1{\cdot}05 \times £3000 &= (1{\cdot}05)^3 \times £3000 \\ &= £3472{\cdot}875. \end{aligned}$$

In this *geometric progression* the first term is the original sum invested, the principal (£3000), and the common ratio is the multiplying factor (1·05); the second term is the amount after one year (£3150), the third term is the amount after 2 years (£3307·50), the fourth term is the amount after 3 years (£3472·875), etc. That is, we can write the GP as:

$$£3000,\ £3000 \times 1{\cdot}05,\ £3000 \times (1{\cdot}05)^2,\ £3000 \times (1{\cdot}05)^3, \ldots .$$

For this case, we can write a general formula for the amount after n years as £$3000 \times (1{\cdot}05)^n$.

In general terms, for any sum £P invested at a rate of $R\%$ p.a. *compound interest*, we can write the amount after n years (£P_n) as

$$\boxed{£P_n = \left(1 + \frac{R}{100}\right)^n \times £P} \text{ or } \boxed{£P_n = £P \times \left(1 + \frac{R}{100}\right)^n}$$

7.22 Exercises

1. Compare this formula for the amount after n years with the formula for the nth term of a GP (see Section 6.9).

2. Use the formula to find to what amount an investment of £500 has grown after 5 years at a rate of 4% p.a. compound interest.
   What is the total increase, that is the total interest received, in this case?

7.23 Depreciation

The ageing process of things like cars or industrial equipment usually implies that they lose part of their value. This decrease in value or *depreciation* can be expressed as a percentage of the original value. For example, if the value, £10 000, of a factory plant has fallen after one year by £1250, we say it has depreciated by, or at a rate of, $12\frac{1}{2}\%$ because $\frac{1250}{10\,000} = \frac{12{\cdot}5}{100} = 12\frac{1}{2}\%$.

If cars depreciate at a rate of 10% per year, a car costing £2000 will have depreciated $\frac{10}{100} \times £2000 = £200$ during the first year, and its value after one year is then £1800.

The multiplying factor of $\frac{90}{100}$, or 0·90, is equivalent to a depreciation of 10%, so the value of this car can be calculated immediately as $0{\cdot}90 \times £2000 = £1800$ after one year. After 2 years its value will be $0{\cdot}90 \times £1800$ or $0{\cdot}9 \times 0{\cdot}9 \times £2000 = £1620$, and so on.

Note. The value of the car is *not* $£2000 - £200 - £200 = £1600$ after 2 years; the *second* decrease is on an amount already reduced by the first year's depreciation.

7.24 Exercises

1. Calculate the value of a television set, bought at £250, after 1 year, 2 years and 3 years, if it depreciates at a rate of 20% p.a.
2. Make up a formula for the value of factory plant, if the rate of depreciation is $R\%$ p.a.
   [Hint: find the multiplying factor representing a depreciation of $R\%$, and use the same approach as in the compound interest formula.]

7.25 Percentage Problems

In problems involving percentages, you should keep in mind the true nature of a 'percentage'.

A percentage is not a sum of money, nor a quantity: it is a *ratio* or a fraction—and it always relates to some other amount. For example, '25%' or 'a quarter' by itself is meaningless. Whenever percentages are given you should immediately ask yourself, 'A percentage of *what*?'

Sometimes the percentage quoted does not refer to the only amount which is mentioned.

If, for example, a hotel bill of £60 includes a 20% service charge, the service charge is *not* 20% of £60. Similarly, if an article carries a Value Added Tax of 10% and is sold at £55, the VAT is not 10% of £55. The percentage in each case relates to the cost before tax or service charge was added.

In problems of this kind, you can name the amount provisionally as, say, £x and use a simple equation, such as:

$$\text{cost price} + \text{profit} = \text{sales price},$$

or

$$\text{cost price} - \text{loss} = \text{sales price},$$

or

$$(\text{retail price without tax}) + \text{tax} = \text{retail price including tax}.$$

For example, if the retail price of an article is £55 and includes a Value Added Tax of 10%, we can write:

$$\begin{aligned} \text{retail price before tax} &= \pounds x; \\ \text{retail price including tax} &= \pounds x + \tfrac{10}{100} \times \pounds x \\ &= (1 + \tfrac{1}{10})\,\pounds x \\ &= \pounds 55\ (\text{given}) \\ \Rightarrow \pounds \tfrac{11}{10}\,x &= \pounds 55 \Rightarrow x = 50. \end{aligned}$$

Therefore the retail price before tax = £50 and VAT = £5.

This 'relative' nature of percentage becomes even more apparent in, for example, a quotation for a car insurance premium.

In this sample quotation, the following allowances are given:

a no claims bonus allowance:	60%
'one driver only' allowance:	$12\frac{1}{2}$%
union membership allowance:	$12\frac{1}{2}$%
'excess of £25' allowance:	10%

The *premium* is then calculated by the insurance company as follows:

basic premium	£65·00
excess: −10%	£ 6·50
	£58·50
union membership: −$12\frac{1}{2}$%	£ 7·31
	£51·19
'owner only driving': −$12\frac{1}{2}$%	£ 6·39
	£44·80
no claims bonus: −60%	£26·88
premium after allowances =	£17·92

Each allowance is calculated as a percentage of the remainder after the other deductions have been made. Clearly, percentages cannot always be added up and *then* deducted. (All the allowances listed here would total 95%!)

Note. The order of the deductions does not matter. This becomes clear if we use multiplying factors, and write the calculation as

$$65 \times \frac{90}{100} \times \frac{87\frac{1}{2}}{100} \times \frac{87\frac{1}{2}}{100} \times \frac{40}{100} = 65 \times \frac{87\frac{1}{2}}{100} \times \frac{40}{100} \times \frac{90}{100} \times \frac{87\frac{1}{2}}{100} = 17{\cdot}92,$$

because of the commutativity of multiplication (compare Exercise 3 of Section 3.4).

7.26 Exercise

Calculate the amount of tax paid on an article which is sold at £92, if this includes a Value Added Tax of 15%.

Unit Eight

The Language of Sets and Solving Problems with Sets

Note. This Unit covers material which is not mentioned in all O-level syllabuses, but it is included because of its inherent interest, and because of the usefulness of the ideas which it introduces.

8.1 Sets

The word *set* was frequently used when we discussed number in Unit 1, and again in Unit 7 when we considered ratios. The idea of a set is very important in mathematics, and it is useful to have a system of special symbols and diagrams which will simplify working with sets and solving problems.

A *set* is simply a collection of things or people, called *elements*. Capital letters such as A, B, C, are used to name such collections. At the beginning of each problem we define a set and name it, either

(*a*) by describing the set in words, or by naming a particular property that identifies the elements of a set, for example,

N is the set of all integers,
Q is the set of all rational numbers,
A is the set of blue-eyed people,

or,

(*b*) by listing or tabulating all the elements of a set; this, of course, is only practicable if the set is relatively small.

The brackets { } are used to enclose the listed elements of a set, or its description by means of a property of its elements.

We could, for example, define the set A of all positive odd integers less than 10 by simply listing all of them:

$$A = \{1, 3, 5, 7, 9\} \text{ or } \{\text{all odd positive integers less than } 10\},$$

or the set B of all the vowels in the alphabet:

$$B = \{a, e, i, o, u\} \text{ or } \{\text{all the vowels of the alphabet}\}.$$

Sets can be visualized, and their handling simplified, by using *Venn diagrams* invented by Leonhard Euler (1707–83) and improved by John Venn (1834–1923). A closed shape or curve is drawn to enclose all the elements of a set—we need not trouble about its precise shape. The letter naming the set is usually placed alongside the curve; it may be written inside it, as long as this creates no confusion.

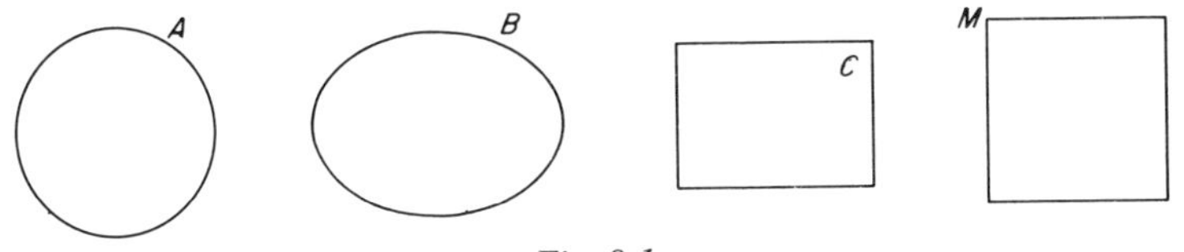

Fig. 8.1

8.2 Elements of Sets

A special symbol $\in$ is used to indicate membership of a set. If, for example, H is the set of all human beings, the statement (Peter is a human being), or (Peter belongs to the set H), or (Peter is a member of the set H) can be abbreviated to:

$$\text{Peter} \in H.$$

Venn diagrams indicate membership of a set by placing the element inside the enclosure representing the set, for example,

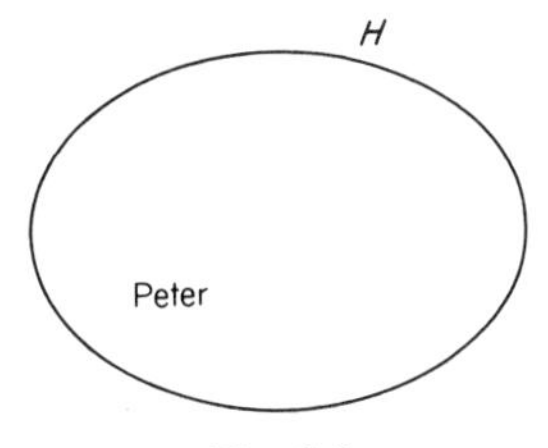

Fig. 8.2

8.3 Subsets

We often need to define a smaller set within a set, which we can call a *subset*; all the members of the subset are members of the set. For example, the set of all men is a subset of the set of all human beings; the set of all positive integers is a subset of the set of all integers.

The symbol $\subset$ indicates the relation 'is a subset of'; if the set of all human beings is denoted by H and the set of all men by M, then we can write

$$M \subset H.$$

The Venn diagram of a subset is a closed shape or curve entirely enclosed within the main set, like the shaded region in Fig. 8.3.

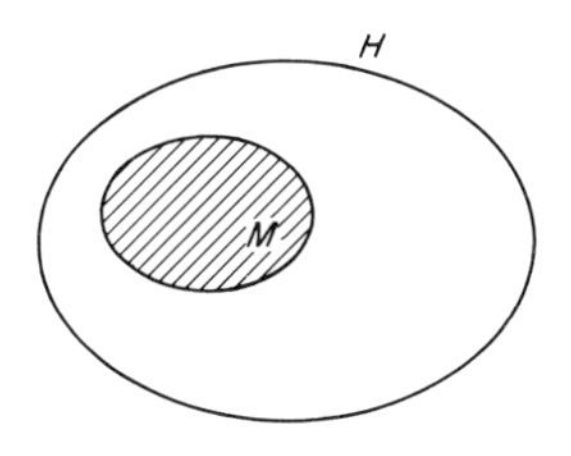

Fig. 8.3

The idea of a subset is purely relative; we can only speak of a subset in relation to a main set. Any set can be thought of as the subset of some larger set; for example, the set of all human beings is a subset of the set of all living creatures. Usually we restrict ourselves to some larger set embracing all the sets under discussion, the *universal set*, $\mathscr{E}$. For example, in a discussion on sets of numbers, the universal set is the set of all numbers.

Fig. 8.4 shows a Venn diagram illustrating a set and a relevant universal set.

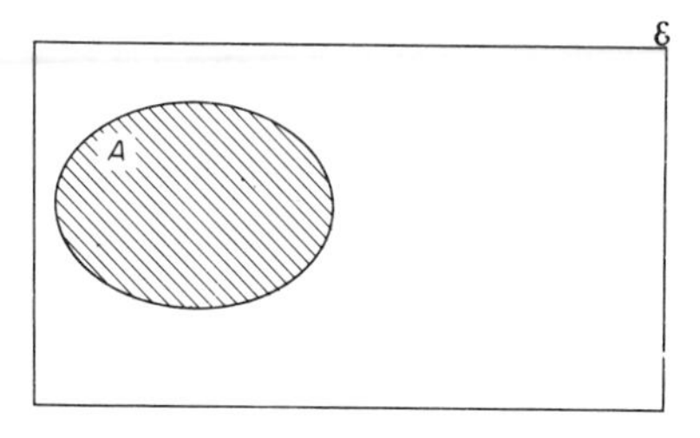

Fig. 8.4

8.4 Complement

Since enclosure of an element within the Venn diagram of a set represents membership of that set, a position outside the enclosure indicates non-membership. The set of all elements of the universal set which are not elements of a set A is called the *complement* of A. It is usually symbolized by a dash (′) after the appropriate letter:

$$A' = \text{the complement of } A.$$

For example, if A is the set of all positive numbers in a universal set of all real numbers, then A' is the set of all real numbers that are *not* positive, i.e. all the negative numbers and zero.

In a Venn diagram of a set A, the complement of A, or A', is the region outside the enclosure representing A, that is, the shaded region in Fig. 8.5.

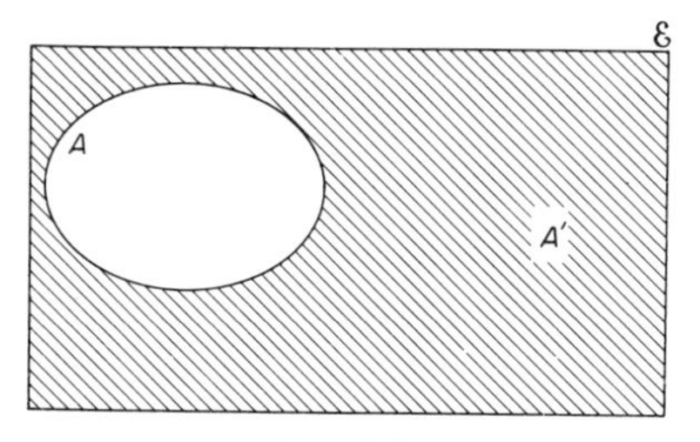

Fig. 8.5

8.5 Intersection

If two sets A and B have elements in common, that is elements which belong both to A and to B, they are said to *intersect*. Fig. 8.6 is a Venn diagram illustrating two sets which overlap or intersect.

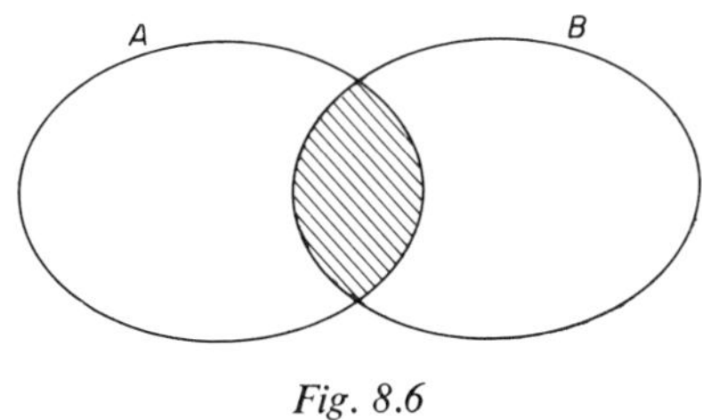

Fig. 8.6

The set of *all* elements which are common to both A and B is called the *intersection* of A and B which, in a Venn diagram, is the whole of the overlapping region (the shaded region in Fig. 8.6).

The intersection of A and B is written

$$A \cap B.$$

For example, if A is the set of all positive even numbers less than 10, or

$$A = \{2, 4, 6, 8\},$$

and B is the set of all positive multiples of 4 which are less than 25, or

$$B = \{4, 8, 12, 16, 20, 24\},$$

then
$$A \cap B = \{4, 8\}.$$

Non-intersecting sets, that is sets which have no members in common, are referred to as *disjoint*.

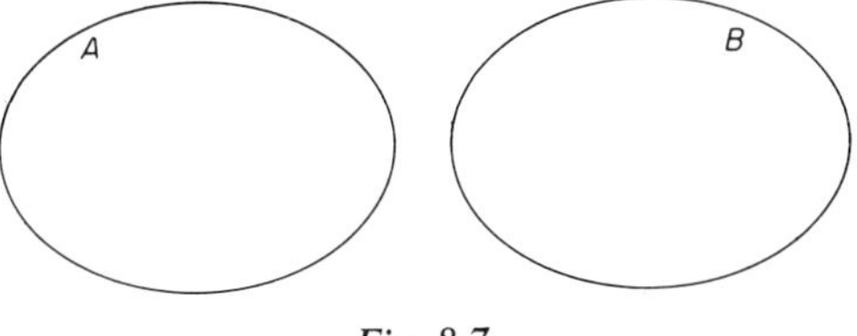

Fig. 8.7

Alternatively, we can say that the intersection of A and B is *empty*. A special symbol, ϕ, is used for the *empty set*, and we can write $A \cap B = \phi$.

8.6 Union

If an element can belong to the set A or to the set B, or is a member of both A and B, we say it belongs to the *union* of A and B, that is, to the set of all elements of A or of B or of both. It is represented by the whole of the shaded region in Fig. 8.8.

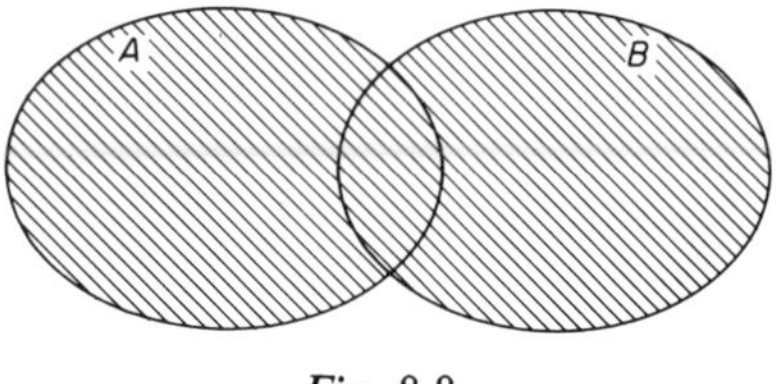

Fig. 8.8

The symbol for union is $\cup$, placed between the letters indicating the sets, so that

$A \cup B$ is 'the union of A and B', and
$A \cup B \cup C$ is 'the union of A, B and C'.

Example

In a class of children, A is the set of all ten-year-olds and B the set of all girls. Then $A \cup B$ is the set of all children who are ten years old, together with all girls in the class (the set $A \cap B$ is the set of all ten-year-old girls).

If $A = \{$Peter, John, Margaret, Mary, Simon$\}$
$B = \{$Sarah, Margaret, Mary, Elizabeth, Kate, Diana$\}$

then,

$A \cup B = \{$Peter, John, Simon, Sarah, Margaret, Mary, Elizabeth, Diana, Kate$\}$.

Obviously, Margaret and Mary are not counted twice in the union; the number of elements in the union of A and B, written $n(A \cup B)$, is therefore not necessarily the sum of the number of elements of A and the number of elements of B. In this example,

$$n(A) = 5,$$
$$n(B) = 6,$$
$$n(A \cup B) = 9.$$

In speaking of sets, the word *or* is not used in an exclusive sense; the set (A or B), that is $A \cup B$, includes the set (both A and B) that is $A \cap B$. Consider, for example, a group of students taking courses in French (the set F) or in German (the set G); 'the set of those taking French or German' ($F \cup G$) includes those studying both French and German ($F \cap G$).

If we want to refer to those who only study one of these languages, we must mention the word *only* specifically: 'the set of all students taking French only', for example.

8.7 Summary

Fig. 8.9 summarizes the notation of sets and Venn diagrams.

The set A	$A = \{ - - - - - - - - - \}$	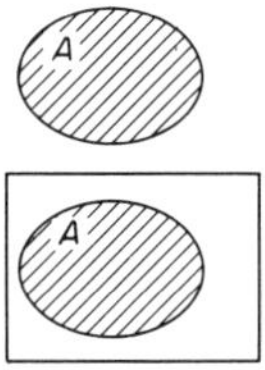
4 is an element of N	$4 \in N$	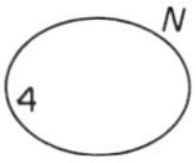
The complement of A	A'	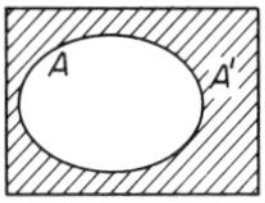
B is a subset of A	$B \subset A$	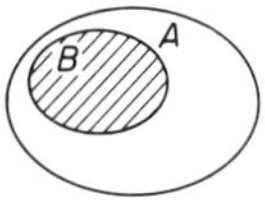
The intersection of A and B	$A \cap B$	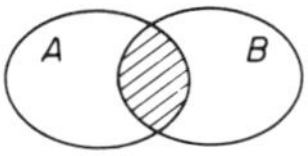
A and B are disjoint (the intersection of A and B is empty)	$A \cap B = \emptyset$	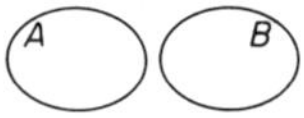
The union of A and B	$A \cup B$	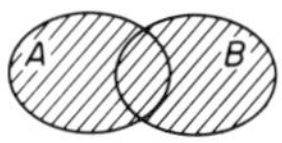 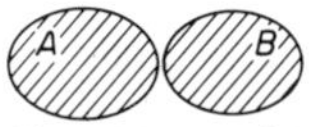 (if A and B are disjoint) 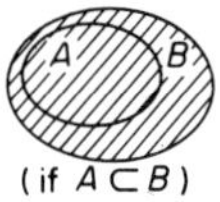 (if $A \subset B$)

Fig. 8.9

8.8 Exercises

1. A is the set of students in a college studying mathematics, B the set of students in the same college studying English. Describe the following sets in words, and in each case draw a Venn diagram, shading the area representing the set:

$$A';$$
$$B';$$
$$A \cup B;$$
$$A \cap B;$$
$$A' \cap B;$$
$$A \cap B';$$
$$A' \cap B';$$
$$(A \cap B)'.$$

2. If P is the set of pretty girls,
   F the set of blonde girls,
   B the set of blue-eyed girls,
   express in symbolic notation, and draw Venn diagrams of, the following sets:

(*a*) the pretty blondes;
(*b*) the set of pretty, blue-eyed blondes;
(*c*) the set of those pretty girls who are neither blonde nor have blue eyes.

8.9 Solving Problems with Venn Diagrams

The example in Section 8.6 showed that the number of elements in the union of two sets A and B, $n(A \cup B)$, is not necessarily the same as the sum of the numbers of elements in each set, $n(A) + n(B)$.

In that example,
$$n(A) + n(B) = 11,$$
$$n(A \cup B) = 9.$$

But we can find the relation between $n(A \cup B)$ and $n(A) + n(B)$ if we consider the number of the elements in the intersection, $n(A \cap B) = 2$. When we added the number of elements of A and the number of elements of B, we found that we had counted those in the intersection (Margaret and Mary) twice. To find the number of elements in the union, or $n(A \cup B)$, we must therefore subtract the number of elements of the intersection, or

$$n(A \cup B) = n(A) + n(B) - n(A \cap B).$$

For example, in a class of boys everyone is a member of the football team (F: eleven boys) or the basketball team (B: five boys); 3 boys play in both teams. Then

$$n(F) = 11,$$
$$n(B) = 5,$$
$$n(F \cap B) = 3.$$

Since all the boys in the class are members of one team or the other, the *number of boys in the class* is

$$n(F \cup B) = n(F) + n(B) - n(F \cap B) = 11 + 5 - 3 = 13.$$

It follows that, if two sets A and B are disjoint, that is $A \cap B = \phi$ and $n(A \cap B) = 0$, then $n(A \cup B) = n(A) + n(B)$.

This result suggests a different method of solving problems with sets. We draw a Venn diagram and name all the *disjoint regions* separately, indicating each region by a small letter to avoid confusion; Fig. 8.10 represents the last example.

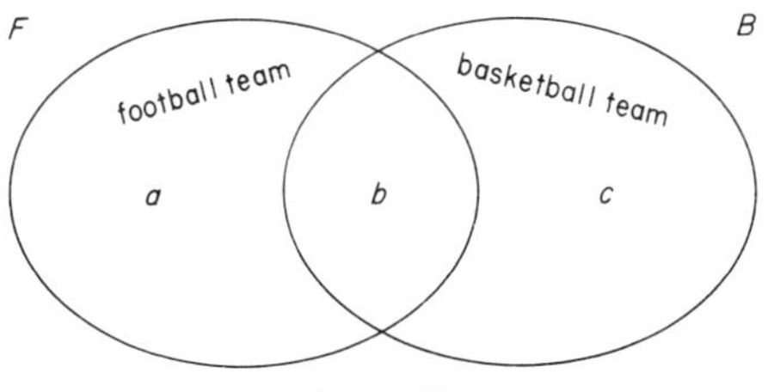

Fig. 8.10

The regions a and b then represent the football team (a: those who play football only, b: those who play football as well as basketball); the regions b and c represent the basketball team (c: those who play basketball only, b: those who play basketball as well as football).

Since the regions a, b and c are disjoint,

$$n(a) + n(b) = 11.$$

We are told that $n(b) = 3$, so $n(a) = 8$.

Similarly, $n(b) + n(c) = 5,$
$n(b) = 3,$
$n(c) = 2.$

Therefore the number of boys in the class is

$$n(a) + n(b) + n(c) = 8 + 3 + 2 = 13.$$

This method is especially recommended for problems involving more than two sets; for such problems, always draw a Venn diagram and name all the disjoint regions separately.

Example

In a sixth form 28 students attend a course in mathematics, 30 a course in physics and 34 a course in geography. Of these, 12 study both mathematics and physics, 7 both mathematics and geography and 11 both physics and geography. These figures include 5 students whose course includes mathematics, physics and geography. There are 38 students in this sixth form who study neither mathematics nor physics nor geography. How many students are there in the sixth form?

We draw a Venn diagram (Fig. 8.11) naming the various disjoint regions, including the region h representing those who do not study mathematics, physics or geography (the whole sixth form is the universal set).

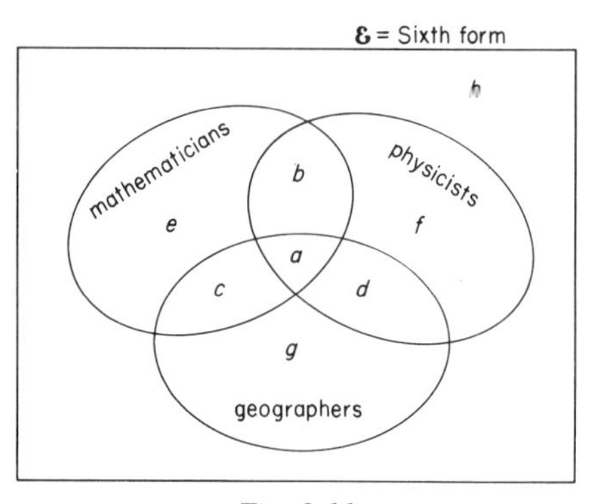

Fig. 8.11

We begin by using the information which gives immediately the number in any of the disjoint regions:

The number of those taking mathematics, physics and geography, or $n(a)$, $= 5$.

The number of those taking mathematics and physics (represented by the regions a and b) is $n(a) + n(b) = 12$; so that $n(b) = 7$.

The number of students taking mathematics and geography, represented by the regions c and a, is $n(c) + n(a) = 7$; so that $n(c) = 2$.

The number of students taking physics and geography, represented by the regions a and d, is $n(a) + n(d) = 11$; so that $n(d) = 6$.

The only regions not known so far are e, f and g, which represent respectively those students who take mathematics only, physics only and geography only.

The number of students taking mathematics only can be found by considering the total number of students taking mathematics, represented by the regions a, b, c and e; then we subtract from this total the number of those who study another subject as well, represented by a, b and c:

$$n(a) + n(b) + n(c) + n(e) = 28$$
$$n(a) + n(b) + n(c) = 14$$

Then $$n(e) = 28 - 14 = 14.$$

Similarly,

$$n(a) + n(b) + n(d) + n(f) = 30$$
$$n(a) + n(b) + n(d) = 18$$

Then $$n(f) = 30 - 18 = 12,$$

and

$$n(a) + n(c) + n(d) + n(g) = 34$$
$$n(a) + n(c) + n(d) = 13$$

Then $$n(g) = 34 - 13 = 21.$$

We now know the numbers of all members of the various disjoint regions. The whole of the sixth form, or

$$= n(a) + n(b) + n(c) + n(d) + n(e) + n(f) + n(g) + n(h)$$
$$= 5 + 7 + 2 + 6 + 14 + 12 + 21 + 38$$
$$= 105.$$

8.10 Exercises

Use Venn diagrams to solve the following problems:

1. In a group of 41 people who all speak French or German or both, 19 speak French and 32 speak German.

   How many of these 41 people speak German only?
   How many speak French only?

2. A survey of a small town of 20 000 households provided the following data:

   60% of the households have a refrigerator;
   35% of the households have a telephone;
   15% of the households have a colour television set;
   18% of all households have both a refrigerator and a telephone;
   12% of all households have both a telephone and a colour television set;
   10% of all households have both a refrigerator and a colour television set;
   8% of all households have a refrigerator, a telephone and a colour television set.

   How many of these households have only a telephone?
   How many have only a refrigerator?
   How many have only a colour television set?
   How many households have neither a telephone, nor a refrigerator, nor a colour television set?

   [Hint: Calculate all the numbers first from the percentages given, and then solve the problem with a Venn diagram. Alternatively, you could consider a 'representative group' of 100 households, saying, 'of every hundred households there are 60 households with a refrigerator,' etc., and then solve the problem with Venn diagrams for these 100 households; remember that these answers should be multiplied by 200, since there are 200 such groups of households (20 000 = 200 × 100).]

Part Three

Algebra and Graphs

Unit Nine

The Processes of Algebra

9.1 Abbreviation and Generalization

Ordinary, everyday language is often long-winded and can be ambiguous; in mathematics this can become so serious that we find we need to use a kind of shorthand, a notation using symbols, so that we can avoid confusion, and make statements which are perfectly clear and precise. Compare, for example, the mathematical way of stating the commutative law for addition, '$m + n = n + m$, for all numbers m and n', with its expression in words, 'The result of adding one number to another is the same as adding the latter to the former'. (Try to describe in words the associative law for addition, $(a + b) + c = a + (b + c)$, mentioned in Exercise 3 of Section 2.3—this will illustrate even more clearly how complicated it can be to make mathematical statements verbally.)

We have already occasionally used *letters* instead of numbers in general results, or in the examples of the commutative and associative laws for addition. The statement $3 + 4 = 4 + 3$ only tells us something about the commutative law for the addition of the numbers 3 and 4; if we want to *generalize* such a statement, that is, to say that it applies to all numbers, we can use symbols, which stand for any number, in place of the particular numbers in the example. Any symbol will do; but almost invariably we use letters of the alphabet.

The branch of mathematics which generalizes results of arithmetic, and uses letters to symbolize numbers, is called *algebra.*

9.2 Algebraic Notation

In algebra, the familiar notation is used to indicate addition, subtraction, multiplication, division, signed numbers, powers, and so forth. For example, $a + a = 2 \times a$; $a^2 = a \times a$.

The multiplication sign, $\times$, is often omitted, or is replaced by a single dot. For example, $a \times b$ is usually written as ab or $a . b$ and, similarly, xyz or $x . y . z = x \times y \times z$ and $3y = 3 \times y$.

Sometimes a product is written as a mixture of symbols and ordinary numerals, such as $3x^2$ or $5xy$. The numbers are then called *numerical coefficients*, or just *coefficients*: in the product $3x$, 3 is the coefficient of x, and 5 is the coefficient of xy in the product $5xy$.

Remember that in algebra as generalized arithmetic, *the letters are used for numbers,* and are in no sense abbreviations for other things.

9.3 The Laws of Algebra

Some of the general rules for arithmetic which we have met so far, are:

$a + b = b + a$ (commutative law for addition);
$(a + b) + c = a + (b + c)$ (associative law for addition);
$a \times b = b \times a$ or $ab = ba$ (commutative law for multiplication);
$a \times (b + c) = a \times b + a \times c$ or $a(b + c) = ab + ac$ (distributive law for multiplication).

The last of these laws is especially important in algebra, and also illustrates the beautiful precision of symbolic notation compared with ordinary words. The sentence, 'I have four lots of 5 pounds and 25 pence' could mean:

'I have $4 \times$ £5 + 25p = £20·25'.

Equally, it could be understood as:

'I have $4 \times$ £5·25 = $4 \times$ £5 + $4 \times$ 25p = £21'.

Mathematical notation avoids this kind of ambiguity: $4 \times a + b$ *always* means $4 \times a + b$, that is, $4a + b$. If not only a but also b is to be multiplied by 4, we write the quantity quite differently, using brackets to indicate the exact meaning:

$$4 \times (a + b), \text{ or } 4(a + b).$$

We can work this out by using repeated addition:

$$\begin{aligned} 4(a + b) &= (a + b) + (a + b) + (a + b) + (a + b) \\ &= a + a + a + a + b + b + b + b \\ &= 4a + 4b. \end{aligned}$$

Of course, this kind of reasoning can be extended to cases where there are 3 or more numbers in the brackets, for example,

$$4(5 + 6 + 7) = 4 \times 5 + 4 \times 6 + 4 \times 7,$$
and
$$a(b + c + d) = ab + ac + ad,$$
and
$$k(l - m + n) = kl - km + kn.$$

9.4 The Distributive Law for Division

The distributive law for division can be stated in general terms as:

For all numbers a, b and c (when c is not zero, which we can write as $c \neq 0$),

$$(a + b) \div c = a \div c + b \div c.$$

You should test this law, using simple examples such as:

$$(4 + 8) \div 2 = 12 \div 2 = 6 \quad \text{or} \quad (4 \div 2) + (8 \div 2) = 2 + 4 = 6;$$
$$(9 + 6) \div 3 = 15 \div 3 = 5 \quad \text{or} \quad (9 \div 3) + (6 \div 3) = 3 + 2 = 5.$$

The distributive law for division follows directly from the distributive law for multiplication. Look, for example, at the quantity $(a + b) \div 4$; this can be written as $\frac{1}{4}(a + b) = \frac{1}{4}a + \frac{1}{4}b = (a \div 4) + (b \div 4)$.

In general, if a, b and n are *any* three numbers ($n \neq 0$),

$$(a + b) \div n = \frac{1}{n}(a + b) = \frac{1}{n}a + \frac{1}{n}b = \frac{a}{n} + \frac{b}{n}.$$

Where the 'division line' is used, brackets can be omitted if no confusion can arise. We write, for example,

$$\frac{a + b}{4} = \frac{a}{4} + \frac{b}{4} \text{ and } \frac{a + b}{n} = \frac{a}{n} + \frac{b}{n}.$$

9.5 Exercises

1. Test the distributive laws for multiplication and division for the following numbers:

$5 \times (6 + 8)$ [Hint: are 5×14 and $5 \times 6 + 5 \times 8$ equal?];

$8 \times (100 + 20 + 4)$

[compare the procedure for multiplication of larger numbers, Section 3.1];

$(6 + 8) \div 7$;
$(400 + 80 + 6) \div 4$

[compare long division and rational numbers, Sections 3.8 and 4.3].

2. Multiplying out the factors of an expression so as to get rid of the brackets is called *expanding* the expression, and the result is called an *expansion*; for example, $3a + 3b + 3c$ is the expansion of $3(a + b + c)$. Expand the following, using the general rules for multiplication of signed numbers (see Section 3.4):

$4(a - 2b + c)$;
$^{-}6(x - y)$;
$^{-}(k - l + m)$ [Hint: use $^{-}1(k - l + m)$];
$^{-}2x(a - 3b - 2c)$.

3. Some calculations can be made a great deal easier if we re-arrange the numbers concerned; for example,

$$4 \times 98 = 4 \times (100 - 2) = 400 - 8 = 392.$$

Using this method, calculate:

9×95;
4×996.

Make up and work out some more examples of your own.

9.6 Further Expansions

The distributive law is useful too when both parts of the multiplication are sums or differences of numbers; for example,

$$\begin{aligned} & (3+2)\times(10+3) \\ =\ & 3\times(10+3)+2\times(10+3) \\ =\ & 3\times10+3\times3+2\times10+2\times3 \\ =\ & 30+9+20+6=65. \end{aligned}$$

In a case like this, we can simply add the numbers in the brackets, and then multiply:

$$\begin{aligned} (3+2)\times(10+3) &= 5\times(10+3) \\ &= 5\times13=65. \end{aligned}$$

When we analysed long multiplication (see Section 3.1) we used the distributive law, when we multiplied each part (or *term*) of one sum by each term of the other; for example:

$$\begin{aligned} 12\times36 &= (10+2)\times(30+6) \\ &= 10\times30+10\times6+2\times30+2\times6=432. \end{aligned}$$

In general, if an expression such as $(a+b)\times(c+d)$, or $(a+b)(c+d)$, is expanded, each term of the first factor, $(a+b)$, is multiplied by each term of the second, $(c+d)$.

Examine the following examples, to see just how this is done.

$$\begin{aligned} (a+b)(c+d) &= a\times c+a\times d+b\times c+b\times d \\ &= ac+ad+bc+bd; \end{aligned}$$

$$\begin{aligned} (a-b)(c-d) &= a\times c+a\times(^-d)+(^-b)\times c+(^-b)\times(^-d) \\ &= ac-ad-bc+bd \text{ (using the laws for operations on signed numbers).} \end{aligned}$$

In expressions such as $3x^2+2x$, we speak of 'a term in x^2', 'a term in x', etc.

9.7 Exercise

Expand the following:

$$(a+b)(c-d);$$
$$(3k+l)(m+2n).$$

9.8 Long Multiplication

A vertical arrangement for algebraic multiplication may be helpful, as in long multiplication of ordinary numbers. 'Like' terms are written vertically one above the other so as to be convenient for the addition part of the calculation.

Examples

1. Calculate $(2x + 3y)(x + 2y)$.

$$\begin{array}{rl} & x + 2y \\ \times & 2x + 3y \\ \hline & 2x \times x + 2x \times 2y \\ & \qquad 3y \times x + 3y \times 2y \\ \hline & 2x^2 + 7xy + 6y^2 \end{array}$$

or

$$\begin{array}{rl} & x + 2y \\ \times & 2x + 3y \\ \hline & 2x^2 + 4xy \\ & \qquad 3xy + 6y^2 \\ \hline & 2x^2 + 7xy + 6y^2 \end{array}$$

(Remember that $x \times y = y \times x$.)

2. Calculate $(x - y)(x + 2y)$.

$$\begin{array}{rl} & x - y \\ \times & x + 2y \\ \hline & x^2 - xy \\ & \qquad 2xy - 2y^2 \\ \hline & x^2 + xy - 2y^2 \end{array}$$

9.9 Exercise

Using this vertical arrangement, expand the following:

$(x + 3)(x + 2)$;
$(x - 4)(x + 5)$;
$(x - 5)(x - 4)$;
$(3x + 2)(2x + 4)$;
$(a + 3)(a + 3)$ or $(a + 3)^2$;
$(x - 2)(x - 2)$ or $(x - 2)^2$.

9.10 Some Special Products

Exercise 9.9 shows up some interesting patterns, which we will look at more closely.

If we need to know the square of a term which is itself the sum of two quantities, a and b, we can expand $(a + b)^2$, or $(a + b)(a + b)$, thus:

$$\begin{array}{rl} & a + b \\ \times & a + b \\ \hline & a^2 + ab \\ & \qquad ab + b^2 \\ \hline & a^2 + 2ab + b^2 \end{array}$$

or,

the square of the sum of any two quantities is equal to the sum of their squares plus twice their product, that is,

$$(a + b)^2 = a^2 + 2ab + b^2$$

You should memorize this result: it will be useful whenever you need to square the sum of two numbers; for example, to expand $(2x + 3y)^2$:

Square both numbers, and add the squares,
$(2x)^2 + (3y)^2 = 4x^2 + 9y^2$;
to this add twice the product of the two numbers
$(= 2 \times 2x \times 3y = 12xy)$;
then $(2x + 3y)^2 = 4x^2 + 9y^2 + 12xy$
$= 4x^2 + 12xy + 9y^2$.

It makes no difference if one or both of the two quantities is negative; for example, to expand $(x + {}^-4)^2$ or $(x - 4)^2$:

Square both numbers, and add the squares,
$(x)^2 + ({}^-4)^2 = x^2 + 16$;
to this add twice the product of the two numbers
$(= 2 \times x \times ({}^-4) = {}^-8x)$
and $(x - 4)^2 = x^2 + 16 + ({}^-8x)$
$= x^2 - 8x + 16$.

We could write this in general terms as:

$$\begin{array}{r} a - b \\ \times \quad a - b \\ \hline a^2 - ab \quad\quad \\ - ab + b^2 \\ \hline a^2 - 2ab + b^2 \end{array}$$

or,

the square of the difference of any two quantities is equal to the sum of their squares minus twice their product, that is,

$$(a - b)^2 = a^2 - 2ab + b^2$$

9.11 Exercises

1. Expand the following:

$(x + y)^2$;
$(2x + 3)^2$;
$(3k + 1)^2$;
$(2k + 3l)^2$;

$$(x - 2y)^2;$$
$$(2x - y)^2;$$
$$(4x - 3y)^2.$$

2. These algebraic 'patterns' can also be used to provide an easy way of calculating squares of numbers; for example,

$$99^2 = (100 - 1)^2 = 100^2 - 200 + 1 = 9801.$$

Calculate, in this way,

13^2 [Hint: $(10 + 3)^2 = \ldots$];
102^2;
$10{\cdot}5^2$ [Hint: $(10 + \frac{1}{2})^2 = \ldots$];
1003^2.

9.12 The Difference Between Two Squares

We can see another interesting pattern in the product of the sum and the difference of two quantities, $(a + b)(a - b)$:

$$\begin{array}{rl} & a - b \\ \times & a + b \\ \hline & a^2 - ab \\ & \quad + ab - b^2 \\ \hline & a^2 \qquad\quad - b^2 \end{array}$$

The product in this case is the difference of the two squares:

$$\boxed{(a + b)(a - b) = a^2 - b^2}$$

Examples

$$(2a - 3)(2a + 3) = (2a)^2 - 3^2 = 4a^2 - 9.$$
$$102 \times 98 = (100 + 2)(100 - 2) = 100^2 - 2^2 = 9996.$$

9.13 Exercise

Using this 'pattern', write down the following expansions:

$$(a - 3)(a + 3);$$
$$(2x + 5)(2x - 5);$$
$$(3m + 2n)(3m - 2n);$$
$$101 \times 99;$$
$$100{\cdot}5 \times 99{\cdot}5.$$

9.14 The Laws of Indices

The process of multiplying a number by itself, i.e. of *squaring* a number, is just a special case of the ordinary process of multiplication. We have already met and used the special *index notation*, for example, 5^2 as an abbreviation of 5×5, or a^2 as an abbreviation of $a \times a$ or $a.a$. The 2 in this raised position is called an *index* (plural *indices*). Index notation is a mathematical tool which is particularly useful and versatile, and you will find it well worth while to master the rules which govern its use.

The *index notation* is used as a shorthand for the further repeated multiplication of a number by itself.

So, for example, $2^3 = 2 \times 2 \times 2,$
$2^4 = 2 \times 2 \times 2 \times 2,$

and, of course, $2^1 = 2.$

It seems superfluous to write the index ^{1}, in this case, and we usually omit it altogether, but you must always remember in calculations that $2 = 2^1$, $3 = 3^1$, $x = x^1$, and so on.

In general terms,

$$a^1 = a,$$
$$a^2 = a \times a,$$
$$a^3 = a \times a \times a,$$
$$\cdots\cdots\cdots$$
$$a^n = \underset{(1)}{a} \times \underset{(2)}{a} \times \underset{(3)}{a} \ldots \times \underset{(n)}{a}$$

Terms such as a^2, a^3, a^n, are known as *powers* of a. We speak of a^2 as 'a squared' or 'a raised to the second power' or 'a to the second power', or very often simply 'a to the second'; similarly of a^4 as 'a raised to the fourth power', or 'a to the fourth power' or 'a to the fourth', and so on.

We can deduce the general rule for the multiplication of powers of the same number by simply writing down the product in full instead of in the abbreviated index notation.

For example, $2^2 \times 2^2 = (2 \times 2) \times (2 \times 2) = 2 \times 2 \times 2 \times 2 = 2^4,$

$$a^2 \times a^3 = (a \times a) \times (a \times a \times a) = a \times a \times a \times a \times a = a^5,$$
$$a \times a^3 = a^1 \times a^3 = a \times (a \times a \times a) = a^4.$$

In all cases, *to multiply powers of a number, add the indices*. Clearly, this only applies to powers of the *same* number; for example, $2^3 \times 3^4$ *cannot* be abbreviated any further. If a, m and n stand for any three numbers,

$$\boxed{a^n \times a^m = a^{n+m}}$$

Similarly, for the division of powers of the same number the indices are subtracted, since division is the inverse of multiplication. For example, in the division $3^5 \div 3^3$ the question is asked: 'by what power of 3 is 3^3 to be multiplied to produce 3^5?' or '$3^{\boxed{?}} \times 3^3 = 3^5$ or $3^{\boxed{?}+3}$'. Clearly, $\boxed{?} = 2$.

Alternatively, we can demonstrate this by writing out each power as a product:

$$\begin{aligned} 3^5 \div 3^3 &= (3 \times 3 \times 3 \times 3 \times 3) \div (3 \times 3 \times 3) \\ &= (3 \times 3) \times (3 \times 3 \times 3) \div (3 \times 3 \times 3) \\ &= (3 \times 3) \times 1 = 3^2. \end{aligned}$$

Using the division line and the 'cancelling' process,

$$\frac{3^5}{3^3} = \frac{\cancel{3} \times \cancel{3} \times \cancel{3} \times 3 \times 3}{\cancel{3} \times \cancel{3} \times \cancel{3}} = 3^2.$$

Similarly, $a^9 \div a^2 = a^7$, $b^{10} \div b^9 = b^1 = b$, etc.

In general, *when dividing powers of the same number, subtract the indices,* or, for any three numbers a, m and n

$$\boxed{\frac{a^m}{a^n} = a^{m-n}}$$

Sometimes the terms of a multiplication or division are themselves products of powers of various numbers, for example,

$$a^2b^3 \times a^4b^2$$

or

$$x^5y^2 \div x^2y.$$

We can re-arrange the terms, using the commutative law for multiplication, so as to simplify such expressions.

Examples

1. $a^2b^3 \times a^4b^2 = a^2 \times a^4 \times b^3 \times b^2 = a^6 \times b^5 = a^6b^5.$

2. $\dfrac{x^5y^2}{x^2y} = \dfrac{x^5}{x^2} \times \dfrac{y^2}{y} = x^3 \times y = x^3y.$

3. $$\begin{aligned} \frac{a^4b^3 + a^5b^6}{a^2b} &= \frac{a^4b^3}{a^2b} + \frac{a^5b^6}{a^2b} \\ &= \frac{a^4}{a^2} \times \frac{b^3}{b} + \frac{a^5}{a^2} \times \frac{b^6}{b} \\ &= a^2 \times b^2 + a^3 \times b^5 \\ &= a^2b^2 + a^3b^5. \end{aligned}$$

9.15 Exercises

1. Expand the following:

$$x^3 (x^2 + 2x + 3);$$
$$ab^2 (a^2 + 3ab + b^2).$$

2. Simplify:

$$\frac{a^5}{a^2};$$
$$\frac{a^4b^2}{ab^2};$$
$$\frac{x^4y^3 + 3x^2y^4}{xy^2}.$$

9.16 Negative Indices

If we accept that we can divide powers of numbers by subtracting indices, we are immediately led to the notion of a *negative index*. When we subtract a positive number from a smaller positive number, the result is a negative number; for example, $^+3 - {}^+5 = {}^-2$.

The result of the division $4^3 \div 4^5$, using the law of indices, is $4^{3-5} = 4^{-2}$. If we use the notation of the division line, we can see what a negative index means at once:

$$\frac{4^3}{4^5} = \frac{4^3 \div 4^3}{4^5 \div 4^3} = \frac{1}{4^2}, \text{ or } 4^{-2} = \frac{1}{4^2}.$$

The negative sign is *not* an indication that the number is negative, but simply that a certain power of a number has been divided by a larger power.

The zero power

Another interesting idea emerges from the subtraction of indices. When a number is divided by itself the result is 1, for example, $3 \div 3 = 1$ and $3^4 \div 3^4 = 1$. But, using the law of indices, $3^4 \div 3^4 = 3^{4-4} = 3^0$; therefore, $3^0 = 1$. We can write this for any numbers, say n and m:

$$n^m \div n^m = n^0, \text{ and } n^m \div n^m = 1.$$

Therefore for any number n (other than 0),

$$\boxed{n^0 = 1}$$

9.17 Exercise

Write down the value of each of the following:

$$2^{-4};\ 3^{-3};\ 4^{-2};\ 8^0.$$

9.18 Multiplication and Division of Indices

If we write out in full what is meant by the abbreviated form, $(5^2)^3$, we discover another law of indices:

$$(5^2)^3 = 5^2 \times 5^2 \times 5^2 = 5^{2+2+2} = 5^{3\times 2} = 5^6.$$

Indices are multiplied when a power of a number is itself raised to another power.

In general terms, $(a^n)^m = a^{m \times n}$

If we want to express $(a^n)^m$ in words, we usually say, 'a to the power n, all to the power m'.

You will remember using the symbol $\sqrt{\ }$ as the square root sign (Section 5.24), and the process of finding a square root as the inverse of squaring a number. For example $\sqrt{4}$, or $\sqrt[2]{4}$ or the square root of 4, is the number which, multiplied by itself (squared), produces 4.

Therefore $\sqrt[2]{4} = {}^{+}2$ or ${}^{-}2$, since $({}^{+}2)^2 = ({}^{+}2) \times ({}^{+}2) = 4$
and $({}^{-}2)^2 = ({}^{-}2) \times ({}^{-}2) = 4$.

We can think of the process of finding a square root as factorizing a number into two equal factors; for example, $\sqrt{36} = \sqrt{(6 \times 6)} = 6$.

The idea of *roots*, that is, of equal factors of a number, can be extended further. An index inserted in the root sign shows how many equal factors we are using.

For example, $\sqrt[3]{8}$—the cube root of 8—is the number which, when 'cubed', equals 8:

$$\sqrt[3]{8} = 2, \text{ since } 2^3 = 8.$$

The cube root is found by factorizing the number into 3 equal factors:

$$\sqrt[3]{8} = \sqrt[3]{(2 \times 2 \times 2)} = 2.$$

Similarly,

$$\sqrt[3]{27} = \sqrt[3]{(3^3)} = 3,$$
$$\sqrt[3]{125} = \sqrt[3]{(5^3)} = 5.$$

This process of factorizing a number into 2, 3 or more equal factors involves dividing the index by 2, 3, 4, etc. For example,

$$\sqrt[2]{3^6} = \sqrt[2]{(3^3 \times 3^3)} = 3^{6 \div 2} = 3^3.$$

In general, for any three numbers a, n and m,

$$\sqrt[m]{(a^n)} = a^{n \div m}$$

This would be read as 'the mth root of a to the power n equals a to the power n divided by m'.

We can summarize the laws of indices derived in Sections 9.15 to 9.18:

(*a*) $a^n \times a^m = a^{n+m}$;
(*b*) $a^n \div a^m = a^{n-m}$;
(*c*) $(a^n)^m = a^{m \times n}$;
(*d*) $\sqrt[m]{(a^n)} = a^{n \div m}$;
(*e*) $a^0 = 1$;
(*f*) $a^{-n} = \dfrac{1}{a^n}$.

9.19 Fractional Indices

Look again at the term $\sqrt[m]{(a^n)}$, which occurs in the statement of the fourth law of indices. If m is a factor of n, there is no problem: the power of a resulting from the simplification of this term will be a whole number. For example, $\sqrt[3]{(a^{12})} = a^{12 \div 3} = a^4$. But if m is not a factor of n, then the resulting power of a will be a fraction; for example, $\sqrt[3]{a^2} = a^{2 \div 3} = a^{\frac{2}{3}}$. This is not really the complication it may appear. Look, for instance, at the term $a^{\frac{1}{2}}$. We could write this as $a^{1 \div 2}$, or $\sqrt{a^1}$, or just $\sqrt{a}$, the square root of a. Similarly, $a^{\frac{1}{3}}$ is $a^{1 \div 3}$, or $\sqrt[3]{a}$, the cube root of a. Now consider the number written as $10^{\frac{2}{3}}$; we could write this as $10^{2 \div 3}$, or $\sqrt[3]{10^2}$, that is, the cube root of 100 (which lies between 4 and 5).

Fractional indices, whether positive or negative, obey all the laws of indices, and are handled in every way like whole-number (integral) indices. For example, $(a^6)^{\frac{1}{3}} = a^{6 \times \frac{1}{3}} = a^2$.

9.20 Exercises

1. Simplify the following:

$(a^4)^{\frac{1}{2}}$;
$(x^{-2})^3$;
$\sqrt[3]{(y^6x^3)}$.

2. Simplify and evaluate:

$(2^4)^{\frac{1}{2}}$;
$\sqrt[3]{(2^3 \times 3^6)}$;
$25^{\frac{1}{2}}$ [Hint: first express 25 as a power];
$27^{\frac{2}{3}}$;
$8^{-\frac{2}{3}}$.

9.21 Logarithms

You have already seen in Unit 6 how to use a commercial slide rule to multiply and divide numbers. The sliding principle of this instrument, as of our 'simple' slide rule, is essentially one of addition and subtraction. The linear scale of the simple slide rule showed the whole numbers spaced along it at equal intervals, and we could add or subtract by sliding it to right or left; on the logarithmic scale the consecutive *powers* of numbers are equally spaced, and it is the indices of these powers which are added or subtracted when we move the slide, that is, the powers themselves are multiplied or divided. Recognizing this, we can examine the laws of indices using the commercial slide rule.

This is most easily done by considering a specific example, say, the calculation of 4×8, using the slide rule. For this we use the familiar procedure: first set the scale of your slide rule with the rider line on 4 on the A-scale; then align 1 on the B-scale with the rider, and lastly move the B-scale two whole 'steps' to the right. Now read off the product on the A-scale, lying opposite to 8 on the B-scale—that is, 32.

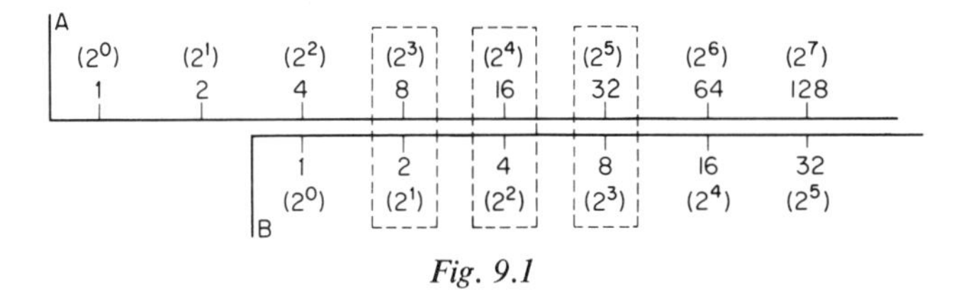

Fig. 9.1

You should notice that while you went through this calculation, $4 \times 8 = 2^2 \times 2^3 = 2^5$, your movements were exactly the same as when you used your simple slide rule to calculate $2 + 3 = 5$. (You will see that your slide rule setting also shows you that $2^2 \times 2^4 = 2^6$, $2^2 \times 2^5 = 2^7$, and so on—each time adding 2 to the power of 2 on the scale.)

As an exercise, use your commercial slide rule to calculate $243 \div 27$ (that is, $3^5 \div 3^3$) and compare your procedure with that of subtraction using the simple slide rule.

The logarithmic scale also illustrates the existence and value of fractional indices. If consecutive whole-number powers are equally spaced out along it, then the points which lie in the intervals between them must represent fractional powers. Fig. 9.1 shows a logarithmic scale based on powers of 2. We will look now at a logarithmic scale with the base 10 (Fig. 9.2), since the logarithm tables in common use are based on 10.

On this scale, $10^{\frac{1}{2}}$ is placed half-way between 10^0 and 10^1, and $10^{1\frac{1}{2}}$ half-way between 10^1 and 10^2, $10^{\frac{1}{4}}$ half-way between 10^0 and $10^{\frac{1}{2}}$, and so on.

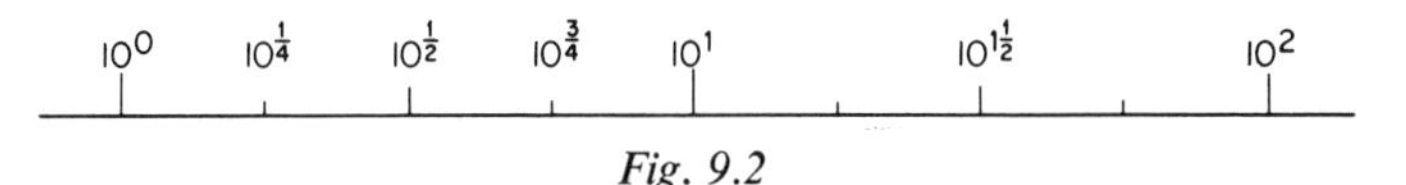

Fig. 9.2

These indices, integral as well as fractional, are also called *logarithms*; as logarithms, fractional indices are usually given in their point notation. For example, $10^{\frac{1}{2}}$ is written as $10^{0\cdot5}$, and $10^{\frac{1}{4}}$ as $10^{0\cdot25}$. The logarithm of 100 is 2, since $10^2 = 100$; 3 is the logarithm of 1000, since $10^3 = 1000$; $10^{0\cdot5}$, or $10^{\frac{1}{2}}$, or $\sqrt{10}$, calculated to 4 significant figures, is 3·162, so that 0·5 is the logarithm of 3·162.

In the *table of logarithms*, all numbers between 1 and 10 are expressed as powers of 10, or powers to the base 10. A table of logarithms appears near the end of this book.

By expressing numbers as powers of 10 (logarithms to the base 10), calculations involving multiplication and division can be done by the much simpler operations of adding and subtracting indices. Using logarithms is an alternative to using the slide rule; it takes rather longer and provides a greater degree of accuracy. However, the slide rule provides a sufficient degree of accuracy for most purposes.

Note. Logarithm tables can be compiled to any base, and in fact other bases are in use. However, common logarithms are to the base 10 and we shall call these simply logarithms.

9.22 Finding the Logarithm of a Number Between 1 and 10

Four-figure tables of logarithms give the four significant figures of the fractional part of the logarithm, which is called the *mantissa*. For example, $2 = 10^{0 \cdot 3010}$ and the table of logarithms only gives the figures 3010 (the mantissa) as the logarithm of 2. In some tables the unit point is actually printed (·3010), but most tables just show the digits 3010. Tables of logarithms use the same arrangement that we met earlier (see Section 5.20). Again, there are three major vertical divisions: a first column for the first two significant figures of the number concerned, a range of columns determined by the third significant figure and another range of columns determined by the fourth significant figure. Let us find, for example, the logarithm of 2·467 (in other words, what power of 10 is equivalent to the number 2·467).

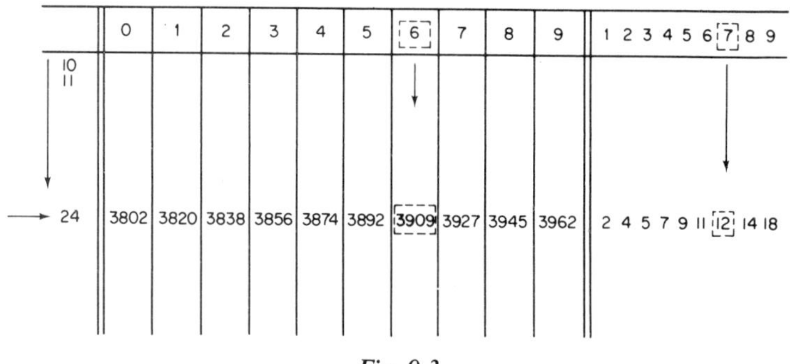

Fig. 9.3

Look down the first column for 24, and so determine the row you need; in this row select the column of the second vertical arrangement headed by the number 6 and find the figures 3909. A minor adjustment has to be made which depends on the fourth figure (7), and you will find this in the last set of columns in the same row, in the column headed by 7; the adjustment shown here is 12 which is added to 3909 giving 3921.

The logarithm of 2·467 is therefore 0·3921. We write log 2·467 = 0·3921, or alternatively $2 \cdot 467 = 10^{0 \cdot 3921}$.

9.23 Exercises

1. Express the following numbers as powers of 10:

$$3{\cdot}432 = 10^{\cdots};$$
$$5{\cdot}321;$$
$$8{\cdot}642.$$

2. Write down the logarithms of these numbers in the alternative notation, for example,

$$\log 3{\cdot}432 = \ldots, \text{ etc.}$$

9.24 The Logarithm of a Number Greater than 10

We can use the standard index notation (see Section 4.21) to find the logarithms of numbers greater than 10.

For example, $4362 = 4{\cdot}362 \times 10^3$.

Using logarithm tables, $\log 4{\cdot}362 = 0{\cdot}6397$ or $4{\cdot}362 = 10^{0{\cdot}6397}$.

Using the laws of indices:

$$4362 = 10^{0{\cdot}6397} \times 10^3 = 10^{3{\cdot}6397} \Rightarrow \log 4362 = 3{\cdot}6397.$$

9.25 Exercise

Express the following numbers in standard index notation and find their logarithms:

$352{\cdot}9$ [Hint: $352{\cdot}9 = 3{\cdot}529 \times 10^2$, so that $\log 352{\cdot}9 = 2{\cdot}\ldots$];
52 460;
12·38.

9.26 Characteristic and Mantissa

The examples show clearly that the integral part of the logarithm, the number on the left-hand side of the unit point (usually called the *characteristic* of the logarithm), is the value of the power of 10 in the standard index notation. It can be found by inspection and can immediately be written down.

All numbers between 10 and 100 (having 2 digits on the left of the unit point) can be expressed in the form $a \times 10^1$, in standard index notation; for example, $99{\cdot}7 = 9{\cdot}97 \times 10^1$. The logarithms of all these numbers therefore have a characteristic of 1.

All numbers between 100 and 1000 (having 3 digits before the unit point) can be expressed in the form $a \times 10^2$ in standard index notation; for example, $994{\cdot}6 = 9{\cdot}946 \times 10^2$. The logarithms of all these numbers therefore have a characteristic of 2.

The characteristic of the logarithm of a number is therefore 1 less than the number of digits before the unit point of the number. For example, the characteristic of the logarithm of $\overbrace{348}^{\text{3 digits}}{\cdot}9$ is 2, or $\log 348{\cdot}9 = 2{\cdot}\ldots$

The characteristic of the logarithm of $\overbrace{86}^{\text{2 digits}}{\cdot}31$ is 1, or $\log 86{\cdot}31 = 1{\cdot}\ldots$; the characteristic of the logarithm of $\overbrace{56\ 280}^{\text{5 digits}}$ is 4, or $\log 56\ 280 = 4{\cdot}\ldots$, etc.

You will have realized too that the mantissae of the logarithms of numbers are the same if the significant figures of the numbers are the same.

For example,
$$4326 = 10^3 \times 4{\cdot}326 = 10^3 \times 10^{0{\cdot}6361} = 10^{3{\cdot}6361},$$
$$432{\cdot}6 = 10^2 \times 4{\cdot}326 = 10^2 \times 10^{0{\cdot}6361} = 10^{2{\cdot}6361},$$
$$43{\cdot}26 = 10^1 \times 4{\cdot}326 = 10^1 \times 10^{0{\cdot}6361} = 10^{1{\cdot}6361}.$$

We therefore find the logarithm of a number by determining

(*a*) the characteristic, by inspection;
(*b*) the mantissa, from the significant figures of the number, by using logarithm tables.

9.27 The Logarithms of Numbers Less than 1

If numbers between 0 and 1 are expressed in standard index notation the power of 10 is always a negative number.

For example,
$$0{\cdot}4326 = 4{\cdot}326 \times 10^{-1},$$
$$0{\cdot}043\ 26 = 4{\cdot}326 \times 10^{-2},$$
$$0{\cdot}004\ 326 = 4{\cdot}326 \times 10^{-3},$$
etc.

Since $4{\cdot}326 = 10^{0{\cdot}6361}$

$$0{\cdot}4326 = 4{\cdot}326 \times 10^{-1} = 10^{0{\cdot}6361} \times 10^{-1} = 10^{-1+0{\cdot}6361}.$$

For simplicity's sake, we do not attempt to add up the negative characteristic and the positive mantissa; the expression is written either as $10^{-1+0{\cdot}6361}$ or as $10^{\bar{1}{\cdot}6361}$ (read as: '10 to the power bar 1 point 6361').

The negative sign before a number makes the whole of that number negative; for example: $-12 = -(10 + 2) = -10 - 2$;
$$-1{\cdot}5 = -(1 + 0{\cdot}5) = -1 - 0{\cdot}5.$$

But when working with *logarithms* you must remember that the *mantissa is always positive*: the negative sign placed over the characteristic only makes the characteristic negative, and has no effect on the mantissa. So if we want to add, for example, the logarithms $\bar{2}\cdot8321$ (read as 'bar 2 point 8321') and $\bar{1}\cdot7426$, we may write:

$$\begin{array}{rl} & {}^{-}2 + 0\cdot8321 \\ + & {}^{-}1 + 0\cdot7426 \\ \hline & {}^{-}3 + 1\cdot5747 = {}^{-}2 + 0\cdot5747 \text{ or } \bar{2}\cdot5747. \end{array}$$

To subtract the logarithm $\bar{2}\cdot1748$ from $1\cdot3421$ we may write:

$$\begin{array}{rl} & 1 + 0\cdot3421 \\ - & {}^{-}2 + 0\cdot1748 \\ \hline & {}^{+}3 + 0\cdot1673 \text{ or } 3\cdot1673 \text{ [Note: } 1 - ({}^{-}2) = {}^{+}3]. \end{array}$$

If we want to subtract the logarithm $\bar{1}\cdot5631$ from $\bar{2}\cdot3742$, we see at once that the mantissae are such that we cannot carry out a straightforward subtraction; but we can get round this by rewriting the second logarithm thus:

$$\begin{aligned} \bar{2}\cdot3742 &= \bar{2} + 0\cdot3742 \\ &= \bar{3} + 1\cdot3742. \end{aligned}$$

Then we can carry out the subtraction in the usual way:

$$\begin{array}{rl} & {}^{-}3 + 1\cdot3742 \\ - & {}^{-}1 + 0\cdot5631 \\ \hline & {}^{-}2 + 0\cdot8111 \text{ or } \bar{2}\cdot8111. \end{array}$$

If you find it easier, you may write:

$$\begin{array}{rl} & \bar{2} + 0\cdot3742 \\ - & \bar{1} + 0\cdot5631 \\ \hline & \bar{2} + 0\cdot8111 \end{array} \quad \text{or} \quad \begin{array}{rl} & \bar{2}\cdot3742 \\ - & \bar{1}\cdot5631 \\ \hline & \bar{2}\cdot8111 \end{array}$$

and make the calculation using ordinary carry-over procedures.

9.28 Exercises

1. Express the following numbers in standard index notation and then find their logarithms:

   0·042 31;
   0·6742;
   0·000 436 7.

2. Add the following pairs of logarithms:

   $\bar{1}\cdot6532 + 2\cdot3219$;
   $\bar{1}\cdot3933 + \bar{3}\cdot8246$.

3. Subtract the following pairs of logarithms:

3·3246 − 2·5783;
2·4825 − 3·5812;
$\bar{2}$·4248 − 1·3242;
$\bar{1}$·3284 − $\bar{2}$·6241.

9.29 Conversion of a Power of 10 into an 'Ordinary Number'

To convert a power of ten into an 'ordinary number'—in other words to find the number whose logarithm we know—we may adopt the reverse procedure with the table of logarithms. For example, to find the number whose logarithm is 2·3878, or the number that equals $10^{2\cdot3878}$, we first determine the significant figures of the required number from the table of logarithms. Scanning the rows and columns for the mantissa, 0·3878, we find the nearest mantissa less than 0·3878 to be 0·3874 in the row labelled 24, and in the column headed by 4 (Fig. 9.4).

	0	1	2	3	4	5	6	7	8	9	1	2	3	4	5	6	7	8	9
24	3802	3820	3838	3856	3874	3892	3909	3927	3945	3962	2	4	5	7	9	11	12	14	18

Fig. 9.4

Now we know the first three figures to be 244; from the last set of columns we find the fourth significant figure (2) corresponding to an adjustment of 4 (3874 + 4 = 3878). The four significant figures are therefore 2442. The characteristic determines the position of the unit point; since $10^2 < 10^{2\cdot3878} < 10^3$, $100 < 10^{2\cdot3878} < 1000$. By placing the unit point after the third significant figure, the required number is found to be 244·2, a number between 100 and 1000. (Also, using our 'rule of thumb': 3 digits before the unit point $\Leftrightarrow$ characteristic is 2.)

Make sure you fully understand this procedure: you could find the logarithm of a number, for example 3782, in the usual way, and then, having found the logarithm to be 3·5777, try to get back to the number by retracing your steps in reverse order. Try this with one or two other numbers, until you are quite sure you have mastered the procedure in both directions.

If the characteristic of a logarithm is negative, we use the same method.

Example

Find the number whose logarithm is $\bar{2}$·3112.

We determine the significant figures of the number using logarithm tables. Since the mantissa is 3112, the significant figures of the required number are 2047 (by the procedure shown above). The simplest method of placing the unit point in the case of a negative characteristic is to consider the characteristic as the *index* of the standard form of the required number, in this example $2{\cdot}047 \times 10^{-2}$. The required number is then:

$$2{\cdot}047 \times 10^{-2} = 0{\cdot}020\,47.$$

(Compare Section 9.27.)

A 'rule of thumb' emerges when we look at more examples. The negative characteristic indicates the position of the first significant figure after the unit point; in the example, the second place after the unit point. That is, if the characteristic is $\bar{1}$, the number is 0· . . .; if it is $\bar{2}$, the number is 0·0 . . .; if it is $\bar{3}$, the number is 0·00 . . ., etc.

9.30 Exercise

Find the numbers whose logarithms are:

$$2{\cdot}3246;$$
$$0{\cdot}6824;$$
$$\bar{1}{\cdot}3672;$$
$$\bar{3}{\cdot}4218.$$

9.31 Calculating with Logarithms

Since logarithms are just indices, all calculations with logarithms are based on the laws of indices. It is just because addition and subtraction are relatively simple compared with multiplication and division, that logarithms are an invaluable tool in calculation, especially where a large number of quantities have to be multiplied. We do this by expressing the numbers as powers of 10, converting the final result back into an 'ordinary number'.

Note. Since logarithm tables concern only the significant figures of a number, the results of calculations by logarithms are always approximations; this also explains the slight variations between different editions of the tables.

To multiply two numbers. Find their logarithms, *add these logarithms* and convert the result to an 'ordinary number'. For example,

$$83{\cdot}67 \times 64{\cdot}29 = 10^{1{\cdot}9226} \times 10^{1{\cdot}8081} = 10^{3{\cdot}7307} = 5379.$$

To divide a number by another number. Find their logarithms, *subtract these logarithms* and convert the result to an 'ordinary number'. For example,

$$259{\cdot}2 \div 88{\cdot}53 = 10^{2{\cdot}4136} \div 10^{1{\cdot}9470} = 10^{2{\cdot}4136-1{\cdot}9470} = 10^{0{\cdot}4666} = 2{\cdot}928.$$

To square a number. Find the logarithm of the number, *multiply the logarithm* by 2 and convert the result into an 'ordinary number'. For example,

$$(67{\cdot}13)^2 = (10^{1{\cdot}8270})^2 = 10^{2\times 1{\cdot}8270} = 10^{3{\cdot}6540} = 4508.$$

This method can also be used for other powers. For example,

$$(4{\cdot}362)^3 = (10^{0{\cdot}6397})^3 = 10^{3\times 0{\cdot}6397} = 10^{1{\cdot}9191} = 83{\cdot}01.$$

To find the square root of a number. Find its logarithm, *divide the logarithm* by 2, and convert the answer to an 'ordinary number'. For example,

$$\sqrt[2]{83{\cdot}67} = \sqrt[2]{(10^{1{\cdot}9226})} = 10^{1{\cdot}9226\div 2} = 10^{0{\cdot}9613} = 9{\cdot}147.$$

Find the cube root of a number by an analogous method: find its logarithm, divide it by 3, and convert the result to an ordinary number'. Similarly,

$$\sqrt[4]{652{\cdot}4} = \sqrt[4]{(10^{2{\cdot}8145})} = 10^{2{\cdot}8145\div 4} = 10^{0{\cdot}7036} = 5{\cdot}053.$$

Particular care is needed when a logarithm with a negative characteristic is divided. For example,

$$\sqrt[3]{0{\cdot}7362} = \sqrt[3]{10^{\bar{1}{\cdot}8670}} = 10^{\bar{1}{\cdot}8670\div 3}.$$

Since the logarithm $\bar{1}{\cdot}8670 = {}^-1 + 0{\cdot}8670$, the division must be done in two parts: a division of the negative characteristic, which *must* result in a *whole number*, and a division of the positive mantissa. If, as in this case, the characteristic is not a multiple of the divisor, we can express the negative characteristic as the sum of two numbers; the first a negative number, the nearest multiple of the divisor (in this case 3), the second a positive number. We can express $^-1$ as $^-3 + 2$, so that

$$\frac{\bar{1}{\cdot}8670}{3} \text{ or } \frac{^-1 + 0{\cdot}8670}{3} = \frac{\overbrace{^-3 + 2{\cdot}8670}}{3} = \frac{^-3}{3} + \frac{2{\cdot}8670}{3} = {}^-1 + 0{\cdot}9557$$
$$= \bar{1}{\cdot}9557,$$

and $\sqrt[3]{10^{\bar{1}{\cdot}8670}} = 10^{\bar{1}{\cdot}8670\div 3} = 10^{\bar{1}{\cdot}9557} = 0{\cdot}9030.$

9.32 Exercises

1. Divide the following logarithms by 5:

$\bar{1}{\cdot}2045$;
$\bar{2}{\cdot}3260$;
$\bar{3}{\cdot}1465$.

[Hint: in each case express the negative characteristic as a sum of two numbers: a negative number which is a multiple of 5, and a positive number.]

2. Using logarithms, calculate:

$$8{\cdot}324 \times 14{\cdot}58;$$
$$3{\cdot}46 \times 89{\cdot}24 \times 6{\cdot}452;$$
$$92{\cdot}45 \times 0{\cdot}0362;$$
$$65{\cdot}23 \div 8{\cdot}936;$$
$$3{\cdot}874 \div 74{\cdot}85;$$
$$(58{\cdot}41)^3;$$
$$(0{\cdot}0726)^3;$$
$$\sqrt[3]{24{\cdot}76};$$
$$\sqrt[3]{0{\cdot}4641};$$
$$\sqrt[3]{0{\cdot}0326}.$$

3. Calculate some of these using your slide rule; compare the degree of accuracy achieved by the two methods.

9.33 A Conventional Arrangement

We often use an arrangement of two columns, one for the numbers and the other for the logarithms, instead of writing out the logarithms as powers of 10. You will find this quite straightforward if you first write down the numbers involved in the calculation in the left-hand (number) column; then look up their logarithms in your tables, and write these in the right-hand (logarithm) column. Carry out the required operations on these logarithms, and write down the result in the right-hand column; then convert this logarithm back into an 'ordinary number', and write this, your final result, in the left-hand column. The following examples illustrates how this way of setting out a calculation is simpler and clearer than writing out all the numbers in index notation, especially for complicated calculations.

Examples

1. Calculate 36·14 × 14·92.

numbers	logarithms
36·14	1·5580
×	+
14·92	1·1738
539·3	2·7318

Note. The result of the calculation is the last number in the 'numbers' column (underlined twice).

2. Calculate $\dfrac{324{\cdot}5 \times 9{\cdot}432}{(8{\cdot}327)^2 \times 4{\cdot}619}$.

numbers	logarithms	
324·5	2·5112	
×	+	
9·432	0·9746	
	3·4858	←
8·327	0·9205	
	×2	
$(8{\cdot}327)^2$	1·8410	(subtracting)
×	+	
4·619	0·6645	
	2·5055	←
9·557	0·9803	

3. Calculate $\sqrt{\dfrac{0{\cdot}005\,672 \times 2{\cdot}618}{0{\cdot}081\,46}}$.

	numbers	logarithms
	0·005 672	$\bar{3}{\cdot}7538$
	×	+
	2·618	0·4179
		$\bar{2}{\cdot}1717$
	÷	−
	0·081 46	$\bar{2}{\cdot}9109$
		$\bar{1}{\cdot}2608$ ($=\bar{2} + 1{\cdot}2608$)
(Take square root)	$\sqrt[2]{}$	÷2
	0·4270	$\bar{1}{\cdot}6304$

9.34 Exercises

Use the column arrangement for the following calculations:

1. $\dfrac{36{\cdot}49 \times 7{\cdot}924}{9{\cdot}104}$

2. $\sqrt{\left(\dfrac{74{\cdot}3 \times 1{\cdot}046}{64{\cdot}07 \times 3{\cdot}26}\right)}$

Unit Ten

Factorization and its Applications

10.1 Factorization of Algebraic Expressions

As you know, there is a simple procedure for multiplying two or more numbers, but no such simple procedure exists for factorizing a number, that is expressing it as the product of two or more numbers. We have to use trial and inspection, and sometimes divisibility tests, as the only methods for factorization.

We have seen how to use the distributive property of multiplication and the laws of indices for the multiplication of algebraic expressions such as:

$$(a + b)(a^2 - b) = (a \times a^2) + (a \times {}^-b) + (b \times a^2) + (b \times {}^-b)$$
$$= a^3 + a^2b - ab - b^2.$$

For the opposite process, 'unravelling' this product into factors, no simple procedure exists; but in the next few sections we will establish some methods which can be used to factorize algebraic expressions.

Note. Remember that some numbers—prime numbers—cannot be factorized. Similarly, some algebraic expressions cannot be factorized either.

10.2 Finding a Common Factor

If an algebraic expression, say $a + b + c$, is multiplied by a number such as 3 or a, this number 3 or a will be a factor of all the terms in the product:

$$3(a + b + c) = 3a + 3b + 3c,$$
$$a(a + b + c) = aa + ab + ac = a^2 + ab + ac.$$

We can take this in the opposite sense: if a number appears as the factor of each term of an expression (a *common factor*), then this number is a factor of the whole expression.

For example, $6b + 6c + 6d = 6(b + c + d)$;

or $x^2 + 3x = xx + 3x = x(x + 3)$.

A simple rule in factorizing any expression is first to look for a common factor and to 'take out' this common factor. If there are more common factors, 'take out' the greatest common factor. For example, in factorizing

$$6x^2 + 12x + 18yx:$$

3 is a common factor,
2 is a common factor,
therefore 6 is a common factor;
x is a common factor,
therefore $6x$ is a common factor.

There are no other common factors; $6x$ is then the greatest common factor, and

$$6x^2 + 12x + 18yx = 6x(x + 2 + 3y).$$

10.3 Exercise

Write as products—that is, factorize—the following expressions, 'taking out' the greatest common factor:

$$3a + 6b;$$
$$2x^2 + 6xy;$$
$$3x^3 + 12x^2 + 6x.$$

10.4 'Grouping' and Taking Out a Common Factor

Sometimes only a few terms in an expression have a factor in common, for example: $ac + ad + bc + bd$. By grouping the terms and taking out the common factor in each group, a common factor for the whole expression may appear:

$$(ac + ad) + (bc + bd) = a(c + d) + b(c + d).$$

Here $(c + d)$ is a common factor of both composite terms. By dividing both terms by $(c + d)$, that is taking out the common factor $(c + d)$, we find

$$a(c + d) + b(c + d) = (c + d)(a + b).$$

Note. You can always check a factorization by multiplying out, or expanding, the result. Here,

$$(c + d)(a + b) = ac + bc + ad + bd.$$

Example

Factorize $xy - y^2 + yz - xz$.

$$\begin{aligned} &(xy - y^2) + (yz - xz) \\ &= y(x - y) + z(y - x) \\ &= y(x - y) - z(x - y) \\ &= (y - z)(x - y). \end{aligned}$$

[Notice that $y - x = -(x - y)$.]

Checking by expanding:

$$(y - z)(x - y) = yx - y^2 - zx + zy = xy - y^2 + yz - xz.$$

10.5 Exercise

Using the above method, factorize:

$$ac + ad + 2bc + 2bd;$$
$$km - kn + lm - ln;$$
$$ac + 4a + 2c + 8;$$
$$2xy - 4x + 3y - 6.$$

10.6 Factorizing Quadratic Expressions: Familiar Products

Expressions such as $x^2 + 5x + 4$ and $y^2 + 2y + 6$ are called *quadratic expressions*; the highest power of the algebraic number is 2, that is x^2, y^2. There are no terms in x^3, y^3 or higher powers of x or y.

The factors of a quadratic expression can be written down immediately if we can recognize one of the familiar patterns:

$$a^2 - b^2 = (a + b)(a - b),$$
$$a^2 + 2ab + b^2 = (a + b)^2,$$
$$a^2 - 2ab + b^2 = (a - b)^2.$$

The *first pattern*, the difference of two squares, is clearly recognizable (sometimes after a common factor has been taken out). For example,

$$k^2 - m^2 = (k + m)(k - m);$$
$$9x^2 - 4y^2 = (3x)^2 - (2y)^2 = (3x + 2y)(3x - 2y);$$
$$12x^2 - 3y^2 = 3(4x^2 - y^2) = 3[(2x)^2 - y^2] = 3(2x + y)(2x - y).$$

In the *second pattern*, there are two positive squares, the squares of two numbers, and the third term is twice the product of these two numbers. If we find two squares in the expression, and can identify the numbers of which they are the squares, we can check whether the remaining term is indeed twice the product of these two numbers. For example, in the expression

$$x^2 + 6x + 9,$$

there are two squares, x^2 and 9, which are respectively the squares of x and 3; the remaining term $6x$ is indeed $2 \times x \times 3$.

Therefore, $$x^2 + 6x + 9 = (x + 3)^2.$$

If we can recognize the *third pattern*, for example,

$$x^2 - 6x + 9,$$

then again we find two squares, and the negative sign of the product tells us that *one* of the two numbers we are trying to identify is negative. There are then two possibilities: the two numbers are either x and $^-3$, or ^-x and 3;

$$-6x = 2 \times x \times {}^-3, \text{ or } -6x = 2 \times {}^-x \times 3,$$

and $$x^2 - 6x + 9 = (x-3)^2 \text{ or } (3-x)^2.$$

10.7 Exercise

Look for a 'pattern' and then factorize the following:

$$x^2 + 8x + 16;$$
$$x^2 - 100;$$
$$a^2 - 10a + 25;$$
$$x^2 + 6xy + 9y^2;$$
$$20x^2 - 45y^2.$$

10.8 Factorizing Quadratic Expressions: Trial and Inspection

Sometimes we cannot recognize any familiar pattern in a quadratic expression, for example $x^2 + 5x + 6$. But we can work out a method for factorizing such expressions if we analyse the multiplication which builds up a product like this.

Consider for example the multiplication $(x + 3)(x + 2)$: each term of the first factor is multiplied by each term of the second factor (see Section 9.6):

$$(x + 3)(x + 2)$$
$$= x \times x + x \times 2 + 3 \times x + 3 \times 2$$
$$= x^2 + 5x + 6.$$

In general terms,

$$(x + a)(x + b)$$
$$= x \times x + x \times b + a \times x + a \times b$$
$$= x^2 + (a + b)x + ab.$$

Here we see that the coefficient of x is the *sum* of the two terms a and b, $a + b$; the last term is the *product* of these terms, ab.

In our first example, we can see that the *sum* of the two numbers 2 and 3 is the coefficient of the term in x, $5x$, and the *product* of 2 and 3 is the last term in the expression $x^2 + 5x + 6$.

So we can identify the two numbers in this quadratic expression by asking ourselves the question: 'Which two numbers have a sum 5 and a product 6?'

Starting with the product 6 we find two possibilities:

$6 = 1 \times 6$, but $6 + 1 = 7$, so these are not the numbers we need;
$6 = 2 \times 3$ and $2 + 3 = 5$.

The required two numbers are therefore 2 and 3, and

$$x^2 + 5x + 6 = (x + 2)(x + 3).$$

Examples

1. Factorize $x^2 + 8x + 12$.

   $12 = 12 \times 1$ and $12 + 1 = 13$;
   $12 = 2 \times 6$ and $2 + 6 = 8$;
   $12 = 3 \times 4$ and $3 + 4 = 7$.

   The required number pair must be 2 and 6, and

   $$x^2 + 8x + 12 = (x + 2)(x + 6).$$

2. Factorize $x^2 - x - 6 = x^2 + (^-1)x + (^-6)$.

   The product of the two numbers required is $^-6$ (the negative product indicates that one of the numbers is negative and the other is positive) and the required sum is $^-1$. Of the possible pairs of numbers whose products are all $^-6$, namely $^+6$ and $^-1$, $^-1$ and $^+6$, $^+2$ and $^-3$, $^-2$ and $^+3$, we choose $^-3$ and $^+2$, since $^-3 + {}^+2 = {}^-1$;

   therefore: $$x^2 - x - 6 = (x - 3)(x + 2).$$

3. Factorize $x^2 - 7x + 10$.

   Again, the sum of the two numbers required is negative ($^-7$), but their product is positive ($^+10$); so both numbers must be negative. Of the possible pairs of negative numbers whose products are all $^+10$, namely, $^-10$ and $^-1$, $^-2$ and $^-5$, we choose $^-2$ and $^-5$, since $^-2 + {}^-5 = {}^-7$,

   and $$x^2 - 7x + 10 = (x - 2)(x - 5).$$

10.9 Exercise

Using this trial-and-inspection method, factorize the following quadratic expressions:

$$x^2 + 9x + 18;$$
$$a^2 + 5a - 14;$$
$$m^2 - m - 20;$$
$$x^2 - 8x + 15.$$

(Remember always to *check* your answer by expanding it.)

10.10 Harder Factorizations by Trial and Inspection

In all our examples so far, the coefficient of the quadratic term has been 1; but we can use the same method of trial and inspection to factorize, for example, $2x^2 + 15x + 7$. Here the term $2x^2$ indicates that the factors are of the form

$$(2x + \ldots)(x + \ldots).$$

The product of the two missing terms is 7, and the two possibilities are 1 and 7, or 7 and 1.

If we try out these possibilities, we find that

$$(2x + 7)(x + 1) = 2x^2 + 9x + 7,$$

and

$$(2x + 1)(x + 7) = 2x^2 + 15x + 7.$$

The second is therefore the required factorization.

This method of inspection and trial is still useful, although more elaborate, if both the coefficient of x^2 and the numerical term are numbers with several factors, for example, $6x^2 + 17x + 12$. If we look at the coefficient of x^2, we see that the factors can be of the form $(2x + \ldots)(3x + \ldots)$

or of the form

$$(6x + \ldots)(x + \ldots).$$

The product of the two missing numbers is 12; the possibilities are therefore 1 and 12, 12 and 1, 3 and 4, 4 and 3, 2 and 6, 6 and 2.

The possibilities 2 and 6, and 6 and 2 can be eliminated at once, since the coefficient of x is an odd number ($6 \ldots + 2 \ldots$ must always be an even number).

Trying out the possibilities,

$$\begin{aligned}
&(2x + 1)(3x + 12),\\
&(2x + 12)(3x + 1),\\
&(6x + 1)(x + 12),\\
&(6x + 12)(x + 1),\\
&(2x + 3)(3x + 4),\\
&(2x + 4)(3x + 3),\\
&(6x + 3)(x + 4),\\
&(6x + 4)(x + 3),
\end{aligned}$$

we will find that only $(3x + 4)(2x + 3)$ will produce $17x$ as the middle term, and this is therefore the required factorization. As soon as you meet the required middle term, you can stop the trial process, since there is only one possible complete factorization; therefore try the most likely possibilities first.

10.11 Exercise

By trial and inspection, factorize the following quadratic expressions:

$$\begin{aligned}
&2y^2 + 5y + 2;\\
&3x^2 - 7x + 2;\\
&6y^2 + 5y - 4.
\end{aligned}$$

10.12 Applications of Factorization

Factorization is an important first step in solving quadratic equations (which you will meet later) and in simplifying complicated expressions, especially those which take the form of a division, or fraction, for example: $\frac{x^2 + 3x}{x^3 + 5x}$. We call such expressions *algebraic fractions*. Most simplifications use the principle of division of the numerator and denominator by a *common factor* (i.e. 'cancelling', see Section 4.4); the first step in simplifying this kind of expression, therefore, must be the factorization of the numerator and denominator.

Examples

1. Simplify $\frac{x^2 + 3x}{x^3 + 5x}$.

Factorizing numerator and denominator:

$$\frac{x^2 + 3x}{x^3 + 5x} = \frac{x(x + 3)}{x(x^2 + 5)}.$$

'Cancelling' x, that is, dividing both numerator and denominator by x:

$$\frac{x(x + 3)}{x(x^2 + 5)} = \frac{x + 3}{x^2 + 5}.$$

2. $$\frac{x^2 + x - 20}{2x^2 - 11x + 12} = \frac{(x + 5)(x - 4)}{(2x - 3)(x - 4)} = \frac{x + 5}{2x - 3}.$$

Note. In this result x cannot be cancelled since it is not a factor of the numerator (i.e. of every term of the numerator), nor of the denominator.

10.13 Exercise

Simplify the following expressions:

$$\frac{3a^2 + 10a + 8}{a^2 - 4};$$

$$\frac{2x - 6}{2x^2 - 4x - 6}.$$

10.14 Expressions that Cannot be Factorized

We must stress again that not every algebraic expression can be factorized and, therefore, not all complicated expressions can be simplified.

When you have had some practice, you will soon be able to establish fairly quickly whether an expression can be easily factorized or not. A good routine is to try the various methods of factorization in succession:

(*a*) take out any common factor;
(*b*) look for familiar products ('patterns');
(*c*) use the method of trial and inspection if the numbers are reasonably small.

10.15 Operations on Algebraic Fractions

In algebraic fractions the numerators and denominators are algebraic expressions which represent numbers. Therefore we can handle algebraic fractions using the same methods as for the familiar rational numbers.

Two algebraic fractions can be *added* or *subtracted* if they have the same *denominator*, and the resulting algebraic fraction will have the sum (or difference) of the numerators as its numerator, and the common denominator as its denominator. For example,

$$\frac{4x+1}{2x^2+1}+\frac{5x+3}{2x^2+1}=\frac{4x+1+5x+3}{2x^2+1}=\frac{9x+4}{2x^2+1}.$$

If the denominators are different, algebraic fractions can only be added or subtracted if one or both are replaced by equivalents with the same denominator.

Just as with rational numbers, we find equivalents by multiplying both numerator and denominator by the same number. For example,

$$\frac{5}{2x^2}+\frac{7}{x}=\frac{5}{2x^2}+\frac{2x\times 7}{2x\times x}=\frac{5}{2x^2}+\frac{14x}{2x^2}=\frac{14x+5}{2x^2}.$$

A common denominator must be found before we can add or subtract the fractions. The *simplest* common denominator is the *lowest common multiple* of the denominators, sometimes called the *lowest common denominator*.

The lowest common multiple of algebraic expressions is found (as for ordinary numbers, see Exercise 3 of Section 3.17) by factorizing each expression; the LCM is the product of *all* factors appearing in the two expressions, to the highest power at which they occur. For example, the LCM of $5x^2y\,[=(5)\,(x^2)\,(y)]$ and $2xy^2\,[=(2)\,(x)\,(y^2)]$ is $5.2.x^2.y^2 = 10x^2y^2$.

Similarly, the LCM of $x^2 - 4\,[=(x+2)\,(x-2)]$
and $x^2+4x+4\,[=(x+2)\,(x+2)$ or $(x+2)^2]$
is $(x+2)^2\,(x-2)$.

Examples

1. Calculate $\dfrac{4}{5x^2y}+\dfrac{3}{2xy^2}$.

LCM of the denominators $= 5.2.x^2.y^2 = 10x^2y^2$;

$$\frac{4}{5x^2y}+\frac{3}{2xy^2}=\frac{2y\times 4}{2y\times 5x^2y}+\frac{5x\times 3}{5x\times 2xy^2}=\frac{8y+15x}{10x^2y^2}.$$

2. Calculate $\dfrac{x}{(x-2)} - \dfrac{4}{(x-2)^2}$.

LCM of the denominators $= (x-2)^2$;

$$\frac{x}{(x-2)} - \frac{4}{(x-2)^2} = \frac{(x-2)\times x}{(x-2)(x-2)} - \frac{4}{(x-2)^2}$$

$$= \frac{x^2-2x}{(x-2)^2} - \frac{4}{(x-2)^2} = \frac{x^2-2x-4}{(x-2)^2}.$$

If the denominators have no factor in common, the LCM is the product of the two (or more) denominators.

Example

Calculate $\dfrac{2}{x-3} + \dfrac{5}{x+2}$.

LCM of the denominators $=(x-3)(x+2)$;

$$\frac{2}{x-3} + \frac{5}{x+2} = \frac{(x+2)(2)}{(x+2)(x-3)} + \frac{(x-3)(5)}{(x-3)(x+2)}$$

$$= \frac{2x+4+5x-15}{(x+2)(x-3)} = \frac{7x-11}{x^2-x-6}.$$

To *multiply* algebraic fractions, multiply numerators and multiply denominators, just as with rational numbers.

Sometimes it helps simplification if we first factorize the numerators of both algebraic fractions, if this is possible.

Example

$$\frac{x^2+3x}{x^2-4} \times \frac{x+2}{x+3} = \frac{x(x+3)}{(x+2)(x-2)} \times \frac{x+2}{x+3}$$

$$= \frac{x(x+3)(x+2)}{(x+2)(x-2)(x+3)} = \frac{x}{x-2}.$$

For *division* of algebraic fractions, follow the same procedure that you used to divide rational numbers (dividing by a fraction is equivalent to multiplying by its inverse).

Example

$$\frac{15y^2}{8x} \div \frac{5y}{2x^2} = \frac{15y^2}{8x} \times \frac{2x^2}{5y}$$

$$= \frac{3.\cancel{5}.\cancel{y}.y.\cancel{2}.x.\cancel{x}}{2.2.\cancel{2}.\cancel{x}.\cancel{5}.\cancel{y}} = \frac{3xy}{4}.$$

10.16 Exercises

Calculate the following (giving the result in its simplest form):

$$\frac{3}{a} + \frac{2}{b};$$

$$\frac{5}{y^2} - \frac{7}{2y};$$

$$\frac{2}{a^2 - b^2} + \frac{5}{a + b};$$

$$\frac{x - 2}{x^2 + 2x - 15} + \frac{x}{x^2 - x - 6};$$

$$\frac{3m^2n}{15l^2k} \times \frac{5lk^2}{12mn^2}.$$

10.17 Long Division

You are now familiar with the division of numbers as repeated subtraction, and also as inverse multiplication (Sections 3.8 and 3.10); we have also considered long division of numbers as repeated subtraction of the divisor or of multiples of the divisor. Now we look at the division of algebraic expressions in the same way.

For example, for the division $(3a + 3) \div (a + 1)$, we ask the usual kind of question, 'by what number is $a + 1$ to be multiplied to produce $3a + 3$?', or, 'how many times can $a + 1$ be subtracted from $3a + 3$?' The answer in this case is evidently 3, whichever method we use.

The idea of repeated subtraction is particularly useful in the division of algebraic expressions, even if the divisor is not a factor of the dividend; for example, $(3a + 5) \div (a + 1)$. If we subtract $a + 1$ three times, we are left with a remainder, 2. We can write this division in the same way as long division of numbers:

$$\begin{array}{r}
3 \\
a + 1 \;\overline{)\,3a + 5} \\
\underline{3a + 3} \\
2
\end{array}$$

or, briefly, $(3a + 5) \div (a + 1) = 3$, remainder 2; or

$$(3a + 5) \div (a + 1) = 3 + \frac{2}{a + 1}.$$

(Compare the division of numbers: $17 \div 3 = 5$, remainder 2, or $17 \div 3 = 5\frac{2}{3}$.)

As in long division of numbers, we try first to subtract the largest possible multiple of the divisor. Consider, for example

$$(2x^2 + 8x + 6) \div (x + 3)$$

We can subtract $(x + 3)$, the divisor, $2x$ times, thus:

$$\begin{array}{r l}
 & 2x \\
x + 3 & \overline{)2x^2 + 8x + 6} \\
 & \underline{2x^2 + 6x} \\
 & 2x + 6
\end{array}$$

and we have a remainder, $2x + 6$, from which $x + 3$ can be subtracted twice more. The complete division then is:

$$\begin{array}{r l}
 & 2x + 2 \\
x + 3 & \overline{)2x^2 + 8x + 6} \\
 & \underline{2x + 6x} \\
 & 2x + 6 \\
 & \underline{2x + 6} \\
 & 0
\end{array}$$

The remainder is zero. Therefore $x + 3$ is a factor of $2x^2 + 8x + 6$. The same method can be used for division of any expression.

Example

Calculate $(x^3 - 4x^2 - x + 2) \div (x^2 + x + 3)$.

$$\begin{array}{r l}
 & x - 5 \\
x^2 + x + 3 & \overline{)x^3 - 4x^2 - x + 2} \\
 & \underline{x^3 + x^2 + 3x} \\
 & {}^-5x^2 - 4x + 2 \\
 & \underline{{}^-5x^2 - 5x - 15} \\
 & x + 17
\end{array}$$

We cannot subtract the divisor any more, so we have a remainder, $x + 17$; $x^2 + x + 3$ is *not* therefore a factor of $x^3 - 4x^2 - x + 2$.

In a long division such as $(x^4 - 1) \div (x - 1)$ you will find it convenient to leave open spaces for any 'missing' powers of x, that is, those whose coefficients are zero:

$$\begin{array}{r l}
 & x^3 + x^2 + x + 1 \\
x - 1 & \overline{)x^4 + 0x^3 + 0x^2 + 0x - 1} \\
 & \underline{x^4 - x^3} \\
 & x^3 \\
 & \underline{x^3 - x^2} \\
 & x^2 \\
 & \underline{x^2 - x} \\
 & x - 1 \\
 & \underline{x - 1} \\
 & 0
\end{array}$$

(Note that $0 - ({}^-x^3) = {}^+x^3$.)

Again, the remainder is zero. Therefore $x - 1$ is a factor of $x^4 - 1$.

10.18 Exercises

Calculate, using long division:

$$(x^2 + 5x + 6) \div (x + 3);$$
$$(y^2 - 12y + 27) \div (y - 3);$$
$$(6b^2 - 11b + 2) \div (3b + 2);$$
$$(9x^2 - 16) \div (3x + 4);$$
$$(x^3 - 9x^2 + 2x + 5) \div (x^2 - 5x - 8).$$

10.19 The Remainder Theorem

(This section can be omitted at the first reading.)

As you know from your checking procedure, we can always re-arrange the numbers involved in a division into a multiplication, possibly including an addition; for example,

$$\text{if } 12 \div 3 = 4 \text{ then } 4 \times 3 = 12;$$

$$\text{if } 17 \div 3 = 5, \text{ remainder } 2, \text{ then } (5 \times 3) + 2 = 17.$$

In general terms, if dividend ÷ divisor = quotient, remainder . . ., then quotient × divisor + remainder = dividend. This equality (the Remainder Theorem) is based on a simple re-arrangement of the terms, and no new information is introduced; that is, it is an *identity* and provides a method for finding the remainder without actually performing the division. For example, if we need to find the remainder of the division,

$$(x^3 - 4x^2 + 2x + 4) \div (x - 3),$$

using Q for the Quotient of this division and R for the Remainder, we can write:

$$Q \times (x - 3) + R \equiv x^3 - 4x^2 + 2x + 4.$$

The sign $\equiv$ means that this is true whatever the value of x, as it is an identity resulting from a re-arrangement of terms. If we make x take the value $^+3$ (so that $x - 3 = 0$), the identity becomes

$$Q \times (3 - 3) + R \equiv 3^3 - 4(3)^2 + 2(3) + 4$$
$$\Rightarrow Q \times 0 + R = 27 - 36 + 6 + 4$$
$$R = 1.$$

To find the remainder resulting from an algebraic division using the Remainder Theorem:

(*a*) Establish what value must be given to the algebraic number x (or whatever symbolic letter is used) to make the divisor equal to zero;
(*b*) substitute this value in the dividend;
(*c*) the numerical value of this dividend is then the remainder.

Example

Find the remainder on dividing $x^3 - 3x^2 - 4x + 7$ by $x + 2$.

(*a*) The divisor $x + 2$ is zero if $x = {}^-2$;
(*b*) substituting ${}^-2$ for x in the dividend:
$({}^-2)^3 - 3({}^-2)^2 - 4({}^-2) + 7 = {}^-5$;
(*c*) the remainder is ${}^-5$.

The Theorem might appear at first sight to be of little practical use; but note that, if the remainder of a division is 0, the divisor is a factor of the dividend. The Remainder Theorem can therefore be used as a simple test in factorization.

For example, we can use the Remainder Theorem to test whether $(a - 3)$ is a factor of $a^3 - 27$:

(*a*) $a - 3 = 0$ if $a = 3$;
(*b*) substituting 3 for a in $a^3 - 27$:
$3^3 - 27 = 0$;
(*c*) the remainder is zero; therefore $a - 3$ is a factor of $a^3 - 27$. You can find the other factor by carrying out the division.

10.20 Exercises

1. Find the remainder on dividing:

$$x^3 - 3x^2 + 2x + 2 \text{ by } x - 4;$$
$$3x^3 - 2x^2 + 3x + 6 \text{ by } x + 3.$$

2. Test whether: $x - 2$ is a factor of $x^3 - 42x + 76$;
$x + 3$ is a factor of $2x^3 - 3x^2 + 54$.

3. For what value of k is $x - 4$ a factor of $x^3 + 3x^2 - 5x + k$?
[Hint: put $x = 4$ in the expression $x^3 + 3x^2 - 5x + k$, and find the remainder. What value must k then have, to make this remainder zero?]

Unit Eleven

Equations

11.1 Equations, and Solving Equations

You must already have begun to appreciate the conciseness and precision of the symbolic notation used in mathematics, as compared with ordinary language; this is especially valuable when we use symbols to express information concerning numbers or quantities. For example, we could describe the size of a man's share of an inheritance by saying: 'An inheritance of £18 000 is to be shared between a man and his two brothers. His share is twice that of his younger brother, and £3000 less than that of his elder brother'.

We could, however, condense the information in this cumbersome statement into an *equation*; we could symbolize the size of the man's share in £s by the symbol S, the *unknown* in this situation, and describe the share of the estate by writing down the relation:

$$(S) + (\tfrac{1}{2}S) + (S + 3000) = 18\,000.$$

The part of this kind of statement which is written before the 'equals' sign (the equation sign, $=$) is called the Left-Hand Side (LHS), and the part after the equation sign the Right-Hand Side (RHS), of the equation.

The same information about the man's share of the inheritance can be given in a much simpler equation:

$$S = 6000.$$

The process of putting a complicated equation into its simplest form, the form which fixes the value of the unknown, is called *solving an equation.* It depends on the *balancing principle* of equations: if two numbers are equal, the same mathematical operation performed on both numbers will produce two further numbers, which are also equal.

Suppose that the symbol N stands for the number 12, or $N = 12$, then

$$\begin{aligned} 2 \times N &= 2 \times 12; \\ N \div 3 &= 12 \div 3; \\ N + 7 &= 12 + 7; \\ N - 8 &= 12 - 8, \text{ and so on.} \end{aligned}$$

If we use the mathematical symbol '$\Rightarrow$' as an abbreviation for 'if . . . then . . .', or 'implies', we can write this as:

$$\begin{aligned} N = 12 &\Rightarrow 2 \times N = 2 \times 12; \\ N = 12 &\Rightarrow N \div 3 = 12 \div 3; \text{ etc.} \end{aligned}$$

Sometimes we can solve an equation, that is, put it into its simplest form, just by using the balancing principle and choosing an appropriate multiplication, division, addition or subtraction. For example,

$$M + 3 = 5.$$

Subtract 3 from both sides of the equation:

$$M + 3 - 3 = 5 - 3$$
$$\Rightarrow M = 2.$$

Here we chose as the appropriate operation the process of 'subtracting 3' from both sides, so as to eliminate the number 3 from the LHS.

Similarly, we could solve the equation $T - 4 = 1$, by choosing the appropriate operation of 'adding 4' to both sides:

$$T - 4 + 4 = 1 + 4 \Rightarrow T = 5.$$

Adding the same number to both sides, or subtracting the same number from both sides, looks just like transferring numbers from one side to the other, and changing their sign; for example,

$$x + 3 = 4 \Rightarrow x = 4 - 3;$$

or $$y - 2 = 5 \Rightarrow y = 5 + 2.$$

The appropriate operation in solving an equation such as

$$5k = 20$$

is 'division by 5':

$$5k = 20 \Rightarrow 5k \div 5 = 20 \div 5 \Rightarrow k = 4.$$

When you are multiplying or dividing both sides of an equation, always remember to carry out the operation on the *whole* of the LHS and the *whole* of the RHS of the equation, that is, on *every* term of *both* sides.

Division of the equation $3x - 5 = 4$ by 3 produces the equation

$$(3x - 5) \div 3 = 4 \div 3$$

or $$\frac{3x - 5}{3} = \frac{4}{3} \Rightarrow \frac{3x}{3} - \frac{5}{3} = \frac{4}{3}$$

$$\Rightarrow x - \tfrac{5}{3} = \tfrac{4}{3} \Rightarrow x = \tfrac{4}{3} + \tfrac{5}{3} \Rightarrow x = \tfrac{9}{3} = 3.$$

You will find it easiest to eliminate numbers or symbols by addition or subtraction *before* eliminating by multiplication or division.

Examples

1. $$S + \tfrac{1}{2}S + S + 3000 = 18\,000$$
$$\Rightarrow 2\tfrac{1}{2}S + 3000 - 3000 = 18\,000 - 3000$$
$$\Rightarrow 2\tfrac{1}{2}S = 15\,000$$
$$\Rightarrow \frac{2\tfrac{1}{2}S}{2\tfrac{1}{2}} = \frac{15\,000}{2\tfrac{1}{2}}$$
$$\Rightarrow S = 6000.$$

2. $5x + 3 = 33$
$\Rightarrow 5x + 3 - 3 = 33 - 3$
$\Rightarrow 5x = 30$
$\Rightarrow 5x \div 5 = 30 \div 5$
$\Rightarrow x = 6.$

3. $5y = 3y + 10$
$\Rightarrow 5y - 3y = 3y + 10 - 3y$
$\Rightarrow 2y = 10$
$\Rightarrow 2y \div 2 = 10 \div 2$
$\Rightarrow y = 5.$

4. $\frac{1}{3}(2x - 3) - \frac{1}{4}(3x - 5) = \frac{1}{6}$
$\Rightarrow 12\{\frac{1}{3}(2x - 3) - \frac{1}{4}(3x - 5)\} = 12 \times \frac{1}{6}$
$\Rightarrow 4(2x - 3) - 3(3x - 5) = 2$
$\Rightarrow 8x - 12 - 9x + 15 = 2$
$\Rightarrow {}^{-}x + 3 = 2$
$\Rightarrow {}^{-}x + 3 - 3 = 2 - 3$
$\Rightarrow {}^{-}x = {}^{-}1$
$\Rightarrow {}^{-}1({}^{-}x) = {}^{-}1({}^{-}1)$
$\Rightarrow x = 1.$

Note. In line 2 of this example, we multiplied both sides of the equation by 12. We chose this number because it is the LCM of 3, 4 and 6, and this meant that we were able at the next stage to eliminate all the fractional coefficients in the equation.

11.2 Exercises

1. (*a*) Express the following information as an equation:
   The sum of four consecutive numbers is 26. What is the smallest number? [Hint: if the first number is N, the next is $N + 1$, then $N + 2$, and so on. Therefore, $N + (N + 1) + \ldots + \ldots = 26$.]
   (*b*) Starting with the smallest angle of a triangle, the angles increase in size by 15°. Find the size of each angle. (The sum of the angles of a triangle is always 180°.)

2. Solve the following equations:

$5t + 6 = 9t - 2;$
$5(6s - 3) - 4(s + 5) = 3(9 - 2s) + 2(6s + 5) + 2s;$
$\frac{1}{3}x + 5 = 2;$
$\frac{2}{5}z + 1 = 4;$
$\frac{1}{3}(y + 4) = \frac{1}{5}(2y + 1).$

Checking

You can always check your solution of an equation by substituting it for the unknown in the original equation. You could, for example, check the solution $y = 5$ of the equation $5y = 3y + 10$ (Example 3 in Section 11.1); substitute 5 for every y in the equation, thus:

$$5(5) = 3(5) + 10 \Rightarrow 25 = 25.$$

Make it a habit to check equations in this way whenever you have worked out a solution.

Note. Several different letters have been used to stand for numbers; probably x is most commonly used, but sometimes other letters seem more appropriate, for example, S for 'share' in the example in Section 11.1. The particular symbol chosen really does not matter at all. You could just as easily choose an asterisk, *, or a picture symbol, ❁, to represent the unknown quantity; but letters are usually used because they are convenient for printers of books, and easily written by hand.

11.3 Quadratic Equations

Equations which involve quadratic expressions of the unknown (defined in Section 10.6) are called *quadratic equations*; for example,

$$x^2 + 12x + 35 = 0$$

is a quadratic equation in x.

Quadratic equations can be solved in various ways, and sometimes one method may be simpler than others, so you will need to be able to use several different methods.

Whatever the method, your first step should always be to see that the equation is put in its simplest form, perhaps by division of both sides by a common factor, or simply by adding 'like terms' (terms which are in the same power of the unknown, for example all the terms in x^2, all the terms in x, and so on).

The following examples illustrate how this is done:

$$4x^2 + 8x = {}^{-}4$$
$$\Rightarrow x^2 + 2x = {}^{-}1$$

(dividing both sides by 4);

or,

$$2y^2 + 3y = y^2 + 4y + 12$$
$$\Rightarrow y^2 - y - 12 = 0$$

(transferring all terms to the LHS, and adding like terms).

11.4 Solving Quadratic Equations by Taking the Square Root of Both Sides of the Equation (Method 1)

We can sometimes simplify a quadratic equation just by taking the square root of both sides of it; but remember that the square root has two values, one positive and one negative (see Section 5.24), and the quadratic equation has two solutions—although, as we shall see, there are some situations in which the negative solution is meaningless. Solutions of an equation are sometimes called the *roots* of the equation, and this term can be used even where square roots are not directly involved at all.

Examples

1. Solve the equation $x^2 = 16$.

$$x^2 = 16$$
$$\Rightarrow \sqrt{x^2} = \sqrt{16}$$
$$\Rightarrow x = {}^{+}4 \text{ or } x = {}^{-}4 \text{ (because } {}^{+}4 \times {}^{+}4 = {}^{-}4 \times {}^{-}4 = {}^{+}16\text{).}$$

2. Solve the equation $y^2 + 3 = 28$.

$$y^2 + 3 = 28$$
$$\Rightarrow y^2 = 25$$
$$\Rightarrow \sqrt{y^2} = \sqrt{25}$$
$$\Rightarrow y = {}^{+}5 \text{ or } y = {}^{-}5.$$

3. If the area of a square is 36 cm², what is the length of the side of the square?

Let us call the length of the side k cm. The area of the square is then k^2 cm²; we know, however, that the area of the square is 36 cm². Therefore, $k^2 = 36$ and the value of k is found by solving this equation.

$$k^2 = 36$$
$$\Rightarrow k = 6$$

(the negative value, $^{-}6$, is meaningless in this particular example, in which the answer is a length). The length of the side is 6 cm.

This method of taking square roots of both sides of the equation can only be used if the square root of the *whole* of each side of the equation can be found. It is a simple method if the equation contains the unknown only in the quadratic term, as in the examples we have given but is *not* suitable for equations such as $x^2 + 16x = 25$.

11.5 Exercises

1. Solve the equations:

$$a^2 - 9 = 0;$$
$$b^2 - 2 = 34.$$

2. If the area of a square is 1·44 m², what is the length of its side?

11.6 Solving Quadratic Equations by Factorization (Method 2)

A fundamental theorem of algebra states that if the product of two numbers is zero, then one or both of these numbers must be zero, or

$$a.b = 0 \Rightarrow a = 0 \text{ or } b = 0$$

(or both $a = 0$ *and* $b = 0$). This follows from the general principle (which we have called the 'balancing principle') that the same operation can be performed on both sides of the equation without altering its validity.

For example,

$$\begin{aligned} 3x &= 0 \\ \Rightarrow \quad x &= \tfrac{0}{3} \quad \text{(dividing both sides by 3)} \\ \Rightarrow \quad x &= 0. \end{aligned}$$

A quadratic equation can be solved by:

(*a*) first reducing the RHS to zero, using the balancing principle;
(*b*) then expressing the LHS as a product, in other words, by factorizing the LHS;
(*c*) putting each of the factors equal to zero in turn.

Examples

1. $x^2 - 5x = {}^-6$

$\Rightarrow x^2 - 5x + 6 = 0$ (reducing the RHS to zero by adding 6 to both sides, or 'transferring' $^-6$ to the LHS and changing its sign)

$\Rightarrow (x - 2)(x - 3) = 0$ (factorizing the LHS)

$\Rightarrow$ either $x - 2 = 0$ **1**

or $x - 3 = 0$ **2**

$\Rightarrow$ either $x = 2$ (adding 2 to both sides of **1**)

or $x = 3$ (adding 3 to both sides of **2**).

This method is useful if the quadratic expression factorizes easily.

2. $a^2 - 4 = 0$

$\Rightarrow (a + 2)(a - 2) = 0$ (factorizing LHS)

$\Rightarrow$ either $a + 2 = 0$ **1**

or $a - 2 = 0$ **2**

$\Rightarrow$ either $a = {}^-2$ (subtracting 2 from both sides of **1**)

or $a = {}^+2$ (adding 2 to both sides of **2**).

(You could solve this equation equally well using Method 1, by taking square roots of both sides; compare Section 11.4).

11.7 Exercise

Solve by factorizing:

$$x^2 + 5x = 36;$$
$$y^2 + 4y = {}^-3;$$
$$m^2 + m = 6;$$
$$3t^2 + 10t + 3 = 0;$$
$$5v^2 + 13v = 6;$$
$$3z^2 = 27.$$

11.8 Practical Problems Involving Quadratic Equations

Some problems arising in practical situations result in quadratic equations.

Examples

1. The cost of hiring a minibus for a party of children to have a day's outing is £11·20. If 2 more children had been able to go, each child would have had to pay 10 pence less. How many children are there in the party?

Let the number of children be represented by the letter c. Then each child would pay £$\frac{11{\cdot}20}{c} = \frac{1120}{c}$ pence. If there had been 2 more children, each child would have had to pay $\frac{1120}{(c+2)}$ pence.

Therefore $\frac{1120}{c} - \frac{1120}{(c+2)} = 10$ (since $\frac{1120}{c}$ is 10 more than $\frac{1120}{c+2}$).

We now solve the equation

$$\frac{1120}{c} - \frac{1120}{c+2} = 10.$$

The LCM of c and $(c + 2)$ is $c(c + 2)$; we therefore multiply both sides of the equation by $c(c + 2)$:

$$c(c+2)\left[\frac{1120}{c} - \frac{1120}{(c+2)}\right] = 10\,c(c+2)$$

(compare Example 4 in Section 11.1)

$$\Rightarrow 1120(c+2) - 1120c = 10c^2 + 20c$$
$$\Rightarrow 1120c + 2240 - 1120c = 10c^2 + 20c$$
$$\Rightarrow 10c^2 + 20c - 2240 = 0$$
$$\Rightarrow c^2 + 2c - 224 = 0$$
$$\Rightarrow (c+16)(c-14) = 0$$
$$\Rightarrow c = {}^-16 \text{ or } 14.$$

Clearly $^{-}16$ cannot be the answer, which must be a number of children; so we take the value $c = 14$, and conclude that there were 14 children in the party.

Check: $\frac{1120}{14}\text{p} = 80\text{p}$;

$$\frac{1120}{16}\text{p} = 70\text{p}, \text{ that is 10p less.}$$

2. A man takes a country walk over difficult ground, covering a distance of 7 km; he walks at a constant speed for the first 3 km, and then 1 km/h faster for the rest of the way. The whole walk takes him 5 hours. What was his average speed for the first 3 km?

First of all, we note that $\frac{\text{distance}}{\text{speed}} = \text{time}$. Let his speed for the first 3 km be represented by s km/h. Then the time he takes for the first part of the walk is $\frac{3}{s}$ hours, and his time for the remainder is $\frac{4}{s+1}$ hours.

Therefore $$\frac{3}{s} + \frac{4}{s+1} = 5.$$

Now we solve this equation:

$$\frac{3}{s} + \frac{4}{s+1} = 5$$

$$\Rightarrow 3(s + 1) + 4s = 5s(s + 1)$$

(multiplying both sides by $s(s + 1)$)

$$\Rightarrow 3s + 3 + 4s = 5s^2 + 5s$$
$$\Rightarrow 5s^2 - 2s - 3 = 0$$
$$\Rightarrow (5s + 3)(s - 1) = 0$$
$$\Rightarrow s = {}^{-}\tfrac{3}{5} \text{ or } 1.$$

Again, $^{-}\frac{3}{5}$ is an inappropriate solution, and so we take the value $s = 1$, concluding his speed for the first 3 km was 1 km/h.

11.9 Exercise

A boat cruises along a river from a point P to a point Q and back again to P, a total distance of 6 km. The current is flowing at 4 km/h, and the journey there and back takes 1 hour. If the boat travels throughout at its maximum speed, what is this maximum speed?

[Hint: let the maximum speed be m km/h. Then the boat will travel at a speed of $(m + 4)$ km/h with the current and $(m - 4)$ km/h against the current. The distance each way is 3 km.]

Therefore $\left(\text{since } \frac{\text{distance}}{\text{speed}} = \text{time}\right)$, $\frac{3}{m-4} + \frac{3}{m+4} = 1$, adding the time for each part of the journey and putting this sum equal to 1 hour.

Then solve the equation $\frac{3}{m-4} + \frac{3}{m+4} = 1$.]

11.10 Solving Quadratic Equations by 'Completing the Square' (Method 3)

If the unknown also appears in a term not involving a square, for example the second term of $x^2 + 6x + 16$, we can make the LHS *into a perfect square*—i.e. into the form $(\ldots \pm \ldots)^2$—by adding to both sides an appropriate number, a process called *completing the square*. We can then take the square root of both sides, as before.

Example

$$\begin{aligned} x^2 + 6x &= 16 \\ \Rightarrow x^2 + 6x + 9 &= 16 + 9 \\ \Rightarrow (x + 3)^2 &= 25 \\ \Rightarrow \sqrt{(x + 3)^2} &= \sqrt{25} \\ \Rightarrow x + 3 &= {}^{+}5 \text{ or } x + 3 = {}^{-}5 \\ \Rightarrow x &= 2 \text{ or } x = {}^{-}8. \end{aligned}$$

In this example, we added the number 9 to both sides. We choose the number to be added by close inspection of the term in x. Remember that we found that the expansion of the perfect square $(a + b)^2$ was $a^2 + 2ab + b^2$ (Section 9.10); in other words:

$$(\boxed{\text{1st term}} + \boxed{\text{2nd term}})^2$$
$$= \boxed{\text{first term}}^2 + 2.\boxed{\text{first term}}.\boxed{\text{second term}} + \boxed{\text{second term}}^2.$$

The 'missing term' in the expansion of the perfect square which begins $x^2 + 12x + \ldots$ is found by writing down $12x$ as the product

$$2\boxed{x}.\boxed{6}.$$

The missing number in this example is therefore 6^2 or 36, and

$$x^2 + 12x + 36 = (x + 6)^2.$$

We can establish a simple procedure for completing the square if we follow through the steps in our example, $x^2 + 12x + \ldots$.

(*a*) We find the 'first term' by taking the square root of x^2: the 'first term' is: $\sqrt{x^2} = x$;

(*b*) we divide $12x$ by 2 and also by the 'first term', x, and find the 'second term': $12x \div 2x \Rightarrow 6$;

(*c*) we square the 'second term', $6^2 = 36$, and add this to $x^2 + 12x$: the result is $x^2 + 12x + 36$, which is a perfect square, $(x + 6)^2$.

11.11 Exercises

1. What must be added to the following expressions to make them into perfect squares?

$$z^2 + 4z \ldots;$$
$$b^2 + 7b \ldots;$$
$$4t^2 + 8t \ldots;$$
$$9v^2 + 6v \ldots;$$
$$x^2 + 2ax \ldots;$$
$$y^2 + ay \ldots.$$

2. Solve the following quadratic equations by completing the square:

$$x^2 + 2x = 3;$$
$$y^2 - 8y = 9;$$
$$9t^2 + 12t = 5;$$
$$m^2 + 4m = 0.$$

[Hint: look back at the example at the beginning of Section 11.10.]

11.12 Harder Cases of Completing the Square

In all these examples, the coefficient of the quadratic term has been a 'square number', such as 1, 4 or 9. If the coefficient of x^2 is *not* a square number, as, for example, in the equation $3x^2 + 8x = 3$, we can reduce it to 1 by dividing both sides of the equation by this coefficient.

$$3x^2 + 8x = 3$$
$$\Rightarrow \quad x^2 + \tfrac{8}{3}x = 1.$$

Completing the square of the LHS,

$$x^2 + \tfrac{8}{3}x + (\tfrac{4}{3})^2 = 1 + (\tfrac{4}{3})^2$$
$$\Rightarrow \quad (x + \tfrac{4}{3})^2 = 1 + \tfrac{16}{9} = 2\tfrac{7}{9} = \tfrac{25}{9} = (\tfrac{5}{3})^2$$

$$\left.\begin{array}{lrcll} \Rightarrow \text{ either} & (x + \tfrac{4}{3}) & = & \tfrac{5}{3} & \quad \mathbf{1} \\ \text{or} & (x + \tfrac{4}{3}) & = & {}^-\tfrac{5}{3} & \quad \mathbf{2} \end{array}\right\} \text{(taking square roots of both sides)}$$

$\Rightarrow$ either $x = \tfrac{1}{3}$ (subtracting $\tfrac{4}{3}$ from both sides of **1**)

or $x = {}^-\tfrac{9}{3} = {}^-3$ (subtracting $\tfrac{4}{3}$ from both sides of **2**).

11.13 Exercise

Solve by completing the square the following equations:

$$6x^2 + x = 2;$$
$$5x^2 - 4x = 1.$$

11.14 The General Solution of the Quadratic Equation (Method 4)

As we have said, the choice of the letter or symbol to represent the unknown in an equation is quite arbitrary and immaterial. For example, the equations $5t^2 + 2t - 20 = 0$ and $5x^2 + 2x - 20 = 0$ are essentially the same, and have the same solutions. The solutions of the equation depend entirely on the numbers involved, that is, the coefficients of x^2 and x, and the numerical (or *constant*) term; we find the solutions of the equation from these numbers alone.

The method of solving quadratic equations by completing the square (Section 11.10) involves a series of steps and calculations which can all be combined into one single formula.

If we use the letters a, b and c to denote respectively the coefficients of x^2, x and the constant term, we can write any quadratic equation in the form

$$ax^2 + bx + c = 0.$$

The formula for the solutions of this equation is then

$$\boxed{x = \frac{-b \pm \sqrt{(b^2 - 4ac)}}{2a}}$$

whatever the values of a, b and c.

Note. The derivation of this formula is rather long-winded, although not really complicated; you can leave it until a second reading if you like. It proceeds quite naturally from the method of completing the square. Try to solve the equations $x^2 + 3x + 5 = 0$ and $3y^2 + 5y + 7 = 0$ by the method of completing the square, analysing each step of your reasoning as you go; you will see that the proof we give just expresses these steps in terms of the general coefficients a, b and c.

Remember that a, b and c in the formula are *signed numbers*. For example, if b is a negative number, ^-b is positive. Similarly, if a is positive while c is negative, then ac is negative, and ^-4ac is therefore positive, and so on.

$$ax^2 + bx + c = 0$$

$$\Rightarrow \quad x^2 + \frac{bx}{a} + \frac{c}{a} = 0 \text{ (dividing both sides by } a\text{)}$$

$$\Rightarrow \quad x^2 + \frac{bx}{a} + \left(\frac{b}{2a}\right)^2 + \frac{c}{a} = \left(\frac{b}{2a}\right)^2 \text{ (completing the square by adding } \left(\frac{b}{2a}\right)^2 \text{ to both sides)}$$

$$\Rightarrow \quad \left(x + \frac{b}{2a}\right)^2 = \left(\frac{b}{2a}\right)^2 - \frac{c}{a} \text{ (subtracting } \frac{c}{a} \text{ from both sides)}$$

$$\Rightarrow \quad \left(x + \frac{b}{2a}\right)^2 = \frac{b^2 - 4ac}{4a^2} \text{ (simplifying the RHS)}$$

$$\Rightarrow \quad x + \frac{b}{2a} = \pm\sqrt{\frac{(b^2 - 4ac)}{4a^2}} \quad \text{(taking the square root of both sides)}$$

$$\Rightarrow \quad x = \frac{-b}{2a} \pm \frac{\sqrt{(b^2 - 4ac)}}{2a} \quad \left(\text{subtracting } \frac{b}{2a} \text{ from both sides}\right)$$

$$\Rightarrow \quad x = \frac{-b \pm \sqrt{(b^2 - 4ac)}}{2a}.$$

Example

$$3x^2 + 5x + 2 = 0$$

$$\Rightarrow x = \frac{-5 \pm \sqrt{(5^2 - 4.3.2)}}{2.3} = \frac{-5 \pm \sqrt{(25 - 24)}}{6}$$

$$\Rightarrow x = \frac{-5 + 1}{6} = \frac{^-2}{3} \text{ or } x = \frac{-5 - 1}{6} = {}^-1.$$

Before using the formula, you must put the equation into the appropriate form; the right-hand side is made equal to zero (by the balancing principle), and the terms of the left-hand side are placed in the right order, with the quadratic term (x^2) first, then the term in x, followed by the constant.

For example,

$$2x - 3 = 5x^2$$

$$\Rightarrow -5x^2 + 2x - 3 = 0$$

(subtracting $5x^2$ from both sides).

You may need to eliminate algebraic fractions before the quadratic form becomes recognizable; compare, for example, the problems in Section 11.8.

Example

Solve the equation $x = \dfrac{x + 4}{x - 1}$.

$$x = \frac{(x + 4)}{x - 1}$$

$\Rightarrow x(x - 1) = x + 4$ (multiplying both sides by $x - 1$)

$\Rightarrow x^2 - x = x + 4$

$\Rightarrow x^2 - 2x - 4 = 0$ (subtracting $x + 4$ from both sides)

$$\Rightarrow x = \frac{-(-2) \pm \sqrt{(-2)^2 - (4 \times 1).(-4)}}{2} \quad \text{(using the formula)}$$

$$\Rightarrow x = \frac{2 \pm \sqrt{4 + 16}}{2}$$

$\Rightarrow$ either $x = 1 + \sqrt{5} = 3{\cdot}236$ (using four-figure tables)

or $x = 1 - \sqrt{5} = {}^-1{\cdot}236$.

Note. $\sqrt{20} = \sqrt{4 \times 5} = \sqrt{4} \times \sqrt{5} = 2\sqrt{5}$. Therefore

$$\frac{2 \pm \sqrt{20}}{2} = \frac{2 \pm 2\sqrt{5}}{2} = 1 \pm \sqrt{5}.$$

11.15 Exercises

1. Solve the following equations using the formula:

$$x^2 + 2x - 5 = 0;$$
$$2t^2 - 7t + 3 = 0;$$
$$2m^2 - 3m - 21 = 0;$$
$$2x - 1 = x^2;$$
$$\frac{3}{x-1} = 5x.$$

2. Express the following information in the form of an equation and solve this equation using the formula:
   (*a*) The area of a rectangle is 10 cm^2, and one side of the rectangle is 2 cm longer than the adjacent side. Find the lengths of these sides. [Hint: if one side is x cm long, the length of the other side is $(x + 2)$ cm. Therefore $x(x + 2) = 10$.]
   (*b*) The height, h metres, to which a body will rise in t seconds when projected vertically upwards with a velocity of u metres per second is given by the formula $h = ut - 5t^2$.
   If a ball is projected vertically upwards at a velocity of 40 m/s, how long will the ball take to reach a height of 5 m? [Hint: when these values for u and h are substituted, there will be a simple quadratic equation in t. This equation has two roots; is the ball at this height at two different times? Why?]

11.16 Summary

Summarizing the various methods of solving quadratic equations:

Method 1 (Section 11.4): Solving quadratic equations by taking the *square root of both sides*: the simplest method if the equation contains the unknown in the quadratic term only; for example,

$$x^2 = 9;$$
$$y^2 = 20;$$
$$z^2 - 25 = 0.$$

Method 2 (Section 11.6): Solving quadratic equations *by factorizing the left-hand side*: a method you can always use, if the LHS of the equation factorizes easily; for example,

$$x^2 - 6x + 9 = 0;$$
$$z^2 + 5z + 6 = 0.$$

Method 3 (Section 11.10): Solving quadratic equations by *completing the square*: a suitable method if Method 2 cannot be used, if the quadratic term is a perfect square (e.g. x^2, $4y^2$, $9z^2$, etc.), and if the middle term can be 'split' easily (that is, if it is an even number) for example,

$$x^2 + 4x - 5 = 0;$$
$$9y^2 + 6y - 20 = 0.$$

Method 4 (Section 11.14): Solving quadratic equations *by formula*: a method that can be used to solve any quadratic equation that is solvable, and which *should* be used if the LHS does not factorize easily.

You should memorize this invaluable formula for the solution of quadratic equations; sometimes it is the only method which is really suitable. However, if the quadratic expression factorizes easily, using the formula is rather like taking a 'sledgehammer to crack a nut'. In quadratic expressions such as $x^2 - 6x + 9 = 0$, or $x^2 - 8x + 15 = 0$, the LHS can be factorized easily, and the solutions found immediately; but in a case such as $3x^2 + 13x + 5 = 0$, where the LHS does not factorize into simple factors, the formula method saves a great deal of effort.

The solutions of a quadratic equation are the same whichever method is used to solve it.

Examiners often indicate the choice of method; for example, the instruction 'find solutions correct to 3 significant figures' is a clear hint that the LHS does not factorize easily, so that the solutions are not whole numbers or simple rational numbers, and that the formula method should be used.

Quadratic equations usually have two solutions; if so *both* solutions must be calculated.

11.17 Exercise

Examine the following quadratic equations; decide what method you will use to solve them and find their solutions:

$$x^2 - 4x + 4 = 0;$$
$$y^2 - 49 = 0;$$
$$7y^2 + 12y + 1 = 0;$$
$$m^2 + 10m + 21 = 0.$$

11.18 Conditions for Quadratic Equations to have Two Real Solutions, One Real Solution, No Real Solutions

(This section and the following exercise can be left till a second reading, if you prefer.)

(*a*) Try to solve the following equation:

$$x^2 + x + 1 = 0.$$

Even if we use the formula, we find a situation which is difficult to accept: $\sqrt{(1-4)}$ or $\sqrt{{}^{-}3}$. Amongst the numbers we have met so far, the *real numbers*, there is no number whose square is ${}^{-}3$.

We say that this equation has *no real solutions* or *no real roots*; and this will happen whenever (in terms of the general equation $ax^2 + bx + c = 0$) $b^2 - 4ac$ is negative or, in mathematical language,

$$b^2 - 4ac < 0.$$

(*b*) The equation $x^2 - 4x + 4 = 0$ can be solved by factorizing the LHS, which is in fact a perfect square: $(x-2)(x-2) = 0$, or $(x-2)^2 = 0$. Using the formula for the solution of quadratic equations, we find that the roots of the equation are $x = \dfrac{4 \pm \sqrt{16-16}}{2} = \dfrac{4 \pm 0}{2}$. In other words, $x = 2$ or $x = 2$; the two roots are the same. The equation is said to have *two coincident roots*, or *one root* or *one solution*.

This is always the case if, in the quadratic equation $ax^2 + bx + c = 0$

$$b^2 - 4ac = 0.$$

(*c*) The quadratic equation $ax^2 + bx + c = 0$ has two different real solutions if the number $\sqrt{b^2 - 4ac}$ has two different real values; in other words, if $b^2 - 4ac$ is positive, or

$$b^2 - 4ac > 0.$$

Since this book only deals with real numbers, we will simply say that equations have *no solutions* or *no roots*, *one solution* or *one root*, and *two solutions* or *two roots*.

11.19 Exercises

1. Without solving the equations, decide whether the following quadratic equations have two different roots, one root or no roots:

$x^2 + 6x + 10 = 0$ [Hint: $a = 1$, $b = 6$, $c = 10$; therefore $b^2 - 4ac = 36 - 4.1.10 = {}^{-}4$];

$y^2 + 6y + 9 = 0$;

$2m^2 - 7m - 20 = 0$.

2. For what values of k does the equation

$$x^2 + 4x + k = 0$$

have (*a*) two different roots;
(*b*) one root only;
(*c*) no roots?

11.20 Simultaneous Equations

So far, all the equations we have solved have involved only one unknown each, and one equation has been sufficient to find its value in each case. If we need to find two unknowns, we require two different pieces of information. These can be given in two ways: either in the form of two equations, each involving one unknown, or as two equations, each involving both unknowns. For example, we know the number of votes cast for and against a resolution discussed in a committee from the statement:

'The motion was passed with a majority of 6 votes; all the 20 members of the committee voted'.

If the number of votes in favour is denoted by f, and the number of votes against by a, then all this information is given concisely by the equations:

$$f + a = 20,$$
$$f - a = 6.$$

Two such equations, both involving the *same unknowns*, are called *simultaneous equations*.

Two simultaneous equations in two unknowns are solved by reducing them to one equation in one unknown, and solving this equation; the second unknown is then easily calculated. This elimination of an unknown can be carried out in different ways.

11.21 Solving Simultaneous Equations by Substitution (Method 1)

This method uses the familiar fact that the LHS and RHS are equal numbers; therefore the one can be replaced by, or substituted for, the other.
Consider the example of the voters in committee:

$$f + a = 20, \qquad \textbf{1}$$
$$f - a = 6. \qquad \textbf{2}$$

Since $f = a + 6$ (equation **2**) we can substitute $(a + 6)$ for f in equation **1**:

$$(a + 6) + a = 20$$
$$\Rightarrow 2a + 6 = 20$$
$$\Rightarrow 2a = 14$$
$$\Rightarrow a = 7.$$

Now we can substitute 7 for a in either equation to find the value of f:

$$f = 7 + 6 = 13 \text{ (using equation } \textbf{2}\text{)}$$
$$\text{or } f + 7 = 20 \Rightarrow f = 13 \text{ (using equation } \textbf{1}\text{).}$$

Thus we can eliminate one of the unknowns (so reducing the two simultaneous equations in two unknowns to one equation in one unknown) by

(*a*) first expressing the value of one unknown in terms of the other, using one equation;
(*b*) then substituting this value in the other equation.

Example

Solve the simultaneous equations:

$$3x + y = 17, \qquad \mathbf{1}$$
$$4x - 3y = 14. \qquad \mathbf{2}$$

Using equation **1**, $y = 17 - 3x$.
Substituting $(17 - 3x)$ for y in equation **2**,

$$4x - 3(17 - 3x) = 14$$
$$\Rightarrow 4x - 51 + 9x = 14$$
$$\Rightarrow 13x = 65$$
$$\Rightarrow x = 5.$$

Substituting 5 for x in equation **1**,

$$3(5) + y = 17$$
$$\Rightarrow 15 + y = 17$$
$$\Rightarrow y = 2.$$

Therefore, the solutions are $x = 5$ and $y = 2$.

11.22 Exercises

By substitution solve the simultaneous equations:

1. $16t + 3v = 14,$
   $v - 10 = {}^{-}4t.$

2. $9a - 7b = 33,$
   $7a - 9b = 15.$

11.23 Solving Simultaneous Equations by Addition or Subtraction (Method 2)

We can use the 'balancing principle' to eliminate one of the unknowns from two simultaneous equations by simple addition or subtraction.

In the example we have already used,

$$f + a = 20, \qquad \mathbf{1}$$
$$f - a = 6, \qquad \mathbf{2}$$

we can add to both sides of the equation **1** *either* the number $(f - a)$ *or* the number 6, which equation **2** tells us are in fact the same thing. If we add $(f - a)$ to the LHS of equation **1**, and 6 to the RHS of the same equation, we find we have constructed a new equation,

$$(f + a) + (f - a) = 20 + 6$$
$$\Rightarrow 2f = 26$$
$$\Rightarrow f = 13.$$

In effect, we have just added the two equations:

$$\begin{array}{rl} & f + a = 20 \\ + & \underline{f - a = \ \ 6} \\ & 2f \quad\; = 26 \text{ (because } a + {}^{-}a = 0) \\ & \Rightarrow f \quad = 13. \end{array}$$

We can find the other unknown, a, either by substituting 13 for f in either equation, or by subtracting equation **2** from equation **1**:

$$\begin{array}{rl} & f + a = 20 \\ - & \underline{f - a = \ \ 6} \\ & 2a = 14 \text{ (because } a - {}^{-}a = a + a = 2a) \\ & \Rightarrow a = 7. \end{array}$$

Two simultaneous equations are not always so conveniently arranged that addition or subtraction alone will eliminate one of the unknowns.

Example

$$3p + 2q = 22, \qquad \mathbf{1}$$
$$p + q = 9. \qquad \mathbf{2}$$

Here we can use an appropriate multiplication which will transform one of the equations so that addition or subtraction does eliminate one of the unknowns. In this example, we can multiply both sides of equation **2** by 3:

$$3(p + q) = 3 \times 9 \Rightarrow 3p + 3q = 27.$$

Leaving equation **1** unchanged: $3p + 2q = 22,$
and subtracting: $q = 5.$

Substituting 5 for q in equation **2**,

$$p + 5 = 9 \Rightarrow p = 4.$$

Sometimes we must transform both equations.

Example

$$5x + 3y = 19 \qquad \mathbf{1}$$
$$2x + 7y = 25 \qquad \mathbf{2}$$

Multiplying both sides of equation **1** by 2: $10x + 6y = 38$
Multiplying both sides of equation **2** by 5: $10x + 35y = 125$
Subtracting the first equation from the second: $29y = 87$
$\Rightarrow y = 3.$

Substituting 3 for y in equation **1**: $5x + 9 = 19$
$\Rightarrow 5x = 10$
$\Rightarrow x = 2.$

11.24 Exercises

1. Solve the following pairs of simultaneous equations by addition or subtraction:

   (*a*) $2p + q = {}^-7,$
   $p + 3q = 4.$
   (*b*) $6x - 5y = 9,$
   $4x - 3y = 7.$

2. Express the following information in terms of equations, and solve the equations (that is, find the price of sugar and flour):
   '5 kg of sugar and 4 kg of flour costs 96 pence; the same total price is paid for 8·5 kg of sugar and 2 kg of flour'.
   [Hint: let the price of sugar be s pence per kg and the price of flour f pence per kg. Then one of the equations you need is: $5s + 4f = 96$. Now find the other equation and solve the two simultaneous equations.]

11.25 Solving Simultaneous Equations when One is Quadratic

If one of the simultaneous equations is a *quadratic equation*, the equations must be solved by substitution.

Example

$$x^2 + y^2 = 13, \qquad \mathbf{1}$$
$$2y - x = 4. \qquad \mathbf{2}$$

From equation **2**, $x = 2y - 4$.
Substituting $(2y - 4)$ for x in equation **1**,

$$(2y - 4)^2 + y^2 = 13$$
$$\Rightarrow 4y^2 - 16y + 16 + y^2 = 13$$
$$\Rightarrow 5y^2 - 16y + 3 = 0.$$

Factorizing the LHS,

$$(5y - 1)(y - 3) = 0$$
$$\Rightarrow y = \tfrac{1}{5} \text{ or } y = 3.$$

Substituting $\frac{1}{5}$ for y in equation **2**,

$$\tfrac{2}{5} - x = 4 \Rightarrow {}^-x = 3\tfrac{3}{5} \Rightarrow x = {}^-3\tfrac{3}{5}.$$

Substituting 3 for y in equation **2**,

$$6 - x = 4$$
$$\Rightarrow x = 2.$$

Therefore $\qquad x = 2$ and $y = 3$,

or $\qquad x = {}^{-}3\frac{3}{5}$ and $y = \frac{1}{5}$.

Note. There will usually be two pairs of solutions if one of the equations is quadratic.

11.26 Exercises

Solve the simultaneous equations:

1. $x^2 - y^2 = 0$,
   $x + 2y = 8$.

2. $3x^2 - xy = 24$,
   $x + y = 4$.

3. $x^2 + xy + 1 = 0$,
   $y - x - 3 = 0$.

11.27 Transformation of Formulae

A formula is an equation expressing the relation between two or more variables. Most of the equations we have so far have involved only two variables, usually represented by the symbols x and y. An example of a formula which relates three variables is the Gas Equation,

$$\frac{pV}{T} = K,$$

where V is the volume of the gas at a certain pressure p and a certain temperature T, and K is a constant. A formula relating four variables is the simple interest formula (see Section 7.16),

$$I = \frac{n \times r \times P}{100}$$

relating the simple interest, I, to the number of years concerned, the interest rate as a percentage per annum and the principal.

An appropriate value of each of these variables can be found if particular values of all the other variables are known. This value is found by substituting the known values in the formula. For example, to calculate the simple interest over 5 years on a sum of £600, at a rate of 4% p.a., we substitute these values in the formula:

$$I = \frac{(5 \times 4 \times 600)}{100} = 120,$$

that is, the simple interest in these circumstances is £120.

Similarly, to find what sum would yield a simple interest of £400 over 4 years at a rate of 5%, we substitute these values for I, n and r in the same formula:

$$£400 = \frac{£(4 \times 5 \times P)}{100}$$
$$\Rightarrow 40\,000 = 20P$$
$$\Rightarrow 2000 = P \text{ or } P = 2000,$$

that is, the sum required is £2000.

The same formula can be used to find the numerical value of any of the variables corresponding to known values of the other variables.

The most convenient form of the equation, however, is the one where all the 'known' variables appear on the right-hand side of the equation and the 'unknown' on the left-hand side, making the unknown the *subject of the formula.*

If the formula is not given in this convenient form, we can use the balancing principle of equations to transform it to an equivalent formula in which the unknown variable is the subject. For example, we can take the simple interest formula expressed as usual with I as its subject, and transform it into an equivalent formula whose subject is P:

$$I = \frac{n \times r \times P}{100}$$
$$\Rightarrow 100 \times I = n \times r \times P$$
$$\Rightarrow \frac{100 \times I}{n \times r} = P \quad \text{or} \quad P = \frac{100 \times I}{n \times r}.$$

We *change the subject of a formula* by choosing and carrying out a series of operations on both sides of the equation so that the new subject is left standing alone on the left-hand side, and all the other variables and constants are collected on the right-hand side.

Examples

1. If $A = P\left(1 + \frac{r}{100}\right)$, to change the subject from A to r we eliminate P, 1 and 100 from the RHS:

$$A = P\left(1 + \frac{r}{100}\right)$$
$$\Rightarrow \frac{A}{P} = \left(1 + \frac{r}{100}\right)$$
$$\Rightarrow \frac{A}{P} - 1 = \frac{r}{100}$$
$$\Rightarrow 100\left(\frac{A}{P} - 1\right) = r,$$
$$\text{or } r = 100\left(\frac{A}{P} - 1\right);$$

r is now the subject of this formula.

2. $T = 2\pi\sqrt{\frac{l}{g}}$; change the subject of this formula from T to g.

$$T = 2\pi\sqrt{\frac{l}{g}}$$
$$\Rightarrow T^2 = 2^2\pi^2\left(\sqrt{\frac{l}{g}}\right)^2 \text{ (squaring both sides)}$$
$$\Rightarrow T^2 = 4\pi^2\frac{l}{g}$$
$$\Rightarrow T^2g = 4\pi^2l \text{ (multiplying both sides by } g\text{)}$$
$$\Rightarrow g = \frac{4\pi^2l}{T^2} \text{ (dividing both sides by } T^2\text{).}$$

3. If $\frac{1}{u} + \frac{1}{v} = \frac{1}{f}$, express v in terms of u and f.

$$\frac{1}{u} + \frac{1}{v} = \frac{1}{f}$$
$$\Rightarrow \frac{1}{v} = \frac{1}{f} - \frac{1}{u} \text{ (subtracting } \tfrac{1}{u} \text{ from both sides)}$$
$$\Rightarrow \frac{1}{v} = \frac{u-f}{fu} \text{ (combining the two fractions on the RHS)}$$
$$\Rightarrow \frac{v}{1} = \frac{fu}{u-f} \text{ (turning both sides `upside down')}$$
$$\Rightarrow v = \frac{fu}{u-f}.$$

11.28 Exercises

Transform the following formulae as indicated:

1. $f = \frac{9}{5}c + 32$; change the subject from f to c. (Do you recognize this formula? It is used to change thermometer readings from Celsius to Fahrenheit.)
2. $E = RC$; change the subject from E to R. (If C represents an electric current, E a potential difference and R a resistance, this formula is a statement of Ohm's law.)
3. $c = 2\pi r$; change the subject from c to r. (This is the relation between the length of the circumference of a circle and its radius.)
4. $s = \frac{1}{2}(u + v)t$; change the subject from s to u.

Unit Twelve

Co-ordinates and Linear Graphs

12.1 Maps

The problem of finding a village or a road on a large map can involve a good deal of searching, but the task can be made easier by a grid which is superimposed on the map, dividing it into squares, or sometimes rectangles, of manageable size. Each square is usually identified by a combination of a letter and a number, or of two numbers, one of which refers to a vertical division of the map into columns, and the other to a horizontal division into rows (see Fig. 12.1).

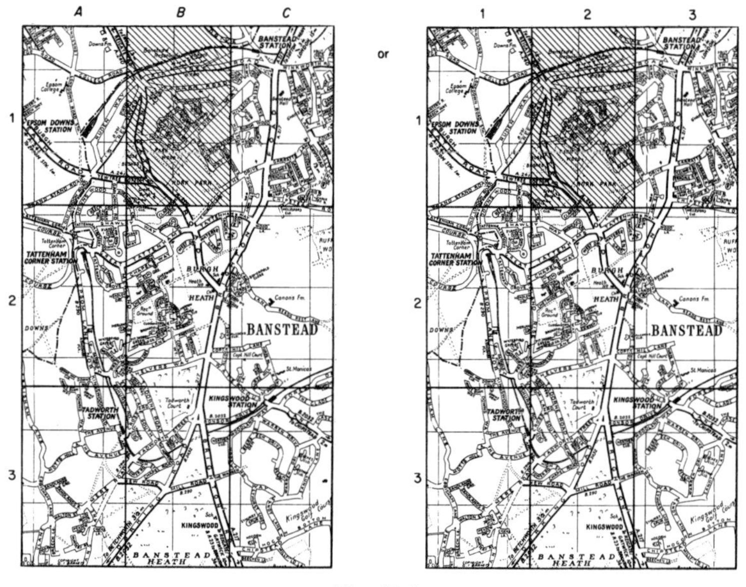

Fig. 12.1

If a pair of numbers is used, we must know which of the two numbers refers to the columns and which to the rows. If we arrange that the first number refers to the columns and the second number to the rows, we can identify the shaded rectangle on the maps in Fig. 12.1 by the 'coding' (B, 1) or (2, 1). The *number pair* is said to be *ordered*; the first number plays a role which is different from that of the second. This means that the order of stating the numbers matters; for example, (2, 1) does not represent the same rectangle as (1, 2). If we know the coding of a street, however, we still do not know its precise location, as any user of a London street map will confirm: one ordered number pair refers to the *whole area* of a square or rectangle.

12.2 Co-ordinates

A method of naming the position of a point very precisely was introduced by René Descartes (1596–1650).

First, we require two lines of reference, which can be simply drawn as two perpendicular lines, or *axes*, usually called the x-axis and the y-axis, as in Fig. 12.2.

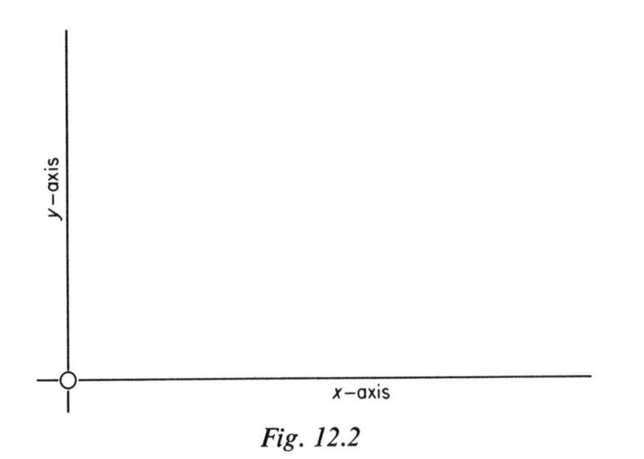

Fig. 12.2

The position of a point is then given by two numbers, *co-ordinates*, which refer to the distances of the point from these two axes. By convention the first number, the x-co-ordinate, always indicates the distance from the y-axis and the second number, the y-co-ordinate, indicates the distance from the x-axis.

Example

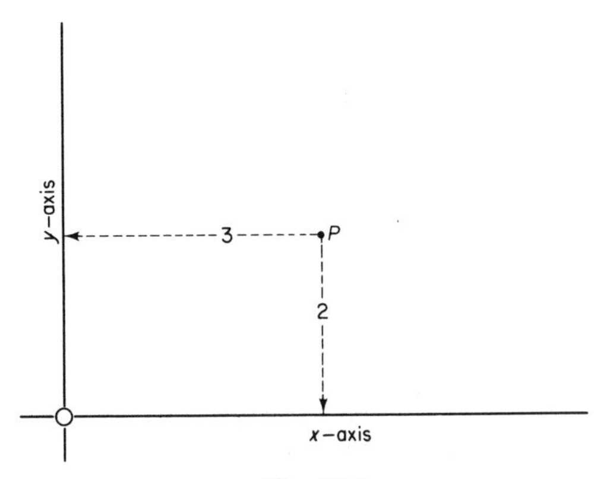

Fig. 12.3

If we choose the centimetre (cm) as the unit of measurement, the co-ordinates of the point P in Fig. 12.3 are 3 and 2. This is usually written (3, 2), which means 'the point which lies 3 units to the right of the y-axis and 2 units above the x-axis'. This arrangement is quite arbitrary although it has been adopted by all mathematicians, and we too shall observe this convention.

12.3 Exercises

1. Draw two perpendicular axes and mark a few points like the point P above. Label each point with a letter and find the co-ordinates of each point by measuring (correct to the nearest mm).
2. In what way does the system of co-ordinates differ from the sub-division into squares as used on maps?

12.4 Graph Paper

We can avoid measuring with a ruler if we use ready-printed squared paper, or graph paper, on which sub-divisions of the unit are also marked.

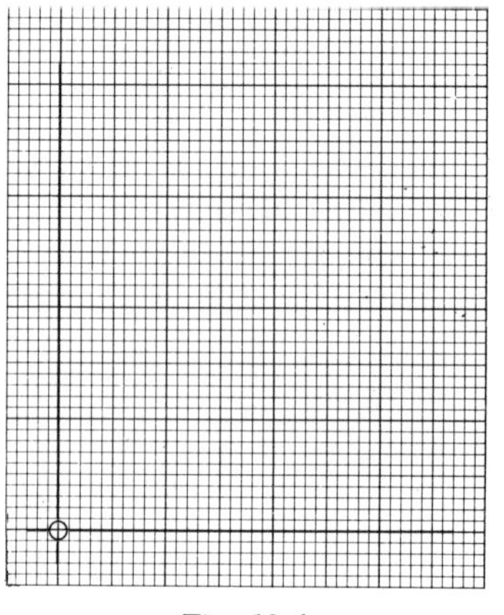

Fig. 12.4

Fig. 12.4 shows graph paper which uses metric units: the heavier lines mark the distances from both axes in centimetres. The two sets of parallel lines on the graph paper are guide lines for measurement. The resulting squares, or rectangles, may look rather like the grid superimposed on maps, but remember when using co-ordinates that the numbers do not refer to the whole area of the squares or rectangles between the lines of the graph paper, but to the distances of these lines from the axes. In Fig. 12.5, for example, we use the lines of the graph paper and the number scale along the axes to locate immediately the point P, the only point represented by (3, 2).

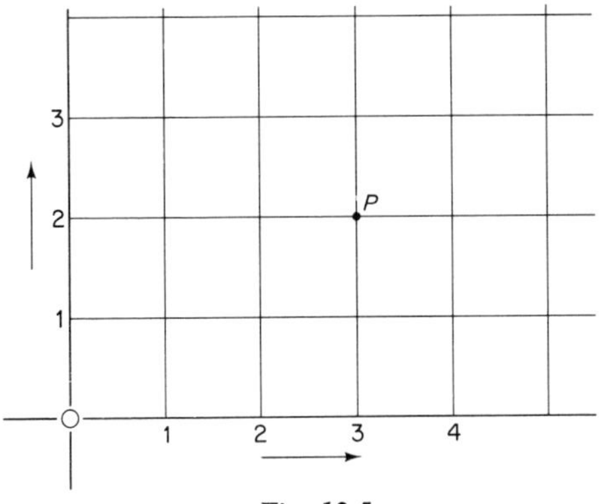

Fig. 12.5

12.5 Quadrants

The two axes divide the plane of the paper into four parts called *quadrants*.

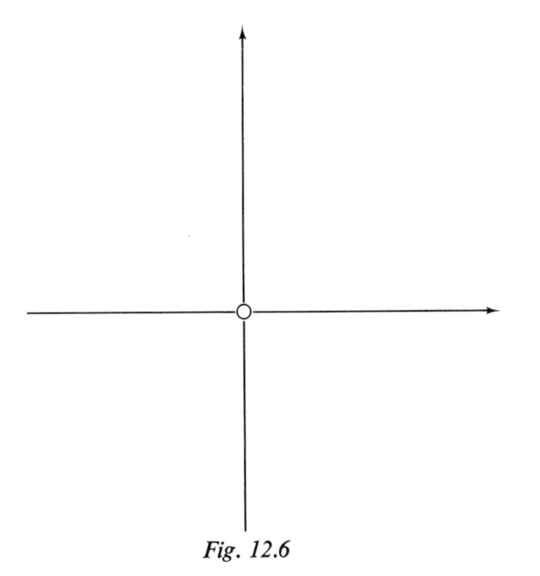

Fig. 12.6

So far, we have only considered points in the right-hand top quadrant, that is, points with positive co-ordinates. The system can be extended to points in the other quadrants if we use negative numbers for co-ordinates, in the same way as when we extended the number line (see Section 1.4).

A negative value of the x-co-ordinate then indicates that we should measure the distance to the left of the y-axis, and a negative y-co-ordinate that we should measure below the x-axis; as usual, the negative sign reverses the usual direction in which distances are measured. (The arrows on the axes always indicate the positive direction.) For example, $(^+2,\ ^-5)$ gives the position of the point which is '2 units to the right and 5 units down', that is, the point P in Fig. 12.7; $(^-3,\ ^+4)$ fixes the point '3 units to the left and 4 units up', namely the point Q.

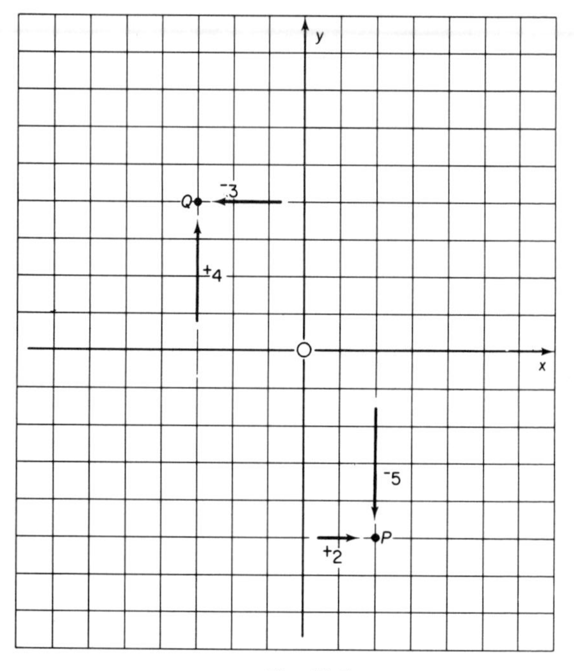

Fig. 12.7

Origin. The point where the two axes intersect is called the *origin*; it represents the number pair (0, 0), that is its distance from the y- and x-axes is zero.

12.6 Exercises

1. On metric graph paper draw a pair of perpendicular axes, one along a heavy line down the middle of the paper and one along a heavy line across the middle of the paper. Using the centimetre as the unit of measurement, mark or 'plot' the following points:

$$P: (^+2,\ ^+1);$$
$$Q: (^+3,\ ^-3\tfrac{1}{2});$$
$$R: (^-4,\ ^+3);$$
$$S: (^-2,\ ^-5\tfrac{1}{2}).$$

Use neatly pencilled crosses, not blobs, to mark the points.

2. On the same graph paper mark a number of points at random, and then read off and write down their co-ordinates. (Do not forget the conventional order.)

   Make up similar exercises until you are thoroughly familiar with the use of axes and co-ordinates.

12.7 Scale

It is often convenient to write a number scale along the axes; this makes plotting points and reading co-ordinates much easier, and also indicates what units have been chosen. For instance, Fig. 12.8 shows every positive and negative whole number marked within a certain range (the unit here is 0·5 cm).

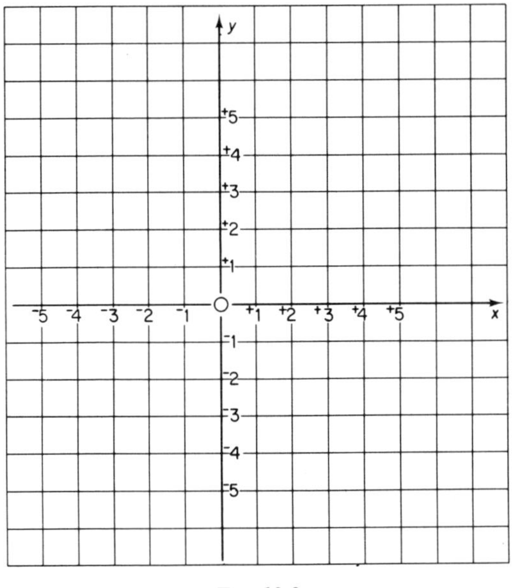

Fig. 12.8

In Fig. 12.9 only multiples of 5 are marked (the unit here is 2 mm).

The unit of measurement for the x-co-ordinates need not be the same as that for the y-co-ordinates; we might, for example, use a 'contracted scale' for the y-co-ordinate in a problem involving large values of y, but very much smaller values of x. The scale in Fig. 12.10 indicates that the x-co-ordinate is measured in centimetre units, and the y-co-ordinate in millimetres.

12.8 Relations and Ordered Pairs

In human society such relations as 'is the husband of', 'is the wife of', 'is the fiancé of', and so on, apply to millions of people. Each describes the relation

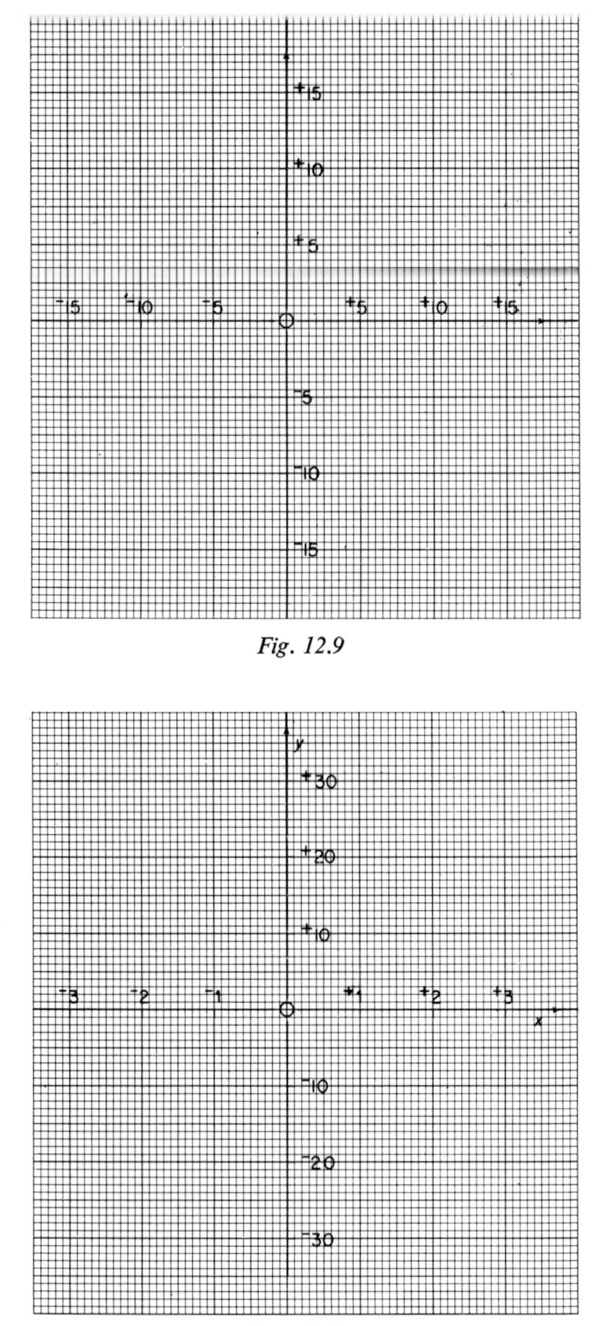

Fig. 12.9

Fig. 12.10

which exists between two people; for example, Mr Paul Smith is the husband of Mrs Margaret Smith.

Similarly, in mathematics, there are links between numbers, called *relations* or *mappings*, or sometimes *functions*; for example, 'is a quarter of', 'is three times as much as', 'is three more than', 'is five less than'. Each of these relations concerns two numbers, a number pair, but may apply to many such pairs. For example, the relation 'is twice as much as' is true for the pairs

2 and 1,
4 and 2,
6 and 3,
$\frac{1}{2}$ and $\frac{1}{4}$,
and between $\frac{2}{3}$ and $\frac{1}{3}$.

These are *ordered* pairs too, that is, the order in which the numbers are given is important. 4 is twice as much as 2, but the reverse is not true.

We can illustrate a relation in a *table*, that is, a list of number pairs (or groups) for which the relation is true. A typical table is the following one which expresses the relation 'is three times as much as'. Here, the second number of the pair is three times the first number; for simplicity, we have only included values of the first number between 0 and 5.

0	0
1	3
2	6
3	9
4	12
5	15

We can also express a mathematical relation by an *equation*: if x stands for the first number of the number pair and y for the second, then the relation 'is three times as much as' is expressed by the equation:

$$y = 3x.$$

This takes a little further our earlier notion of an equation as a piece of information about particular unknowns. For example, if Mr Jones is aged x years and his wife is aged y years, their ages could be given by the two simultaneous equations:

$$x + y = 45,$$
$$x - y = 5.$$

Each of these equations can be regarded as an incomplete piece of information concerning two particular numbers or 'unknowns'.

An equation such as $x + y = 45$ describes a whole range of ordered number pairs; the two numbers of each pair add up to 45. We can call x and y *variables*; x and y can vary, they can have many values, but the value of y always depends on the value of x, and vice versa.

If $x = 1$ then $y = 44$, or $x = 1 \Rightarrow y = 44$,
$x = 4 \quad \Rightarrow y = 41$,
$x = 20 \quad \Rightarrow y = 25$, etc.

Consider now a typical example of an equation giving the relation between two variables. The formula for converting temperatures given in degrees Celsius (x°C) to temperatures in degrees Fahrenheit (y°F) is:

$$y = \tfrac{9}{5}x + 32.$$

Any thermometer reading given in degrees Celsius can be converted to the corresponding measurement in degrees Fahrenheit by simple substitution in this equation; for example, a temperature of 25°C is $[\tfrac{9}{5}(25) + 32]$°F, or 77°F.

The table of this equation, given here for a small range of values of x, can be used as the conversion table from degrees Celsius to degrees Fahrenheit.

x	y
0	32
5	41
10	50
15	59
20	68
25	77
30	86

12.9 Exercises

1. Mrs Johnson is precisely 25 years old when her son is born. This age difference will be the same at any time of their lives. Express this relation as an equation, and make up a table showing the ages of Mrs Johnson and her son for a limited number of years.
2. Make up a table for the relation given by each of the following equations, in each case using integral values of x between 0 and 6:

$$y = 5x;$$
$$y = 5x + 2;$$
$$y = 5x - 3.$$

12.10 Graphs

One of the greatest advances in the history of mathematics was made when the idea of a numerical relation was linked to that of the number pair of the co-ordinate plane. A relation can be visually represented in this way by a collection of points, each representing one ordered pair of numbers to which the relation applies. For example, the relation 'is twice as much as', expressed by the equation $y = 2x$, is visually represented by all the points whose second co-ordinate is twice the first; some of these points are (1, 2), (2, 4), (3, 6), (0·1, 0·2) and ($^-3$, $^-6$). This set is described in mathematical notation by:

$\{(x, y): y = 2x\}$, which means 'the set of all points with co-ordinates (x, y), such that $y = 2x$'.

The set of all points whose second co-ordinate is twice the first co-ordinate is the *graph* of this relation, or the *graph* of the equation $(y = 2x)$ which expresses the relation.

12.11 Exercises

1. Write down some pairs of numbers connected by the relation 'is three more than' (i.e. the second number is 3 more than the first number, or $y = x + 3$), and mark or *plot* on graph paper the points represented by these number pairs. [Hint: take about 5 values of x, positive or negative, and find the corresponding value of y for each by adding 3. Now draw two perpendicular axes with scales covering the values of x and y you have chosen, and plot the points representing the number pairs that you have found.]

   What do you notice about these points?
2. Join the points. Write down some more number pairs which have the same relation, and plot their corresponding points. Are these points on the line(s) joining the previous points?
3. Mark another point on the line, and write down its co-ordinates. Does the relation 'is three more than' apply to the co-ordinates of this point? [That is, is the pair of co-ordinates of this point a member of $\{(x, y): y = x + 3\}$?]
4. Repeat the first three parts of this exercise using the relation 'is half of' $(y = \frac{1}{2}x)$.

12.12 Linear Graphs

In the last exercise, we found that all the points representing the number pairs of each relation lay on a straight line, and that every point on this straight line represented a number pair to which the relation also applied.

These 'straight-line' graphs are called *linear graphs* and their corresponding equations *linear equations*. Later in this Unit we shall see that all equations of the form $y = mx + c$, where m and c are given numbers, are linear equations and their graphs are linear graphs; for example:

$$y = 2x + 3;$$
$$y = 3x \ (\Rightarrow y = 3x + 0);$$

and also

$$2y = 3x + 6 \ (\Rightarrow y = \tfrac{3}{2}x + 3);$$
$$y + x = 4 \ (\Rightarrow y = {}^{-}(1)x + 4).$$

None of these equations has terms in x^2 or y^2, or higher powers of x and y.

Note. Remember that operations based on the balancing principle of equations do not change the information given about the relation. The same information is given by $y + x = 1\frac{1}{2}$ and $2x + 2y = 3$; the same relation between x and y is expressed by $y - x = 2$ and $y = x + 2$. You can confirm this by plotting the graphs of the equivalent equations; you will find that they are the same straight line.

12.13 Plotting a Linear Graph

If we know that a graph is linear (a straight line), then we need only mark or plot two points of that line; we can then draw a straight line through these two points, extending as far as we like in either direction. This is called *plotting a graph*.

Examples

1. Plot the graph of $y = 4x - 3$.

   Two number pairs are found by substituting any two values for x in the equation $y = 4x - 3$. We could choose $x = 1$, and $x = 3$. Then,

$$x = 1 \Rightarrow y = 4(1) - 3 = 1,$$
$$x = 3 \Rightarrow y = 4(3) - 3 = 9.$$

We know then that the points (1, 1) and (3, 9) lie on the graph, and we can 'plot' it by drawing a straight line through these two points (Fig. 12.11).

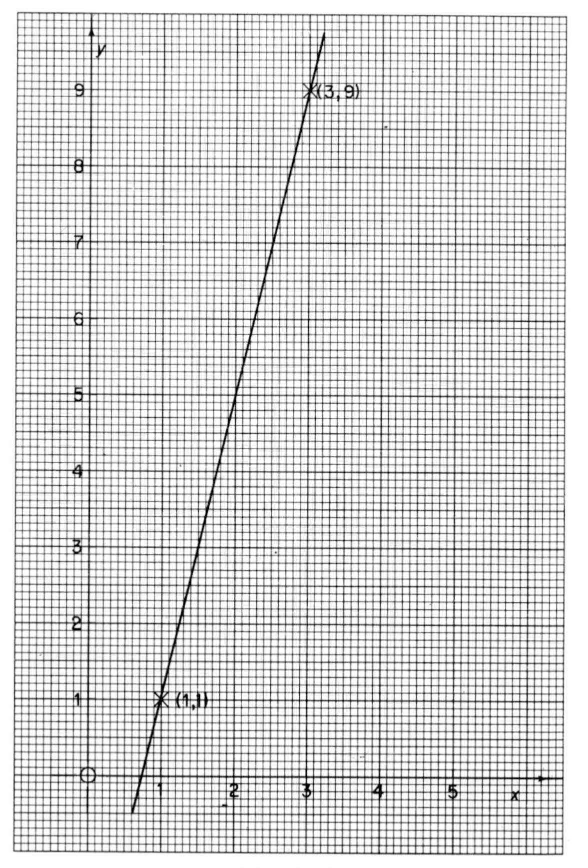

Fig. 12.11

2. Draw a graph of the relation between degrees Celsius and degrees Fahrenheit:

$$y = \tfrac{9}{5}x + 32$$

for the range 0°C (freezing point) to 30°C.

We choose the millimetre as a convenient unit on both y- and x-axes. If we substitute successively the values 0 and 30 for x in the equation $y = \frac{9}{5}x + 32$, we find the two ordered pairs (0, 32) and (30, 86). We can then plot these two points and draw a straight line through them to obtain the graph (Fig. 12.12).

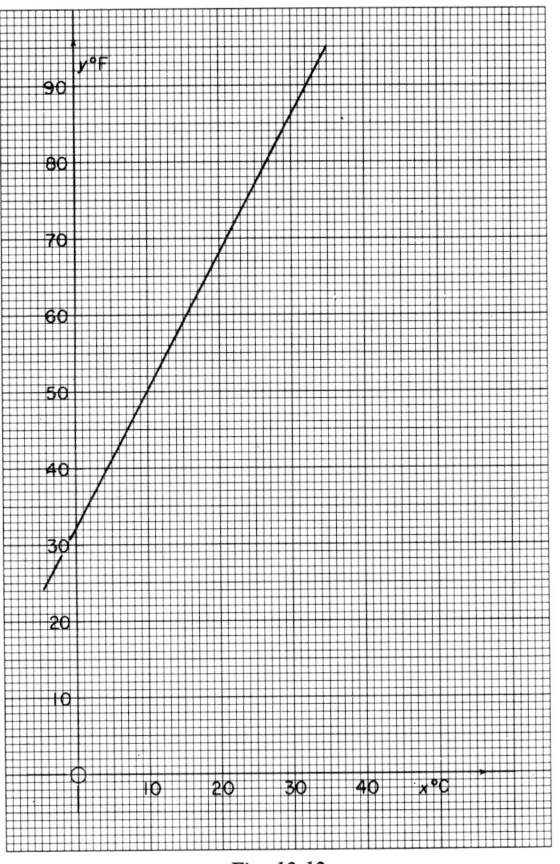

Fig. 12.12

12.14 Reading a Graph

A table of number pairs illustrating a relation or function can only include a limited number of entries; look at the tables which you made up for the exercises in Section 12.9. *A graph represents all possible number pairs* within the limits of the graph paper; the temperature conversion graph in Fig. 12.12, for example, represents all the possible number pairs of which the first co-ordinate lies between 0 and 42 (including all the possible fractions).

A graph can be used in two ways. Knowing a particular co-ordinate, the x-co-ordinate, we can find the corresponding y-co-ordinate; vice versa, knowing a particular y-co-ordinate we can find its corresponding x-co-ordinate.

Example

If we need to convert a thermometer reading of $22\frac{1}{2}$°C to °F from the graph, we can ask ourselves any of the following questions:

> Which point on the graph has $22\frac{1}{2}$ as its first co-ordinate? or
> What is the value of y when $x = 22\frac{1}{2}$? or
> $(22\frac{1}{2}, ?)$.

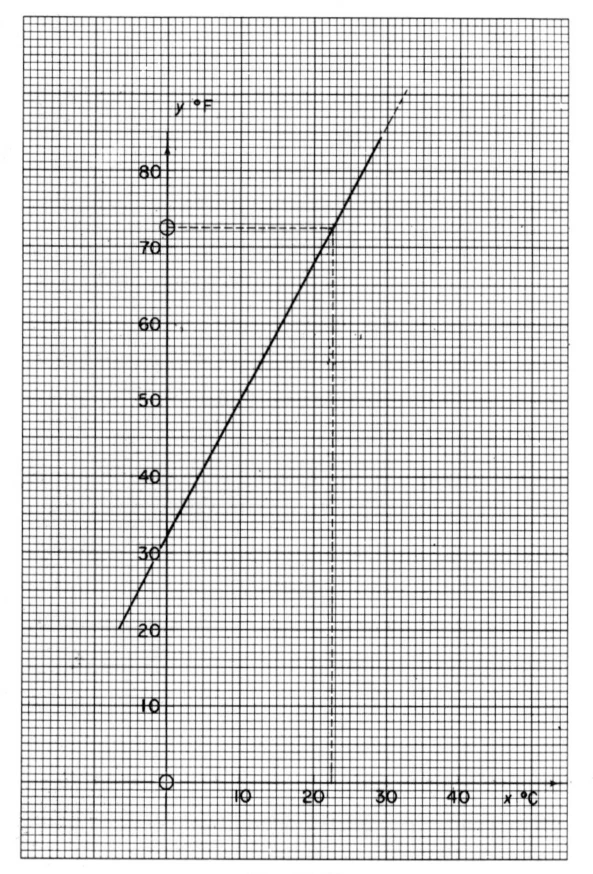

Fig. 12.13

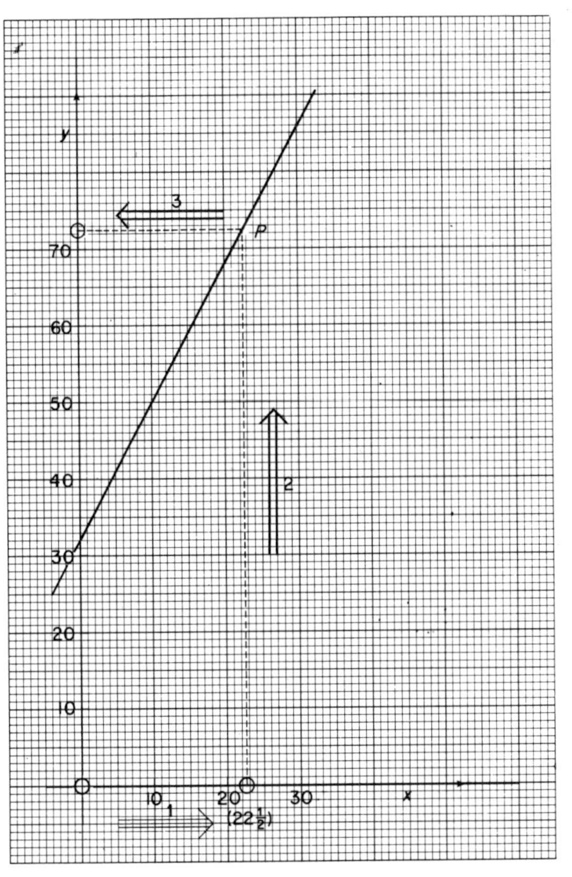

Fig. 12.14

All the points with $22\frac{1}{2}$ as their first co-ordinate lie along the line parallel to the y-axis at a distance of $22\frac{1}{2}$ units from the y-axis (the line $x = 22\frac{1}{2}$). We are concerned with the point where the conversion graph meets this line; the y-co-ordinate of this point of intersection is found by measuring or by reading from the scale. We need not actually draw the line parallel to the y-axis. A simple procedure is:

1. look for the point $(22\frac{1}{2}, 0)$ on the x-axis;
2. from this point, proceed parallel to the y-axis until you meet the graph at the point P;
3. read the y-co-ordinate of P from the scale on the y-axis; in this case, $y = 72\frac{1}{2}$.

You can use a similar procedure to find the x-co-ordinate, given a particular value of the y-co-ordinate. For example, to convert 60°F to °C, using the same conversion graph, we can ask the following questions:

What is the value of x when $y = 60$? or

Which point on the graph has 60 as its second co-ordinate, and what is the x-co-ordinate of this point? or

(?, 60).

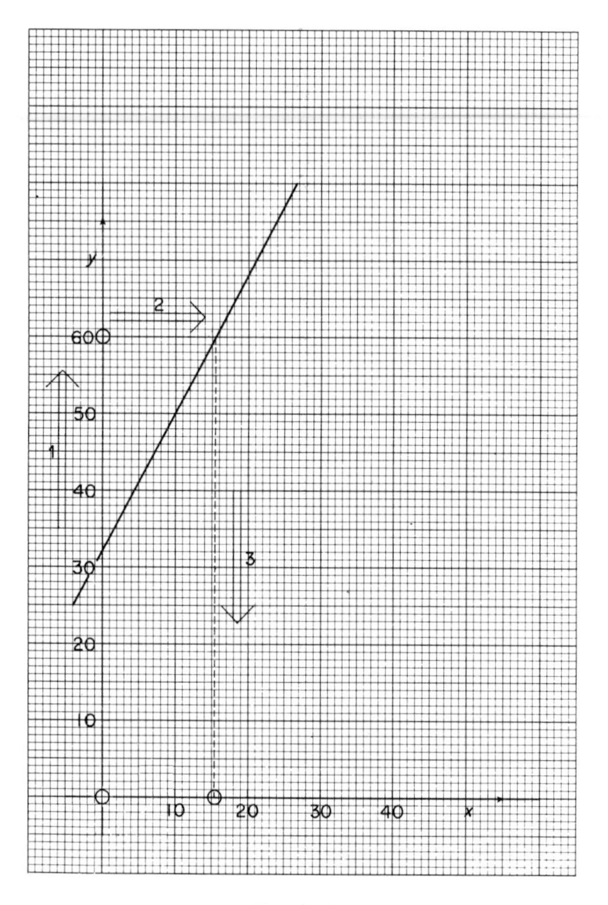

Fig. 12.15

All points with 60 as their second co-ordinate lie 60 units above the x-axis, on the line $y = 60$; again, this set of points includes the point where the graph meets the line $y = 60$. The corresponding x-co-ordinate is found by measuring or reading from the scale on the x-axis; in this case x is just over $15\frac{1}{2}$.

Note. Since reading a graph is in effect a matter of taking measurements along the axes, the results are always approximations. The degree of accuracy depends on the choice of unit for the scales on the axes.

12.15 Exercises

1. (*a*) Draw the graph of $y = \frac{9}{5}x + 32$, from $x = {}^{-}20$ to $x = 45$, using 2 mm as the unit on both axes (1 cm for every 5 units). From this Celsius–Fahrenheit conversion graph find the temperature in °F corresponding to:

   0°C (freezing point);
   24°C (comfortable room temperature);
   32°C (hot weather);
   36°C (normal body temperature);
   41°C (high body temperature).

   (*b*) Using the same graph convert the following to °C:

   40°F;
   65°F;
   80°F;
   102°F.

2. The relation between the lengths of the circumference of a circle (y) and its diameter (x) is expressed by the equation $y = \pi x$ (see Exercise 2 of Section 1.10; this will be discussed in more detail in Section 17.8), where π (the Greek letter 'pi') is a number approximately equal to 3·14.

   Plot the graph of this function from $x = 0$ to $x = 8$, using the centimetre as the unit on both axes to represent the lengths of diameter and circumference respectively in centimetres. (Negative values of x and y have no meaning in this calculation, so you need only consider the top right-hand quadrant.)
   [Hint: first plot two points on the graph as accurately as you can; for example,

   $x = 1, y = \pi \times 1 = 3{\cdot}14$, i.e. the point (1, 3·14);
   $x = 2, y = \pi \times 2 = 6{\cdot}28$, i.e. the point (2, 6·28).

   Then draw a straight line through these two points.]

   From this graph find (correct to the nearest mm):

   (*a*) the circumference of the circle whose diameter is 6 cm, and the circumference of the circle whose diameter is 24 mm;
   (*b*) the diameter of the circle whose circumference is 20 cm, and the diameter of the circle whose circumference is 68 mm.

12.16 The Slope or Gradient of a Graph

(This section can be left until a second reading, if you like.)

We begin this section with an exercise:

1. First, draw the graph of the relation, or function, represented by the equation $y = x$ (this instruction is often abbreviated to 'draw the graph of

the function $y = x$'); then, on the same sheet of graph paper and using the same axes, draw the graphs of the functions $y = 2x$, $y = 3x$, $y = 4x$. You will notice that all of these graphs pass through the point (0, 0), the *origin*.

An interesting feature is that, as the coefficient of x becomes larger, the graph becomes 'steeper'. The graph of $y = 2x$ is steeper than the graph of $y = x$ (that is, $y = 1x$); the graph of $y = 3x$ is steeper than the graph of $y = 2x$, and so on. The close connexion between the steepness, or *slope* or *gradient* of the graph and the coefficient of x in the equation is shown even more clearly by the graphs of functions where the coefficients of x are the same.

2. On one sheet of graph paper plot the graphs of the functions:

$$y = 3x;$$
$$y = 3x + 1;$$
$$y = 3x - 2.$$

All these graphs are parallel lines.

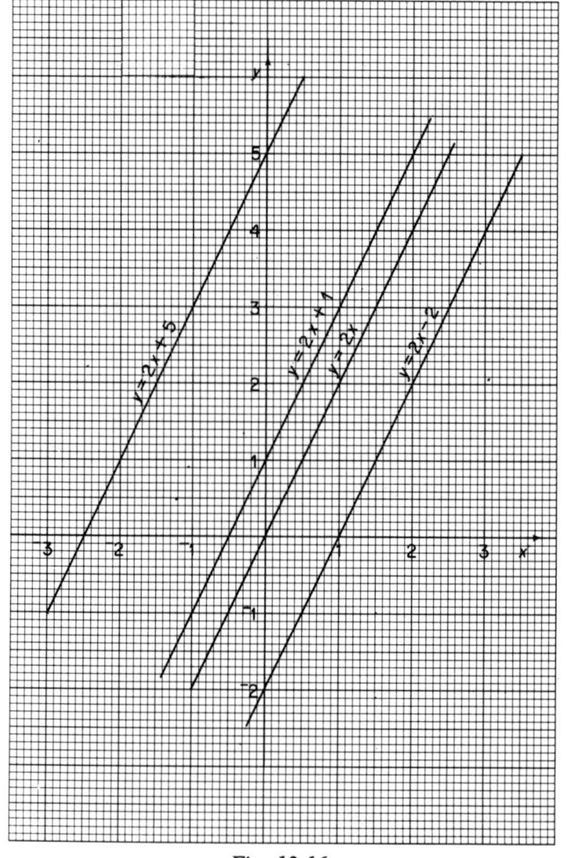

Fig. 12.16

Compare also (see Fig. 12.16) the graphs of

$$y = 2x \text{ [i.e. } y = 2x + 0];$$
$$y = 2x + 1;$$
$$y = 2x + 5;$$
$$y = 2x - 2.$$

The slope or *gradient* of the graph is determined by the coefficient of x, that is, the number m in the general form of the linear equation, $y = mx + c$.

You will be familiar with road signs which express the slope or gradient of a hill by a *ratio* such as 1 in 10, or 1:10. This means that if you walk 10 m along the road—in the diagram from P to Q—you will actually climb 1 m (RQ in Fig. 12.17).

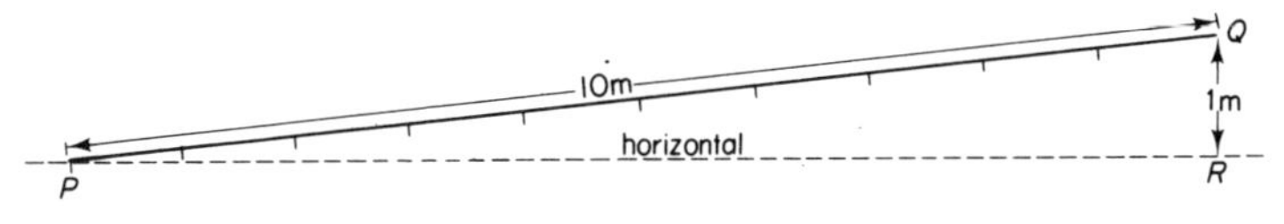

Fig. 12.17

The slope or gradient of a graph is usually given not by the ratio of these two lengths but by the ratio of the lengths of QR and PR. In Fig. 12.18, if SV is vertical and TV horizontal, then the slope or gradient of ST is the ratio of the lengths SV and VT.

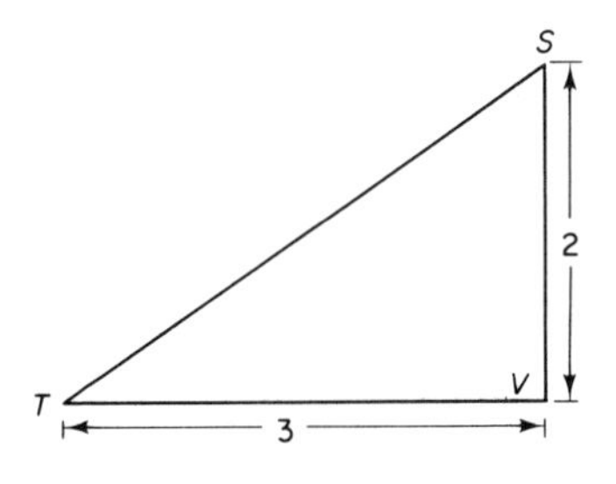

Fig. 12.18

The slope of the line joining the points S and T, or the line ST, is

$$\frac{\text{the length of } SV}{\text{the length of } VT} = \frac{2}{3}.$$

Now consider the graph $y = 2x$, in Fig. 12.19.

Look first at the small section of the graph between the points P (1, 2) and Q (2, 4). As the x-co-ordinate increases from 1 at point P to 2 at Q—that is by 1 unit—the y-co-ordinate increases from 2 at P to 4 at Q, in other words by 2 units. (We can think of the gradient of PQ as 'when we move from P to Q we move two units up and one unit to the right'.) The gradient of the line PQ is

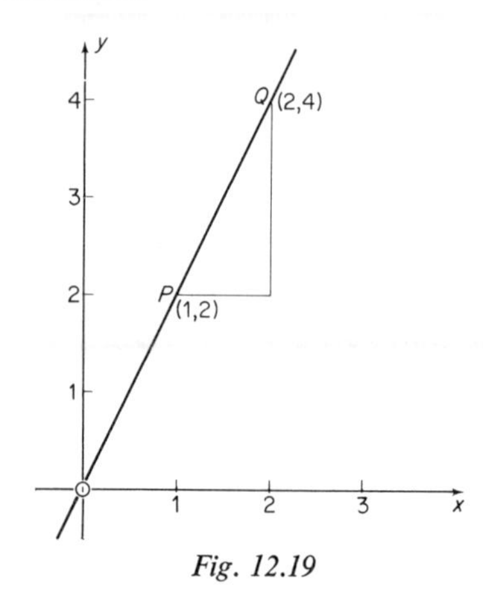

Fig. 12.19

therefore $\frac{2}{1}$ or 2, which is the coefficient of x in the equation $y = 2x$. We could use any other section of the graph, and the result would be the same: the gradient of the graph everywhere is 2.

This is obvious, because the graph is a straight line, and its slope does not change from one section to another. The table of the equation also makes this clear.

Look, for instance, at the table and the graph of the equation $y = 2x + 3$.

x	y
0	3
1	5
2	7
3	9
4	11
5	13

(Each interval: x increases by 1; y increases by 2.)

Each time x increases by 1, y increases by 2, and this relation is unaffected by the value of the constant term (3 in this example) (Fig. 12.20). The gradient in each interval is the same: $\frac{2}{1} = 2$.

We can always transform the equation of a straight-line graph so that the gradient can be immediately 'read off'; we need only re-arrange it so that the coefficient of y is 1, and the gradient is then equal to the coefficient of x. For example,

$$3y = 2x + 6$$

can be divided by 3,

$$y = \tfrac{2}{3}x + 2,$$

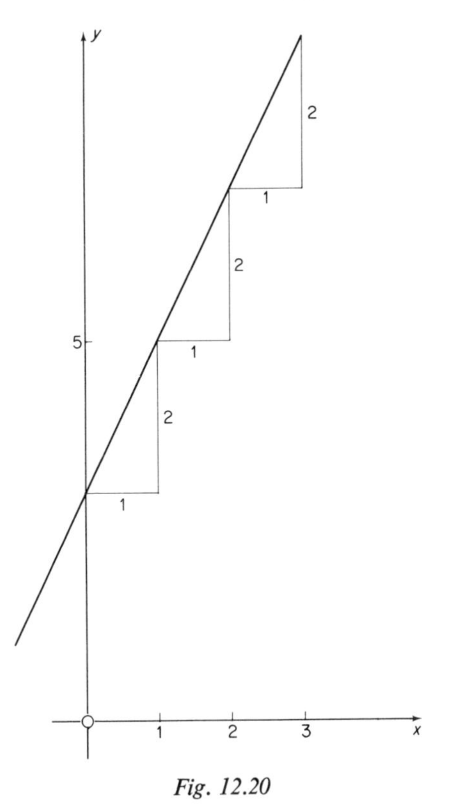

Fig. 12.20

and the gradient of the line is then $\frac{2}{3}$. We can test the validity of this result by drawing up the table of the original equation, or by plotting its graph (Fig.12.21). (As x increases by 3, y increases by 2.)

12.17 Exercises

1. On a sheet of graph paper draw a pair of axes and sketch a series of lines with gradient 3.
2. Repeat this exercise for gradients $\frac{1}{2}$ and $\frac{3}{4}$.
3. State the gradient of the graph of each of the following equations (do not plot the graphs):

$$y = 5x + 2;$$
$$2y = 6x + 7 \text{ [Hint: divide through by 2 first]};$$
$$4y = 5x + 2.$$

Note. When 'reading off' the gradient from a graph, be careful to observe the scales on *both* axes. The gradient of the graph in Fig. 12.22, for example, is $\frac{2}{6}$ (or $\frac{1}{3}$), *not* $\frac{1}{2}$.

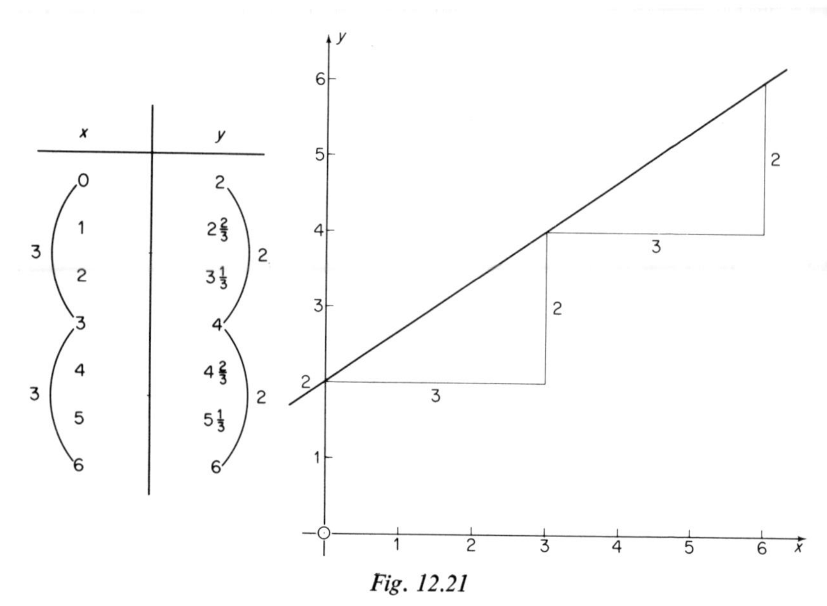

Fig. 12.21

12.18 Negative Gradients

All the examples in the last two sections were chosen so that the value of y increased, or became larger, as the value of x increased. This is not always so: look at the table of the equation $y + 2x = 6$.

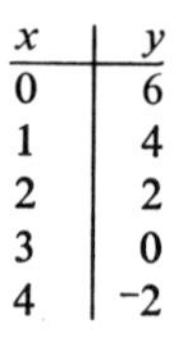

x	y
0	6
1	4
2	2
3	0
4	−2

It is clear that the value of y *decreases* as the value of x *increases*. For

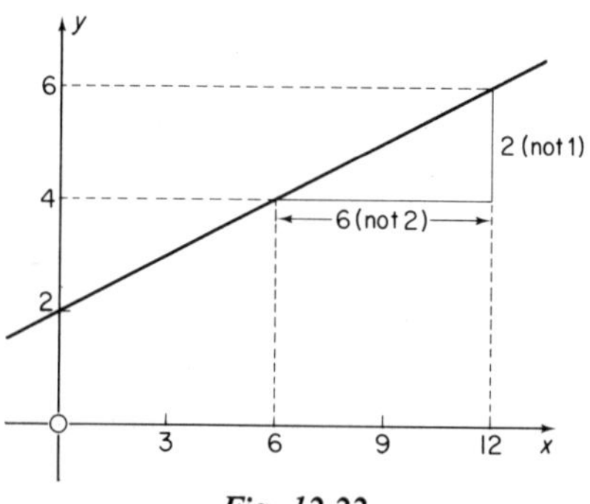

Fig. 12.22

example, as the value of x increases from 1 to 2—an increase of 1—the value of y decreases from 4 to 2, a decrease of 2.

This appears in the graph as a *downward* slope as we move along the graph from left to right (Fig. 12.23).

When we move along the graph in Fig. 12.23 from the point P (1, 4) to the point Q (2, 2), we move 1 unit to the right and 2 units down. The amount of the increase in x is the difference between the two x-co-ordinates $(2 - 1 = 1)$, and the amount of the increase in y is the difference between the y-co-ordinates, $(2 - 4 = {}^{-}2)$, that is, a *decrease* of 2. As one of these numbers is negative, the gradient is a negative number: $\frac{^{-}2}{1} = {}^{-}2$. The equation

$$y + 2x = 6$$

can be re-written in the form

$$y = {}^{-}2x + 6,$$

from which we can immediately read off the gradient as a negative number, $^{-}2$.

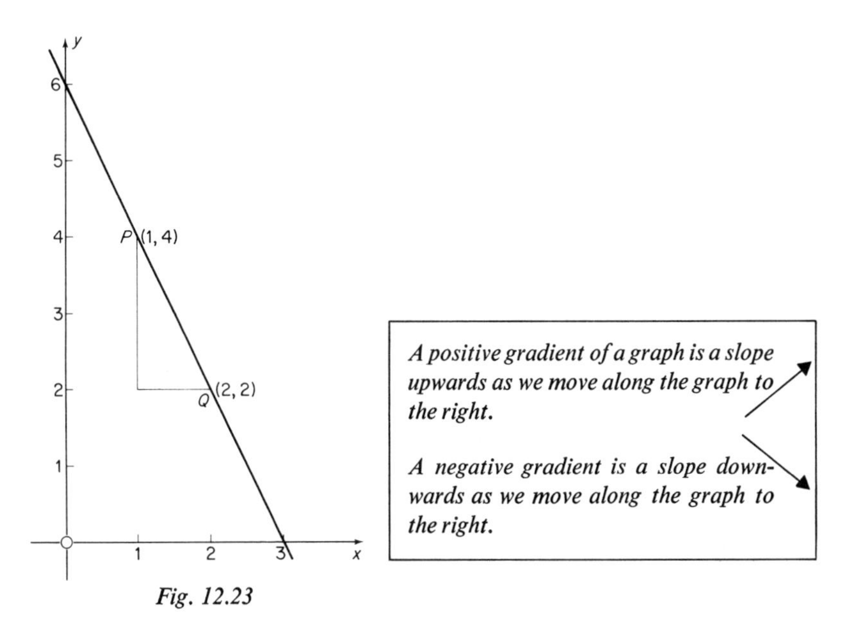

Fig. 12.23

12.19 Exercise

Write down the gradients of the graphs of the following equations (without first plotting the graphs):

$y + 5x = 2$ [Hint: write the equation as $y = -5x + 2$];
$2y - 3x = 7$;
$y = 7 - 4x$.

12.20 Intercept

The points where the graph meets or cuts the axes, in particular the y-axis, are interesting. The distance of this point from the origin is called the *intercept* on the y-axis. At this point the distance of the graph from the y-axis is zero, that is, the x-co-ordinate of this point is 0, and we can find its y-co-ordinate by substituting 0 for x in the equation. Consider the equation $y = 2x + 3$:

$$x = 0 \Rightarrow y = 2(0) + 3 = 3.$$

The graph will therefore cut the y-axis at the point (0, 3).

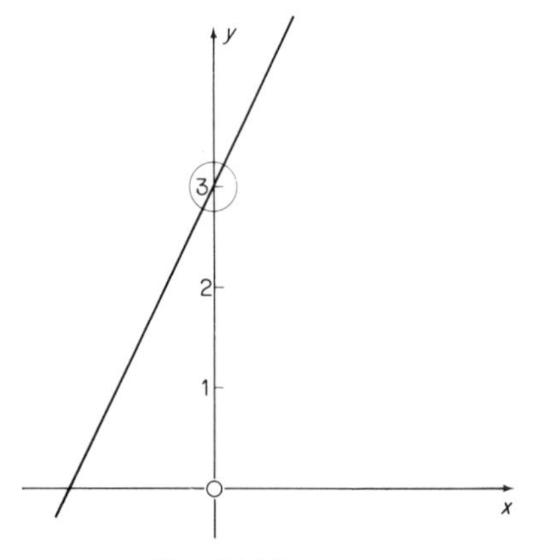

Fig. 12.24

Again, if we transform the equation to the general form $y = mx + c$, we can read off the y-co-ordinate of this point immediately; it is the constant term, c. For example,

$$3y = 2x + 6 \Rightarrow y = \tfrac{2}{3}x + 2$$

and the y-co-ordinate of the point where the graph cuts the y-axis, the intercept, is 2, since

$$x = 0 \Rightarrow y = 2.$$

As an exercise, work out the intercepts on the y-axis for the equations and graphs mentioned in previous exercises.

12.21 The Gradient–Intercept Method of Sketching Graphs

The intercept and the gradient together provide an alternative method for plotting linear graphs. Use the following simple procedure:

(*a*) express the equation in the form $y = mx + c$;
(*b*) mark the point $(0, c)$ on the graph paper;
(*c*) draw the line with gradient m which passes through the point $(0, c)$.

Examples

1. Plot the graph of $5y = 3x + 10$.

$$5y = 3x + 10 \Rightarrow y = \tfrac{3}{5}x + 2$$
$$\Rightarrow \text{gradient} = \tfrac{3}{5}, \text{ intercept is } (0, 2).$$

Mark the point (0, 2).

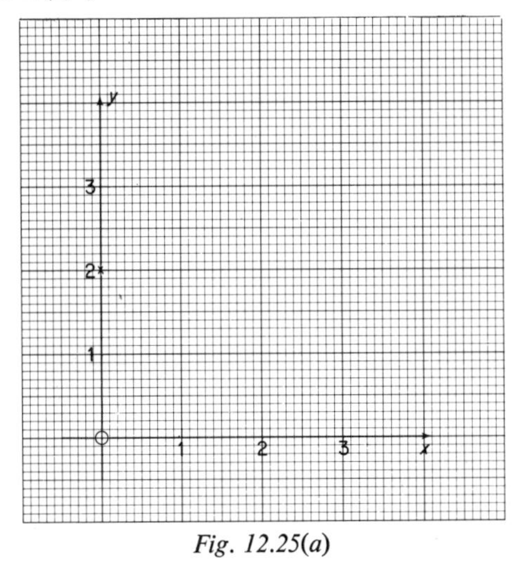

Fig. 12.25(a)

Draw the line passing through the point (0, 2) and a second point 5 units to the right and 3 up.

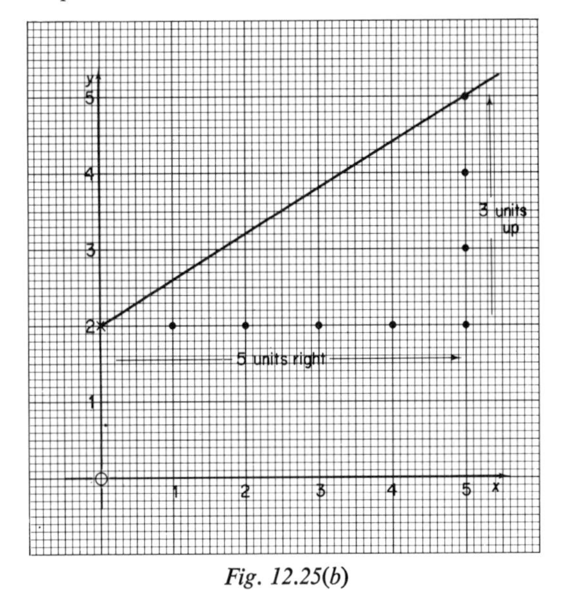

Fig. 12.25(b)

2. Plot the graph of $2y + 3x + 8 = 0$.

$$\begin{aligned} & 2y + 3x + 8 = 0 \\ \Rightarrow\ & 2y = {}^-3x - 8 \\ \Rightarrow\ & y = {}^-\tfrac{3}{2}x - 4 \\ \Rightarrow\ & \text{gradient} = -\tfrac{3}{2}, \\ & \text{intercept is } (0,\ {}^-4). \end{aligned}$$

Mark the point (0, ⁻4).

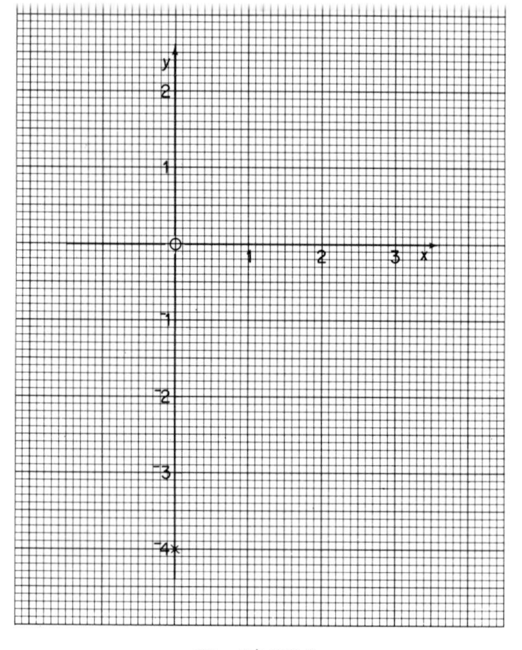

Fig. 12.26(a)

Draw the line passing through the point (0, ⁻4) and a second point 2 units to the right and 3 units *down* (negative gradient) (Fig 12.26 (*b*)).

12.22 Exercise

Sketch the graphs of the following equations, using the gradient–intercept method:

$$y = \tfrac{3}{4}x + 2;$$
$$7y = 5x + 3\tfrac{1}{2};$$
$$5y + 6x = 15.$$

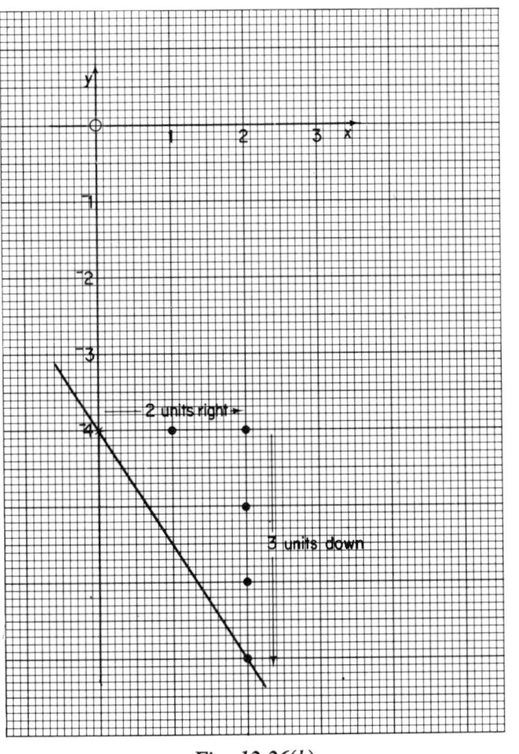

Fig. 12.26(b)

12.23 Solving Simultaneous Equations by Graphical Methods

The graph of the equation $y + x = 10$ represents *all* the pairs of numbers whose sum is 10; the graph of the equation $y = x + 5$ represents all the pairs of numbers of which the second is 5 more than the first. If there is a number pair whose sum is 10, and of which the second number is 5 more than the first (two simultaneous equations in two unknowns), this pair should appear as a point which lies on both graphs or, in other words, as the point where the two graphs intersect. If we draw both graphs on the same sheet of graph paper, using the same axes (Fig. 12.27), we find that these graphs meet at the point $(2\frac{1}{2}, 7\frac{1}{2})$. The point of intersection represents a number pair which satisfies both equations:

$$x = 2\tfrac{1}{2} \text{ and } y = 7\tfrac{1}{2}$$
$$\Rightarrow x + y = 10 \text{ and } y = x + 5.$$

We can therefore solve problems involving simultaneous equations by plotting the graphs of the two equations and finding their point of intersection.

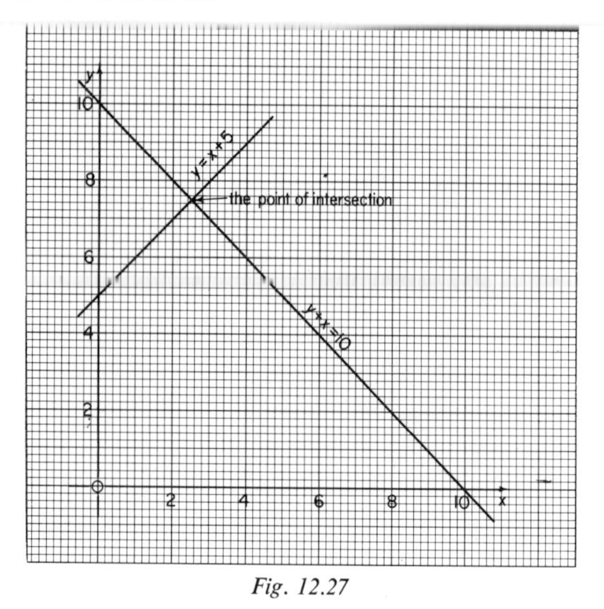

Fig. 12.27

Example

To the cost of an article, a Value Added Tax of 25% is added. The price of the article, including this tax, is £60. Find the VAT on this article and the price of the article without this tax.

If VAT on the article is £y and the price without VAT is £x, we can express this information by the equations

$$y + x = 60$$
$$y = \tfrac{1}{4}x \ (25\% \text{ of } x = \tfrac{1}{4}x),$$

which are drawn out in Fig. 12.28.

The graph of $y + x = 60$ represents all the ordered pairs of numbers whose sums are 60, and the graph of $y = \frac{1}{4}x$ represents all the number pairs where the second number is a quarter of the first.

The point of intersection, (48, 12), represents the *only* number pair which satisfies both conditions: $y + x = 60$ *and* $y = \frac{1}{4}x$.

Therefore, VAT on the article is £12, and the price without tax is £48.

12.24 Exercises

1. Solve the following simultaneous equations by the graphical method:

$$x + y = 5,$$
$$y = 2x + 1.$$

2. Express the following information by simultaneous equations, and solve these equations by the graphical method:

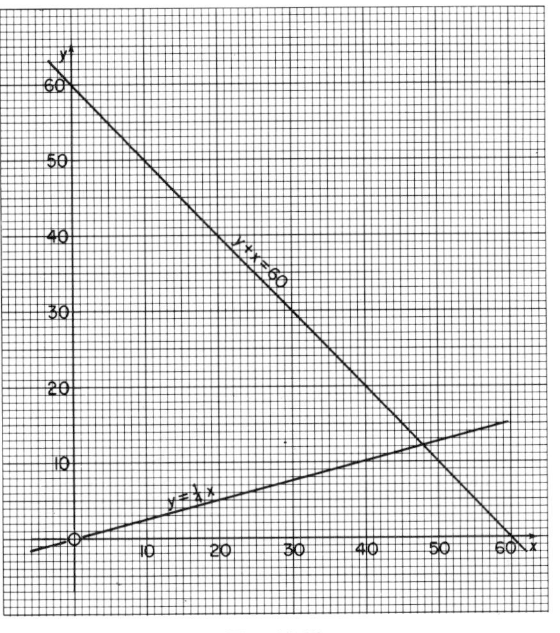

Fig. 12.28

The perimeter (the sum of the lengths of the sides) of a rectangle is 48 cm. The larger side is 10 cm longer than the shorter side. Find the lengths of the sides of this rectangle. [Hint: the perimeter of a rectangle is *twice* the length of the larger side plus *twice* the length of the shorter side.]

12.25 Graphs of Inequalities

You can easily verify our earlier statement that the graph of an equation represents *all* the number pairs within the range of the graph paper. Plot the graph of $x + y = 3$ as an example, and mark any point above, or on the right-hand side of, the graph. Find the co-ordinates of this point, and add them together; you will notice that this sum is *not* 3. In other words, the number pair represented by this point does not satisfy the equation $x + y = 3$.

In Fig. 12.29 we have arbitrarily marked some points P and Q. The sum of their co-ordinates is not 3; in fact, if the co-ordinates of the point P, chosen to be (4, 2), are added, this sum is more than 3, or $4 + 2 > 3$. Indeed, the sum of the two co-ordinates of *every* point above the graph is more than 3.

Compare now the co-ordinates of the points $T(1, 2)$ and $P(4, 2)$; their second co-ordinate is the same. In Fig. 12.29, we have drawn a dotted line through P and T; this is the line whose equation is $y = 2$. The y-co-ordinate of all the points along this line is 2, but the x-co-ordinate of every point on this line to the right of T is more than 1 (the x-co-ordinate of T). Therefore, the sum of the co-ordinates of every point along this line to the right of T is more than 3.

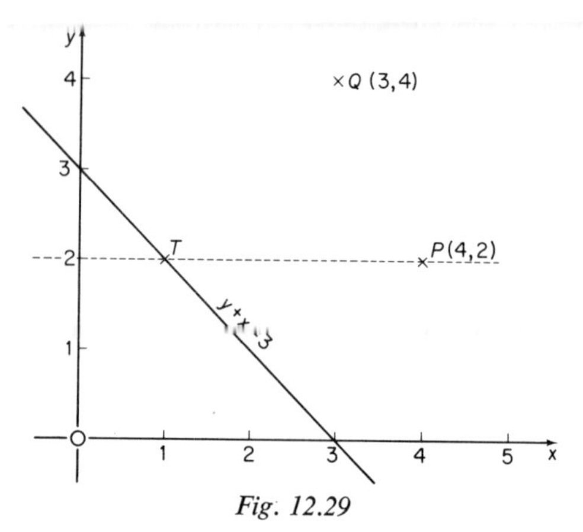

Fig. 12.29

Every point above, or to the right of the graph of the equation $x + y = 3$, that is the shaded region in Fig. 12.30, therefore represents a number pair (x, y) whose sum is more than 3, or $x + y > 3$. This shaded region is the set of all points representing ordered number pairs to which the relation $x + y > 3$ applies, or in mathematical shorthand $\{(x, y): x + y > 3\}$. *The shaded region is therefore the graph of the inequality* $x + y > 3$.

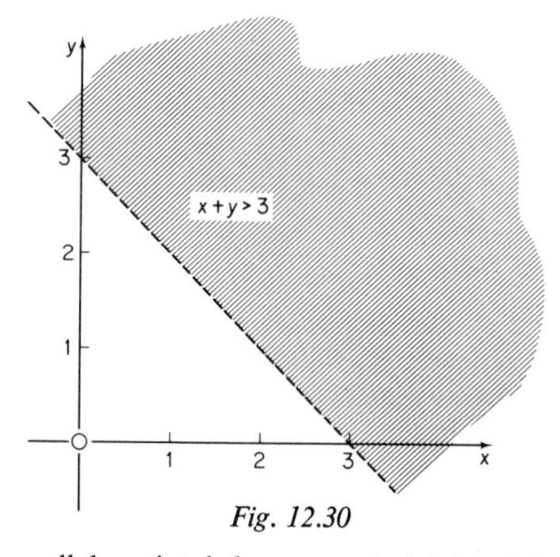

Fig. 12.30

In the same way all the points below, or on the left-hand side of, the graph of $x + y = 3$ have co-ordinates whose sums are less than 3. The graph of the inequality $x + y < 3$ is therefore the shaded region of Fig. 12.31, which represents $\{(x, y): x + y < 3\}$.

The graph of $x + y = 3$ in fact divides the graph paper into 3 parts or sets of points (Fig. 12.32):

the line representing the equation $x + y = 3$, or $\{(x, y): x + y = 3\}$;
the region representing the inequality $x + y < 3$, or $\{(x, y): x + y < 3\}$;
the region representing the inequality $x + y > 3$ or $\{(x, y): x + y > 3\}$.

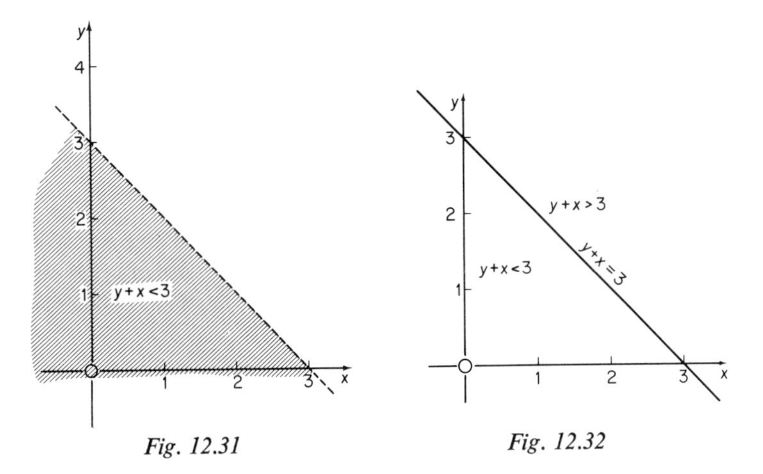

Fig. 12.31 *Fig. 12.32*

Notice that neither the graph of $x + y < 3$ nor the graph of $x + y > 3$ includes the number pairs whose sums equal 3. The 'frontier', or the *bounding line*, of the two regions is therefore indicated by a dotted line, in Figs. 12.30 and 12.31, which is the line $x + y = 3$.

We can indicate the set of all points representing number pairs, (x, y), whose sums are equal to 3 or whose sums are less than 3—together expressed as $x + y \leqslant 3$—by the shaded region *including* the line $x + y = 3$, which is fully drawn.

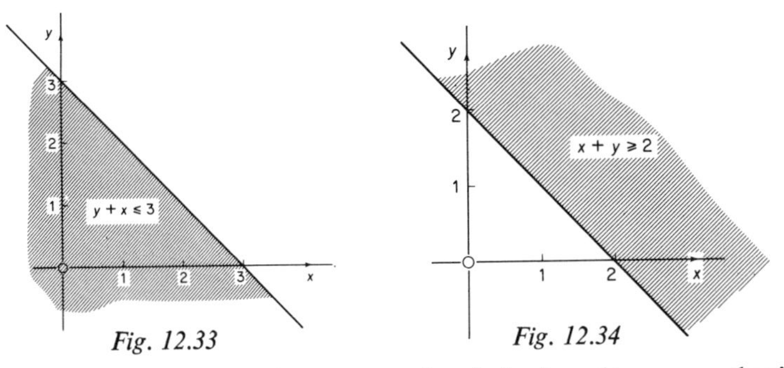

Fig. 12.33 *Fig. 12.34*

We use a similar notation, $\geqslant$, to express the relation 'equal to or more than'; for example, $x + y \geqslant 2$ expresses the relation between the co-ordinates whose sums are equal to 2 or more than 2. The set of all such ordered number pairs, or $\{(x, y): x + y \geqslant 2\}$ is represented by all the points along the line $x + y = 2$ and the points in the region $x + y > 2$.

Note. When you are drawing the graphs of inequalities, you should draw the bounding line of the region first; the bounding line is the graph of the corresponding equation. For example, the equation $x + y = 7$ is the bounding line of the inequalities $x + y < 7, x + y > 7, x + y \leqslant 7$ and $x + y \geqslant 7$.

12.26 Exercises

1. Draw the graphs of the following inequalities (use a different piece of graph paper for each graph):

$$y + 2x < 4;$$
$$3y + x \geqslant 7.$$

Show clearly whether the bounding line is included or not.

2. Draw the graph of $y = 3$. [Hint: this graph is the set of all the points whose y co-ordinates are 3, or, whatever the value of x, the value of y is always 3.] Mark, by shading, the region representing $y > 3$. Shade the region $y < 3$ in a different way.
3. (*a*) Draw the graph of $x < 4$.
   (*b*) On a different piece of graph paper, draw the graph of $x \geqslant 2$.

12.27 Simultaneous Inequalities

When we solved two simultaneous equations by the graphical method, we used the simple fact that two straight lines are either parallel or meet—*intersect*—at just one point. The point of intersection represents the number pair of the solutions.

But the intersection of two *regions* is not a single point; the 'overlap', the set of points which lie in both regions, is itself a region containing an unlimited number of points. For example, the shaded region in Fig. 12.35

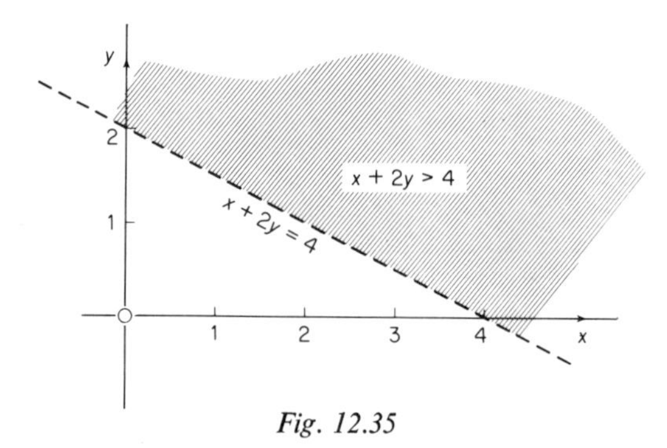

Fig. 12.35

represents all the points satisfying $x + 2y > 4$, or $\{(x, y): x + 2y > 4\}$. Similarly, the shaded region in Fig. 12.36 represents all the points satisfying $x - 2y \geqslant 3$, or $\{(x, y): x - 2y \geqslant 3\}$. If we draw both graphs on the same diagram and axes, the intersection of the two regions is the doubly shaded area in Fig. 12.37.

The points in this region represent number pairs which satisfy both inequalities at once. In graphs of two inequalities involving the *same two variables, two*

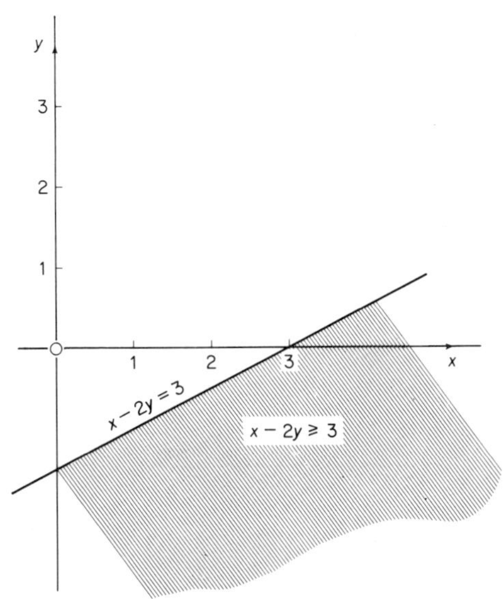

Fig. 12.36

simultaneous inequalities, this doubly shaded region represents all the possible solutions; in our example, it represents all the number pairs, (x, y), satisfying both the inequalities $x + 2y > 4$ and $x - 2y \geqslant 3$.

If we use 'set language' (see Section 8.5) the overlap is the *intersection* of both sets $\{(x, y): x + 2y > 4\}$ and $\{(x, y): x - 2y \geqslant 3\}$ which we can write as

$$\{(x, y): x + 2y > 4\} \cap \{(x, y): x - 2y \geqslant 3\}.$$

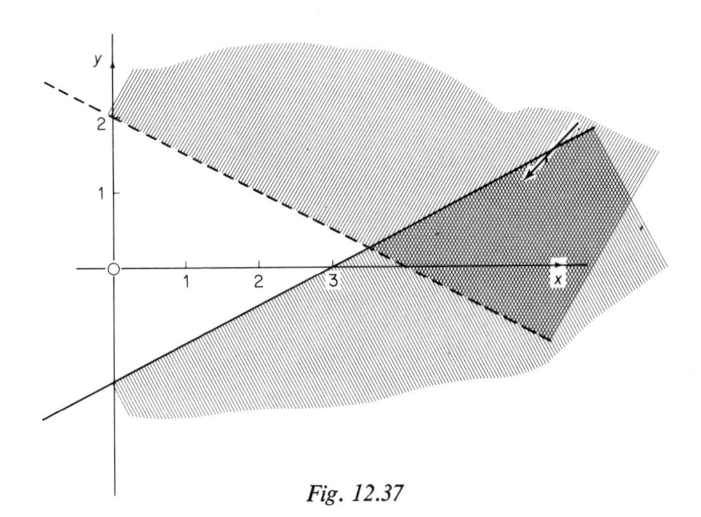

Fig. 12.37

When drawing the graph of two such simultaneous inequalities, we usually only shade in the area of overlap, that is the set of all points which satisfy both inequalities. Sometimes this is called the *solution set*.

The graph of the simultaneous inequalities $x + 2y > 4$ and $x - 2y \geqslant 3$ is then the shaded region in Fig. 12.38.

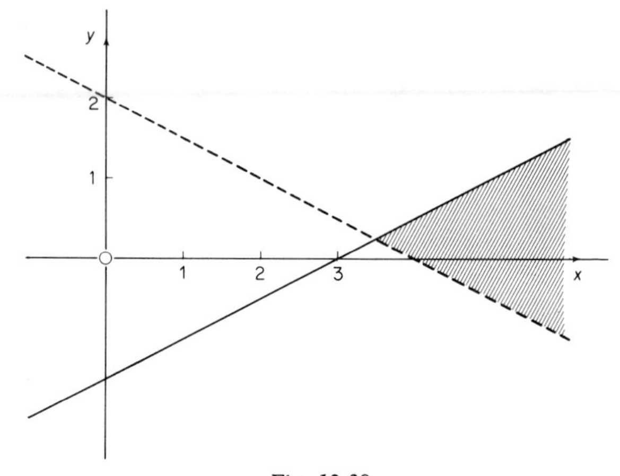

Fig. 12.38

Similarly, in Fig. 12.39, the shaded region is the graph of the simultaneous inequalities $x > 3$ and $x \leqslant 6$.

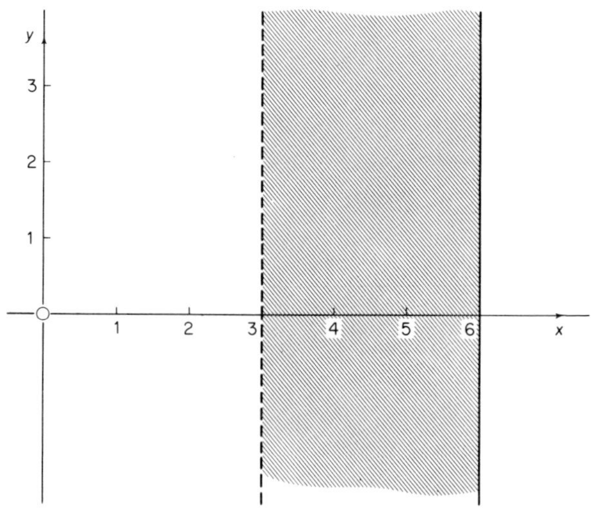

Fig. 12.39

12.28 Exercise

Draw the bounding lines and shade the region representing the simultaneous inequalities:

$$x < 5,$$
$$2x + 3y \geqslant 12.$$

12.29 Linear Programming

The graphical method of solving *simultaneous equations* is often more elaborate than the other available methods (see Unit 11). But for solving problems involving *simultaneous inequalities* in two unknowns, the graphical method is the simplest and most suitable one. Solving simultaneous inequalities is called *linear programming*.

Practical problems often involve three or more inequalities. The graph representing the solution set is then the region where all three (or more) regions overlap, that is, the points included in the graph of each and every inequality. The more inequalities there are, the more restricted the region of overlap becomes.

For example, the region representing the solution set of the simultaneous inequalities

$$y \leqslant 3,$$
$$x - 2y < 1,$$
$$y \geqslant 5 - 2x$$

is the shaded region in Fig. 12.40.

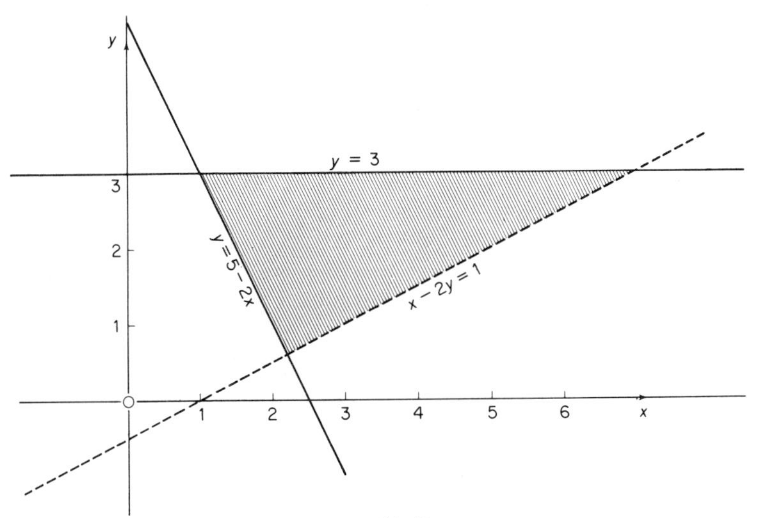

Fig. 12.40

If we then introduce a third inequality concerning the same two variables, the solution set can be even more restricted. If, for example, the inequality $x > 3$ is added to the set of inequalities of this example, the solution set is the shaded region in Fig. 12.41.

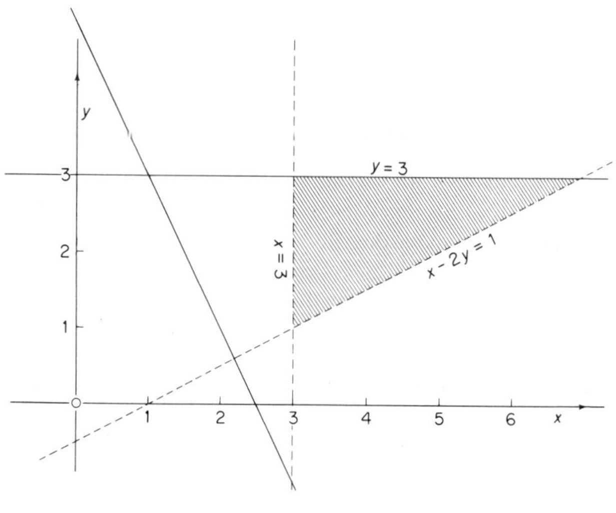

Fig. 12.41

Practical problems involving inequalities in two variables are solved by this graphical method. Computers are usually used when the problem involves more than two variables. In each case, we must first express the relevant conditions as inequalities or equations, whichever are appropriate.

Example

The manager of a charter airline has to decide the type of aircraft to be used to transport 750 passengers, bearing in mind the following:

(*a*) there are 6 *S*-type aircraft available carrying 125 passengers each, and 10 *V*-type aircraft carrying 50 passengers each;
(*b*) there are only 11 pilots;
(*c*) the total amount in fares paid by passengers is £4000; the cost to the company for flying an *S*-type aircraft is £550, and for flying a *V*-type aircraft £200.

If the number of *S*-type aircraft to be used is denoted by x, and the number of *V*-type aircraft by y, the conditions under (*a*) and (*b*) are expressed by the inequalities:

at most 6 *S*-type aircraft: $x \leqslant 6$,
at most 10 *V*-type aircraft: $y \leqslant 10$,
at most 11 pilots (and therefore at most a total of 11 aircraft): $x + y \leqslant 11$,

750 passengers to be transported: $x\,(125) + y\,(50) \geqslant 750 \Rightarrow 5x + 2y \geqslant 30$ (dividing both sides by 25).

When the bounding lines are drawn, all the number pairs satisfying these inequalities simultaneously are represented by the points in the shaded region in Fig. 12.42.

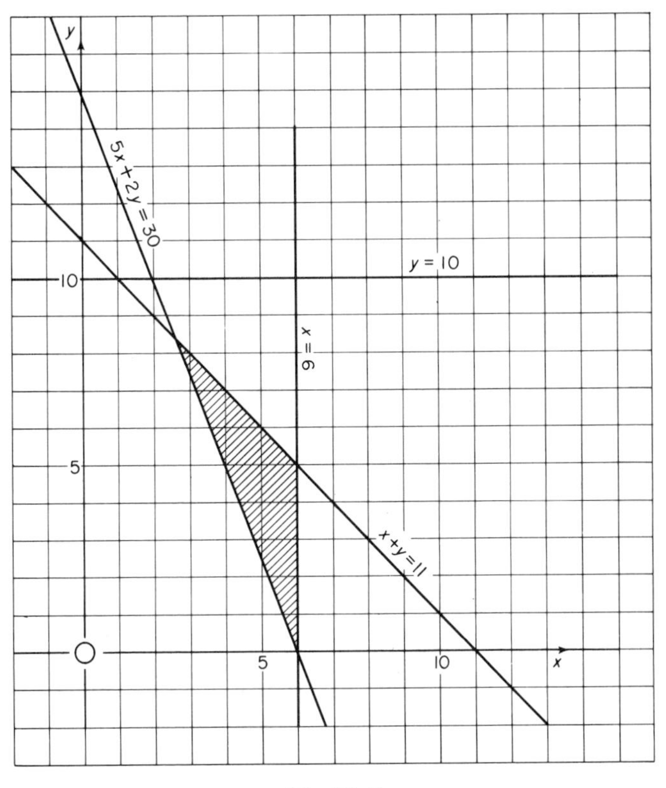

Fig. 12.42

However, a further restriction is implied, although not stated: the number of aircraft can only be a *positive whole number* (no negative numbers, no fractions); therefore, only those points within the shaded region can be considered which represent pairs of integers, indicated in Fig. 12.43 by dots.

The possible solutions are therefore represented by the points in the *solution set*:

(3, 8): that is, 3 *S*-type aircraft and 8 *V*-type aircraft;
(4, 5), (4, 6), (4, 7);
(5, 3), (5, 4), (5, 5), (5, 6);
(6, 0), (6, 1), (6, 2), (6, 3), (6, 4), (6, 5).

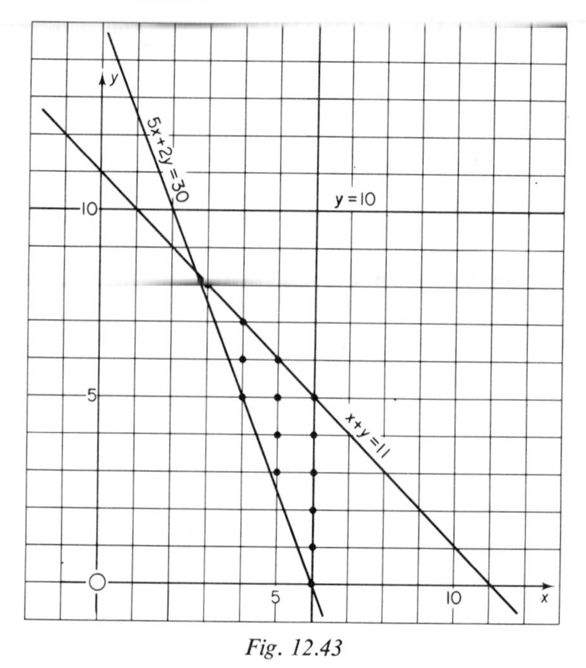

Fig. 12.43

The most profitable solution

Clearly, not all these solutions are equally profitable. If (6, 0) is a solution (that is, 6 *S*-type aircraft, which could carry all 750 passengers), the solutions (6, 1), (6, 2), (6, 3), (6, 4) and (6, 5) represent cases when aircraft are flying with vacant seats. (6, 0) is a more profitable alternative than any of these. Similarly, (5, 3) (5 *S*-type and 3 *V*-type aircraft) is less costly than (5, 4), (5, 5) and (5, 6); (4, 5) is less costly than (4, 6) or (4, 7). We can find which of the solutions (3, 8), (4, 5), (5, 3) and (6, 0) is the most profitable one by calculating the cost in each case and comparing them. If there are many possible solutions, a more systematic method is the consideration of the cost graph.

We know that the cost is £$(550x + 200y)$. For the company to just break even, this must be £4000. So $550x + 200y = 4000 \Rightarrow 11x + 4y = 80$.

If this cost graph is also drawn on the diagram (Fig. 12.44), the region to the left of the line $11x + 4y = 80$ represents all possible values of x and y when the company makes a profit (the cost of flying the aircraft is less than the money received in fares, or $11x + 4y < 80$).

Indeed, the further a point lies to the left of the line $11x + 4y = 80$, the less the cost of flying the aircraft and the greater the company's profits in the circumstances which that point represents.

Of the four points under consideration—(3, 8), (4, 5), (5, 3) and (6, 0)—the point (4, 5) is the farthest away from the bounding line $11x + 4y = 80$, and represents, therefore, the most profitable solution. We can therefore say that the most profitable way for the airline to transport these passengers would be to use 4 S-type and 5 V-type aircraft.

12.30 Exercise

A shoe manufacturer is to decide on the number of ladies' and children's shoes to be made. The relevant data are:

(*a*) each of the 6 skilled workers can make, per hour, 2 pairs of ladies' shoes or 3 pairs of children's shoes;

(*b*) there are sufficient machines and equipment to manufacture, per hour, 7 pairs of ladies' shoes and 12 pairs of children's shoes;

(*c*) the total cost of employing these 6 workers is £7 per hour, whether any shoes are made or not; this has to be met from a profit of 70p on a pair of children's shoes and 100p on a pair of ladies' shoes.

The number of workers required to make 6 pairs of children's shoes in one hour is $\frac{6}{3} = 2$.

By the same reasoning, the number of workers required to make x pairs of children's shoes per hour is $\frac{x}{3}$.

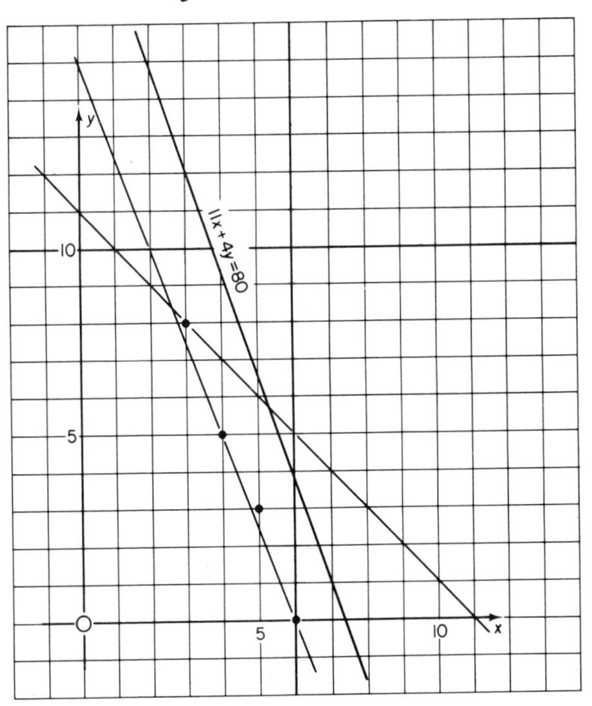

Fig. 12.44

Similarly, the number of workers required to make y pairs of ladies' shoes per hour is $\frac{y}{2}$.

The number of workers required to make x pairs of children's shoes and y pairs of ladies' shoes per hour is $\frac{x}{3} + \frac{y}{2}$.

According to (*a*), the manufacturer has only 6 men at his disposal:

$\frac{x}{3} + \frac{y}{2} \leqslant 6 \Rightarrow 2x + 3y \leqslant 36$ (multiplying both sides by 6). This inequality expresses all the information in condition (*a*).

1. Draw the graph of this inequality.
2. Write down the inequalities expressing condition (*b*), and draw the graphs of these inequalities. [Hint: if the machinery can make only 12 pairs of children's shoes per hour, the number x cannot be more than 12. In other words, x is less than 12, or at most equal to 12.]
3. A minimum number of shoes must be made to cover the cost of employing the men. Using condition (*c*), express this as an inequality. [Hint: the profit on x pairs of children's shoes is $x \times 70$p; the total profit on x pairs of children's shoes and y pairs of ladies' shoes is $x \times 70\text{p} + y \times 100\text{p}$, or $(70x + 100y)$ pence. For there to be any *net* profit, this sum must be more than the labour cost of 700p per hour.]
4. Mark with dots all the points which represent a profitable solution, using only *whole numbers* for simplicity's sake.
5. Which point represents the most profitable solution? Calculate the maximum *net* profit per hour, deducting the labour costs.
   [Hint: there are many solutions, and it is impractical to calculate the cost in each case. As with the problem of the aircraft, the most profitable solution is represented by the point, plotted in Exercise 4, which lies farthest to the left of the bounding line of the cost graph, $7x + 10y > 70$.]

Unit Thirteen

Curves

13.1 Graphs of Non-linear Equations

If we plot the graphs of such simple relations as 'is the square of' or 'is the cube of', we find that not all graphs of equations are straight lines. The relation 'is the square of' is expressed by the equation $y = x^2$. We set up a table of this equation for integral values of x from $x = 0$ to $x = 5$, as follows:

x	0	1	2	3	4	5
y	0	1	4	9	16	25

Fig. 13.1 shows that, when we plot these points on graph paper, they do not all lie on one straight line. The 'gradient' of the graph (see Section 12.16) changes from one place to another along it, instead of remaining the same everywhere, as it does on a linear graph; you can study this by comparing the increase in y for an increase of 1 in the value of x, for different values of x.

The graph of $y = x^2$ is not drawn by joining the plotted points by straight lines. This becomes obvious if we consider 'intermediate' values of x, for

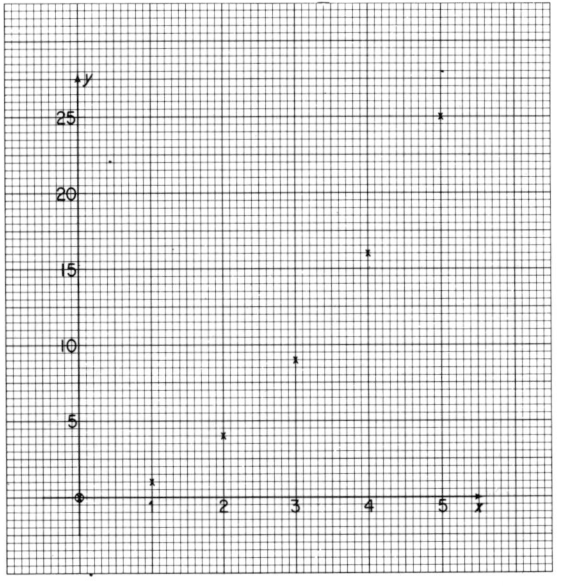

Fig. 13.1

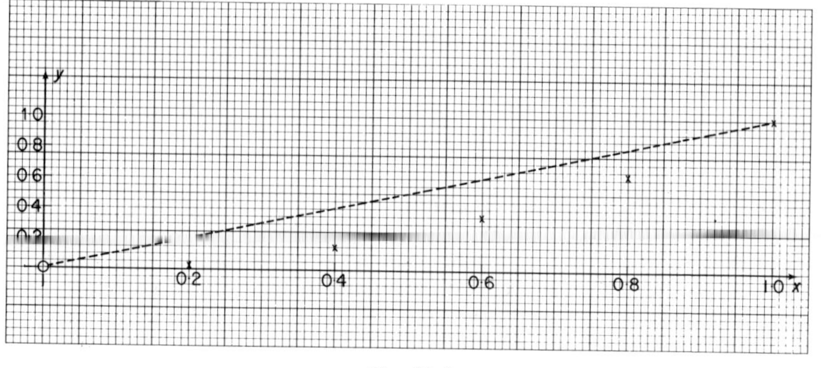

Fig. 13.2

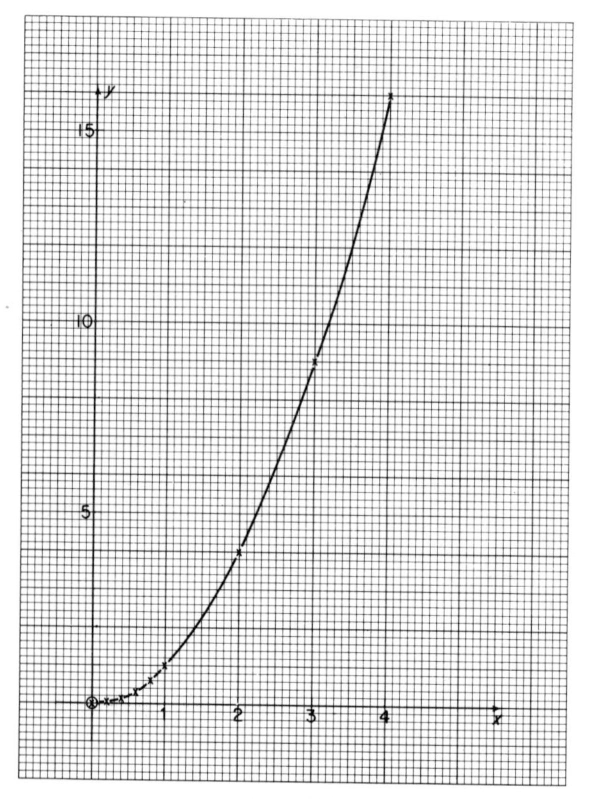

Fig. 13.3

example, between $x = 0$ and $x = 1$—in other words, fractional values of x such as $\frac{1}{5}, \frac{2}{5}, \frac{3}{5}, \frac{4}{5}$, or 0·2, 0·4, 0·6, 0·8.

x	0	0·2	0·4	0·6	0·8	1
y	0	0·04	0·16	0·36	0·64	1

(Take care when working with unit points; for example, $(0{\cdot}2)^2 = 0{\cdot}04$, *not* 0·4.) The corresponding points plotted on an enlarged graph *do not* lie along the straight line joining the points (0, 0) and (1, 1), that is, the dotted line in Fig. 13.2. The graph is in fact a *curved line* or a *curve*, and we cannot use a ruler to draw it. The plotted points must be joined by a smooth curving line, sketched neatly and carefully but freehand. You will find it helpful to plot some intermediate points in the interval where the 'bend' seems sharpest. Figure 13.3 illustrates the graph of $y = x^2$ over the range from $x = 0$ to $x = 4$, sketched after plotting all the points of both tables.

If we extend the graph of $y = x^2$ to include negative values of x, an interesting phenomenon appears.

x	$^-3$	$^-2$	$^-1$	0	$^+1$	$^+2$	$^+3$
y	$^+9$	$^+4$	$^+1$	0	$^+1$	$^+4$	$^+9$

The values of y are all *positive* even when the corresponding value of x is negative (see Section 3.4).

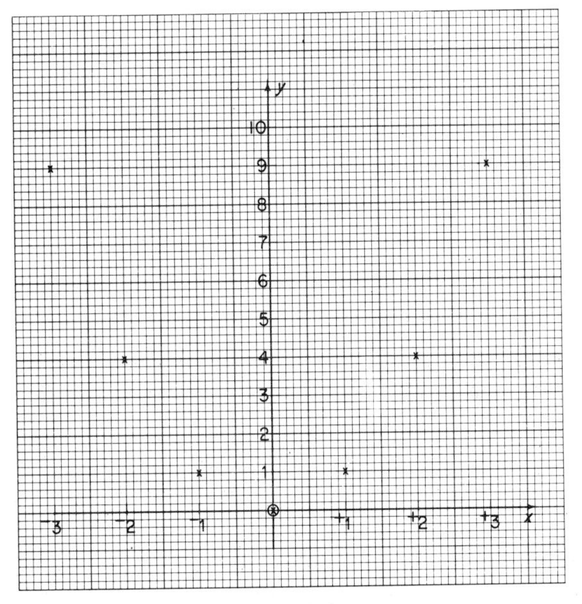

Fig. 13.4

We can see, from the general appearance of the set of points, that the sharpest bend occurs between $x = {}^-1$ and $x = {}^+1$, so we include some intermediate values in this interval. The table of $y = x^2$ from $x = {}^-3$ to $x = {}^+3$, including these intermediate points, is then:

x	$^-3$	$^-2$	$^-1$	$^-0{\cdot}6$	$^-0{\cdot}3$	0	$^+0{\cdot}3$	$^+0{\cdot}6$	$^+1$	$^+2$	$^+3$
y	$^+9$	$^+4$	$^+1$	$^+0{\cdot}36$	$^+0{\cdot}09$	0	$^+0{\cdot}09$	$^+0{\cdot}36$	$^+1$	$^+4$	$^+9$

These points are plotted, and joined by a smooth curve to form the graph of $y = x^2$.

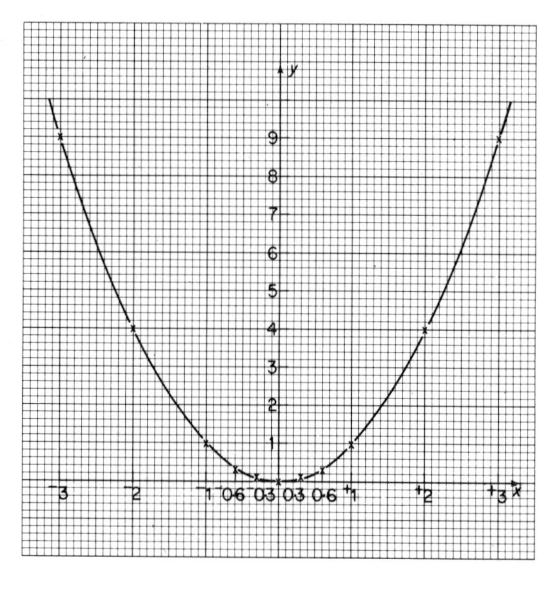

Fig. 13.5

13.2 The Graph of $y = x^3$

We can plot the graph of $y = x^3$, using a similar procedure. First, we set up a table for this equation for values of x in a convenient range, say from $x = {}^-3$ to $x = {}^+3$.

x	$^-3$	$^-2$	$^-1$	0	$^+1$	$^+2$	$^+3$
y	$^-27$	$^-8$	$^-1$	0	$^+1$	$^+8$	$^+27$

This time the value of y is negative if the corresponding value of x is negative; for example, $({}^-2)^3 = {}^-8$. Again, when these points are plotted the sharpest bend appears between the values $x = {}^-1$ and $x = {}^+1$ (Fig. 13.6).

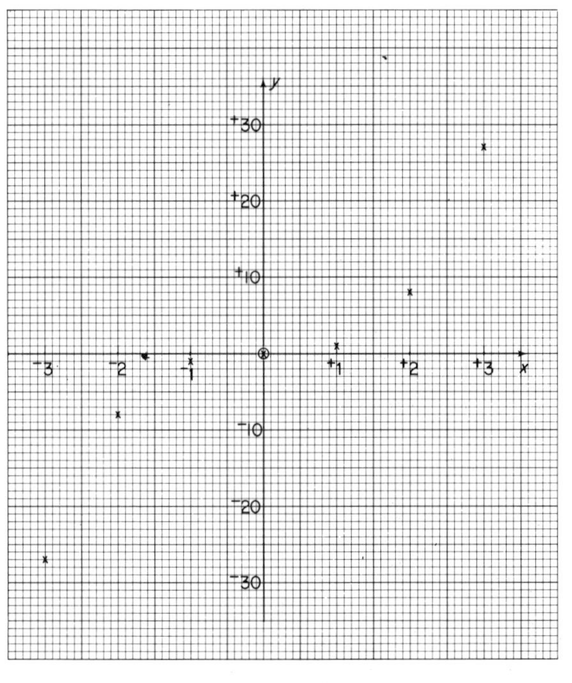

Fig. 13.6

We therefore add some intermediate values to our table in the interval $x = {}^-1$ to $x = {}^+1$.

x	$^-0{\cdot}6$	$^-0{\cdot}3$	0	$^+0{\cdot}3$	$^+0{\cdot}6$
y	$^-0{\cdot}216$	$^-0{\cdot}027$	0	$^+0{\cdot}027$	$^+0{\cdot}216$

All the points are then plotted and the graph sketched by joining these points with a smooth curve, as in Fig. 13.7.

13.3 The General Shape of the Graph of $y = x^n$

We will begin with an exercise.

1. Complete the following table for the equation $y = x^4$.

x	$^-2$	$^-1$	$^-\frac{1}{2}$	0	$^+\frac{1}{2}$	$^+1$	$^+2$
y	$^+16$						

On a sheet of graph paper draw the x- and y-axes. You will be able to tell from the table which is the most suitable position: the y-axis down the middle of the page and the x-axis as low as possible across the page, since there are no negative values of y.

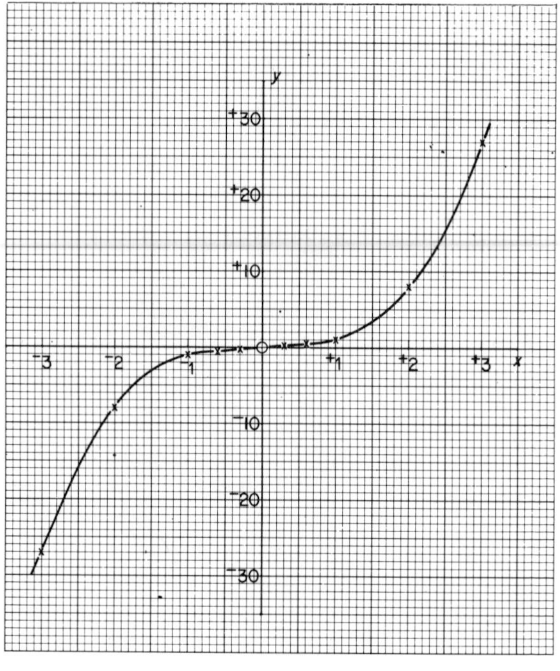

Fig. 13.7

2. Plot the points representing the number pairs of this table.
3. Join the plotted points by a smooth curve.

If we study this graph, comparing it with the graphs of the previous sections, we find we have a U-shaped curve very like the graph of $y = x^2$, except that this graph 'droops' much lower between $x = {}^{-}1$ and $x = {}^{+}1$. Both these graphs are quite different from the graph of $y = x^3$.

We can easily account for this similarity and this difference. Remember what we observed earlier (see Section 3.6) about the sign of powers of negative numbers:

An even power of a negative number is itself a positive number, for example,

$$({}^{-}2)^2 = {}^{+}4, \ ({}^{-}2)^4 = {}^{+}16.$$

An odd power of a negative number is itself a negative number, for example,

$$({}^{-}2)^1 = {}^{-}2, \ ({}^{-}2)^3 = {}^{-}8.$$

The shape of the graph of $y = x^n$ is U-shaped if n is an even number and S-shaped (or an inverted S-shape) if n is an odd number.

13.4 Exercises

1. Sketch on one large sheet of graph paper, using the same axes, the graphs of:

$$y = x;$$
$$y = x^2;$$
$$y = x^3;$$
$$y = x^4;$$
$$y = x^5.$$

   Plot points over an interval from $x = {}^-2$ to $x = {}^+2$. Notice the 'crossing' of the curves at the points $(1, 1)$, $({}^-1, 1)$ and $({}^-1, {}^-1)$.
2. Draw on one sheet of graph paper the graphs of $y = x^2$ and $y = {}^-x^2$. [Hint: ${}^-x^2$ means ${}^-(x^2)$, that is 'the negative of the square of x'.]
3. Similarly, on one sheet of graph paper, sketch the graphs of $y = x^3$ and $y = {}^-x^3$.
   (If you have drawn them properly, each pair of graphs should be like mirror images on either side of the x-axis.)

13.5 The Graph of $y = \frac{1}{x}$ or $y = x^{-1}$

If we study the table of the graph of $y = \frac{1}{x}$, it is soon clear that the shape of its graph differs completely from any other we have so far discussed.

x	$^-5$	$^-4$	$^-3$	$^-2$	$^-1$	$^-\frac{1}{2}$	$^-\frac{1}{3}$	$^-\frac{1}{4}$	$\frac{1}{4}$	$\frac{1}{3}$	$\frac{1}{2}$	1	2
y	$^-\frac{1}{5}$	$^-\frac{1}{4}$	$^-\frac{1}{3}$	$^-\frac{1}{2}$	$^-1$	$^-2$	$^-3$	$^-4$	4	3	2	1	$\frac{1}{2}$

We have already seen (in Exercise 2 of Section 3.13) that we cannot divide by zero, and, therefore, the function $y = x^{-1}$ has no value at the point where $x = 0$. Plotting the points represented by the number pairs of the table, we find again that the graph is not a straight line but a curve—a curiously shaped curve consisting of two parts, which is known in mathematics as a *hyperbola* (Fig. 13.8).

13.6 The Graph of $y = \sqrt{x}$

When we plot the graph of the relation $y = \sqrt{x}$, we must remember that the square root of a positive number has two values, a positive one and a negative one; for example, $\sqrt{4} = {}^+2$ or ${}^-2$. For every value of x, there are two corresponding values of y; this is why $\sqrt{x}$ is often written as $\pm\sqrt{x}$.

Remember too that there is no real number whose square is a negative number; therefore we cannot find any value of y for negative real values of x, and so the least possible value of x is 0.

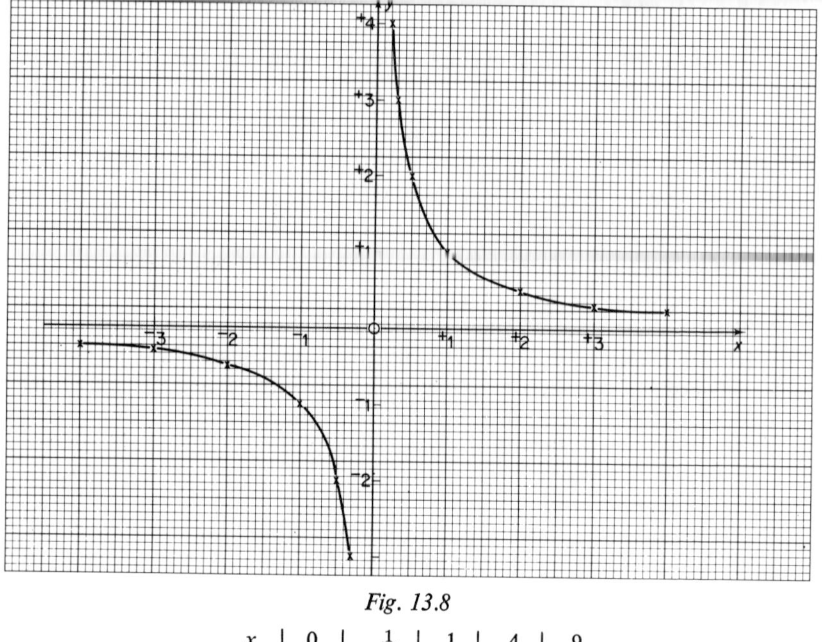

Fig. 13.8

x	0	$\frac{1}{4}$	1	4	9
y	0	$+\frac{1}{2}$ or $-\frac{1}{2}$	+1 or −1	+2 or −2	+3 or −3

The graph of $y = \sqrt{x}$ is shown in Fig. 13.9.

13.7 Exercise

1. Plot the graph of $y = \dfrac{2}{x}$.
2. Compare the graph of $y = \sqrt{x}$ with the graph of $y = x^2$. Can you account for a certain similarity? [Hint: $y = \sqrt{x} \Rightarrow y^2 = x \Rightarrow x = y^2$.]

13.8 Solving Simultaneous Non-linear Equations by Graphical Methods

We can use graphical methods to solve simultaneous equations involving one or two quadratic equations in exactly the same way that we used them to solve two simultaneous linear equations.

Example

The graph of $y = x^2$ represents all the number pairs of which the second number is the square of the first; the graph of the function $y = 4x$ represents all the number pairs of which the second number is 4 times the value of the first.

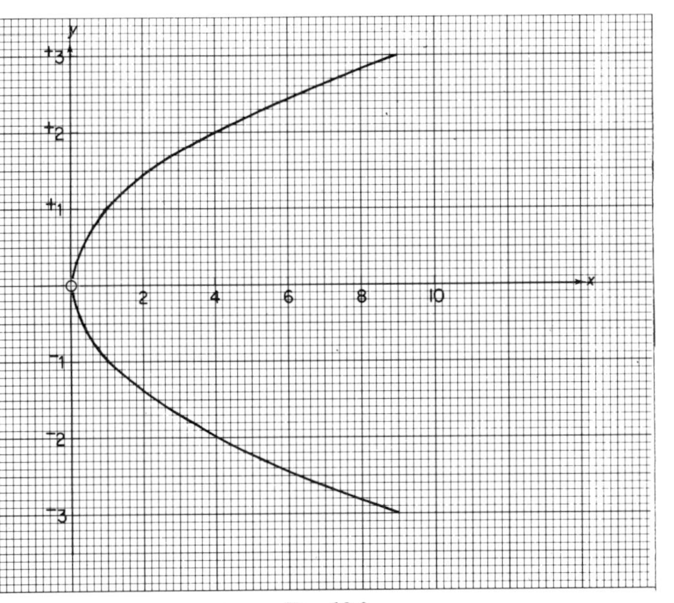

Fig. 13.9

If these two graphs are plotted on the same sheet of graph paper with, of course, the same scales and axes, the point(s) which lie(s) on both graphs (the point(s) of intersection) represent the number pairs which satisfy both equations.

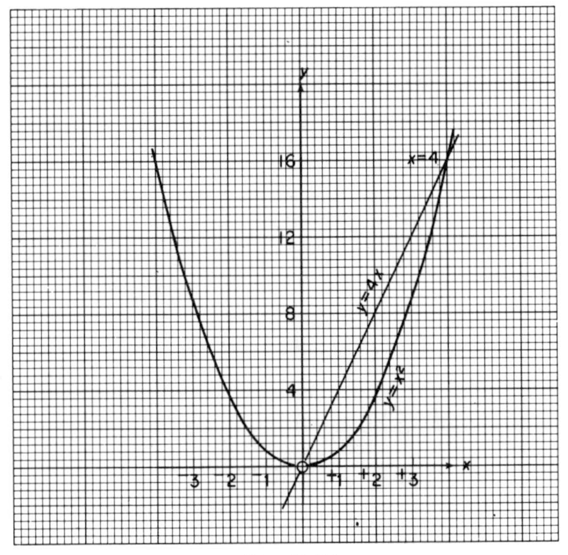

Fig. 13.10

The graphical method gives the solutions (0, 0) and (4, 16); checking by substitution confirms these solutions are valid:

$$0 = 0^2 \text{ and } \quad 0 = 4 \times 0;$$
$$16 = 4^2 \text{ and } 16 = 4 \times 4.$$

Therefore the solutions of the simultaneous equations:

$$y = x^2,$$
$$y = 4x,$$

are $\quad x = 0$ and $y = 0$;

and $\quad x = 4$ and $y = 16$.

If one or both of the simultaneous equations includes a cubic power of x, or an even higher power of x, the procedure is exactly the same.

Example

Solve the simultaneous equations $y = x^3$,

$$y = 3x - 1.$$

The graphs of both equations are plotted on the same sheet of graph paper (using the table for $y = x^3$ as given in Section 13.2).

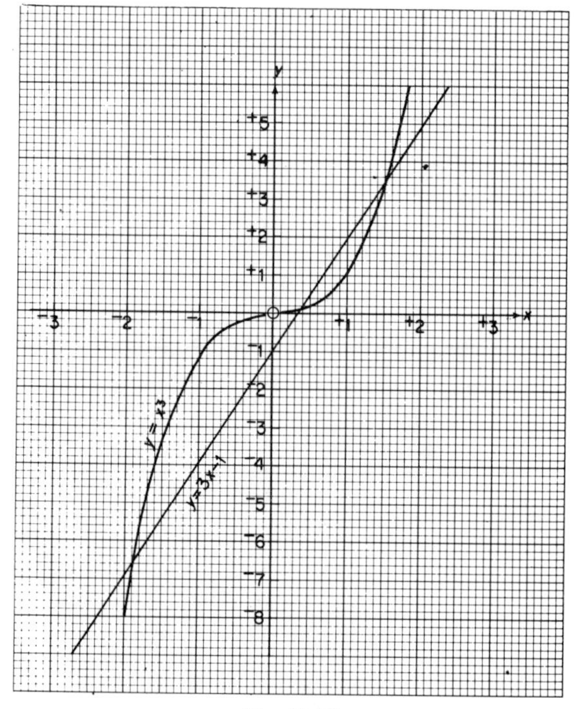

Fig. 13.11

As with any graphical method, the solutions are approximations within the limits of accuracy of the graph. The points of intersection are, as nearly as we can measure, ($^-1{\cdot}9$, $^-6{\cdot}8$), ($0{\cdot}34$, $0{\cdot}03$) and ($1{\cdot}5$, $3{\cdot}4$) and the alternative pairs of solutions are

$$x = {}^-1{\cdot}9,\quad y = {}^-6{\cdot}8;$$
$$x = 0{\cdot}34,\quad y = 0{\cdot}03;$$
$$x = 1{\cdot}5,\quad y = 3{\cdot}4.$$

You can obtain more accurate results by sketching the graph on a larger scale, and plotting more intermediate points.

13.9 Exercise

Solve graphically the simultaneous equations:

$$y = x^2 + 1,$$
$$y + x = 2.$$

[Hint: follow the procedure in the examples above. Draw the graphs of the two equations on the same graph paper with the same axes, and read off the points of intersection.]

13.10 Solving Quadratic Equations by the Graphical Method

Consider now a particular pair of simultaneous equations:

$$y = x^2 + 2x,$$
and
$$y = 2.$$

We draw up the table for $y = x^2 + 2x$ for values of x between $^-3$ and $^+2$.

x	$^-3$	$^-2$	$^-1$	0	$^+1$	$^+2$
x^2	$^+9$	$^+4$	1	0	$^+1$	$^+4$
$2x$	$^-6$	$^-4$	$^-2$	0	$^+2$	$^+4$
$y = x^2 + 2x$	$^+3$	0	$^-1$	0	$^+3$	$^+8$

Drawing the graph of $y = x^2 + 2x$ and, on the same axes, the graph of $y = 2$ (Fig. 13.12) we find the graphs intersect at (approximately) ($^-2{\cdot}75$, 2) and ($0{\cdot}75$, 2).

We can solve other pairs of simultaneous equations which include the same quadratic equation $y = x^2 + 2x$, without needing to plot this equation again.

Examples

1. Solve the simultaneous equations

$$y = x^2 + 2x,$$
$$y = 1.$$

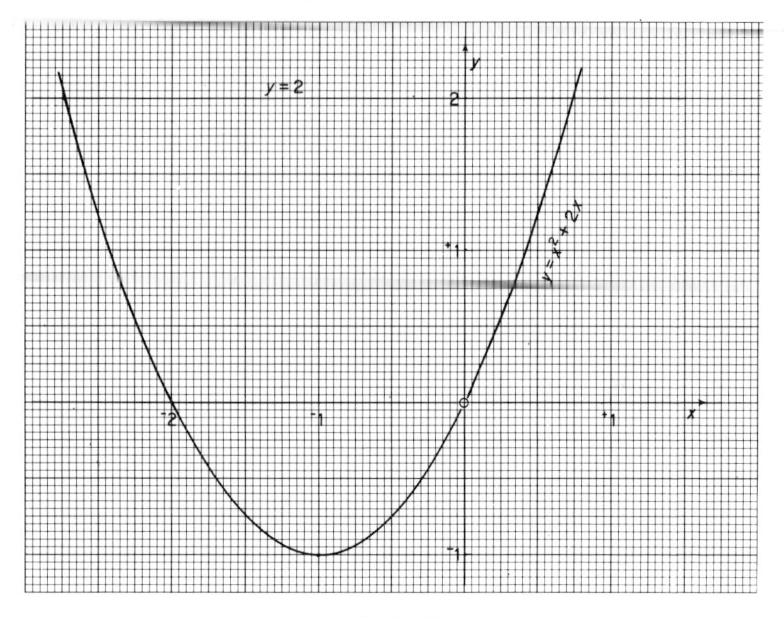

Fig. 13.12

Draw the line $y = 1$ on Fig. 13.12; you can now read off the new points of intersection, $({}^-2{\cdot}4, 1)$ and $(0{\cdot}4, 1)$ (Fig. 13.13) and therefore the solutions:

$$x = {}^-2{\cdot}4,\ y = 1$$

and

$$x = 0{\cdot}4,\quad y = 1.$$

2. Solve the simultaneous equations

$$y = x^2 + 2x,$$
$$y = 0.$$

The line $y = 0$ is the x-axis, and the points of intersection are $({}^-2, 0)$ and $(0, 0)$.

In all these cases, the pairs of simultaneous equations could each have been reduced by substitution to one quadratic equation in one unknown.

The simultaneous equations

$$y = x^2 + 2x \qquad \mathbf{1}$$
$$y = 2 \qquad \mathbf{2}$$

can be reduced to $2 = x^2 + 2x$ (substituting 2 for y in equation **1**)

or $x^2 + 2x - 2 = 0.$

The values of x which satisfy this equation are the x-co-ordinates of the points of intersection, $({}^-2{\cdot}75, 2)$ and $(0{\cdot}75, 2)$. In other words, $x = {}^-2{\cdot}75$ and $x = 0{\cdot}75$ are the solutions of the quadratic equation $x^2 + 2x - 2 = 0$.

Similarly, in Example 1 the values of the x-co-ordinates of the points of intersection are $({}^-2{\cdot}4, 1)$ and $(0{\cdot}4, 1)$ or, in other words, $x = {}^-2{\cdot}4$ and $0{\cdot}4$ are

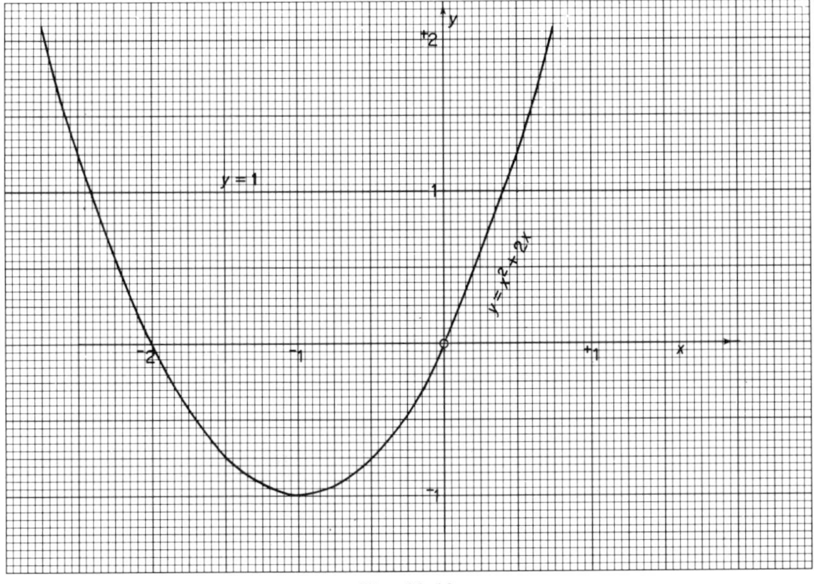

Fig. 13.13

the solutions of the quadratic equation $1 = x^2 + 2x$ or $x^2 + 2x - 1 = 0$.

In Example 2, the points of intersection of the graph of $y = x^2 + 2x$ and the graph of $y = 0$ are the points where the graph of $y = x^2 + 2x$ meets the x-axis. The values of the x-co-ordinates of these two points are the solutions of the quadratic equation $x^2 + 2x = 0$.

Alternatively, if we wish to read off the solutions of the quadratic equation $x^2 + 2x = 2$ from the graph of $y = x^2 + 2x$, we could ask ourselves the question, 'At what point(s) of the graph is the y-co-ordinate 2 and what is the value of x at that point?' or simply, 'For what value of x (on this graph) is the value of y equal to 2?'

Similarly, if we wish to read off the solutions of the equation $x^2 + 2x = 0$ from the graph of $y = x^2 + 2x$, the question then is, 'For what value of x is the y-co-ordinate of the graph zero?'; in other words, 'Where does the graph meet the x-axis, and what is (are) the value(s) of x at this (these) point(s)?'

Notice that the graph we are using is *not* the graph of $x^2 + 2x = 0$ but the graph of $y = x^2 + 2x$; that is, the set of all points representing number pairs (x, y) to which the relation $y = x^2 + 2x$ applies. This graph can be used to solve any equation in this form, such as $x^2 + 2x = 0$, or $x^2 + 2x = 3$. We simply find the appropriate two points (if there are two solutions) of the graph.

Note. If the quadratic equations $x^2 + 2x - 2 = 0$, $x^2 + 2x - 1 = 0$, $x^2 + 2x = 0$ were solved by the methods described in Unit 11, the same solutions would of course be obtained, within the limits of accuracy of the graph.

13.11 Exercises

1. Complete the table and plot the graph of $y = x^2 - x + 2$ from $x = {}^-3$ to $x = {}^+3$.

x	$^-3$	$^-2$	$^-1$	0	1	2	3
x^2		4	1			4	
^-x		2	1				
$^+2$		2	2				
y		8					

On the same diagram draw the graph of $y = 8$.
Use these graphs to solve the quadratic equation $x^2 - x + 2 = 8$.

2. Plot the graph of $y = x^2 - 2x - 5$ from $x = {}^-2$ to $x = {}^+4$.
   (*a*) Read off from this graph the values of x for which $y = {}^-1$.
   (*b*) For what values of x is $x^2 - 2x - 5$ equal to zero? (in other words, solve the equation $x^2 - 2x - 5 = 0$.)
   (*c*) Can you use this graph to solve the equation $x^2 - 2x - 3 = 0$? [Hint: $x^2 - 2x - 3 = 0 \Rightarrow x^2 - 2x - 5 = {}^-2$.]

13.12 Using a Graph to Solve Other Quadratic Equations

In Section 13.10, and again in the last exercise, we used a graph of a quadratic function to solve quadratic equations in which the constant term was different from that of the original function. For example, in the last exercise we used the graph of $y = x^2 - 2x - 5$ to solve the equation $x^2 - 2x - 3 = 0$.

Now we will use this same graph, $y = x^2 - 2x - 5$, to solve quadratic equations in which the term in x is also different, for example, to solve the equation $x^2 - 3x - 6 = 0$.

Using the balancing principle of equations, the left-hand side of this equation can be changed to $x^2 - 2x - 5$ by adding $x + 1$ to both sides:

$$x^2 - 3x - 6 = 0 \Rightarrow x^2 - 2x - 5 = x + 1.$$

Solving the quadratic equation $x^2 - 3x - 6 = 0$ is therefore equivalent to solving the two simultaneous equations

$$y = x^2 - 2x - 5$$
$$y = x + 1$$

(see Section 13.8).

They are solved graphically, as usual, by plotting both graphs on the same diagram and finding the point(s) of intersection; but here, we only need to plot the graph of $y = x + 1$ on the diagram on which $y = x^2 - 2x - 5$ was sketched in Exercise 2 in Section 13.11.

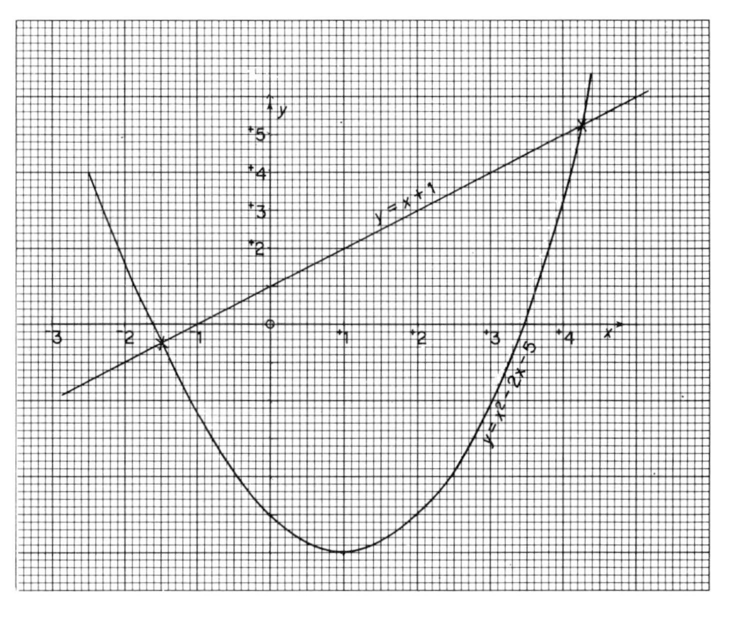

Fig. 13.14

13.13 Exercise

In the same way (by drawing a suitable straight line), find the solutions of the equation $x^2 - 4x - 6 = 0$ from your graph of $y = x^2 - 2x - 5$. [Hint: $x^2 - 4x - 6 + 2x + 1 = x^2 - 2x - 5$.]

13.14 Maxima and Minima

In the previous sections, we used graphs to solve quadratic equations; all our examples concerned one or more points of intersection. For example, we could plot the graph of $y = x^2 - 8x + 15$, and then solve the quadratic equation $x^2 - 8x + 15 = 0$ by asking ourselves the question, 'At what points of the graph is y equal to zero?' When we solve the quadratic equation $x^2 - 8x + 16 = 0$, the question is, 'At what point of the same graph is the value of the y-co-ordinate 1?' ($x^2 - 8x + 15 = 0 \Rightarrow x^2 - 8x + 16 = 1$.)

A graph of a relation between x and y also provides the answer to more general questions such as:

For what values of x is the y-co-ordinate positive?
For what values of x is the y-co-ordinate negative?
What and where is the least value of the y-co-ordinate?

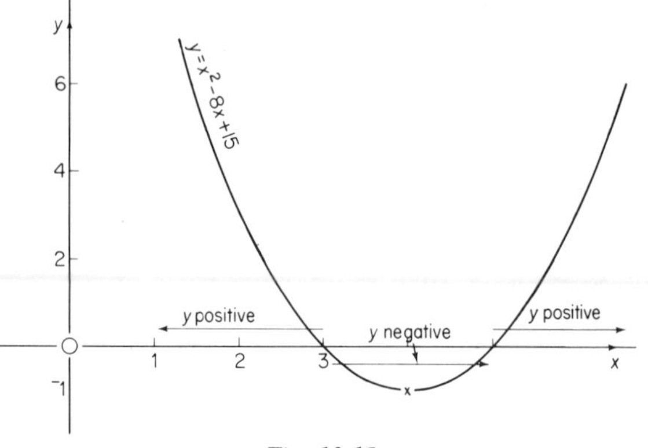

Fig. 13.15

From the graph of $y = x^2 - 8x + 15$ in Fig. 13.15, the range of values of y (or $x^2 - 8x + 15$) can be read off as:

positive: for values of x less than 3 ($x < 3$) and more than 5 ($x > 5$); that is, to the left of $x = 3$ and to the right of $x = 5$, the y-co-ordinates of the graph are positive.

negative: between the values $x = 3$ and $x = 5$ the y-co-ordinates of the graph are negative (below the x-axis).
The statement that '$x > 3$ but $x < 5$' can be written as $3 < x < 5$, that is 3 is less than x but x is less than 5, or x is between 3 and 5.

In this particular graph there is a point which can be called the *minimum*, the point $(4, ^-1)$ where the y-co-ordinate reaches its lowest value ($^-1$). Points on the graph on either side of this point, to the left and to the right, have y-co-ordinates which are more than $^-1$; the further to the left of this point and the further to the right of this point, the greater the value of y. The least value of y (or of $x^2 - 8x + 15$), or the *minimum value* of y, is therefore $^-1$; it reaches this lowest point, the *minimum point*, where $x = 4$.

All graphs which appear U-shaped have such a minimum.

Similarly, graphs which have an 'inverted U-shape', such as the graph of $y = {^-x^2} + 4x - 3$ in Fig. 13.16, are said to have a *maximum point*, a highest point.

At the point (2, 1), the value of y (or of ${^-x^2} + 4x - 3$) is 1; the further to the left or the further to the right of this point, the smaller the value of y. The maximum value of y is 1, and occurs when $x = 2$.

13.15 Exercise

Plot the graph of $y = {^-x^2} + 2x + 4$, from $x = {^-2}$ to $x = {^+4}$ using 2 cm as the unit on the x-axis and 1 cm as the unit on the y-axis. From this graph find:

(*a*) the maximum value of y;
(*b*) the value of x for which $^{-}x^2 + 2x + 4$ has a maximum value;
(*c*) the range of values of x for which $^{-}x^2 + 2x + 4 > 0$.

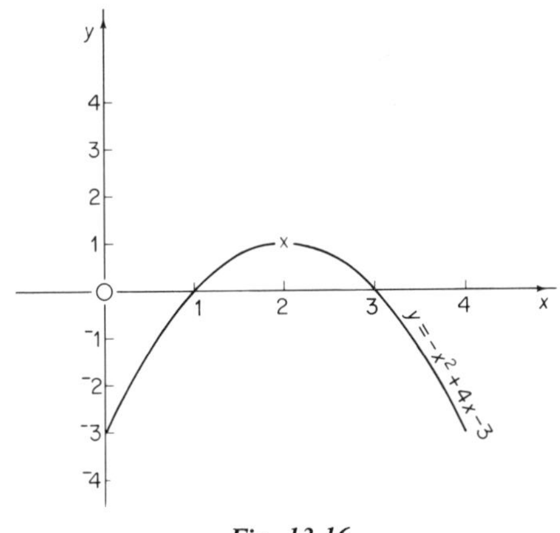

Fig. 13.16

13.16 Summary

We have stressed that equations are mathematical expressions of information and relations. We have concentrated on the techniques of solving various types of equations, and in most cases just called the variables x and y. Remember, however, that these two letters are only 'dummies'; other letters may be used to refer to the two variables which in practical problems stand for two varying quantities.

The exercises which follow all involve expressing some given information as equations. In each case, you should then decide how to interpret the questions that are asked in terms of the variables that have been introduced, or in terms of the graph.

13.17 Exercises

1. One side of a rectangle is 5 cm longer than the adjacent side. Use graphical methods to find the length of each side, if the area of the rectangle is 84 cm^2.

   Now use your graph to find the lengths of the sides of a rectangle whose area is 24 cm^2, and whose larger side is also 5 cm longer than its other side.
2. If a stone is thrown vertically upwards with a velocity of 3·43 m/s, its height (h m) after t seconds is given by the equation $h = 3{\cdot}43t - 0{\cdot}49t^2$. Draw the graph of this relation, taking 2 cm as the unit on the horizontal (t) axis (every 2 cm representing 1 second), and 2 mm as the unit on the

vertical (h) axis (every 2 mm representing 1 m, or every cm representing 5 m).

From your graph find:

(*a*) the highest point the stone will reach, and after how many seconds it will reach this point;
(*b*) after how many seconds the stone will reach a height of 5 m;
(*c*) after how many seconds the stone will reach the ground again, if it is thrown from ground level.

Part Four

Geometry

Unit Fourteen

Lines and Angles

14.1 Introduction

Man has been exploring the properties of shapes since the beginning of civilization, as soon indeed as his need for tools and structures made the understanding of shape a necessity. Levelling devices and other tools used by Egyptian and Babylonian surveyors and builders thousands of years ago show that they were thoroughly familiar with the properties of isosceles and right-angled triangles. The Greeks called this kind of knowledge *geometry*, which means 'Earth measurement', and the name indicates the practical value of the science. The architects of the ancient world applied it to the design and building of their pyramids and temples, and the merchant adventurers of those early days used it to develop a navigation system that could guide them back home from the other side of the ocean.

As a more purely intellectual curiosity developed, the Greek geometers began to interrelate the geometric facts and ask themselves the question 'Why?' From concern with the practical usefulness of the geometric facts, they turned to the reasons for them, and began to relate these facts and reasons into one coherent system. Euclid's attempt to derive all the known geometric properties ultimately from some simple basic facts became the basis and model for all mathematics, and this stood almost unchanged for thousands of years. The limitations of Euclid's geometry did not become clear until the 19th century, when other, non-Euclidean geometries were discovered. Since then, the scope of geometry has greatly expanded; there is now a wide range of branches of mathematics that can be classed as geometry.

14.2 Our Programme

This book will introduce you to geometry as a simple exploration and analysis of shape, following the historical pattern outlined in Section 14.1.

(*a*) Like the early Egyptians, we shall consider the basic familiar shapes, and rediscover the geometric facts, the properties, of shapes.
(*b*) With the curiosity of the Greeks, we will then try to answer the question 'Why?' and establish whether, and how far, these properties are necessarily found in all such shapes. We shall also see how the various properties are interrelated.
(*c*) Then we shall examine how these properties can be used, and how they can be applied in practical constructions.

The examination of shape often consists of comparing two shapes, or two parts of one shape, searching for some *equality*. Two line segments (see Section 14.4), for example, can be compared to see if they have the same length or not.

If two shapes, whether they are two flat shapes or two solids, are identical in every aspect except the positions they occupy, equality is called *congruence*. We speak, for example, of two congruent triangles, two congruent parallelograms; this means that one can be placed on the other so that all their corresponding parts coincide.

When two pairs of intersecting lines can be placed so that the points of intersection coincide as well as the pairs of lines (see Section 14.5), we speak of angles of the same size, or *equal angles*, rather than of 'congruence'.

However, two maps of England, for example, drawn to different scales, are clearly not congruent, but have a particular kind of 'sameness' of shape called *similarity*; the general appearance of two similar shapes is the same, but their sizes may be different.

Sometimes, looking for equality in shape, we find a certain 'sameness' *within* a shape known as *symmetry*.

When a plane symmetrical shape is drawn on paper, the paper can be folded so that the folding line divides the shape into two congruent halves, the one fitting exactly over the other. This shape is then said to have *bilateral symmetry*.

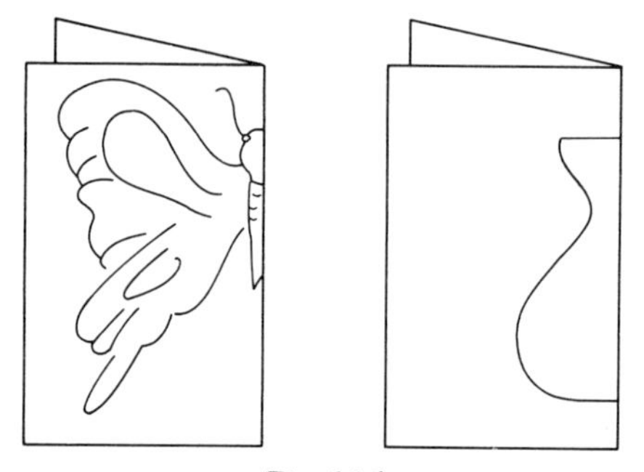

Fig. 14.1

The folding line is the *line of symmetry*, or the *axis of symmetry*. Some shapes have more than one line of symmetry, for example the rectangle (Fig. 14.2).

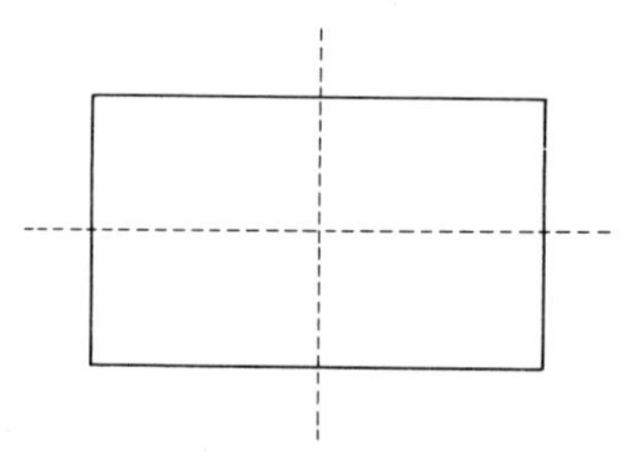

Fig. 14.2

As an exercise, study the symmetry of such shapes as squares, the letters of the alphabet, and so on, by drawing them on tracing paper and folding; establish all their lines of symmetry by folding the paper in different ways.

Alternatively, you can think of bilateral symmetry in terms of mirror images or reflections: one half of the shape is the mirror image or reflection of the other. If you place a mirror on the line of symmetry, the whole of the shape is generated from one half.

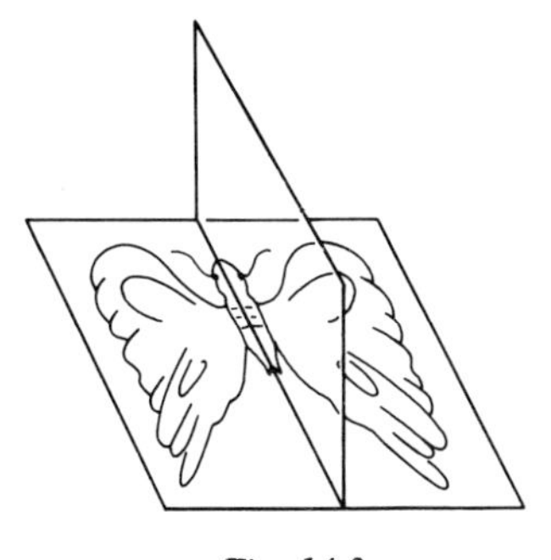

Fig. 14.3

Some shapes have a marked 'sameness' in their composition without having bilateral symmetry at all. They consist of two or more congruent parts and the whole of the shape can be generated from one of these parts, not by folding or reflecting, but by *rotating* the part about a point (Fig. 14.4).

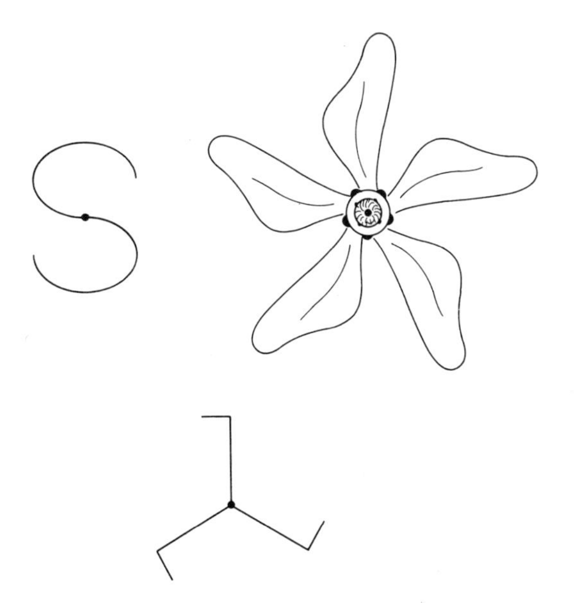

Fig. 14.4

We can call this *cyclic symmetry* or *rotational symmetry* or *point symmetry*.

Even measuring shape (see Units 20 and 21) is a search for equality. When we are measuring the *area* of a flat shape, we are in effect trying to find a number of squares whose total area equals that of the shape; measuring the *volume* of a solid is equating it to the total volume of a number of cubes of a certain fixed size.

The measurement of areas of shapes bounded by curves can only be an approximation; there is a special branch of mathematics that concerns itself with the calculation of these areas and with the 'gradient' of curves. It is called *calculus*; we shall treat it separately in the last part of this book.

Note. *Notation* and *special terms* in geometry are introduced at the appropriate places.

The main geometric facts, that is, the results of our investigations, are called *theorems*, and are all listed at the end of Unit 21—*Summary of Theorems*.

We will discuss *geometric constructions* together with the theorems on which they are based; these also are listed at the end of Unit 21—*Summary of Constructions*.

14.3 Points and Lines

Medieval theologians debated the question 'How many angels can stand on the point of a needle?' The point of a needle was the smallest space they could imagine, and angels were considered not to occupy any space at all, although at any time they were located at a particular position in the universe.

In a more mathematical context, Euclid defined a point as 'that which has no size', emphasizing that points have no length, no area and no volume. We cannot, then, imagine a point, and certainly cannot draw one; we do not need a magnifying glass to see that the points we draw always cover some area, and that the lines we draw always have a certain thickness. We can keep the imperfections of our drawings to a minimum by using sharp pencils and by indicating points by intersections of two strokes rather than by blobs; this is especially important when making a geometrical construction.

However, a point within a given geometric figure has *position*; it is located at a definite place. Many geometric constructions are just methods of finding the exact location of a point.

Notation. By mathematical convention, points are named or referred to by capital letters, such as the point A or the point P.

14.4 Lines and Line Segments

For ordinary purposes we need not trouble ourselves with the philosophical question of the impossibility of drawing a perfect straight line of no thickness. Aware of our limitations, we start from the straight line we know, and accept its property that on a perfectly flat surface, which we call a *plane*, it joins two points along the shortest route between them. A *line* will always be understood to be a *straight line*, unless it is specifically mentioned otherwise. All lines that we draw are in fact *line segments*, pieces of lines of given lengths and with two definite ends: we have to stop somewhere even if only because we reach the edge of our paper. (Compare also the restrictions of graph paper, mentioned in Section 12.14.) When we want to indicate that line segments are intended, we mark the end-points by two short strokes, as in Fig. 14.5.

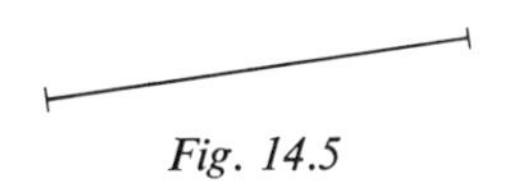

Fig. 14.5

Lines which have one definite end-point but which are unrestricted in the other direction are called *half-lines*.

When the mathematician says that a *line* has no end, or even that it is *infinitely* long, he is not arrogantly assuming divine powers. He is simply saying that no restriction has been set, that no matter how far one proceeds along the line, one can always go further.

A line drawn on a flat surface has a certain position on that plane, and has a certain direction.

Two lines are different when they occupy different positions on the plane. When *two different lines* are drawn on the same plane they either meet or *intersect*, that is have one point in common, or they do not intersect. If they do not intersect, however far they are extended, they are said to be *parallel*: they have the same direction.

We indicate in a drawing that two (or more) lines are parallel by marking each of the lines with an arrow. If a different pair of parallel lines is introduced into the same drawing, we can use double-headed arrows (Fig. 14.6).

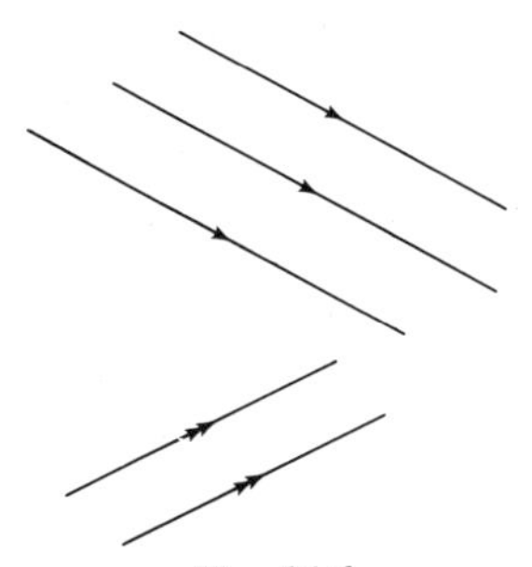

Fig. 14.6

Notation. A line is sometimes referred to by a single lower-case letter, written alongside the line in the diagram; for example, the line a, the line l in Fig. 14.7.

Fig. 14.7

Alternatively we can name the line after two of the points which lie on it, especially if in the diagram these points are the end-points of a line segment. For example, if P and Q are points lying on a line, we speak of the line PQ; if A and B are the end-points of a line segment, we speak of the line segment, or sometimes colloquially the line, AB.

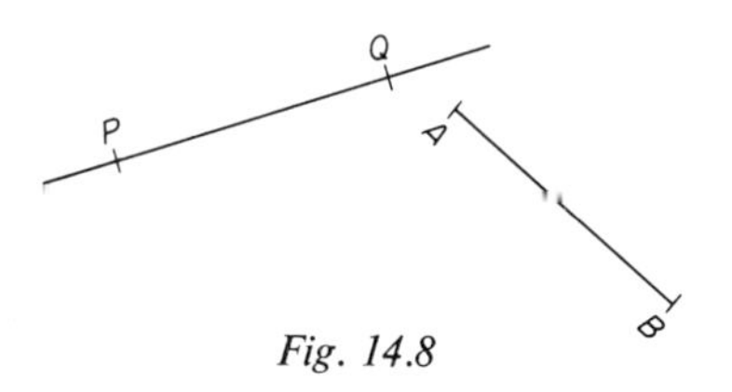

Fig. 14.8

14.5 Angles

When we examine two intersecting lines—let us call them l and m—we concentrate first on the open-ended shape formed by the two half-lines intersecting at their end-points, the *point of intersection.*

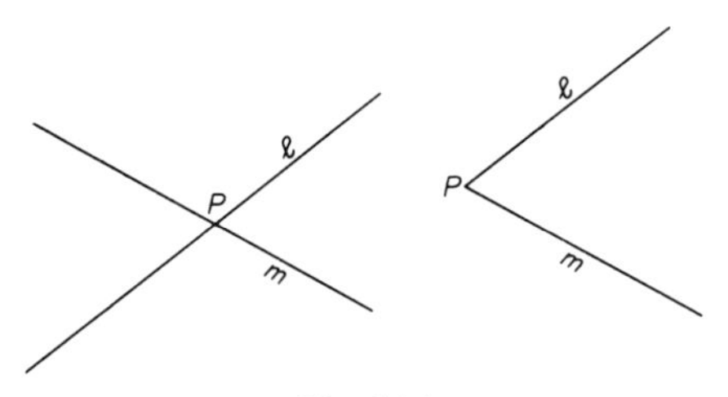

Fig. 14.9

Since the lines l and m intersect, the direction of the line l is different from that of m. This difference of direction is what we refer to when we speak of the *angle between two lines*, not the space or area between the lines, which proceeds ad infinitum. You will find that differences of direction between pairs of intersecting lines vary widely. The direction of the line a in Fig. 14.10, for example, is almost that of b; the difference between their directions is smaller than the difference in the directions of lines l and m.

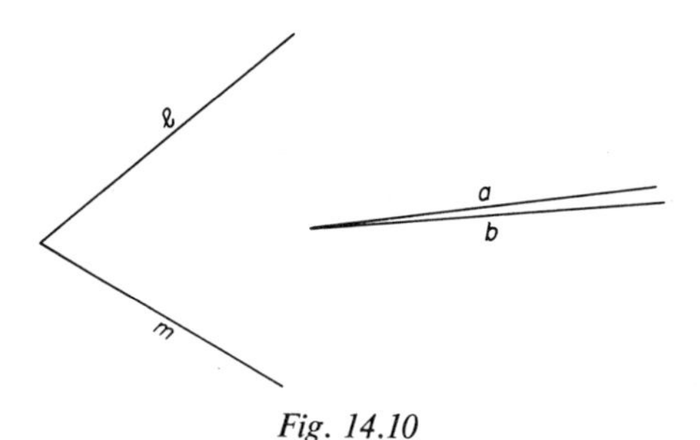

Fig. 14.10

There are in fact innumerable possible differences of direction—*angles*—between pairs of half-lines. You can demonstrate this by joining two meccano strips (two strips of card will do), each strip representing a half-line and the join representing the point of intersection (Fig. 14.11).

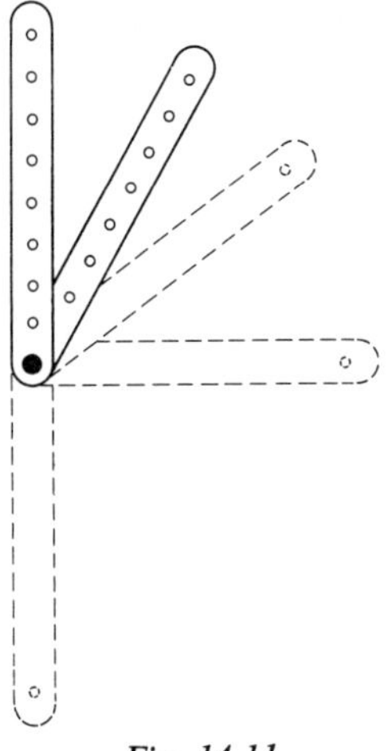

Fig. 14.11

Move the ends of the strips a little wider apart; the difference between the directions of the strips is greater than before. You can repeat this process again and again; indeed, you can think of an angle between two lines as the result of turning one line from an initial position at which it completely overlaps the other. The size of an angle is then measured by the extent of the turn or 'rotation' which this line has made from its initial position.

You can illustrate this in a little more detail by fitting a strip of meccano, or stiff paper or wire, on a card, as the hand of a clock is fixed to a clock face (Fig. 14.12).

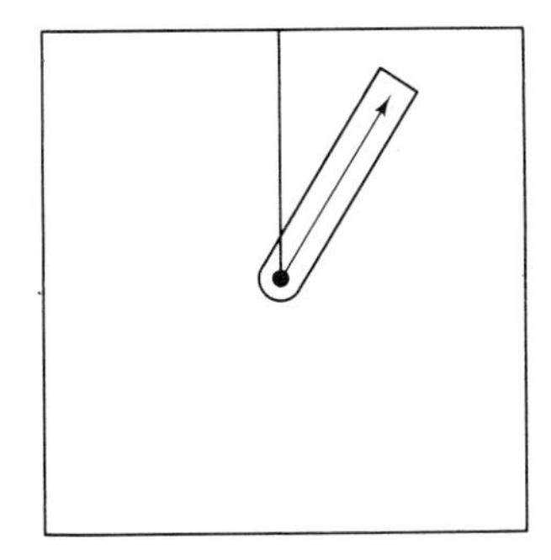

Fig. 14.12

Mark the initial position by drawing a half-line from a chosen point near the centre of the card. Fit one end of the movable arm or pointer at this point to represent the second half-line.

Place the pointer so that it coincides with the line drawn on the card—the 'twelve o'clock position'; then turn it in a clockwise direction. After one complete turn the pointer again covers the line on the card.

All angles can be thought of as resulting from a turn like this, or from a fraction of a complete turn (Fig. 14.13).

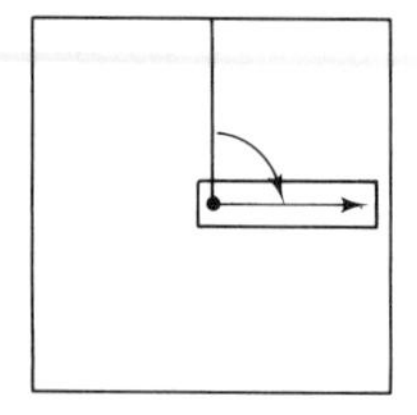

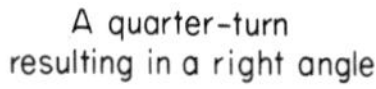

A quarter-turn resulting in a right angle

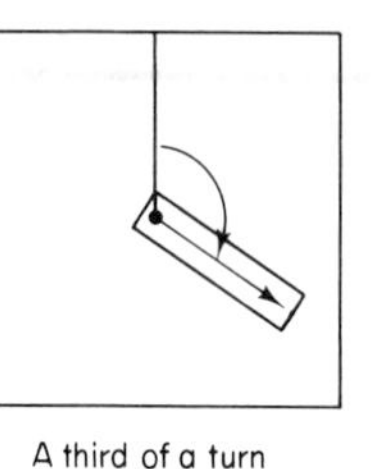

A third of a turn

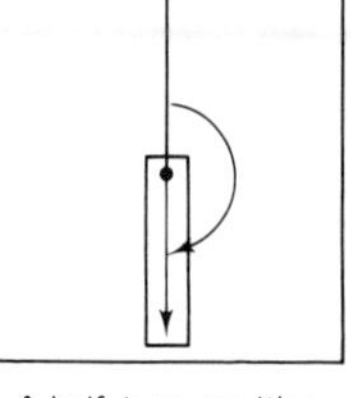

A half-turn, resulting in a straight line

Fig. 14.13

After a quarter of a turn from the initial position we arrive at a familiar angle, the *right angle*, where the direction of one line is said to be *perpendicular* to that of the other; after a half-turn, the two half-lines form one straight line.

Usually the words 'the angle between two half-lines l and m' mean 'the smallest angle between l and m', which is generated by making the turn by the shortest route (Fig. 14.14). The largest angle possible using this convention is then the half-turn, when the two lines l and m form a straight line.

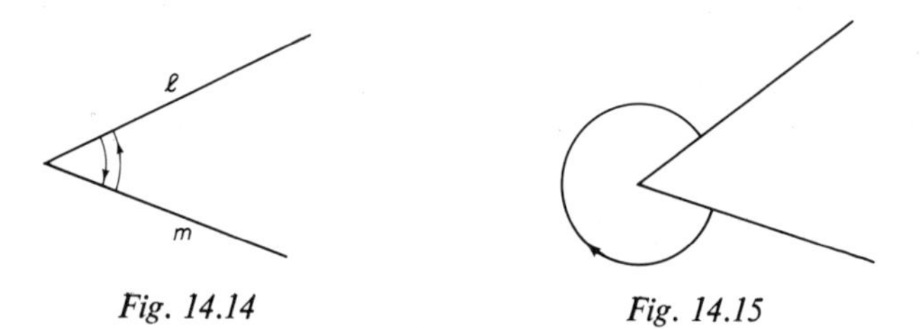

Fig. 14.14 *Fig. 14.15*

Sometimes, however, the larger angle has to be considered, that is, the angle measured by the longer route (Fig. 14.15).

It is even theoretically possible to think of angles greater than one complete turn: 2 rotations, $1\frac{1}{2}$ rotations, and so on.

Notation and terms. If we need to indicate a particular angle in a diagram, we can draw an arc between the appropriate lines (Fig. 14.16).

A quarter-turn, or right angle, is indicated by two perpendicular strokes (Fig. 14.17).

Fig. 14.16 *Fig. 14.17*

Angles are often named by referring to three points: the point of intersection of the two half-lines (the *vertex* of the angle), and a point on each of the lines; for example, in Fig. 14.18 the angle is named the angle ABC, written $\angle ABC$, or, equally validly, the angle CBA, written $\angle CBA$. (You may encounter another notation, $A\hat{B}C$ or $C\hat{B}A$, but we have not used it in this book.)

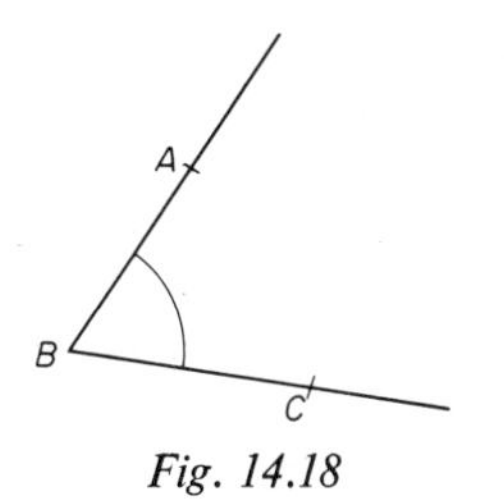

Fig. 14.18

The 'equals' sign (=) is used to indicate that two angles are of the same size; if, for example, $\angle ABC$ is of the same size as $\angle CAB$ we write '$\angle ABC = \angle CAB$'. The half-lines forming the angle are often referred to as the *sides* or *arms* of the angle. An angle smaller than a right angle is called an *acute angle*, an angle greater than a right angle and less than two right angles is called an *obtuse angle*, and an angle greater than two right angles and less than four right angles is called a *reflex angle*.

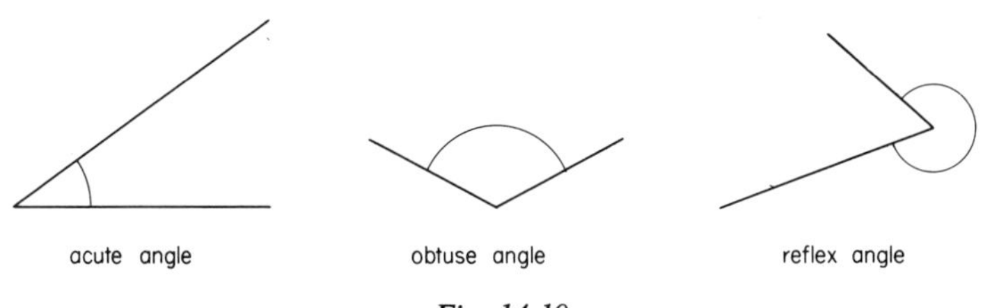

Fig. 14.19

14.6 Measuring Angles

The card with pointer (see Section 14.5) can be used as an instrument for measuring angles, if we add a scale on which the possible sub-divisions of a turn are indicated by marks, rather like the face of a clock (Fig. 14.20).

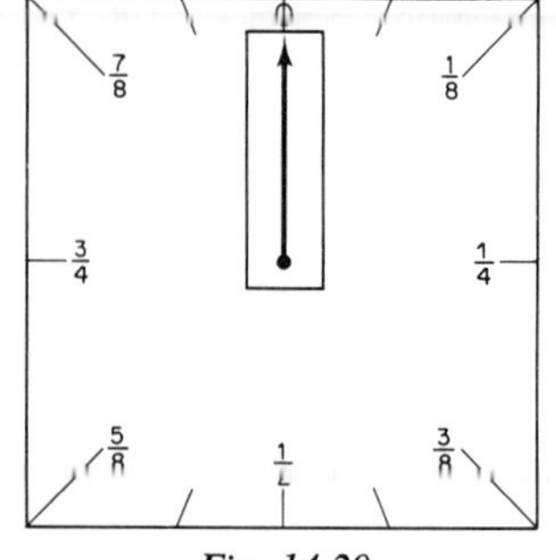

Fig. 14.20

As usual, we have to choose a unit of measurement, and a system of arbitrary sub-divisions of a turn is universally accepted. A complete turn is divided into 360 equal parts, called *degrees* (360 degrees is written 360°). For very precise measurements, each degree is sub-divided into 60 *minutes*, written 60′, and each minute into 60 *seconds*, written 60″. For example, an angle which measures 25 degrees, 40 minutes and 30 seconds is written as 25° 40′ 30″.

A right angle, a quarter-turn, is therefore an angle which measures 90°, or simply an angle of 90°; a straight line includes an angle of 180°; a three-quarter turn is an angle of 270°, and so on.

You will need a *protractor* for measuring angles; this is a flat circular object, made of transparent plastic with markings for degrees.

If you fix a metal pointer to the centre of a circular protractor (for example, a hair clip through the hole at the centre), you have an excellent instrument to practise measuring angles. (Such an instrument can be bought commercially.) If you use an ordinary circular protractor, you will find that the 'initial-position line' is marked by 0° or 360°.

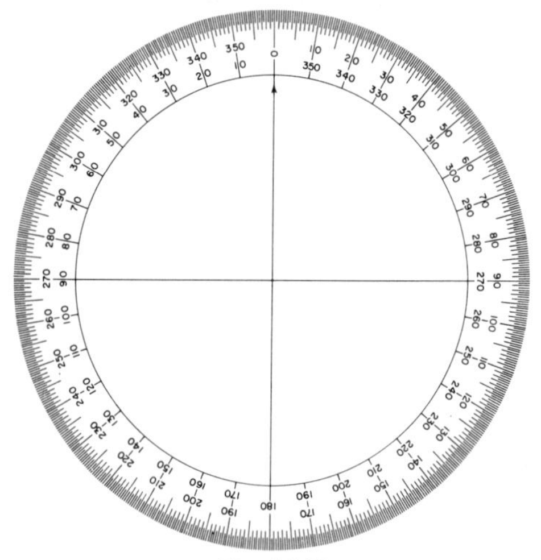

Fig. 14.21

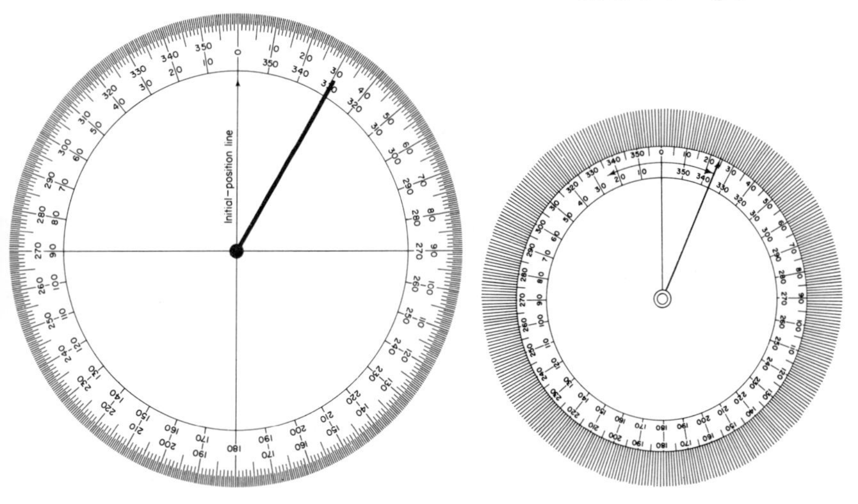

Fig. 14.22

To measure an angle between two half-lines:

(*a*) place the protractor over the angle to be measured so that the centre of the protractor lies as exactly as possible over the vertex of the angle, and the 'initial-position line' is over one of the two lines;

(*b*) turn the pointer until it covers the second line and read off the size of the angle from the scale. Most protractors have two scales, one for clockwise rotation and one for anti-clockwise rotation. Fig. 14.23 illustrates the measurement of an angle of 120°.

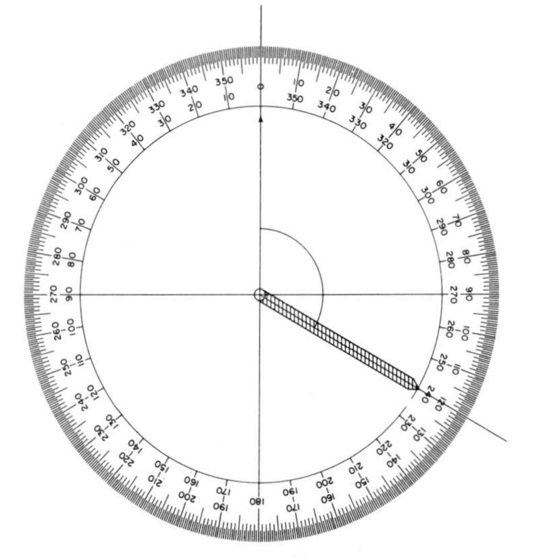

Fig. 14.23

After some practice with the angle indicator, you will be able to dispense with the movable arrow and should use the protractor alone to measure angles. You should use a similar method as with the angle indicator:

(*a*) place the protractor over the angle to be measured so that the centre of the protractor coincides with the vertex of the angle and the 'initial-position line' (marked as 0° or 360°) covers one of the two half-lines;

(*b*) find the angle, measured in degrees, by reading the number on the mark where the second half-line crosses the scale. (You may need to extend the half-lines forming the angle, if they are not drawn sufficiently far.)

To draw an angle of a required size, for example of 72°:

(*a*) draw a half-line, marking clearly the point which is to become the vertex of the angle;

(*b*) place the protractor over this line so that the centre of the protractor coincides with the mark for the vertex and the initial-position line of the protractor covers the half-line (that is, so that it crosses the scale at the zero mark or 360°);

(*c*) mark on the paper the position where the protractor indicates the required angle (72°, in this case);

(*d*) remove the protractor and join this mark to the vertex.

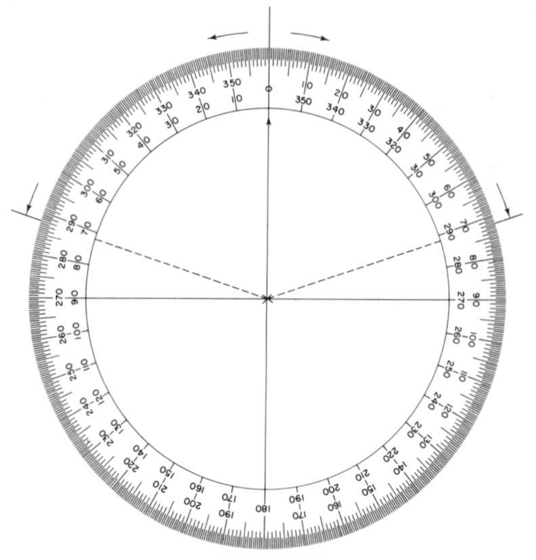

Fig. 14.24

Fig. 14.24 shows that there are two possible positions of the second line, since we can use either the clockwise scale or the anti-clockwise scale.

Note. You will find many kinds of protractors in the shops. A circular instrument is more useful than a semicircular one, especially in the work in Section 14.8 on bearings.

14.7 Exercises

1. Draw some pairs of intersecting lines and measure the angles formed by the half-lines. Write down the size of these angles in degrees, using the conventional notation. (Ordinary protractors only allow measurements to be made to the nearest degree.)
   Also measure the angles found on surfaces of everyday articles such as books, designs on textiles or carpets, tiles of various shapes, and so on.
2. Using a protractor, draw angles of 30°, 45°, 60°, 72° and 120°.

14.8 Bearings

The simplest method of showing someone the way is often to raise an arm and point a finger in the appropriate direction. If the inquirer proceeds far enough along the line indicated by the arm or finger, he will arrive at his destination.

Navigation on the high seas or in the air involves finding and following the right direction, the line between one's present position and the point of destination. (In this section we shall assume that the Earth is flat!)

As long as we have a compass, there is always one direction that can be established almost everywhere on Earth: the direction *due north*. If the line which points due north is once fixed, the direction of any other line on the surface of the Earth can be given as the angle which it makes with the north-pointing line; this angle is called a *bearing*.

Imagine that a person pointing his right arm in the direction of a village—let us call it A—also has a compass available and at the same time points his left arm in the direction due north. His two arms will form an angle which is the *bearing of the village A* from the point where he is standing.

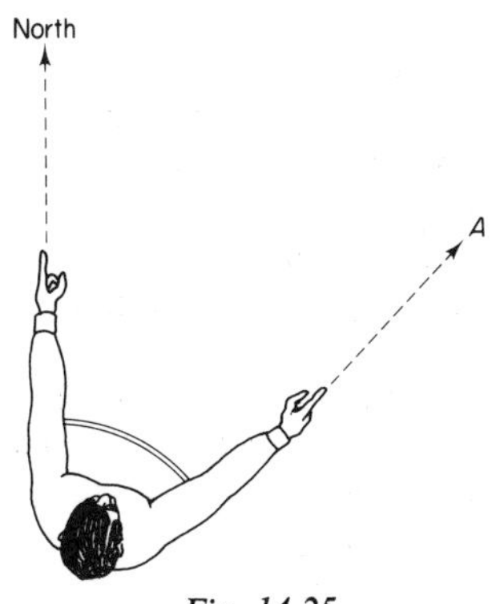

Fig. 14.25

To avoid confusion, we adopt the convention that the bearing is to be measured from the line due north in a *clockwise direction*.

The direction to be followed to reach one's destination depends entirely, of course, on one's starting point. Fig. 14.26 shows that the bearing of the village A is 60° when taken by an observer at P, but 290° when taken from point Q; we speak of the bearing of A from P, and the bearing of A from Q.

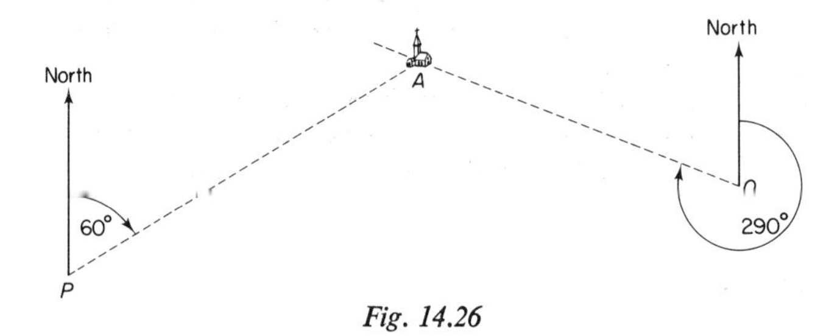

Fig. 14.26

The bearing of a point A from a point P is therefore the angle between the line due north through P and the line AP, measured in a clockwise direction starting from the line due north.

Again by convention, this angle is always given in three figures. For angles smaller than 100°, zeros are added to the left; for example, a bearing of 60° is written as 060°, and a bearing of 7° as 007°. Given in this form, a bearing is called a *three-figure bearing*.

14.9 Exercise

You will understand the idea of a bearing best if you actually take some bearings. Sophisticated instruments are necessary for accurate navigation. But rough bearings can be taken using only a circular protractor, a compass and a piece of string or wire.

Fit a string through the hole at the centre of your protractor (make a hole if there is not one already). Place the protractor on a table—out of doors if possible—and put the compass on the centre of the protractor. Now turn the protractor until the initial-position line points in the same direction as the needle of the compass—that is, due north. Choose a suitable object for your observations, such as a church steeple, a tall tree or a factory chimney, in the distance.

Without moving the protractor, align the string so that it points in the direction of the object you have chosen. You can do this fairly accurately by looking along the line of the string, keeping your eye at the level of the table, while an assistant moves the string according to your directions. You can now read off the bearing from the point where the string crosses the scale (Fig. 14.27).

You have now taken the bearing *of* the church steeple, chimney etc., *from* the point where you were standing—or, more accurately, from the point where the centre of your protractor was located.

Take the bearings of objects lying in various directions.

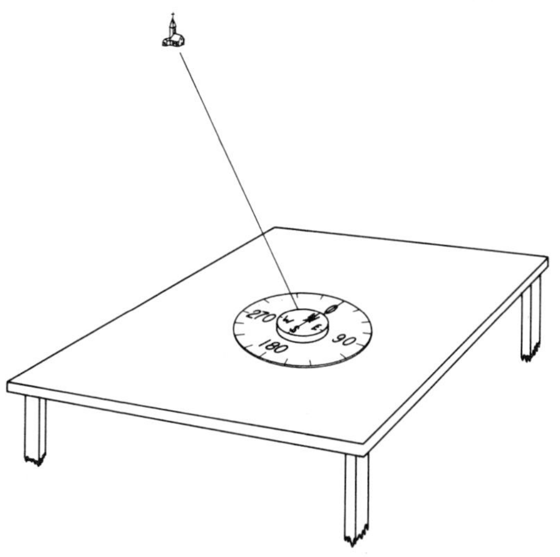

Fig. 14.27

14.10 Navigation

The simpler methods of navigation for ships and aircraft have now been superseded by more sophisticated methods like radar. But these modern methods are also based on a simple principle: the position of a certain point is made known to the navigator by means of a beam of light from a lighthouse or an electronic beam produced by a special device, so that the navigator can detect the direction and take a bearing of this lighthouse or point. Every lighthouse or station sends out its own individual signal, which can be identified by the navigator. Using the bearing of the lighthouse found in this way, he can now draw a line on his chart through the point marking the lighthouse, in the same direction as the bearing. He still does not know how far his ship is from the lighthouse, but he knows that the position of his ship on the chart is somewhere along the line he has drawn.

If at the same time the navigator can take the bearing of another, different point he can then draw a second line on the chart, using the same method. Since the ship must be somewhere along each line, the exact position is given by the only point that lies on both lines, the point of intersection.

Similarly, the exact position of a ship can be established by two observers from two different positions on the shore.

Example

A ship is sighted off the coast by an observer from a point A on a bearing of 140° and simultaneously by another observer from a point B on a bearing of 205°.

The positions of the points A and B are marked on the chart in Fig. 14.28, as well as the direction of the line due north.

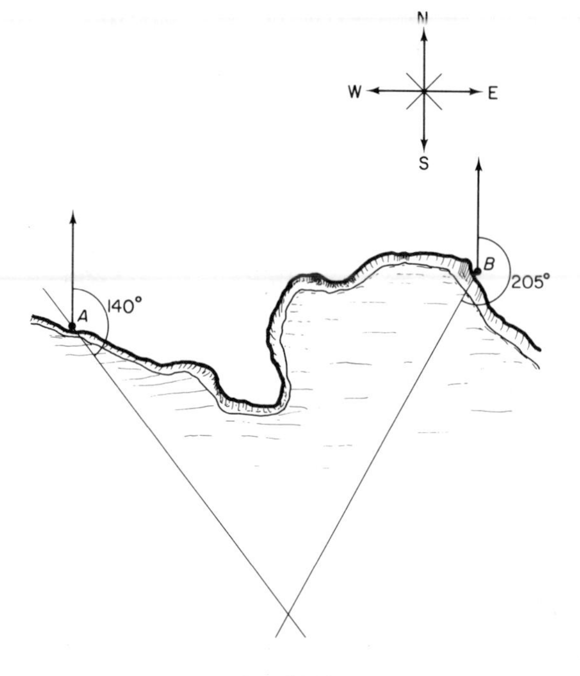

Fig. 14.28

The direction north is indicated on the chart, so that we can draw a line through point A which marks the bearing of 140°, and another through point B marking the bearing of 205°. The position of the ship is now indicated by the point where the two lines intersect.

14.11 Exercise

Try to establish the position of your home on a large-scale map of the area where you live, by taking bearings of two landmarks which are both visible in the distance and recorded on the map.

14.12 Angles at a Point

You will remember that the angle formed by two half-lines can be measured in two different ways, that indeed there are two angles between a pair of lines (Section 14.5). In Fig. 14.29, for example, the angle between the lines l and m, measured from the line l in a clockwise direction, is 45°, while the angle from the line m to the line l in a clockwise direction is 315°.

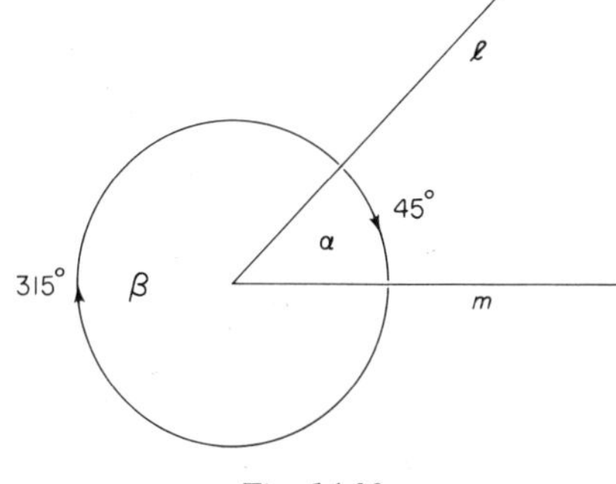

Fig. 14.29

As we measured first one angle and then the other, we made a complete turn; a part-turn of 45° followed by a part-turn of 315° in the same direction is equivalent to a complete turn or rotation.

We can interpret this phrase 'followed by' mathematically, and speak of the *addition of angles*; the two different angles between two half-lines *add up to* 360° or, using mathematical notation,

$$\angle\alpha + \angle\beta = 360°.$$

The same situation arises if three half-lines meet at a point (Fig. 14.30).

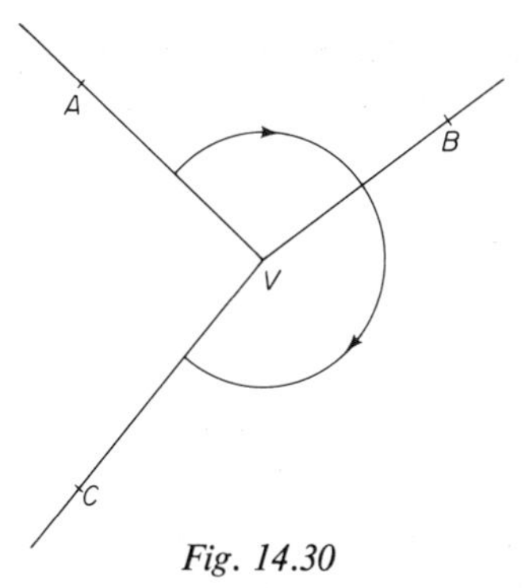

Fig. 14.30

The two angles $\angle AVB$ and $\angle BVC$ have the same, or a common, vertex, the point V, as well as sharing the line VB; they are *adjacent angles* (i.e. neighbouring or side-by-side angles).

Measuring the angle AVB from the line VA to the line VB in a clockwise direction, and then measuring the angle BVC in a clockwise direction from the line VB, we have turned through the reflex angle AVC, or:

$$\angle AVB + \angle BVC = \text{reflex } \angle AVC.$$

Now we can complete the rotation (up to 360°) by measuring the angle CVA in a clockwise direction starting from the line VC, or

$$\angle AVB + \angle BVC + \angle CVA = 360°.$$

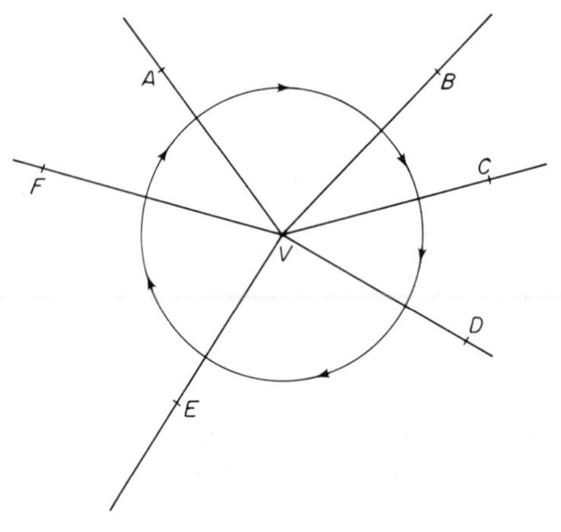

Fig. 14.31

We could add any number of further half-lines, all meeting at V; adding all these angles in the same way, we complete one full turn (Fig. 14.31).

We can express this result by saying,

the sum of the angles at a point is 360°

Terms. Two angles which add up to 180° are called *supplementary angles* (they supplement each other to form a straight line) (Fig. 14.32).

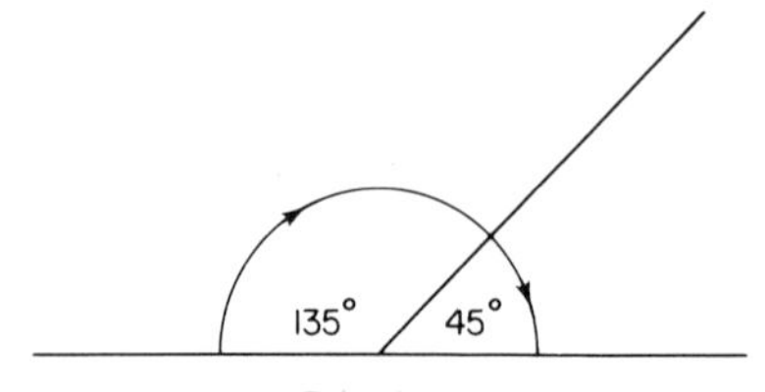

Fig. 14.32

For example, an angle of 135° is the supplement, or supplementary angle, of 45°; similarly, an angle of 45° is the supplement of an angle of 135°.

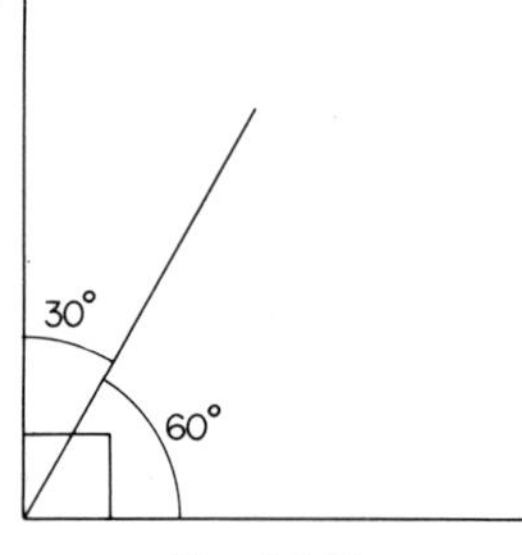

Fig. 14.33

Two angles whose sum is 90° are called *complementary angles* (they complement each other to form a right angle) (Fig. 14.33). For example, an angle of 30° is the complement, or the complementary angle, of 60°, and vice versa.

14.13 Two Intersecting Lines

When two lines intersect they form four angles. We can think of this as a special case of four half-lines meeting at a point, special because two of the pairs of half-lines form straight lines.

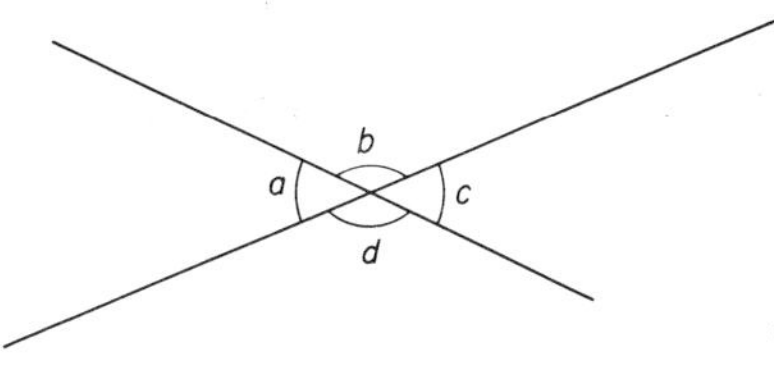

Fig. 14.34

This means that, in Fig. 14.34, $\angle a + \angle b = 180°$, or that angle a is the supplement of angle b, or $\angle a = 180° - \angle b$.

It also follows that angle c is the supplement of angle b, or $\angle c = 180° - \angle b$.

Therefore $\angle a$ is the same size as $\angle c$, or $\angle a = \angle c$.

In the same way, we can show that $\angle b = \angle d$.

Because of their position in this 'cross' situation, the angles a and c are said to be *vertically opposite angles*. Similarly, b and d are vertically opposite angles.

Using these terms, we can say that

> *when two straight lines intersect, the vertically opposite angles are equal*

14.14 Exercises

1. One of the angles formed by two intersecting lines is 45°; find the size of the other three angles.
2. Use a set-square with angles of 90°, 60° and 30°, and a ruler to construct an angle of 120° and an angle of 150°.

Note. A *set-square* is a rigid frame of which two sides form a right angle. Commercial set-squares are normally triangular in shape and made of wood or plastic; some are in the shape of a triangle of which the angles are 90°, 60° and 30°, others have angles of 90°, 45° and 45° (Fig. 14.35).

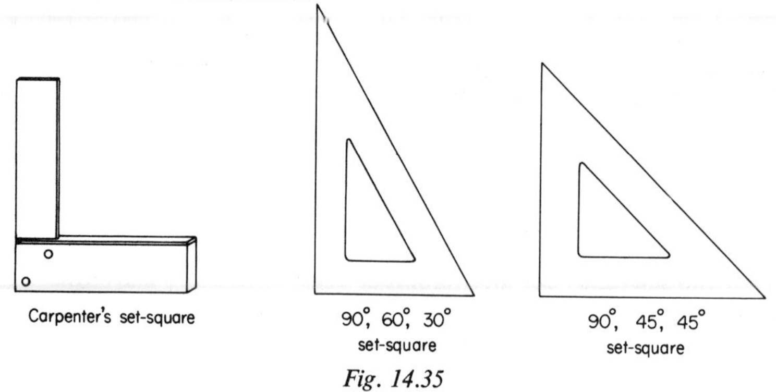

Fig. 14.35

3. Construct an angle of 135°, using a ruler and a set-square with angles of 90°, 45° and 45°.

14.15 Angle Bisectors

The line which divides an angle into two angles of equal size is called the *bisector* of the angle, or the *angle bisector*.

The line l in Fig. 14.36 bisects the angle ABD (= 90°); angles ABC and CBD are both 45°,

$$\angle ABC = \angle CBD = 45°.$$

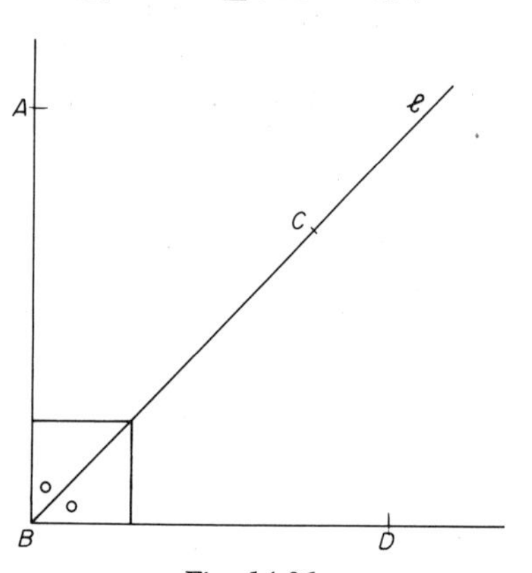

Fig. 14.36

(Equal angles can be indicated in diagrams by putting a mark, such as ○ or ×, or a double arc as in Fig. 14.37, between the sides of the angles.)

A construction of the angle bisector with compasses and ruler will be discussed in Section 16.11. The angle bisector can be found very roughly by drawing the angle on transparent paper, and folding the paper so that the sides of the angle coincide (Fig. 14.37).

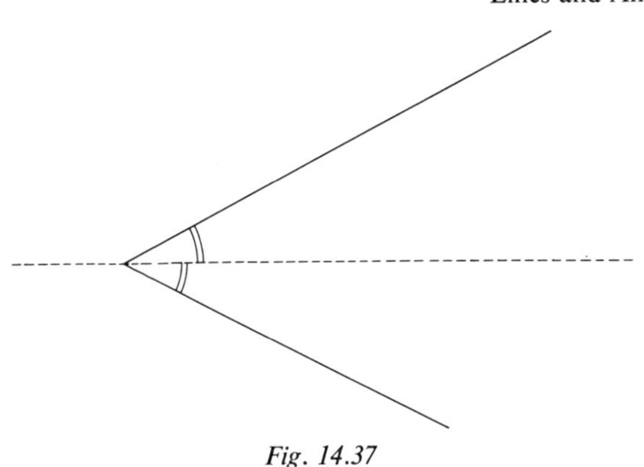

Fig. 14.37

The line of the fold makes an angle with each of the sides of the original angle, and clearly these two angles are equal; so the line of the fold is the bisector of the angle.

14.16 Exercises

1. Draw an angle on a transparent sheet of paper, and find its bisector by the folding method.
2. Use the straight edge of a sheet of paper as a line; 'construct' an angle of 90° by folding, then by folding again, 'construct' an angle of 45°.
3. Draw two intersecting lines on transparent paper.
   (*a*) Fold the paper so that the line of the fold passes through the point of intersection, and the one pair of half-lines covers the other pair. The line of the fold bisects two angles. Show that the four angles so formed are all equal in size.
   (*b*) By folding the sheet of paper again find the bisector of the other pair of vertically opposite angles. Measure the angle between the angle bisectors, using a protractor.

14.17 Bisectors of Supplementary Angles

Theorem 14.1: *The bisectors of the angles formed by two intersecting lines are perpendicular to one another.*

(This is an expression of the results of Exercise 3(*b*) in Section 14.16.)

If two lines l and m and their bisectors form angles A_1, A_2, . . ., A_8 as marked in Fig. 14.38:

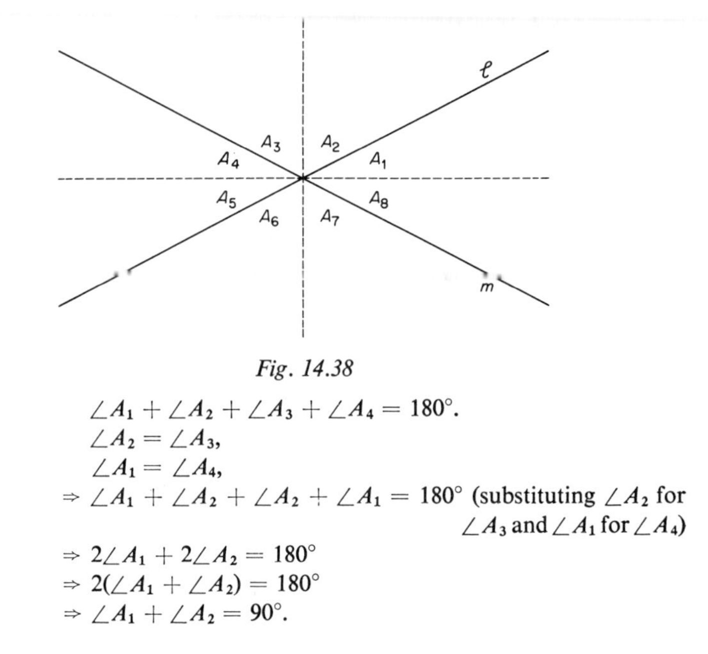

Fig. 14.38

$$\angle A_1 + \angle A_2 + \angle A_3 + \angle A_4 = 180°.$$
$$\angle A_2 = \angle A_3,$$
$$\angle A_1 = \angle A_4,$$
$\Rightarrow \angle A_1 + \angle A_2 + \angle A_2 + \angle A_1 = 180°$ (substituting $\angle A_2$ for $\angle A_3$ and $\angle A_1$ for $\angle A_4$)

$$\Rightarrow 2\angle A_1 + 2\angle A_2 = 180°$$
$$\Rightarrow 2(\angle A_1 + \angle A_2) = 180°$$
$$\Rightarrow \angle A_1 + \angle A_2 = 90°.$$

14.18 Lines Intersecting Two Parallel Lines

If a line is drawn to cross a pair of parallel lines, the situations which appear at the intersections look remarkably similar. Indeed, if these lines are drawn on tracing paper and the paper is cut in two between the parallel lines, the fragments can be so placed that the pairs of lines completely coincide.

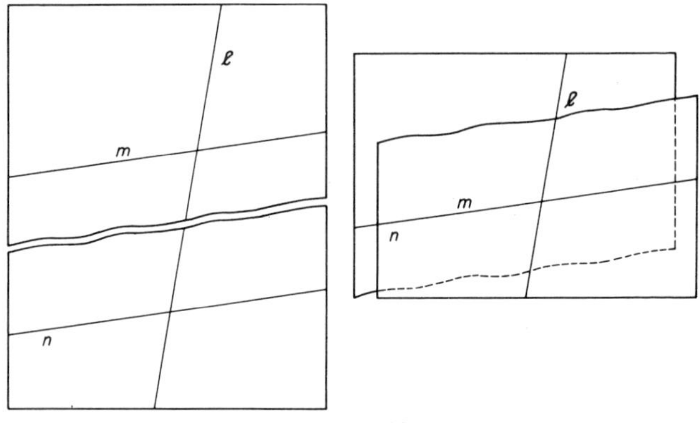

Fig. 14.39

The part of line l on the upper portion coincides with the part of line l on the lower portion and the lines m and n coincide. The angles between these lines are evidently equal.

This discovery should not surprise you. We defined parallel lines as lines which have the same direction; the difference between each of their directions and the direction of a third line will clearly be the same.

Terms. Some special names are used for the angles formed by parallel lines.

The pairs of angles which coincide in our experiment above are called *corresponding angles*; these are the angles which occupy similar positions at the two intersections. For example, the angles marked a and e in Fig. 14.40 are corresponding angles; so are the angles marked b and f, and so on.

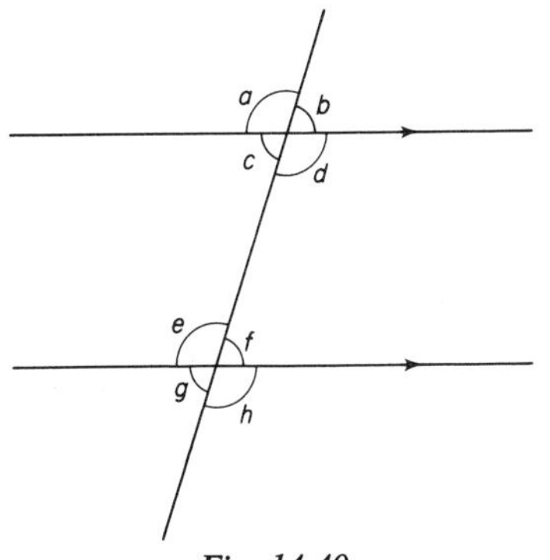

Fig. 14.40

Angles on the 'outside' of the parallel lines are sometimes called *exterior angles* (in Fig. 14.40, the angles marked a, b, g and h) and angles on the inside of the parallel lines *interior angles* (the angles marked c, d, e and f).

The pair of angles marked in Fig. 14.40 as c and f are called *alternate angles*; d and e are also alternate angles. It is easy to show that, for a pair of parallel lines cutting a third line, each pair of alternate angles is equal; you should work this out, for practice.

Using this terminology, we can describe the result of our experiment as follows:

> *when two parallel lines are intersected by a third line, the corresponding angles are equal; the alternate angles are also equal*

In mathematical notation:

Two parallel lines intersect a third line $\Rightarrow$ the corresponding angles are equal.

Similarly,

Two parallel lines intersect a third line $\Rightarrow$ the alternate angles are equal.

It is interesting to find out whether two lines are parallel if corresponding angles are equal (the opposite or *converse* of the first part of the statement above).

We start by marking two points P and Q on a line l and then drawing another line through point P, choosing its direction arbitrarily.

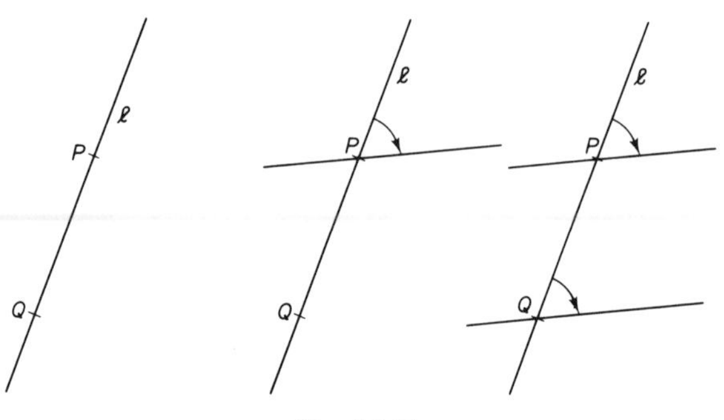

Fig. 14.41

Using a protractor, we measure the angle between these two lines in a clockwise direction starting from the line l. Now, starting at point Q, we draw the same angle, again in a clockwise direction. The two lines drawn in this way through P and Q indeed appear parallel.

Note. If the angle at Q is measured from the line l in an anti-clockwise direction, the lines through P and Q are not necessarily parallel, as Fig. 14.42 shows.

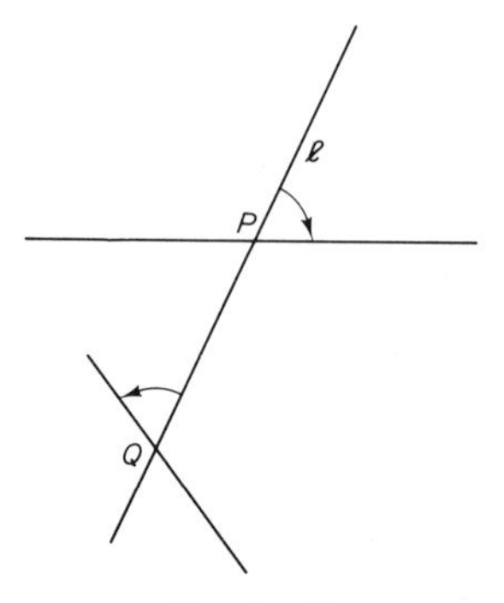

Fig. 14.42

But then the two equal angles are not what we have called *corresponding* angles.

if two lines intersect a third line and the corresponding angles are equal, the lines are parallel

In mathematical notation:

Corresponding angles are equal $\Rightarrow$ the lines are parallel.

We can use this result as the basis of a construction of parallel lines using a ruler and a set-square. The more sophisticated drawing board used by draughtsmen and architects is based on the same principle.

14.19 Constructing Parallel Lines

Construction 14.1: *Parallel lines.*

Hold your ruler firmly on your sheet of paper and place a set-square beside it, as indicated in Fig. 14.43. Now slide the set-square along the ruler;

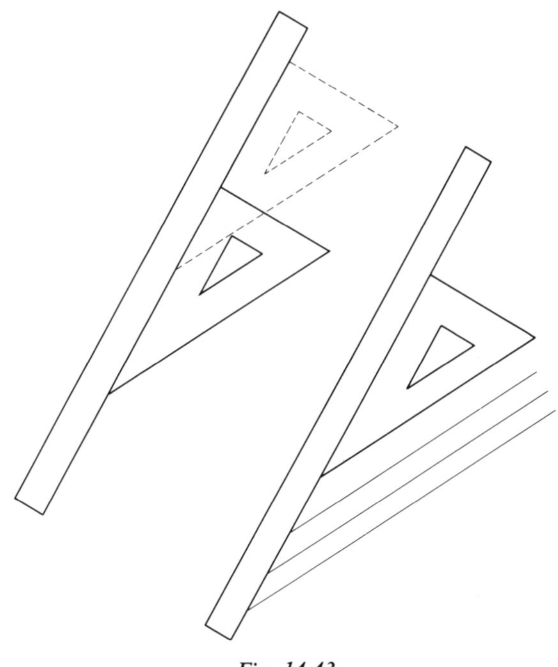

Fig. 14.43

no matter what its position along its line of travel, a side of the set-square always makes *the same angle* with the line of the ruler. If you draw lines along the same side of the set-square at various stages of its movement across the paper, they are all parallel.

14.20 Exercises

1. The construction of parallel lines with a ruler and a set-square can also be used to draw a line parallel to a given line. First place the set-square on the paper so that one of its sides coincides with the given line, and then place the ruler along another of its sides. Then slide the set-square along the ruler as before (Fig. 14.44).

   Draw a straight line on a sheet of paper and, using a ruler and a set-square, construct a line parallel to it.

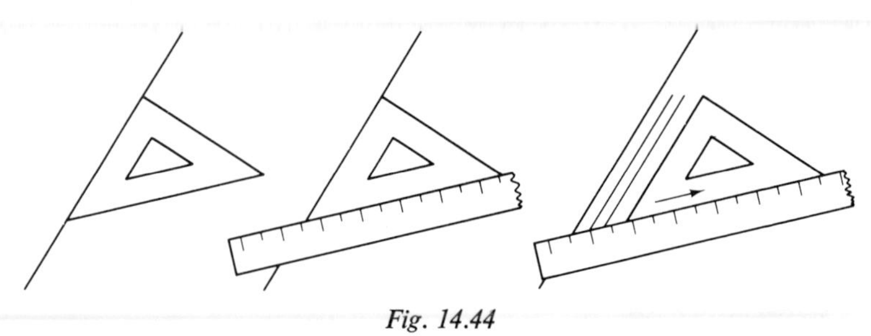

Fig. 14.44

2. Draw a straight line and mark a point P, not on the line. Construct a line through the point P parallel to the line. How many *different lines* can be drawn through this point parallel to the given line?
3. Draw an angle and mark a point Q which is not on one of the sides of the angle.

   Use the construction of parallel lines to 'transfer' the angle to this point Q, that is, so that Q is then the vertex of the angle.

Unit Fifteen
Triangles

15.1 Triangles: Terms and Notation

When three distinct straight lines are drawn on a flat surface, so that none of these lines is parallel to either of the other two, the three lines will intersect at three points (unless they happen to be *concurrent*, that is, all passing through one point). Three such lines are shown in Fig. 15.1 and the three points—or vertices—are marked A, B and C.

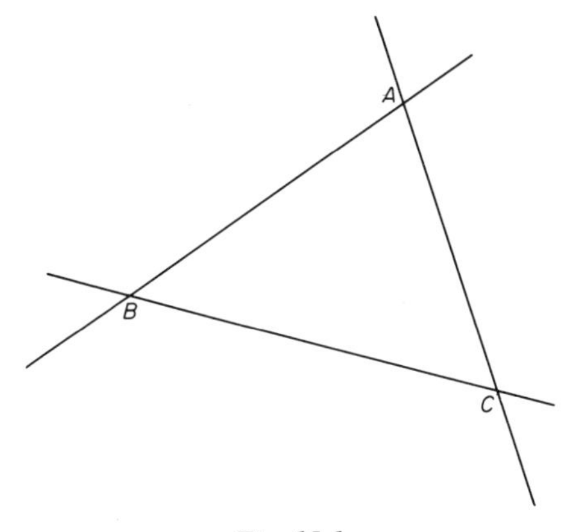

Fig. 15.1

The plane (or flat) closed shape bounded by the three line segments AB, BC and CA is called a *triangle*.

Terms and notation. Triangles are named from the letters indicating the vertices; we speak of the triangle ABC, or write $\triangle ABC$. The same triangle ABC can also be called the triangle BCA ($\triangle BCA$) or $\triangle CAB$; also as $\triangle ACB$, $\triangle CBA$ or $\triangle BAC$ (i.e. moving in a clockwise direction or in an anti-clockwise direction, and starting at any of the vertices).

Enclosed within the triangle are the *angles of the triangle*, formed by the three line segments which are called the *sides of the triangle*.

The notation for angles, introduced in Section 14.5, can be used to name the angles of the triangle; for example, $\angle ABC$ in $\triangle ABC$ is the angle at the vertex B. Often we call this angle simply angle B, or $\angle B$. If there are several angles with the same vertex, we can call them $\angle B_1$, $\angle B_2$, $\angle B_3$, and so forth, and mark them in the diagram by the appropriate subscript numbers $_1$, $_2$, $_3$, etc.

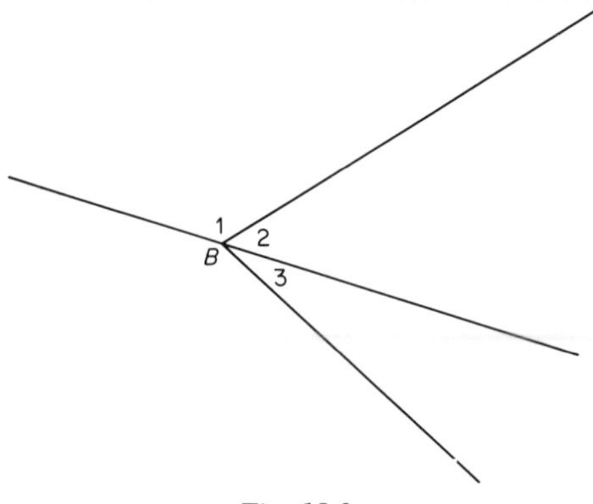

Fig. 15.2

The sides of $\triangle ABC$ are usually called the sides AB, BC and CA, but sometimes it is convenient to indicate a particular side by its position within the triangle. Instead of the side BC in $\triangle ABC$, we can speak of *the side opposite* $\angle A$,

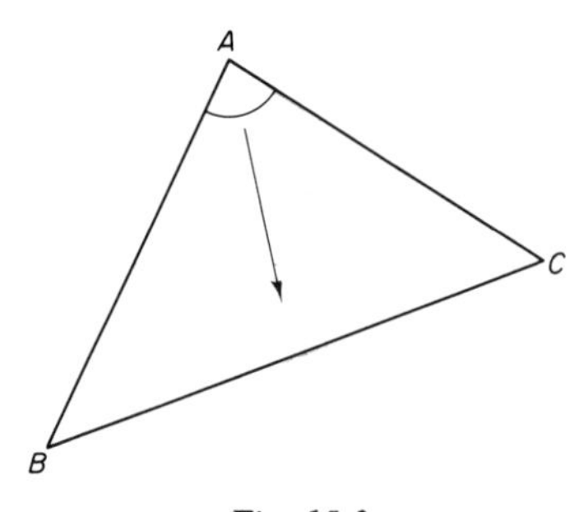

Fig. 15.3

the side AC as the side opposite $\angle B$, and so on. Vice versa, we may call $\angle A$ the angle opposite the side BC, etc.

A special notation for *the length of the side AB* is $|AB|$.

15.2 The Triangle Inequality

The three vertices A, B and C in Fig. 15.1 do not all lie on one straight line. The definition of a triangle in fact implies that they *cannot* do so; whenever three points A, B and C do lie on one straight line, the line segments AB, BC and CA are all part of this one line and not, as is required for a triangle, three *distinct* lines.

This brings us immediately to what is known as the *triangle inequality*. The shortest distance between two points A and B is the length of the line segment AB; if a point C does not lie on this line—in other words, if the points A, B and

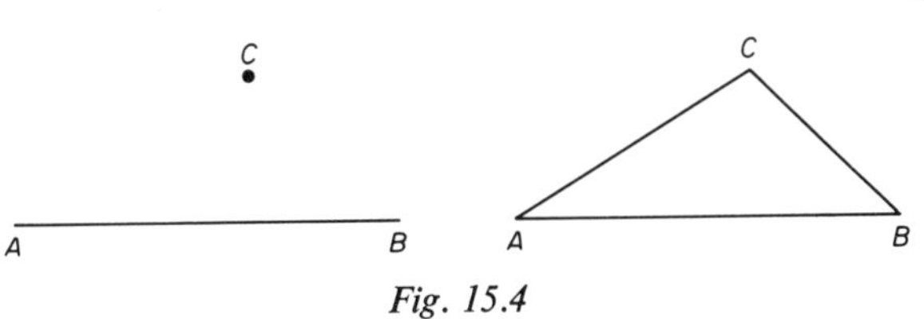

Fig. 15.4

C form a triangle—then the route from A to B via C must always be longer than the direct straight-line route AB (Fig. 15.4).

We can write this as:

(length of segment AB) < (length of segment AC) + (length of segment CB),

or,

$$|AB| < |AC| + |CB|$$
or
$$|AC| + |CB| > |AB|$$

In words,

the sum of the lengths of two sides of a triangle is always greater than the length of the third side

15.3 Exercises

1. Confirm this last result by drawing a number of triangles, and measuring the lengths of their sides.
2. Cut out some strips of paper of various lengths and try to form triangles with these strips as 'sides'. Test the triangle inequality using two strips whose lengths total less than that of the third strip.

15.4 The Angles of a Triangle

Triangles vary widely in their shapes and sizes. Some have interesting characteristics, such as symmetry about a line: they are called *isosceles triangles*, and

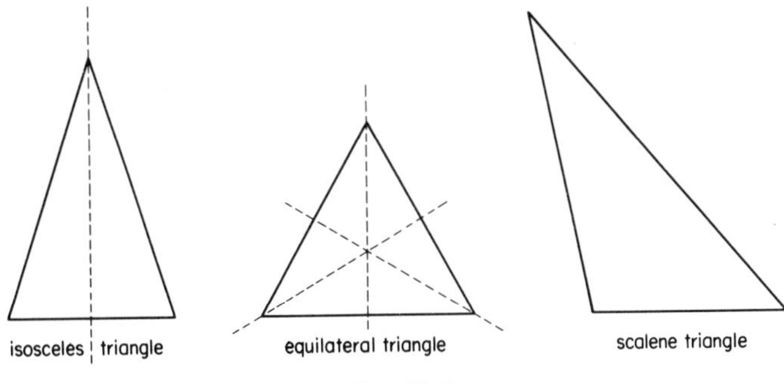

Fig. 15.5

have two sides of equal length. Some triangles have three lines of symmetry: they are called *equilateral triangles* (all three sides have the same length). In other triangles all the sides are of different lengths; they are called *scalene triangles*.

All triangles have an interesting property in common concerning the sum of their angles. You can re-discover this property for yourself by physically adding the angles of the triangle. Draw any triangle on a sheet of paper, and cut it out. Tear or cut off the three corners representing the three angles of the triangle and place them (or paste them) on another piece of paper alongside each other so that all three vertices meet at one point. You will find that

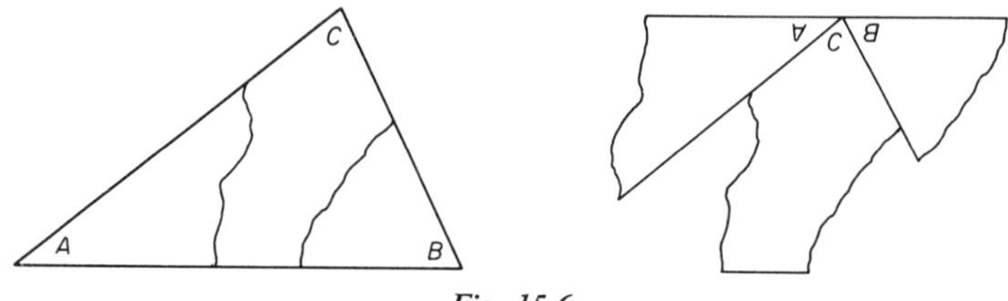

Fig. 15.6

together these three angles will make up a straight line, in other words, an angle of 180°.

the sum of the angles of a triangle is always 180°

This can be shown if we draw a line l through one of the vertices, parallel to the side opposite this vertex, in Fig. 15.7, the line through the point B parallel

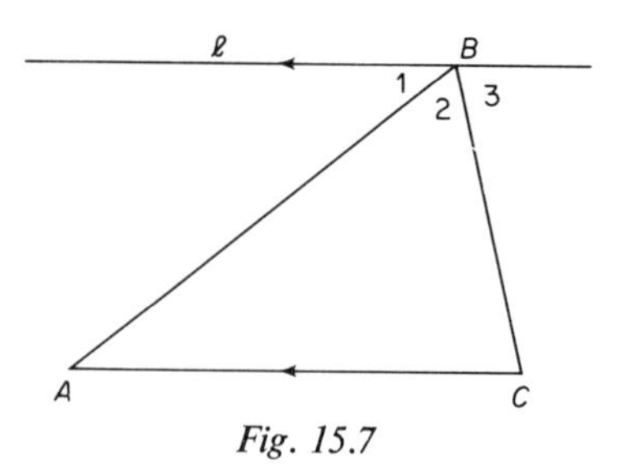

Fig. 15.7

to AC. If the angles are marked B_1, B_2 and B_3 as in the diagram, then $\angle B_1 + \angle B_2 + \angle B_3 = 180°$ (line l is a straight line). But $\angle B_1 = \angle A$ and $\angle B_3 = \angle C$, since these are pairs of alternate angles formed by parallel lines. Substituting $\angle A$ for $\angle B_1$ and $\angle C$ for $\angle B_3$ in $\angle B_1 + \angle B_2 + \angle B_3 = 180°$, we can write:

$$\angle A + \angle B_2 + \angle C = 180°,$$

that is, the sum of the angles of $\triangle ABC$ is 180°.

15.5 Exercises

1. Calculate the third angle, if the other two angles of a triangle are known to be:

   (*a*) 70° and 50°;
   (*b*) 90° and 20°;
   (*c*) 60° and 60°;
   (*d*) 90° and 45°.

2. Can a triangle have angles of 60°, 70° and 75° respectively? Try to draw it.

15.6 The Exterior Angles of a Triangle

The supplementary angles of a triangle are also called the *exterior angles of the triangle.* They are constructed by extending (in mathematical jargon '*producing*') the sides of the triangle beyond the vertices (Fig. 15.8).

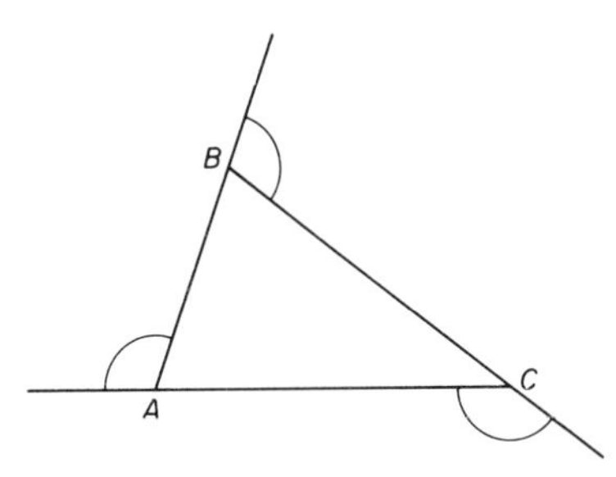

Fig. 15.8

The angles formed by the sides of the triangle *inside* the triangle are usually called 'the angles of the triangle'; but if we need to distinguish them from the exterior angles of the triangle we use the term *interior angles.*

The exterior angle at A, $\angle A_2$ in Fig. 15.9, is formed by the side CA and the side BA produced, the two sides opposite $\angle B$ and $\angle C$ respectively.

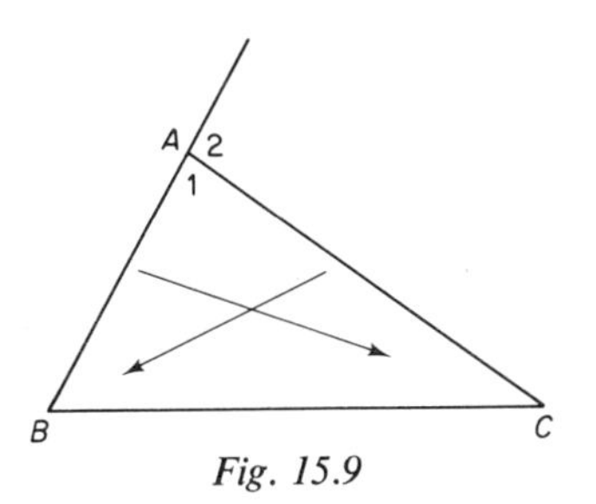

Fig. 15.9

When referring to the angle A_2 we may speak of the angles B and C as the *interior opposite angles.*

Theorem 15.1: *Each exterior angle of a triangle is equal in size to the sum of the interior opposite angles.*

This is immediately clear from the result in Section 15.4 concerning the sum of the angles of a triangle, taken with the definition of an exterior angle.

Using the notation of the triangle ABC in Fig. 15.9,

$$\angle A_1 + \angle B + \angle C = 180° \Rightarrow \angle B + \angle C = 180° - \angle A_1$$
(subtracting $\angle A_1$ from both sides);

also: the exterior angle A, or $\angle A_2$, $= 180° - \angle A_1 \Rightarrow \angle B + \angle C = \angle A_2$.

Theorem 15.1 can also be proved by drawing a line l through one of the

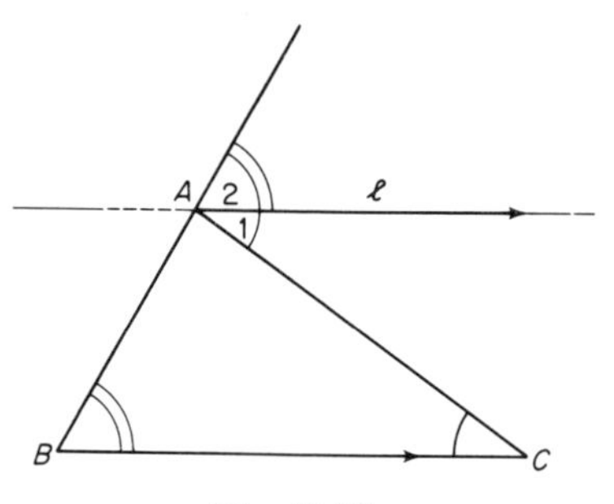

Fig. 15.10

vertices parallel to the side opposite this vertex, so as to divide the exterior angle A into two angles, $\angle A_1$ and $\angle A_2$ (Fig. 15.10).

$\angle B = \angle A_2$ (corresponding angles formed by parallel lines);
$\angle C = \angle A_1$ (alternate angles formed by parallel lines).

Adding,

$\angle B + \angle C = \angle A_1 + \angle A_2 =$ the exterior angle A.

15.7 Exercises

1. Calculate the sizes of the angles indicated in Fig. 15.11 by question marks.
   [Hint: first find the angles that can be read off immediately, such as the supplementary angles; then use the theorem of the angle sum of the triangle.]
2. Draw a number of different triangles, and measure their angles and sides. Confirm that *opposite the largest angle in a triangle* we always find the *largest side*. (A more precise relation between angles and sides will be discussed in Unit 19.)

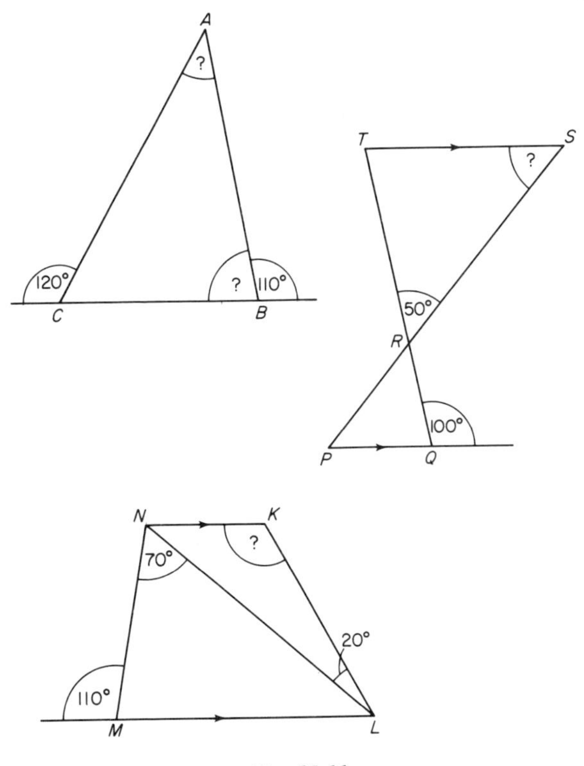

Fig. 15.11

15.8 The Triangle as a Rigid Framework

The triangle is the simplest plane shape enclosed by straight lines; it is also the most fundamental of all constructions. The earliest and most primitive human achievements in engineering, like the enormously larger structures of our own day, were based on the rigidity of a triangular framework. In geometrical terms this 'rigidity' means that three line segments meeting at their end-points can only produce *one triangle*; in other words, a triangle is determined by three line segments which satisfy the triangle inequality (i.e. no side is longer or equal in length to the total length of the other two sides).

You can test this by a simple experiment:

Take three meccano strips, or any three rigid strips, to represent the three sides of the triangle; the holes at the ends represent the end-points of the line segments. Make sure that none of the three strips is longer than (or equal to) the other two strips put together.

Using split pins, join the strips to form a triangle (Fig. 15.12). You will find that, if the split pins fit well, this triangle is *rigid*, unlike for example a framework of four strips. (Try four strips to confirm this.) Place this framework on

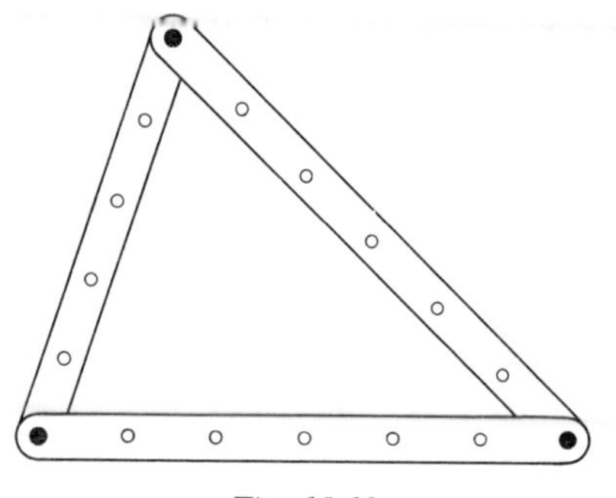

Fig. 15.12

a sheet of paper and draw lines along the inside of the frame, using it as a stencil or template. All triangles drawn in this way will be congruent.

Take the frame apart, and with the same three strips—using the same holes—try to form a triangular framework which is different from the one you made first. Test it by putting the frame over the triangles drawn on the paper.

The only difference that you may observe is that one triangle is the mirror image of another.

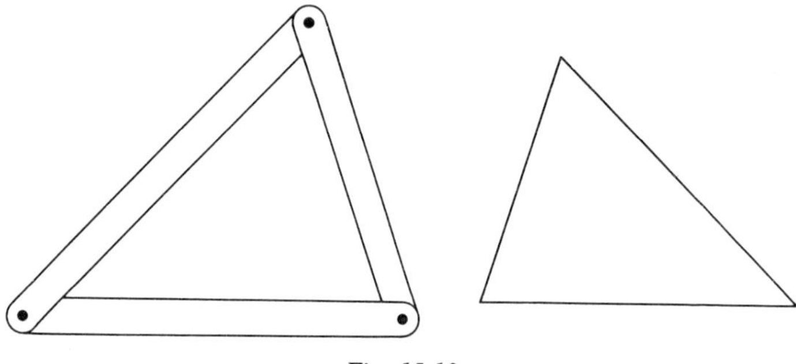

Fig. 15.13

They can, however, be made to fit precisely over one another and are therefore *congruent triangles*.

Construction 15.1: *A triangle of which the lengths of the sides are known.*

Note. In this and subsequent constructions with compasses we will use a characteristic property of the circle, which will be discussed in more detail in Unit 17: all the points on the circumference of a circle are at the same distance from the centre of the circle. Vice versa, all the points at a given distance from a fixed point form a curved line named the circumference of a circle. Any part of this circumference of a circle is called an *arc* of this circle.

The line segment joining the centre to any point of the circumference is named the *radius* of the circle.

Let a, b and c be three line segments, which satisfy the triangle inequality.

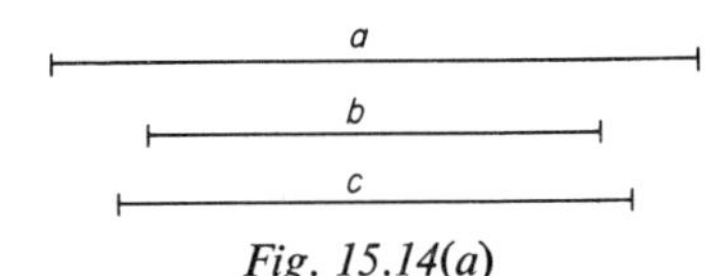

Fig. 15.14(a)

To construct a triangle whose three sides have the same lengths as those of a, b and c:

Draw a line segment, equal in length to one of the three line segments, say a. (The length of a line segment can be transferred by 'measuring' the distance between the end-points with compasses.) 'Measure' the length of the segment b with compasses in the same way; then place the point of the compasses at one of the end-points of a, and mark an arc of the circle on one side of a.

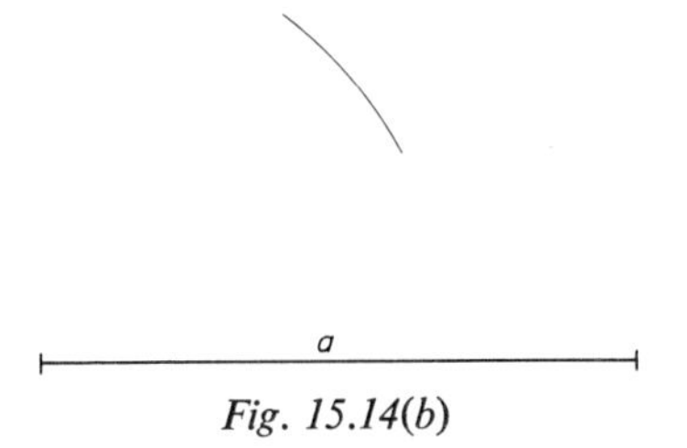

Fig. 15.14(b)

Similarly, mark the length of segment c from the other end-point of a, on the same side of the line.

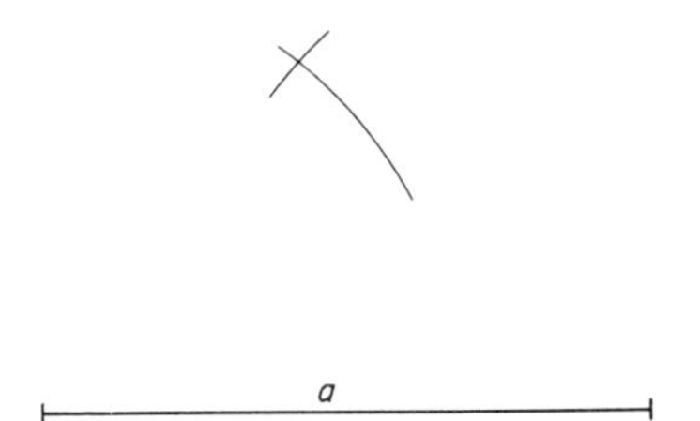
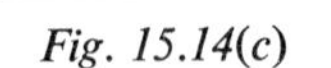

Fig. 15.14(c)

When the point of intersection of the markings is joined to the end-points of a, a triangle is formed.

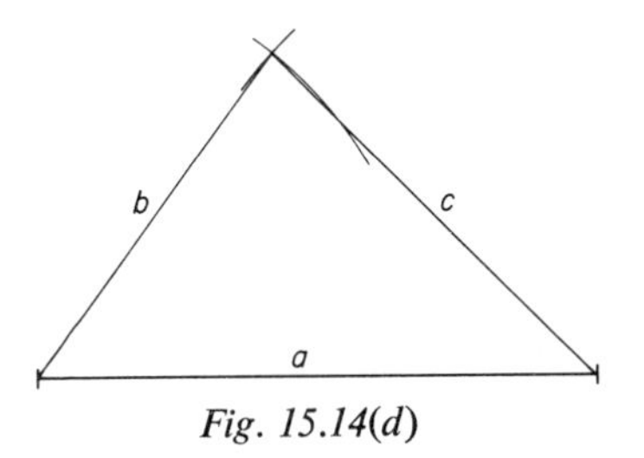

Fig. 15.14(d)

You can draw a triangle on the other side of a in the same way. This triangle is a mirror image of the first and therefore congruent with it.

15.9 Exercise

Construct a triangle whose sides are of lengths 6 cm, 7 cm and 8 cm. Start with the shortest side, and check whether the construction produces the same triangle as the construction which starts with the side of 7 cm (and 8 cm).

15.10 Congruent Triangles

When two triangles fit perfectly over one another, that is, when they are congruent, the pairs of lines that coincide are the *corresponding sides* and the pairs of angles are the *corresponding angles*.

For example, if the triangles ABC and FGH in Fig. 15.15 are congruent, the

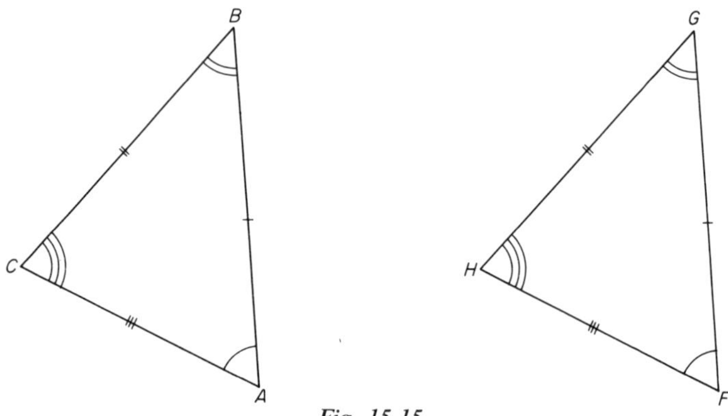

Fig. 15.15

angles BAC and GFH are corresponding angles, the sides BC and GH are corresponding sides, and so on.

When naming congruent triangles, it is a good habit to arrange the letters in the right order, that is, so that the corresponding vertices, sides and angles can be read off immediately. For example, taking the triangles in Fig. 15.15 again,

$$\begin{array}{cccc} \triangle & A & B & C \\ & \updownarrow & \updownarrow & \updownarrow, \\ \triangle & F & G & H \end{array}$$

showing at once that the pairs of corresponding angles are A and F, B and G, and C and H, and that the pairs of corresponding sides are AB and FG, BC and GH, and CA and HF.

We can then write, 'the triangles $\begin{matrix} ABC \\ FGH \end{matrix}$ are congruent', or

$$\text{'}\triangle\text{s } \begin{matrix} ABC \\ FGH \end{matrix} \text{ are congruent'.}$$

The use of the same three letters is often convenient, with a dash attached to each letter to indicate the vertices of the second triangle, as with the triangles ABC and $A'\,B'\,C'$ in Fig. 15.16.

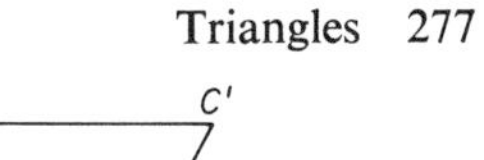

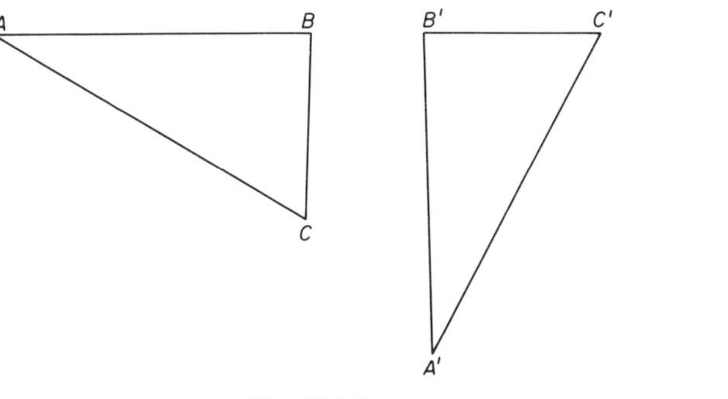

Fig. 15.16

(Notice that the corresponding sides of congruent triangles need not necessarily be parallel, and may not always be very conveniently drawn.)

15.11 Sufficient Conditions for the Congruence of Triangles

When we read that two triangles, say $\triangle s \begin{matrix} ABC \\ A'B'C' \end{matrix}$, are congruent, we are presented with at least seven pieces of information which we can write down as seven equations.

There are three pairs of equal angles:

$$\angle A = \angle A', \ \angle B = \angle B' \text{ and } \angle C = \angle C'.$$

There are three pairs of sides of equal length:

$$|AB| = |A'B'|, \ |BC| = |B'C'| \text{ and } |AC| = |A'C'|.$$

The area of $\triangle ABC =$ the area of $\triangle A'B'C'$.

On the other hand, remember that in Section 15.8 we saw that, if three sides of a triangle are equal to three sides of another triangle, these two triangles must be congruent. In other words, all these seven equations need not be stated before we can decide that two triangles are congruent: it is *sufficient*, for example, to know that all three sides of one triangle are of the same length as the three sides of another triangle.

(*a*) We say that equality of length of three pairs of sides is *a sufficient condition* for the congruence of two triangles. This particular condition is known as *side–side–side*, abbreviated to *S–S–S*.

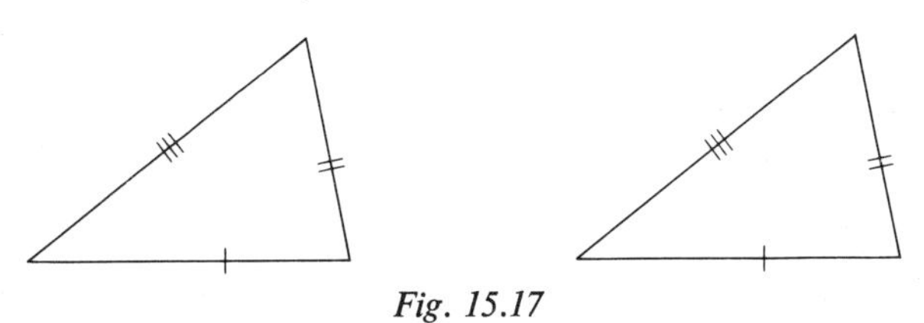

Fig. 15.17

There are other conditions for congruence of triangles which are also sufficient in themselves.

(*b*) Two triangles are congruent if two pairs of sides are of equal length, and if the angles included by these two sides are also equal. This condition is called *side–angle–side* (*S–A–S*).

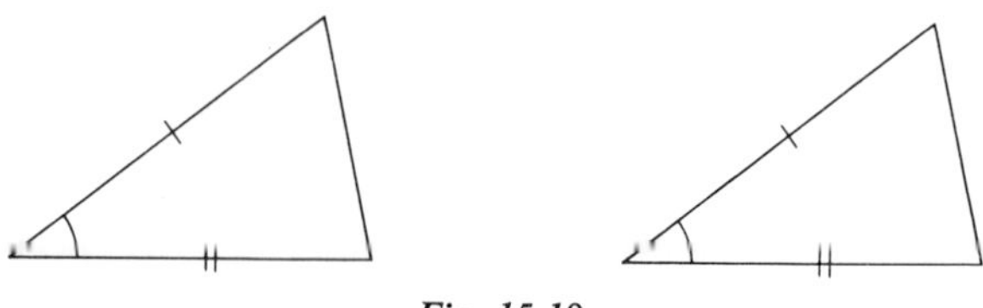

Fig. 15.18

You can demonstrate this by drawing two equal angles on separate sheets of transparent paper,

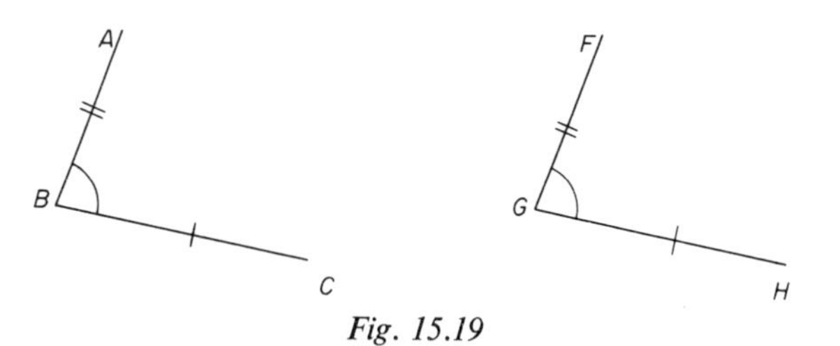

Fig. 15.19

marking the points *A* and *F*, *C* and *H* on the sides of the angles so that the length of *BA* equals the length of *GF*, or $|BA| = |GF|$, and $|BC| = |GH|$.

If you place the angle *ABC* over the angle *FGH*, you can immediately confirm that the points *B* and *G*, *A* and *F*, *C* and *H* can be made to coincide simultaneously, and that the triangles *ABC* and *FGH* are congruent.

You can see that this must necessarily be the case, by considering the meaning of the equality of angles, and the equality of the lengths of line segments. If the angles at *B* and *G* are equal, then the points *B* and *G* and the half-lines forming the angles can always be made to coincide, the half-line *GF* covering the line *BA* and the half-line *GH* covering the half-line *BC*.

Because the lengths of the two line segments *BA* and *GF* are equal (or $|BA| = |GF|$), the points *A* and *F* are bound to coincide too (point *A* and point *F* are now the same distance from point *B*, and *A* and *F* both lie on the same line, on the same side of *B*).

Therefore, whenever two sides of a triangle are of the same length as two sides of another triangle and the angles included by these sides are equal, the two triangles are congruent.

(*c*) A third possible sufficient condition for congruence of triangles is equality of length of one side in each triangle and equality of the pairs of corresponding angles at both ends of these segments: *angle–side–angle* (*A–S–A*).

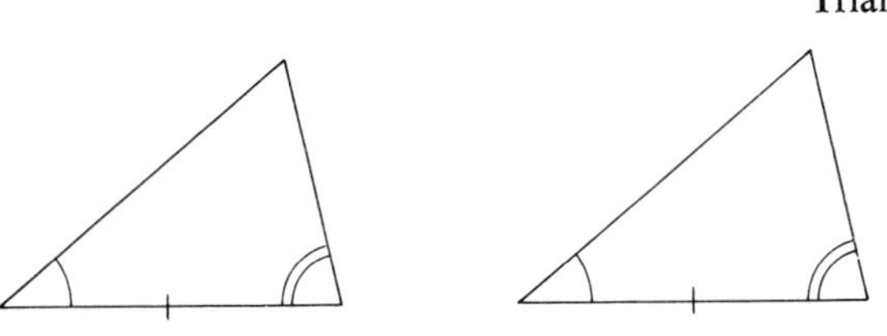

Fig. 15.20

You can test this in the same way as before, by drawing two line segments of equal length on separate sheets of transparent paper and a pair of equal angles at the end-points of each segment, as in Fig. 15.21.

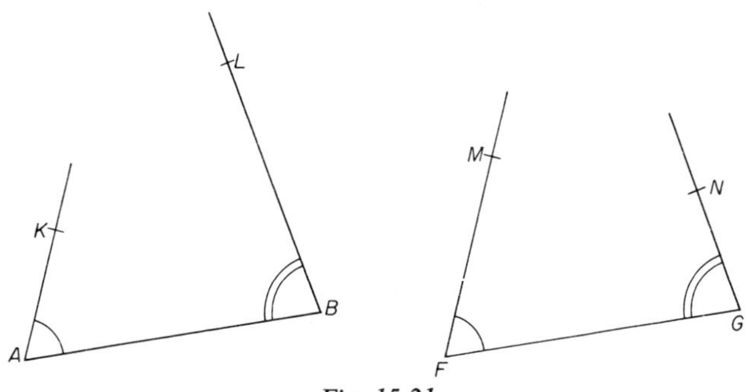

Fig. 15.21

You can now place one sheet over the other so that

the point A coincides with F,
the point B coincides with G,
the line AK coincides with FM,
the line BL coincides with GN.

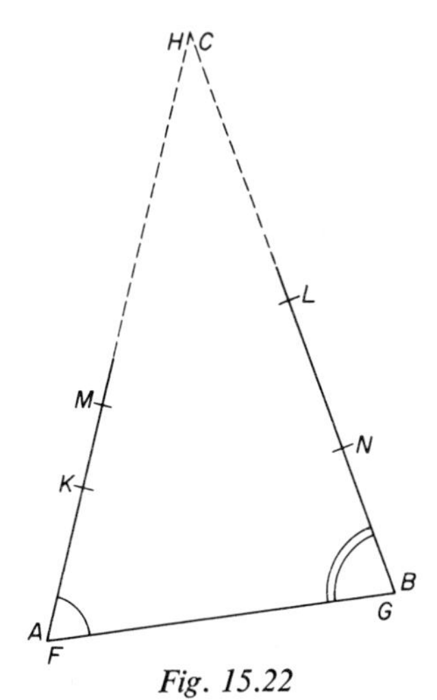

Fig. 15.22

The lines AK and DL, when extended, will meet at the point C and so complete the triangle ABC. They will completely cover the extensions of the lines FM and GN and their point of intersection, H. $\triangle ABC$ will then completely coincide with $\triangle FGH$, and the $\triangle$s ABC and FGH are therefore congruent.

This sufficient condition for the congruence of two triangles (*A–S–A*) will hold whenever two triangles have one side of equal length and any two angles of one triangle are equal to the *corresponding* angles of the other. This follows from the angle-sum property of the triangle; since the sum of the angles of each triangle is 180°, then if two angles of one triangle are equal to two angles of another, the third angle of the one must equal the third angle of the other. For example if, in $\triangle ABC$ and $\triangle XYZ$, in Fig. 15.23,

$$\begin{aligned} \angle A &= \angle X \\ + \quad \angle C &= \angle Z \\ \hline \text{then} \quad \angle A + \angle C &= \angle X + \angle Z, \text{ and} \\ \angle B &= 180^\circ - (\angle A + \angle C) \\ &= 180^\circ - (\angle X + \angle Z) = \angle Y. \end{aligned}$$

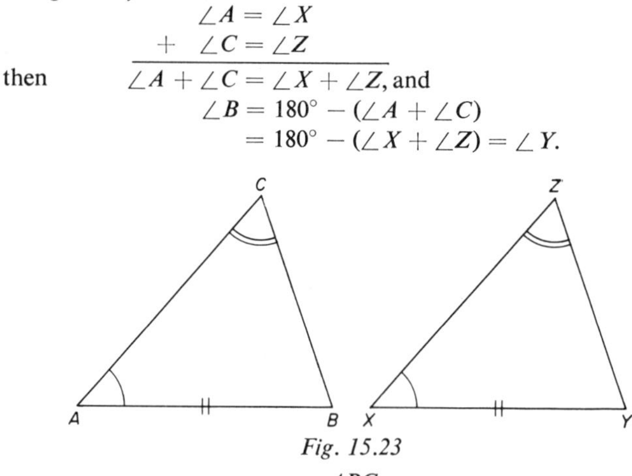

Fig. 15.23

If $|AB| = |XY|$ as well, then $\triangle$s $\begin{matrix} ABC \\ XYZ \end{matrix}$ are congruent (*A–S–A*).

The equal angles must, of course, be *corresponding* angles, i.e. in the same position relative to the sides of equal length. Fig. 15.24 shows two triangles FGH and KLM, in which

$$\begin{aligned} |FG| &= |KL|, \\ \angle F &= \angle M, \\ \angle G &= \angle L. \end{aligned}$$

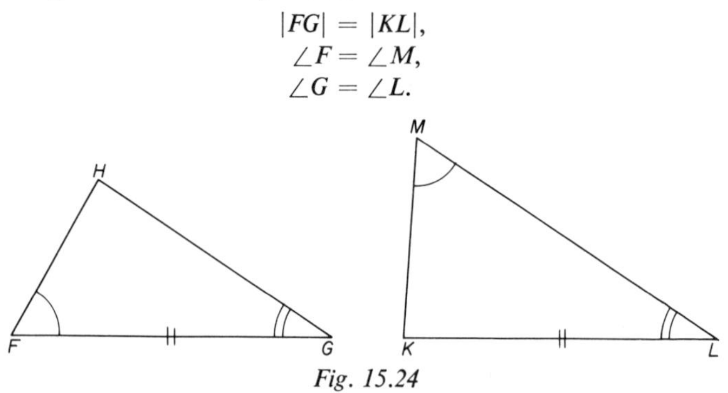

Fig. 15.24

Angles F and M are *not* corresponding angles; they occupy a different position in relation to the equal sides FG and KL. Therefore these triangles are *not* necessarily congruent.

Another *ambiguous* situation arises when the lengths of two sides of a triangle are equal to the lengths of two sides of another triangle and the angles opposite one of the pairs of equal sides are equal; in other words, when the equal angles are not the angles included by the equal sides (not S–A–S but *side–side–angle*).

For example, in $\triangle ABC$ and $\triangle FGH$ in Fig. 15.25,

$$|AC| = |FH|,$$
$$|CB| = |HG|,$$
$$\angle B = \angle G.$$

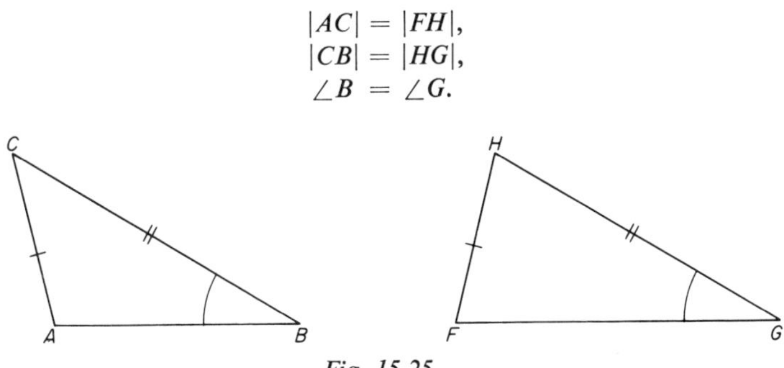

Fig. 15.25

The diagram shows that $\triangle$s $\frac{ABC}{FGH}$ are *not* necessarily congruent. Indeed, if we attempt to construct a triangle FGH (Fig. 15.26) when the angle G is of a given size and the lengths $|GH|$ and $|HF|$ are also fixed, we find *two* possible positions of the point F and therefore *two different* triangles.

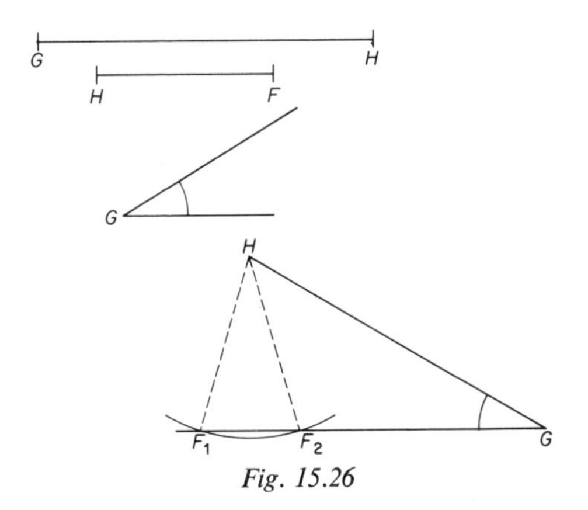

Fig. 15.26

We draw angle G, and mark the point H. The two points of a pair of compasses are adjusted to the given length of the line segment HF, and one of the points is placed on the point H; the other point marks two possible points F_1 and F_2, so that HF_1 and HF_2 both have the required length.

The case *side side angle* is therefore *not* a sufficient condition for the congruence of two triangles.

(*d*) No ambiguity can arise if the equal angle is a right angle. For example, in $\triangle ABC$ and $\triangle FGH$ in Fig. 15.27,

$$\angle B = \angle G = 90°,$$
$$|BC| = |GH|,$$
and
$$|CA| = |HF|.$$

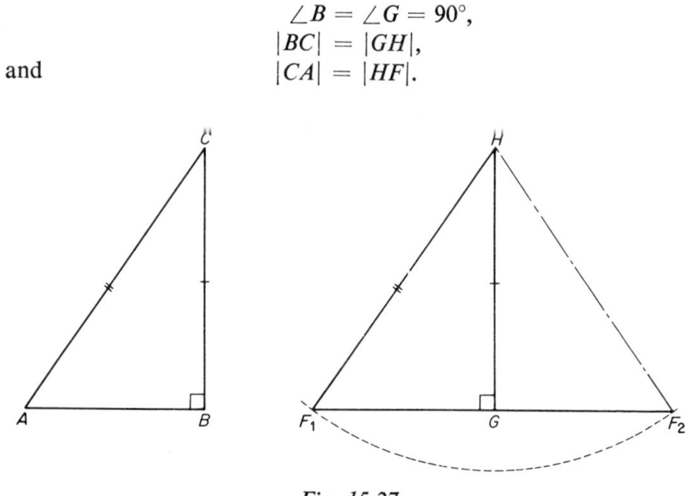

Fig. 15.27

Using the same method of construction as before we find that the two possible points F_1 and F_2 lie on either side of the line GH and, with G and H, form two triangles which are mirror images of each other, both congruent with $\triangle ABC$.

This sufficient condition for congruence of two triangles is known as *right angle–hypotenuse–side* (*R–H–S*). (The *hypotenuse* of a right-angled triangle is the side opposite the right angle.)

The congruence of the two triangles $\begin{matrix} HGF_1 \\ HGF_2 \end{matrix}$ will be discussed in more detail in Section 15.14.

Fig. 15.28 summarizes the various sufficient conditions for the congruence of two triangles.

15.12 Applications of Congruence of Triangles

In each of the four cases mentioned in Section 15.11, a knowledge of only three equalities is sufficient to provide a guarantee of the congruence of a pair of triangles; this knowledge also supplies information about the other four aspects of the triangle. The proof of the congruence of two triangles is often, therefore, a good way to demonstrate the equality of length of two line segments, or two angles, or two areas.

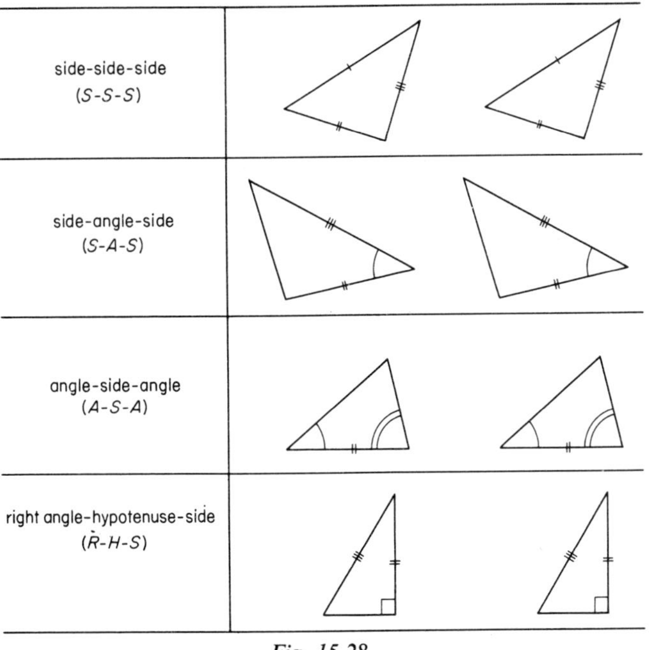

Fig. 15.28

Examples

1. In a plane four-sided shape (a *quadrilateral*) $ABCD$ (Fig. 15.29),

$$\angle A = 60^\circ,$$
$$|AB| = |CB|,$$
$$|CD| = |AD|.$$

We can use these facts to find the size of $\angle C$.

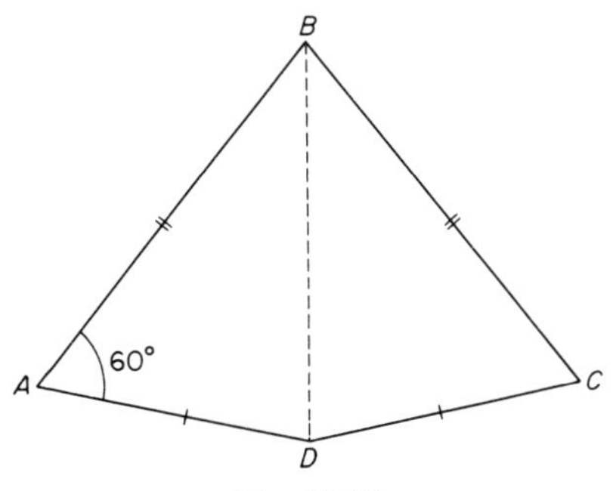

Fig. 15.29

We draw the line BD (the *diagonal*), dividing the quadrilateral into triangles ADB and CDB. Of these we know that

$$\begin{aligned} |AB| &= |CB| \text{ (given)}, \\ |AD| &= |CD| \text{ (given)}, \\ |BD| &= |BD| \text{ (a 'common' side of the two triangles)}. \end{aligned}$$

Therefore $\triangle\text{s}\ \begin{matrix} ADB \\ CDB \end{matrix}$ are congruent (S–S–S) and, because they are congruent, $\angle C = \angle A = 60°$ (corresponding angles).

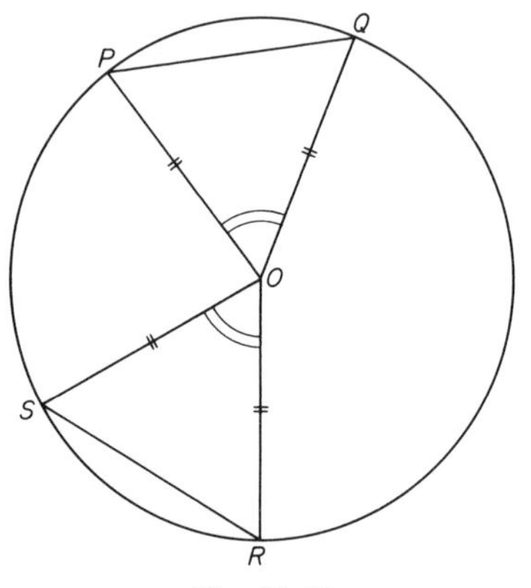

Fig. 15.30

2. In Fig. 15.30, $\triangle OPQ$ and $\triangle ORS$ are formed by the point O, the centre of a circle, and the points P, Q, R and S which are all points on the circle; also

$$\angle POQ = \angle ROS.$$

These facts tell us that the line segments PQ and RS are of the same length, since

$$\begin{aligned} |OP| &= |OR| \text{ (radii)}, \\ |OQ| &= |OS| \text{ (radii)}, \\ \angle POQ &= \angle ROS \text{ (given)}. \end{aligned}$$

$\triangle\text{s}\ \begin{matrix} POQ \\ ROS \end{matrix}$ are congruent (S–A–S), so $|PQ| = |RS|$ (corresponding sides).

A practical application of the condition S–S–S for congruent triangles is the following:

Construction 15.2: *An angle equal to a given angle.*

An angle A and a line segment BC are given (Fig. 15.31). An angle is to be constructed at B which is equal to angle A: in other words, a line through B which forms with BC an angle equal to angle A.

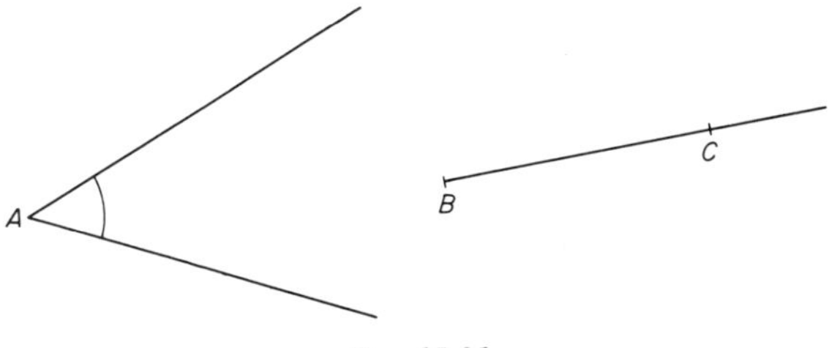

Fig. 15.31

We may, in fact, draw any line intersecting the sides of angle A (at the points D and E in Fig. 15.32), forming a triangle ADE, and then construct on the half-line BC a triangle whose sides have the same lengths as those of $\triangle ADE$ (using Construction 15.1 in Section 15.8).

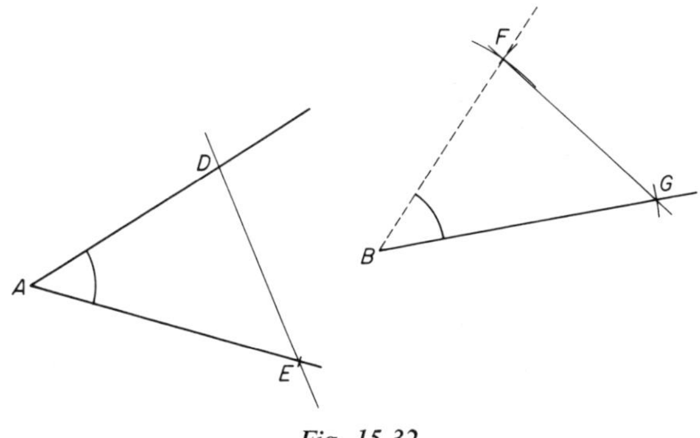

Fig. 15.32

$\triangle$s $\begin{matrix} ADE \\ BFG \end{matrix}$ are congruent (S–S–S) $\Rightarrow \angle A = \angle B$.

There is also a simpler method based on the same principle (Fig. 15.33).

(*a*) Using compasses, mark two points, P and Q, on the sides of $\angle A$ at equal distances from A.

(*b*) Leaving the distance between the points of the compasses unchanged, mark an arc of a circle with B as centre, intersecting the line BC at R, so that $|AP| = |BR|$.

(*c*) Measure off the distance PQ with the points of the compasses, and mark the point S on the arc through R, so that $|RS| = |PQ|$.

(*d*) Join the points B and S.

$\angle SBR$ is the required angle, since $\triangle$s $\begin{matrix} APQ \\ BRS \end{matrix}$ are congruent (S–S–S).

In this example, we named the points P, Q, R and S and drew the lines QP and RS so as to demonstrate the congruence of triangles on which the construction is based; this is not usually necessary in a simple construction.

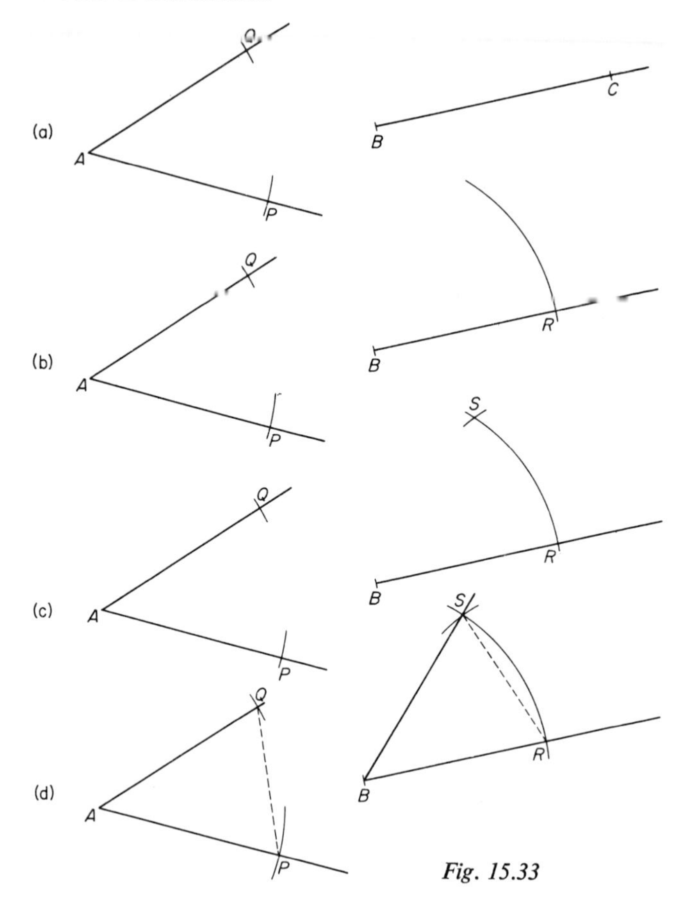

Fig. 15.33

Construction 15.3: *Parallel lines, using compasses and a ruler.*

If a line l and a point P are given, a line can be constructed through the point P parallel to the line l. We use the same principle that we used in Construction 14.1: equal corresponding angles $\Rightarrow$ parallel lines.

Draw a line (any line) through point P which intersects the line l at some point Q. Construct a line m which makes an angle at P of the same size as angle Q, using Construction 15.2 (Fig. 15.34). Then since $\angle P$ and $\angle Q$ are equal corresponding angles, l is parallel to m.

15.13 Exercises

1. Practise Construction 15.2 by drawing an angle and a half-line, and constructing an angle of the same size at one of the end-points of the half-line.
2. Similarly, practise Construction 15.3 by drawing a line, marking a point not on this line and constructing a parallel line through this point.

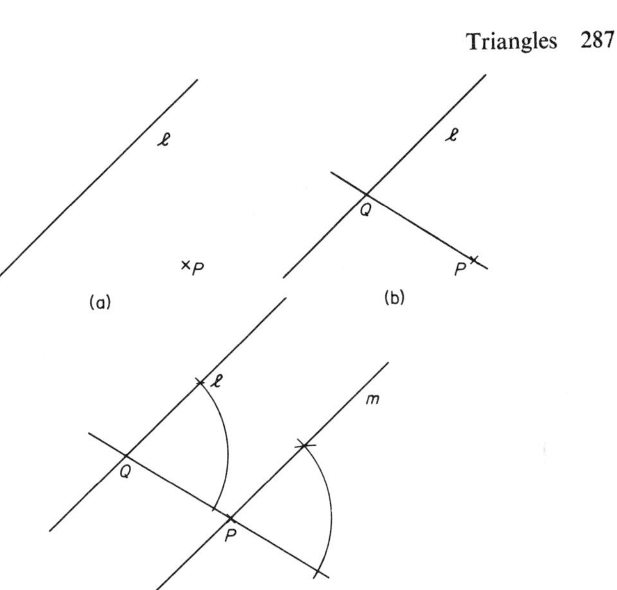

Fig. 15.34

3. Draw a square $ABCD$ on squared paper, and mark a point P somewhere on the line AB (Fig. 15.35).

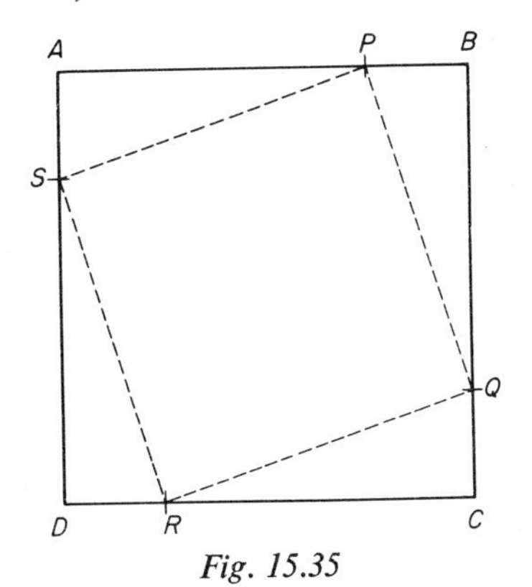

Fig. 15.35

Use the distance AP to mark points Q, R and S on BC, CD and DA, such that $|AP| = |BQ| = |CR| = |DS|$. Join the points P, Q, R and S in order (that is, P to Q, Q to R, etc.).

(*a*) Show that the triangles $\begin{matrix} SAP \\ PBQ \\ QCR \\ RDS \end{matrix}$ are congruent, and that therefore

$|SP| = |PQ| = |QR| = |RS|$.

(*b*) Mark the equal angles in the four congruent triangles and show that $\angle PSR$, $\angle SRQ$, $\angle RQP$ and $\angle QPS$ are all right angles. [Hint: in any right-angled triangle, the sum of the two angles which are not right angles is 90°.]
From (*a*) and (*b*) it follows that *PQRS* is a square.

15.14 The Isosceles Triangle

An isosceles triangle is a triangle which is symmetrical about a line; when drawn on paper it can be folded so that one half of the triangle covers the other half exactly.

This suggests a simple method of constructing an isosceles triangle: fold a piece of transparent paper in half and mark on one of the halves a line segment *AB* so that *A* is a point on the line of the fold.

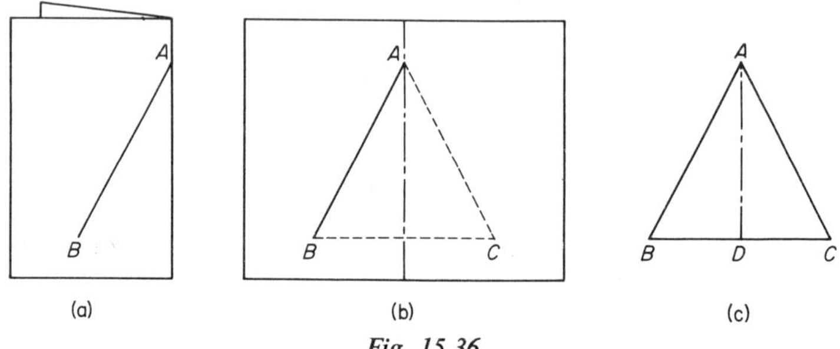

Fig. 15.36

At the point *B*, prick a hole through both halves of the sheet. Open out the sheet of paper and mark the second hole as point *C*; join *C* to the points *A* and *B* by straight lines. When the paper is folded again the line segment *AB* will cover the line segment *AC*, and the two halves of the line segment *BC* will also coincide. If the point where the fold intersects *BC* is marked as *D*, we can show that $\triangle$s $\begin{matrix} BDA \\ CDA \end{matrix}$ are congruent as follows:

Our method of fixing *C* by pricking a hole at *B* ensures that $|BA| = |CA|$, and that the points *B* and *C* are the same distance from the line of the fold, i.e. $|BD| = |CD|$. In other words, the point *D* is the *mid-point* of *BC*.

Both triangles share, or have in common, the line segment *AD*, i.e. $|AD|$ in $\triangle BDA = |AD|$ in $\triangle CDA$.

The $\triangle$s $\begin{matrix} BDA \\ CDA \end{matrix}$ are therefore congruent (*S–S–S*).

15.15 Exercise

Mark all the corresponding equal angles and equal sides of the two congruent $\triangle$s $\begin{matrix} BDA \\ CDA \end{matrix}$ in Fig. 15.36(*c*) and so account for the properties of the isosceles triangle as given in detail in Section 15.16.

15.16 The Properties of the Isosceles Triangle

Two of the sides of an isosceles triangle have the same length;

in Fig. 15.37, $|BA| = |CA|$.

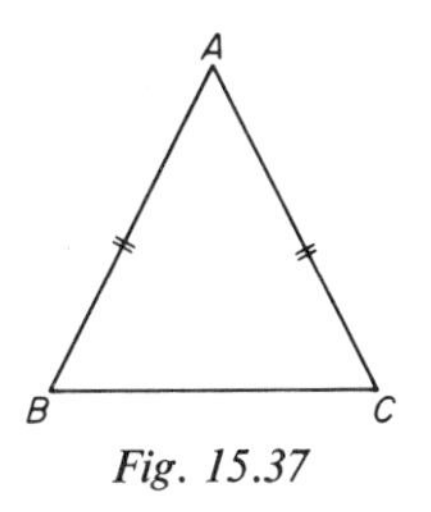

Fig. 15.37

We have already remarked that the Greek word 'isosceles' means 'equal-sided'.

Construction 15.4: *An isosceles triangle.*

If the lengths of the sides of an isosceles triangle are given, we can construct it in the same way as any other triangle (see Construction 15.1 in Section 15.8). As two sides are of equal length, the construction is much simpler.

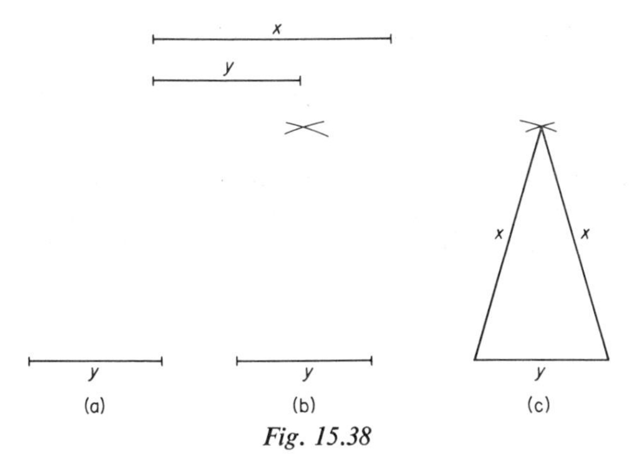

Fig. 15.38

If the length of the line segment x in Fig. 15.38 represents the lengths of the two equal sides of an isosceles triangle, and the length of the line segment y is that of the third side of the triangle (often called the *base* of the isosceles triangle), the isosceles triangle is constructed by

(*a*) drawing the base;
(*b*) marking the lengths of the other two (equal) sides from the end-points of the base, using compasses;
(*c*) joining the end-points of the base to the intersection of these two arcs.

Other ways of constructing isosceles triangles are outlined in Section 15.20.

Two angles – the base angles – of an isosceles triangle are equal;

in Fig. 15.39, $\angle B = \angle C$.

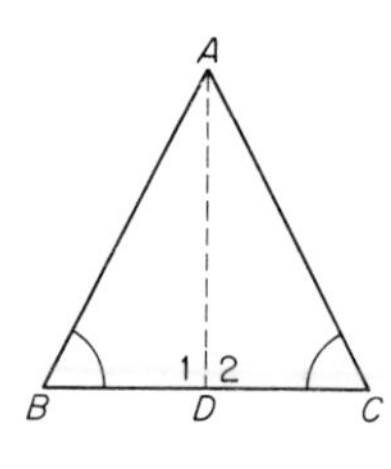

Fig. 15.39

This follows immediately from the congruence of the $\triangle$s $\dfrac{BDA}{CDA}$; the angles B and C are corresponding angles.

It also follows from the congruence of $\triangle$s $\dfrac{BDA}{CDA}$ that $\angle D_1 = \angle D_2$. Moreover, since $\angle D_1 + \angle D_2 = 180°$ (BDC is a straight line),

$$\angle D_1 = \angle D_2 = 90° \text{ (a right angle)}.$$

The line DA is, therefore, *perpendicular* to the line BC.

In any triangle XYZ we can draw a number of lines which have interesting properties:

(*a*) the *median* XM from (or through) the vertex X, that is, the line joining the vertex X to the mid-point M of the opposite side YZ. The medians from the vertex Y and the vertex Z can be drawn in the same way (Fig. 15.40).

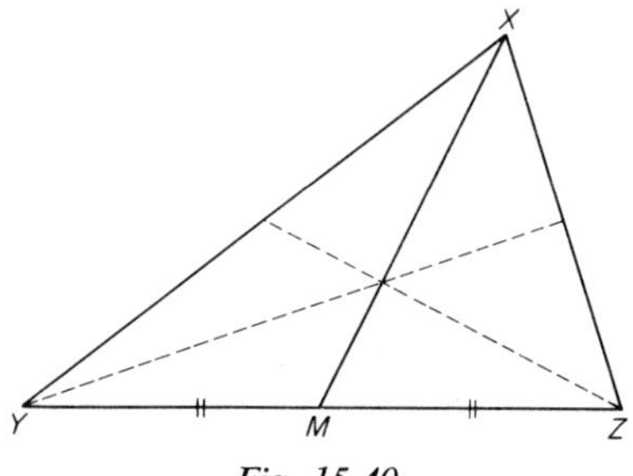

Fig. 15.40

(*b*) the *altitude* of the triangle XYZ relative to the base YZ, that is, the line segment XH through the vertex X and perpendicular to the base YZ in Fig. 15.41.

Similarly, the altitudes of the triangle XYZ relative to XY and XZ as base respectively can be drawn (Fig. 15.41).

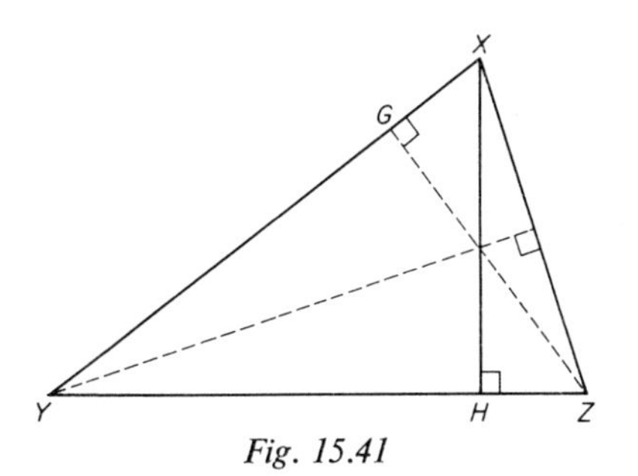

Fig. 15.41

(*c*) the *bisector of angle X*, or the *angle bisector of* $\angle X$, that is, the line which divides the angle X into two equal parts: $\angle X_1 = \angle X_2$ in Fig. 15.42.

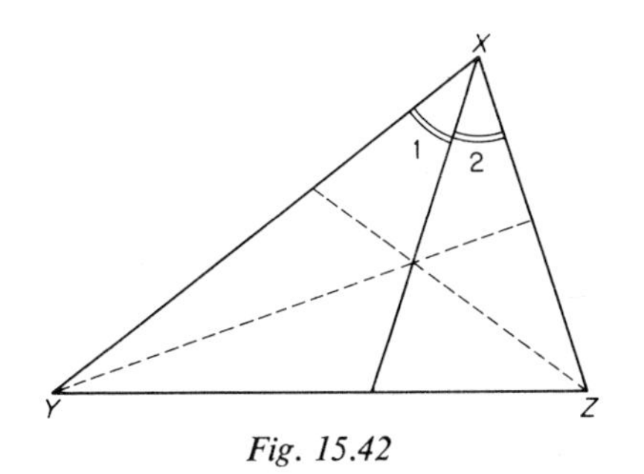

Fig. 15.42

(*d*) the *perpendicular bisector of a side of the triangle*, that is, the line through the mid-point of the side at right angles to it: for example, the line through M in Fig. 15.43.

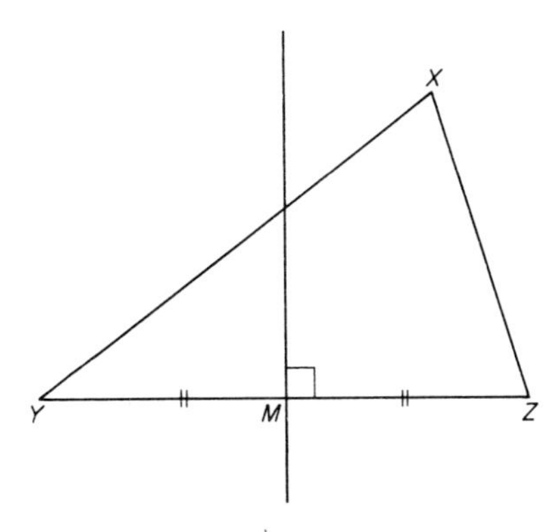

Fig. 15.43

Usually these four lines in a triangle are different lines or line segments (Fig. 15.44).

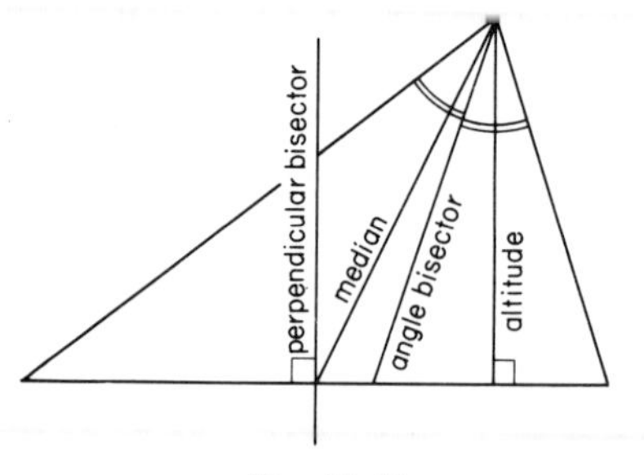

Fig. 15.44

In an isosceles triangle these four lines all coincide: in the isosceles triangle ABC in Fig. 15.39, the median through A is also the altitude relative to BC, the perpendicular bisector of BC and the angle bisector of angle A.

We have already seen (Section 15.16) that the line DA, which was drawn as the *median* of the isosceles triangle ABC, is perpendicular to BC and is therefore the *altitude* of the triangle relative to BC. Since it is perpendicular to BC and passes through the mid-point D of BC, DA is at the same time the *perpendicular bisector* of BC. Also, since $\triangle$s $\begin{matrix} BDA \\ CDA \end{matrix}$ are congruent, the two angles A_1 and A_2 are equal; the line DA is therefore the *bisector of angle A* as well.

15.17 Sufficient Conditions for a Triangle to be Isosceles

Each of the properties described in Section 15.16 is sufficient to establish that a triangle is isosceles; this is demonstrated in the following series of theorems.

Theorem 15.2: *If two sides of a triangle have the same length, the triangle is isosceles and the base angles are equal.*

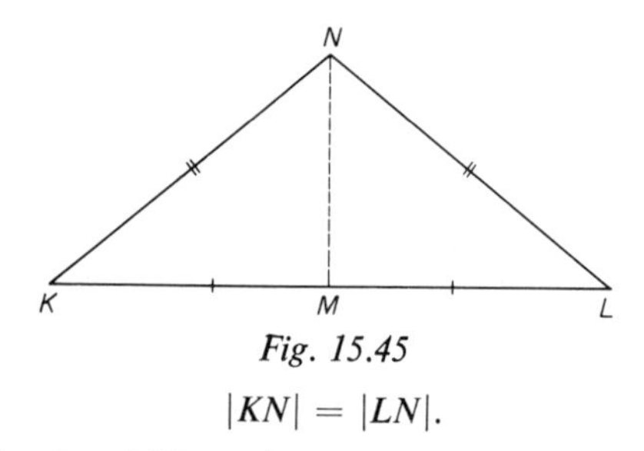

Fig. 15.45

In $\triangle NKL$, $$|KN| = |LN|.$$

Let M be the mid-point of KL, so that

$$|KM| = |ML|.$$

We join the points N and M, so that NM is a common side of $\triangle KMN$ and $\triangle LMN$:

$$|NM| = |NM|.$$

Therefore the triangles KMN and LMN are congruent (S–S–S).

All the other properties of the isosceles triangle then follow from the congruence of these two triangles, in particular that $\angle K = \angle L$.

Theorem 15.3: *If two angles of a triangle are equal, the triangle is isosceles and therefore two sides are of equal length.*

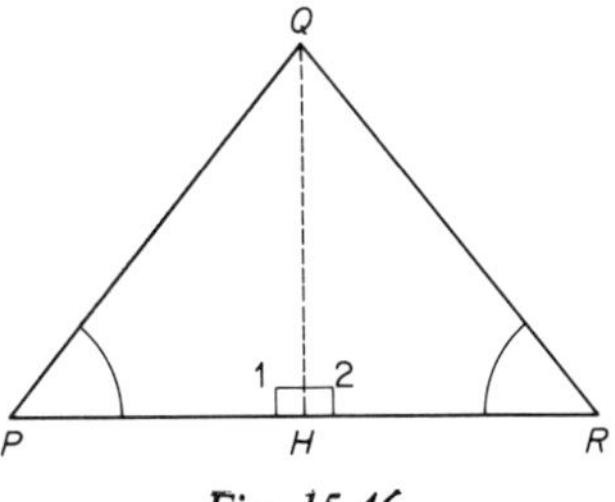

Fig. 15.46

In $\triangle PQR$, $\angle P = \angle R$.

We draw the altitude from Q relative to the base PR, so that

$$\angle H_1 = \angle H_2 = 90°.$$

The line QH divides the $\triangle PRQ$ into two triangles PHQ and RHQ; QH is a common side of both triangles:

$$|QH| = |QH|.$$
$$\angle P = \angle R \text{ (given).}$$

Therefore the triangles $\begin{matrix} PHQ \\ RHQ \end{matrix}$ are congruent (A–A–S), and therefore the two sides PQ and RQ are of equal length.

Theorem 15.4: *If, in a triangle, the altitude from a vertex coincides with the median from that vertex, the triangle is isosceles.*

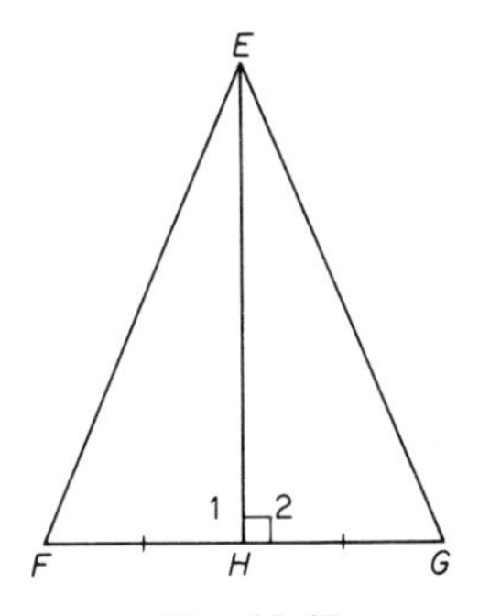

Fig. 15.47

In $\triangle EFG$ (Fig. 15.47), the altitude from the vertex E coincides with the median from E. Let EH be the altitude from E, so that $\angle H_1 = \angle H_2 = 90°$. EH is also given to be the median, so that $|FH| = |GH|$ (H is the mid-point). $\triangle FHE$ and $\triangle GHE$ have the side EH in common: $|EH| = |EH|$.

Therefore $\triangle$s $\begin{matrix} FHE \\ GHE \end{matrix}$ are congruent (S–A–S).

This theorem can be put alternatively as: '*If the perpendicular bisector of a side of a triangle passes through the vertex opposite that side, the triangle is isosceles*'.

Theorem 15.5: *If the bisector of the angle C of a triangle ABC coincides with the median through the vertex C, the triangle ABC is isosceles.*

A proof of this sufficient condition is a little complicated since the congruence of the triangles $\begin{matrix} ADC \\ BDC \end{matrix}$ cannot immediately be shown from the given facts $|AD| = |BD|$, $|DC| = |DC|$ and $\angle C_1 = \angle C_2$; this is the ambiguous case (S–S–A).

Instead, we draw two auxiliary lines DP and DQ perpendicular to AC and BC respectively.

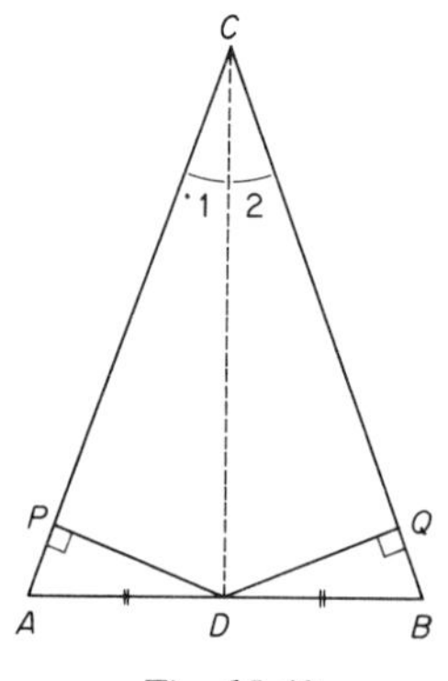

Fig. 15.48

CD is the angle bisector of angle C, so that $\angle C_1 = \angle C_2$;
CD is the median, so that $|AD| = |BD|$.

First, we show the congruence of $\triangle$s $\begin{matrix} DPC \\ DQC \end{matrix}$:

$$\angle C_1 = \angle C_2,$$
$$|DC| = |DC|,$$
$$\angle CPD = \angle CQD = 90°.$$

$\triangle$s $\begin{matrix} DPC \\ DQC \end{matrix}$ are therefore congruent (A–S–A), and it follows that $|PD| = |QD|$.

Now we show that $\triangle s \begin{matrix} ADP \\ BDQ \end{matrix}$ are congruent:

$$\begin{aligned} \angle DPA &= \angle DQB = 90^\circ, \\ |AD| &= |DB| \text{ (given)}, \\ |PD| &= |QD| \text{ (from the congruence of } \triangle s \begin{matrix} DPC \\ DQC \end{matrix}). \end{aligned}$$

$\triangle s \begin{matrix} ADP \\ BDQ \end{matrix}$ are therefore congruent (*R–H–S*); it follows that $\angle A$ and $\angle B$ are equal, and $\triangle ABC$ is therefore isosceles.

15.18 Exercise

Show that if, in a triangle ABC, the altitude relative to the base BC coincides with the median through A, the triangle is isosceles. [Hint: if M is the mid-point of BC, show the congruence of $\triangle s \begin{matrix} BMA \\ CMA \end{matrix}$.]

15.19 Applications

The most useful *applications* of these sufficient conditions for isosceles triangles are based on the two-way implication: 'Whenever two sides of a triangle are equal, the two angles opposite these sides are equal, and vice versa'. In a triangle ABC,

$$|AC| = |BC| \Rightarrow \angle B = \angle A,$$

and

$$\angle B = \angle A \Rightarrow |AC| = |BC|.$$

If these two implications are combined,

$$|AC| = |BC| \Leftrightarrow \angle B = \angle A.$$

Many proofs and constructions use this property of the isosceles triangle.

15.20 Constructions of Isosceles Triangles

(*a*) We have already discussed the construction of an isosceles triangle when the lengths of the sides are given (Section 15.16).

(*b*) If the vertical angle A of an isosceles triangle ABC (the angle opposite the base) and the length of the equal sides BA and CA are given, the triangle ABC is easily constructed by drawing the angle A and marking off the points B and C at the given distance from A (Fig. 15.49).

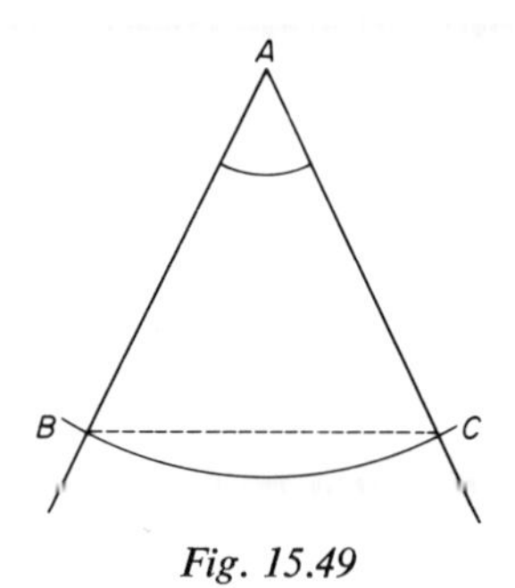

Fig. 15.49

(*c*) Since a triangle with two equal angles is necessarily isosceles, the construction of two equal angles at the end-points of a line segment will always produce an isosceles triangle (Fig. 15.50).

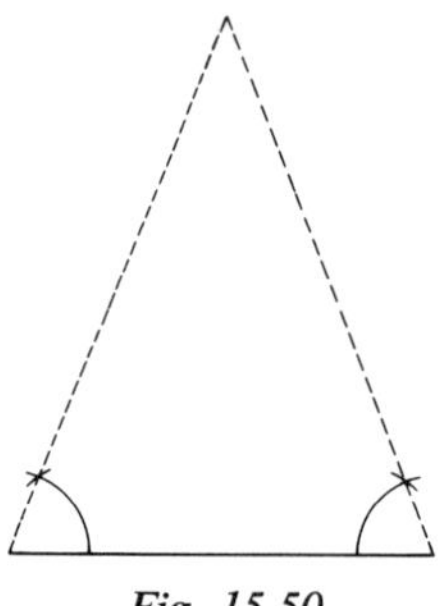

Fig. 15.50

15.21 The Equilateral Triangle

A triangle with three lines of symmetry is called an equilateral triangle.

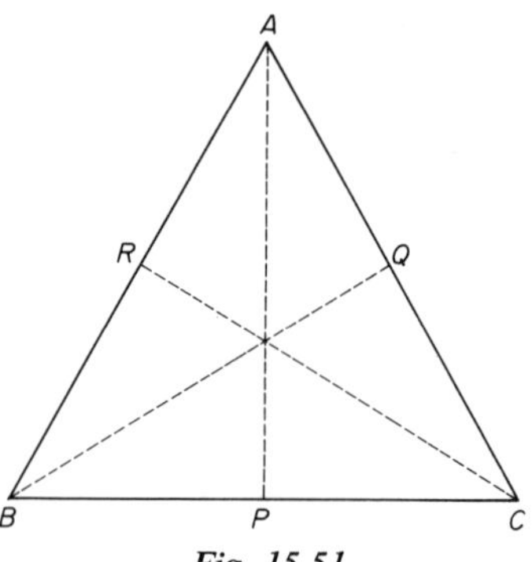

Fig. 15.51

It is an isosceles triangle, whichever way we look at it. Drawn on paper, as in Fig. 15.51, it can be folded along the line AP to show the congruence of

$\triangle$s $\frac{BPA}{CPA}$, implying $\angle B = \angle C$ and $|AB| = |AC|$; it can be folded along BQ to show the congruence of $\triangle$s $\frac{AQB}{CQB}$, implying that $\angle C = \angle A$ and $|BA| = |BC|$; again, it can be folded along CR to show congruence of $\triangle$s $\frac{ACR}{BCR}$, implying that $\angle A = \angle B$ and $|AC| = |BC|$.

Properties of the equilateral triangle

If we study these congruences and the resulting equalities of angles and line segments, we can see that:

(*a*) the three sides of an equilateral triangle are of equal length; hence the name 'equilateral' (Latin for 'equal-sided');
(*b*) the three angles of an equilateral triangle are equal; since the angle sum of a triangle is 180°, *each angle is* 60°.

Either of these properties is sufficient to establish that a triangle is equilateral.

15.22 Sufficient Conditions for a Triangle to be Equilateral

Theorem 15.6: *If the three sides of a triangle are of equal length, the triangle is equilateral and therefore the angles are equal.*
In the $\triangle$ ABC (Fig. 15.52), $|AB| = |BC| = |CA|$.

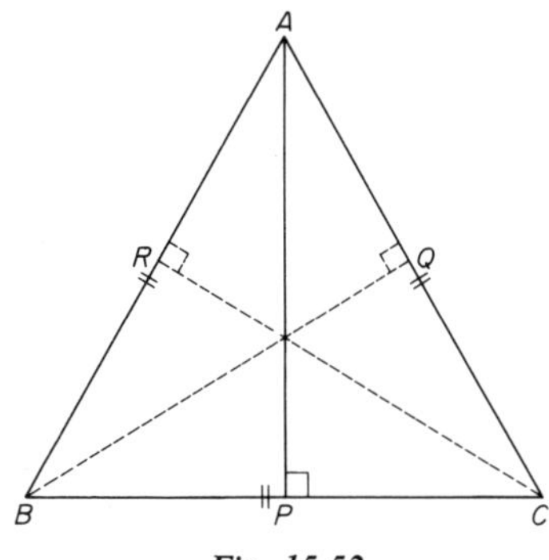

Fig. 15.52

Each angle of the triangle is considered in turn as the vertical angle of an isosceles triangle; the triangle is shown to be symmetrical about the axis through each vertex, and the angles are therefore all equal.

Taking $\angle A$ as the vertical angle of triangle ABC,
$|AB| = |AC| \Rightarrow \triangle ABC$ is isosceles and symmetrical about the axis AP

$$\Rightarrow \angle B = \angle C.$$

Taking $\angle B$ as the vertical angle,
$|BC| = |BA| \Rightarrow \triangle BCA$ is isosceles and symmetrical about the axis BQ

$$\Rightarrow \angle C = \angle A.$$

Taking $\angle C$ as the vertical angle,
$|AC| = |BC| \Rightarrow \triangle CAB$ is isosceles and symmetrical about the axis CR

$$\Rightarrow \angle A = \angle B.$$

So $$\angle A = \angle B = \angle C.$$

Theorem 15.7: *If the three angles of a triangle are equal, the triangle is equilateral and therefore the sides are of equal length.*

15.23 Exercise

1. Prove Theorem 15.7 by considering each pair of angles in turn as the two base angles of an isosceles triangle.
2. Why is it sufficient to know that two angles of a triangle are 60° each to decide that the triangle is equilateral?

15.24 Constructing Equilateral Triangles

Construction 15.5: *An equilateral triangle.*

This construction is based on the general construction of a triangle of which the lengths of the sides are given; the equality of the lengths of the three sides makes this a simple procedure.

For example, to construct an equilateral triangle whose sides are 6 cm long:

(*a*) Draw a line segment PQ which is 6 cm long.
(*b*) Place one of the points of your compasses on P and the other on Q; then, using this distance as radius, mark two arcs of circles with P and Q as centres. These arcs meet at the point S.
(*c*) Join the points P, Q and S.

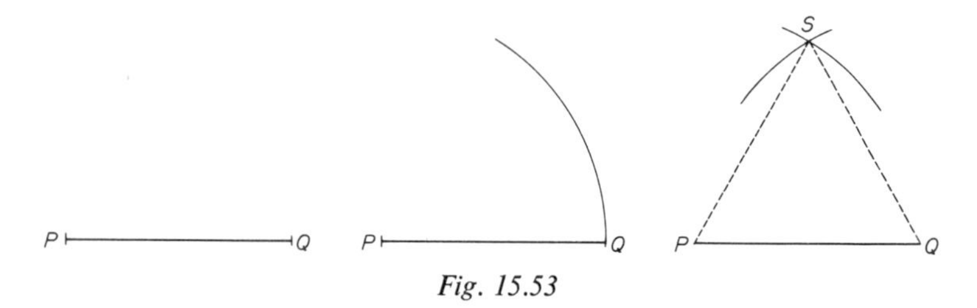

Fig. 15.53

Construction 15.6: *An angle of* 60°.

The construction of an equilateral triangle can also serve to construct an angle of 60°. In this case any length can be used as the side of the equilateral triangle.

15.25 Exercises

1. Construct an equilateral triangle whose sides are 4 cm long. Using each of the three sides of this equilateral triangle in turn as the base of another equilateral triangle, construct three other (congruent) equilateral triangles.
   Continue this process, and so design a 'joined-up' pattern of triangles, a *triangular grid.*
2. Construct an equilateral triangle whose sides are 8 cm long. Cut it out and fold it along the lines of symmetry.

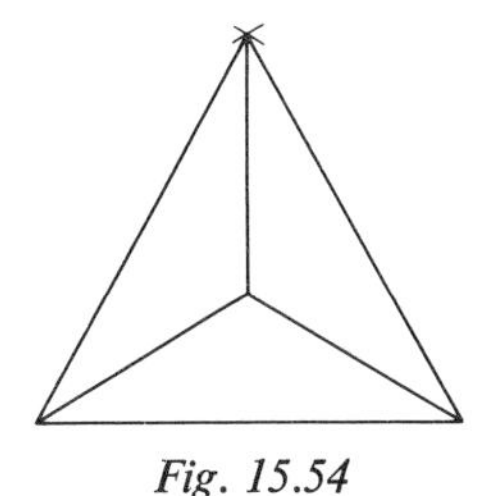

Fig. 15.54

   Cut out the three triangles formed by the 'centre' and the vertices of the triangle.
   (*a*) Test these three triangles for congruence.
   (*b*) Prove that these three triangles must be congruent.
   (*c*) Calculate the angles of these triangles.
3. Construct and cut out another equilateral triangle. Fold it along the lines of symmetry, and cut out the six triangles that are formed by the lines of symmetry and the sides of the triangle.

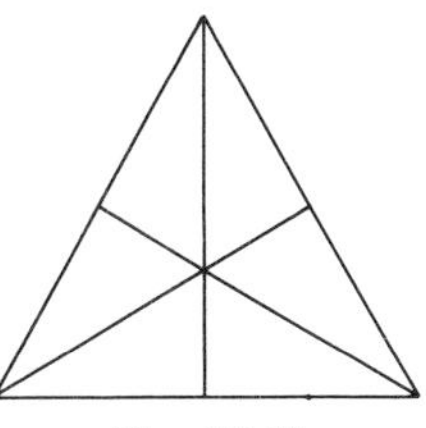

Fig. 15.55

   Prove that these six triangles are congruent.

15.26 The Rotational Symmetry of the Equilateral Triangle

The division of the equilateral triangle into three congruent isosceles triangles, meeting at a central point (see Exercise 2 of Section 15.25), reveals the rotational symmetry of the equilateral triangle (Fig. 15.56).

Any of these three triangles will generate an equilateral triangle when it is rotated.

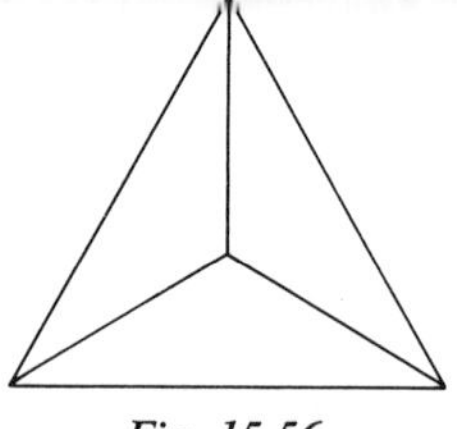

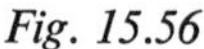

Fig. 15.56

On stiff paper construct an isosceles triangle PQR, whose angles are 120°, 30° and 30°. Cut it out, leaving a little of the paper round the vertex P.

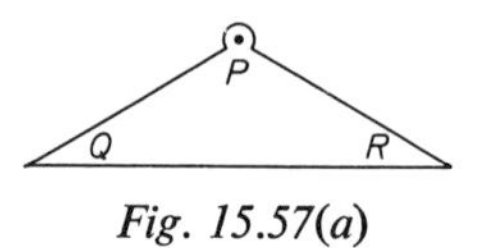

Fig. 15.57(a)

By drawing along its sides, use $\triangle PQR$ as a template to draw $\triangle ABC$. Hold point P in the same position, using a pin, and rotate the triangle PQR through an angle of 120°. Again, draw lines along the template and so construct $\triangle ACD$.

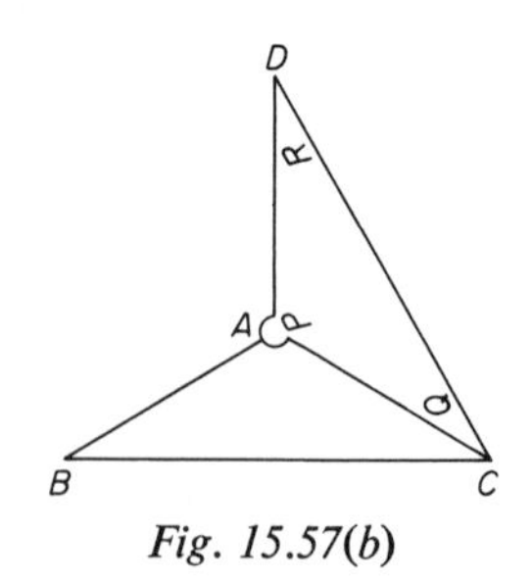

Fig. 15.57(b)

If this process is repeated once more, the equilateral triangle BCD is completed.

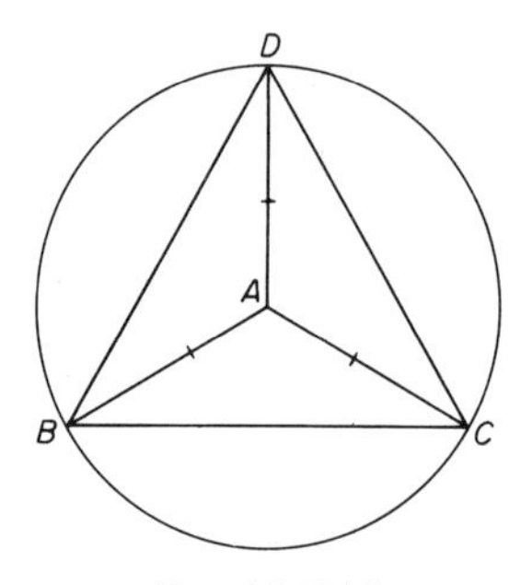

Fig. 15.57(c)

Since the points B, C and D are all at the same distance from A, we can draw a circle of radius $|AB|$ and centre A, which will pass through the points B, C

and D. The point A is called the *central point*, and each of the three angles at A is called the *angle at the centre*.

15.27 Exercise

Use a protractor to draw three angles of 120°, meeting at one point (S). Construct an equilateral triangle with point S as central point.

15.28 The Right-angled Triangle

If one angle of a triangle is a right angle, the triangle is called a *right-angled triangle*.

The side of the triangle opposite the right angle is named the *hypotenuse*; it is always the largest side of the triangle. (*Why?*) Unit 19, on Trigonometry, will deal with the ratios of the lengths of the sides of the right-angled triangle.

One interesting property of the lengths of the sides of the right-angled triangles was recognized long ago by the early Egyptians, and is usually known as Pythagoras' Theorem.

Theorem 15.8 (Pythagoras' Theorem): *In a right-angled triangle, the square on the hypotenuse equals the sum of the squares on the other two sides.*

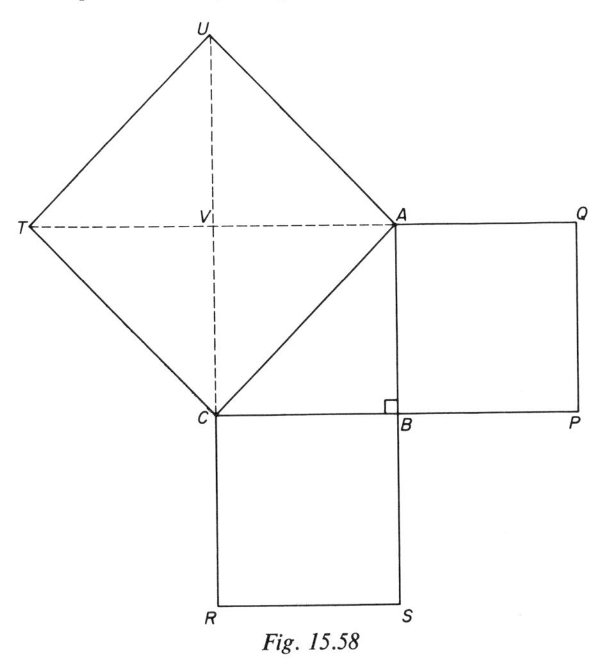

Fig. 15.58

Pythagoras' Theorem primarily concerns the *areas* of the squares on each of the sides of the right-angled triangle.

The simplest case is the triangle mentioned in Section 1.9 of which two sides are of unit length, say 1 dm each (Fig. 15.58).

When we draw squares on each of the sides of the triangle, we can see that the area of the larger square is equal to that of the sum of the areas of the two smaller squares if we cut the larger square into four congruent triangles, indicated in the diagram as $\triangle$s ACV, CTV, TUV and UAV. These four triangles can be placed upon the two squares $ABPQ$ and $BCRS$ and cover them

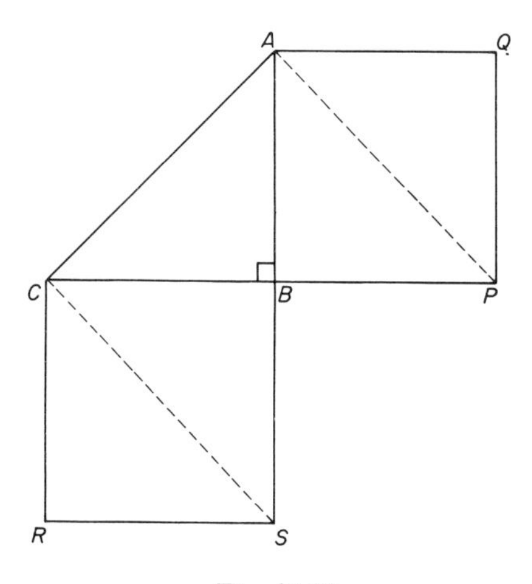

Fig. 15.59

exactly. The area of each of the two smaller squares (each 1 dm²) is covered by two of the triangles and the area of the larger square by four triangles, showing:

$$\text{area } ACTU = \text{area } ABPQ + \text{area } BCRS \,(= 2\,\text{dm}^2)$$

or

$$|AC|^2 = |AB|^2 + |BC|^2.$$

Clearly this relationship will hold whatever the length of $|AB|$ $(= |BC|)$. Pythagoras' Theorem applies indeed to any right-angled triangle; we can show this as follows:

Draw a square $ABCD$ and mark a point P somewhere on AB; mark points Q, R and S on the sides BC, CD and DA respectively so that $|AP| = |BQ| = |CR| = |DS|$. In Exercise 3 of Section 15.13, you were able to show that $PQRS$ is a square.

If the lengths of the sides of each of the (congruent) triangles APS, BQP, CRQ and DSR are p cm, q cm and r cm (Fig. 15.60),

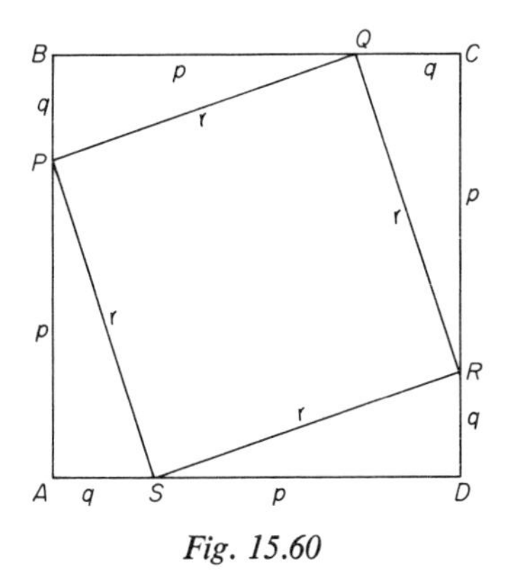

Fig. 15.60

Pythagoras' Theorem states that $r^2 = p^2 + q^2$, that is: (the area of the square of side r cm) = (the area of the square of side p cm) + (the area of the square of side q cm). You can make this clear by drawing the same square $ABCD$ again, and then drawing its component parts, the squares of side p cm and of side q cm and the four triangles, as in Fig. 15.61.

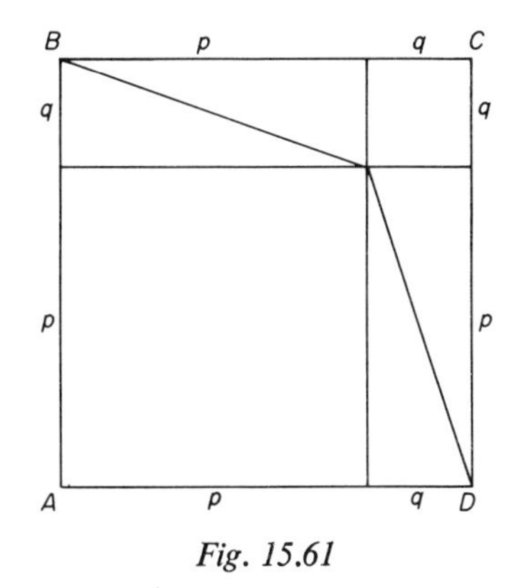

Fig. 15.61

The area of $ABCD$ is the same in both figures, however it is sub-divided; removing the four triangles (by cutting out) from each, the same areas are left: in the first instance one square of side r cm (area = r^2 cm²), in the second instance two squares, one of area p^2 cm² and the other of area q^2 cm². Therefore $r^2 = p^2 + q^2$.

We have confirmed that if a triangle PQR is right-angled at R, and if p, q and r are the lengths of the sides opposite the vertices P, Q and R, then:

$$r^2 = p^2 + q^2.$$

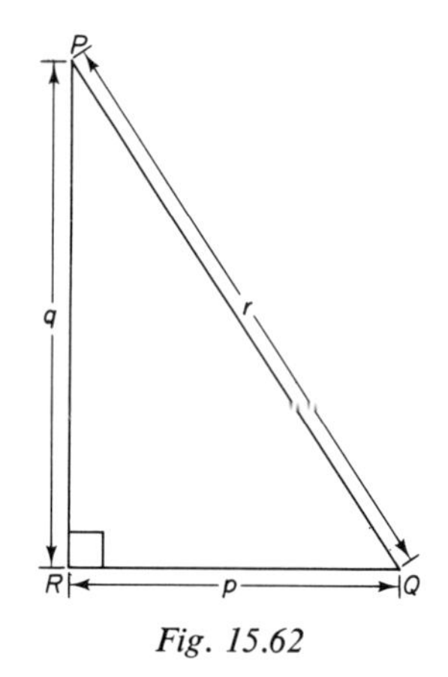

Fig. 15.62

Conversely, any triangle whose sides have these same lengths, p, q and r, is congruent with $\triangle PQR$ (S–S–S) and will therefore be right-angled.

15.29 Pythagorean Triples

Although Pythagoras' Theorem concerns the *areas* of the squares on the sides of a right-angled triangle, its usefulness derives mainly from the relation between the *lengths* of the sides that follows from it.

If the lengths of two sides of a right-angled triangle are known, the length of the third side can be calculated.

Example

$$\triangle ABC \text{ is right-angled at } A;\quad |AB| = 3\text{ cm};\quad |AC| = 4\text{ cm}.$$

By Pythagoras' Theorem: $|CB|^2 = |AB|^2 + |AC|^2$

$$\Rightarrow |CB|^2 = 3^2 + 4^2 = 9 + 16 = 25 = 5^2$$
$$\Rightarrow |CB| = 5.$$

(The negative value of the square root is ignored, since there are no negative lengths!)

These three numbers 3, 4 and 5 are called a *Pythagorean triple*; other Pythagorean triples are:

6, 8 and 10 ($6^2 + 8^2 = 10^2$),
5, 12 and 13 ($5^2 + 12^2 = 13^2$),
etc.

Usually, however, the lengths of the three sides of a triangle are not all whole numbers, nor even all rational numbers (see Section 1.9).

For example,

$$\triangle PQR \text{ is right-angled at } P;$$
$$|QR| = 5 \text{ cm};$$
$$|PQ| = 2 \text{ cm}.$$

By Pythagoras' Theorem:

$$|RQ|^2 = |PQ|^2 + |RP|^2$$
$$\Rightarrow 5^2 = 2^2 + |RP|^2$$
$$\Rightarrow |RP|^2 = 5^2 - 2^2 = 25 - 4 = 21$$
$$\Rightarrow |RP| = \sqrt{21} = 4{\cdot}583 \text{ cm (using four-figure square root tables).}$$

15.30 Exercises

1. Calculate the length of the third side in the right-angled triangles in Fig. 15.63.

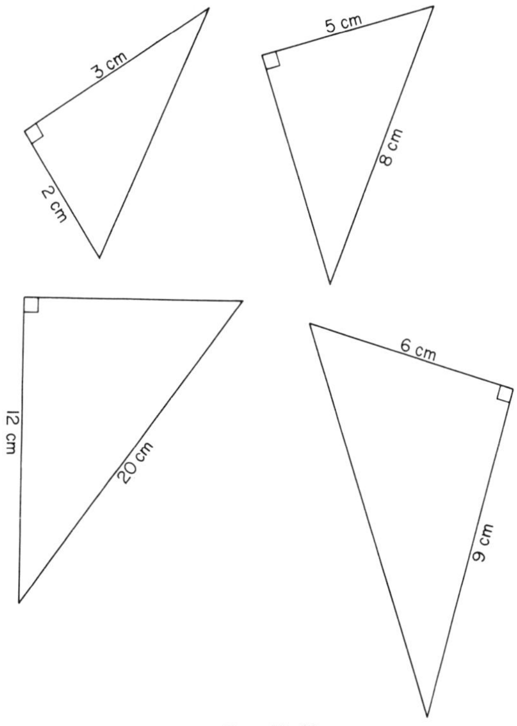

Fig. 15.63

2. The sides of a triangle are 10 cm, 24 cm and 26 cm. Decide whether this triangle is right-angled or not.

3. $\triangle ABC$ is an isosceles triangle; $|AB| = |CB| = 8$ cm and $|AC| = 6$ cm. If D is the mid-point of AC, calculate the length of the line segment BD (the altitude relative to AC).
4. The Egyptians used a 'knotted rope' to construct right angles: knots were laid in a rope at equal distances and the Pythagorean triple 3, 4 and 5 was used to form a right-angled triangle with the rope (Fig. 15.64).

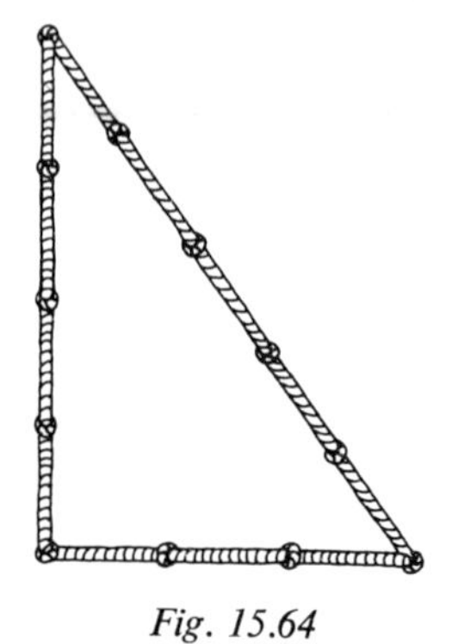

Fig. 15.64

Construct a right-angled triangle with compasses and ruler, using the same principle.

Unit Sixteen
Quadrilaterals and Other Polygons

16.1 The General Quadrilateral

Any plane shape bounded by straight-line segments is called a *polygon* (Greek for 'many-angled'); a plane shape enclosed by four line segments—a four-sided polygon—is named a *quadrilateral* (Latin for 'four-sided').

Like triangles, quadrilaterals come in all shapes and sizes! Some have special features, such as equality of angles or of lengths of sides, and some have symmetrical properties not found in others. These special quadrilaterals such as the square, parallelogram and kite may be more familiar than the general quadrilateral which does not necessarily show any of these characteristics.

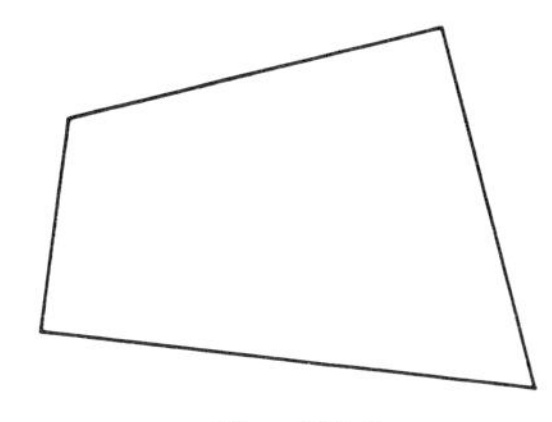

Fig. 16.1

We have already noticed (Section 15.8) a rather negative property of the quadrilateral: four given line segments do not determine a uniquely shaped quadrilateral. If four meccano strips are joined at their end-points, the framework is *not* rigid; they provide an endless variety of possible quadrilaterals.

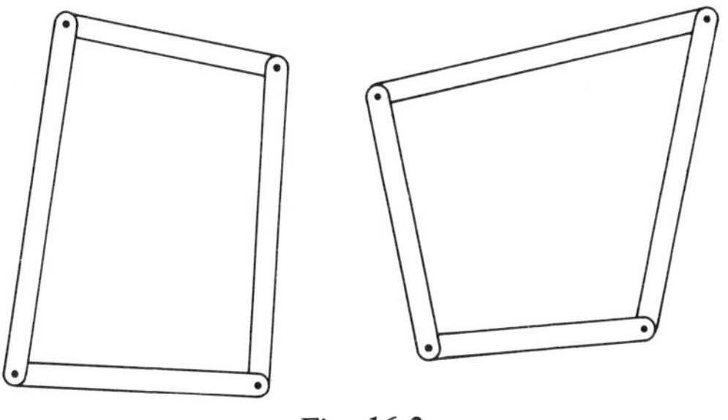

Fig. 16.2

We can, however, make this framework rigid if we add a fifth strip, a strut, to form two triangles (Fig. 16.3).

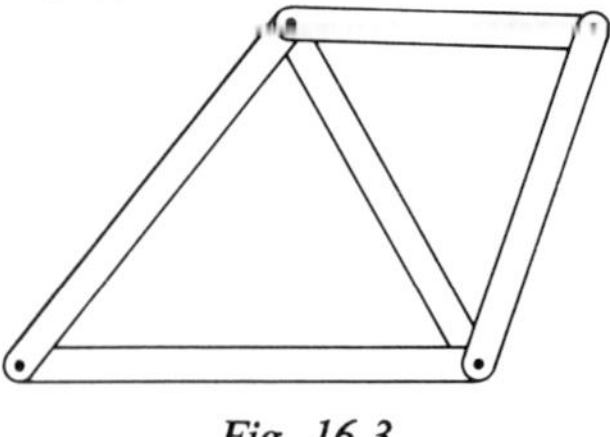

Fig. 16.3

The line which joins two different vertices of a polygon which are not the end-points of the same line segment is called a *diagonal.* Any quadrilateral has two such diagonals, each dividing it into two triangles.

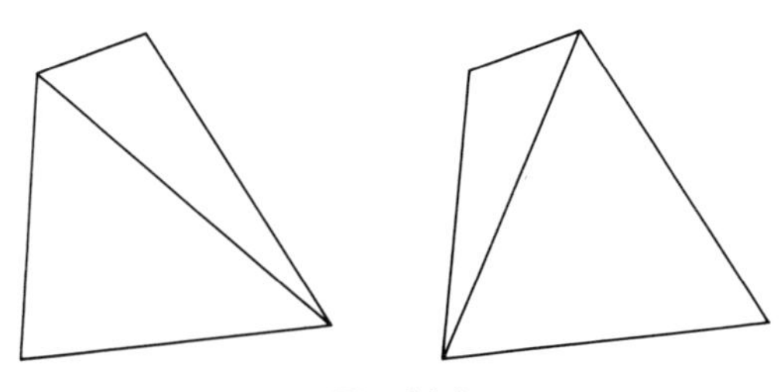

Fig. 16.4

This demonstrates the following property of *all* quadrilaterals:

Theorem 16.1: *The sum of the angles of any quadrilateral is* 360°.
We draw any quadrilateral $ABCD$, and show that

$$\angle A + \angle B + \angle C + \angle D = 360°.$$

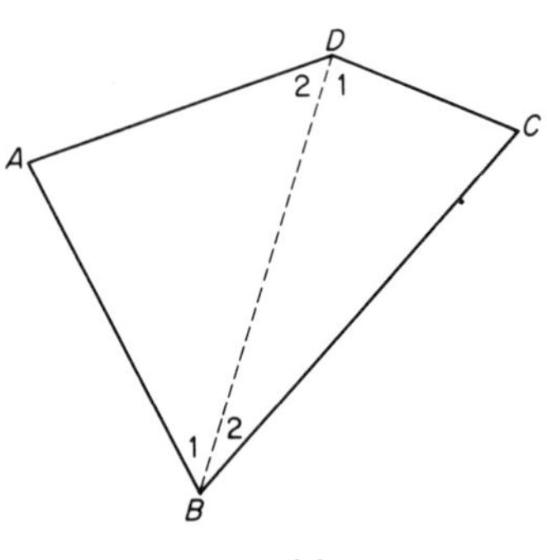

Fig. 16.5

We draw diagonal BD and mark the angles as in Fig. 16.5.

$$\angle A + \angle B_1 + \angle D_2 = 180° \ (ABC \text{ is a triangle});$$
$$\angle B_2 + \angle D_1 + \angle C = 180° \ (BCD \text{ is a triangle});$$

Adding both sides of these equations:

$$\angle A + \underbrace{\angle B_1 + \angle B_2} + \angle C + \underbrace{\angle D_1 + \angle D_2} = 360°$$

$$\Rightarrow \angle A + \angle B + \angle C + \angle D = 360°.$$

You can confirm this result by a simple experiment:

Draw any quadrilateral on a sheet of paper and cut it out. Tear or cut off the corners, and place them together so that all the four vertices meet at a point without overlap (Fig. 16.6).

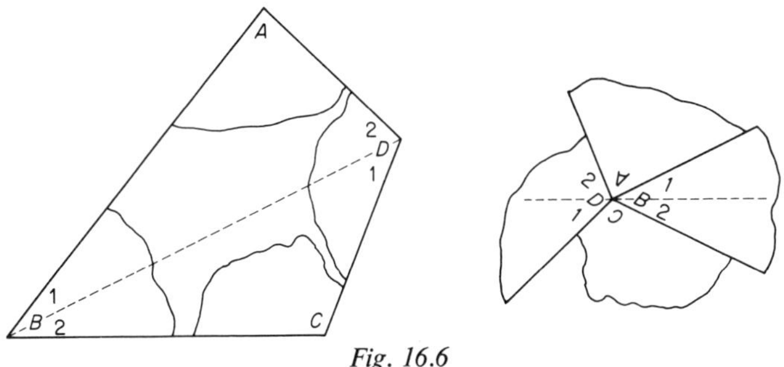

Fig. 16.6

Clearly, the four angles at the vertices form an angle of 360°.

16.2 The Angle Sum of any Polygon

We can establish the angle sum of any *polygon* of more than four sides in this way, that is, by dividing it into triangles by drawing diagonals. As an exercise, sketch any five-sided polygon, a *pentagon*, and find how many different diagonals can be drawn from one of its vertices. You should be able to show that the sum of the 5 angles of the pentagon is the same as the sum of the angles of the 3 triangles formed by the sides of the pentagon and the diagonals through one vertex.

A six-sided polygon, a *hexagon*, can be made from a pentagon by replacing one of the sides of the pentagon by two sides (Fig. 16.7).

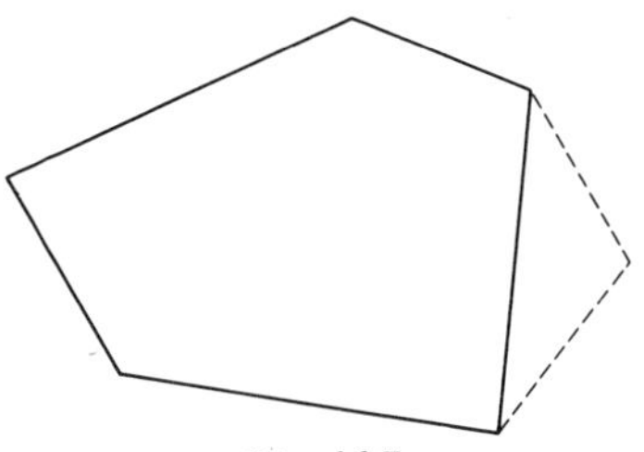

Fig. 16.7

Examine this diagram closely. Clearly, if we increase the number of sides of this pentagon by 1, we also increase by 1 the number of triangles into which the polygon can be divided by its diagonals; and therefore we increase the angle sum of the polygon by 180°.

When we try to draw diagonals from one of the vertices of *any* polygon—say from the vertex A of a polygon $ABCDEFGH$ (Fig. 16.8)—we find that diagonals are produced by joining A to *all but three* of the vertices of the polygon. The exceptions are B and H (the line segments AB and AH are sides of the polygon) and, of course, A itself.

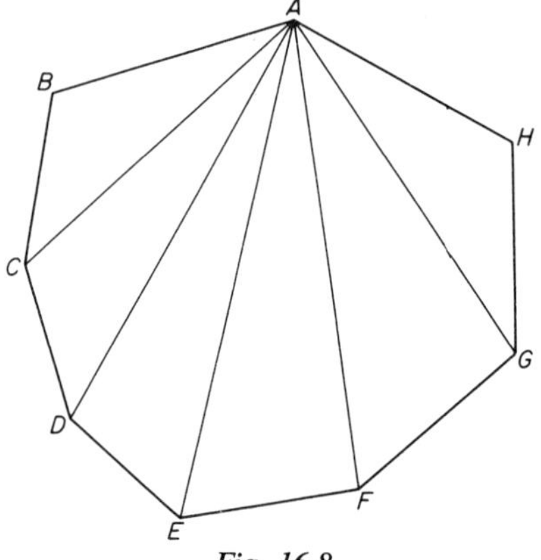

Fig. 16.8

The number of diagonals of a polygon through one of its vertices is therefore three less than the number of its sides; in an n-sided polygon there are $(n - 3)$ diagonals through any one of its vertices. These diagonals divide the polygon into $(n - 2)$ triangles.

This means that we can write down the formula for the angle sum of any polygon:

the sum of the angles of an n-sided polygon is $(n - 2) \times 180°$

16.3 Exercise

Check the formula for the angle sum of the polygon by drawing diagonals through one of the vertices of a five-sided polygon, a six-sided polygon, a seven-sided polygon and an eight-sided polygon.

Calculate the angle sum of each of these polygons.

16.4 The Kite

A kite (Fig. 16.9) is one of the special quadrilaterals mentioned in Section 16.1: it is symmetrical about one diagonal. The other diagonal divides the kite into two isosceles triangles.

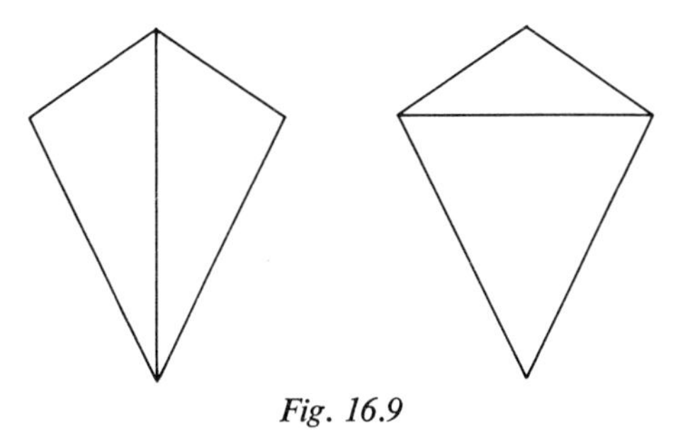

Fig. 16.9

You will appreciate both these aspects of the kite if you 'construct' it by folding a sheet of paper and then drawing two line segments AB and BC so that both A and C are points on the line of the fold (Fig. 16.10).

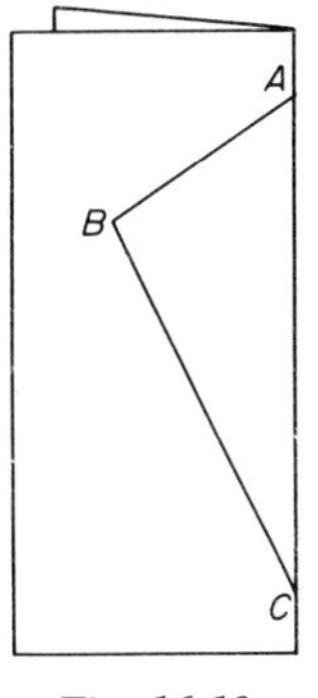

Fig. 16.10

Now prick a hole through the point B, marking a corresponding point D. When the sheet is unfolded and the points A and D and C and D are joined by straight lines, the figure $ABCD$ is a kite (Fig. 16.11).

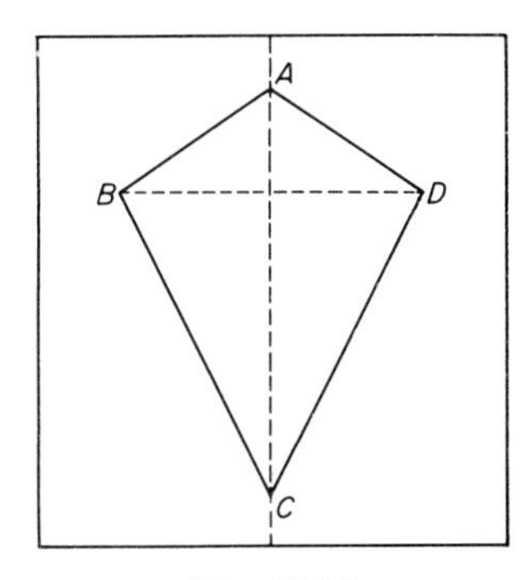

Fig. 16.11

The two triangles $\begin{matrix} ABC \\ ADC \end{matrix}$ are congruent;
$\triangle BDA$ is an isosceles triangle;
$\triangle BDC$ is an isosceles triangle.

16.5 Exercise

In the kite $ABCD$, mark all the pairs of equal angles and of line segments of equal length, using your previous work on congruent and isosceles triangles, and so confirm the properties of the kite listed in Section 16.6.

16.6 Properties of the Kite

(*a*) A kite $ABCD$ (Fig. 16.11) is symmetrical about the diagonal AC, or $\triangle$s $\begin{matrix} ABC \\ ADC \end{matrix}$ are congruent.

(*b*) The diagonal BD divides a kite into two isosceles triangles, $\triangle BDA$ and $\triangle BDC$.

(*c*) Two pairs of adjacent sides are of equal length (Fig. 16.12):

$$|AB| = |AD| \text{ and } |BC| = |DC|.$$

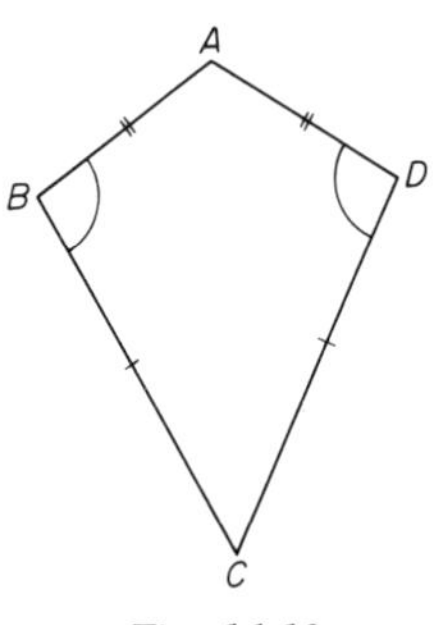

Fig. 16.12

(*d*) Two angles of a kite are equal:

$$\angle B = \angle D.$$

(*e*) The diagonal AC, the axis of symmetry, bisects the angle A and the angle C.

(*f*) The axis of symmetry, AC, divides each of the two isosceles triangles, $\triangle BDA$ and $\triangle BDC$ into two congruent triangles (Fig. 16.13):

$\triangle$s $\begin{matrix} AEB \\ AED \end{matrix}$ are congruent, and $\triangle$s $\begin{matrix} BEC \\ DEC \end{matrix}$ are congruent.

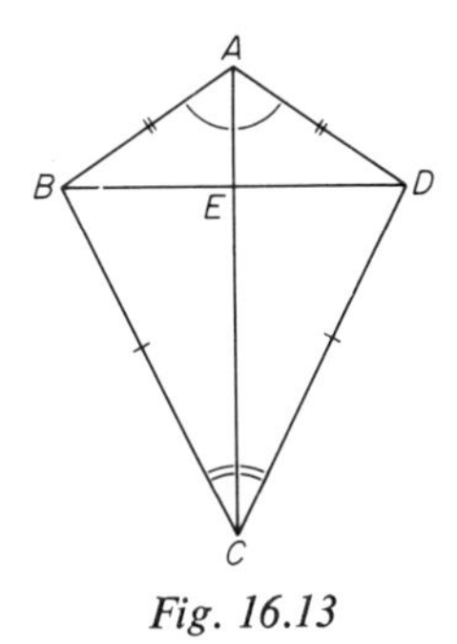

Fig. 16.13

(*g*) The diagonal AC divides the diagonal BD into two line segments of equal length, or $|BE| = |ED|$; that is, one diagonal of a kite is bisected by the other, although the other diagonal is not necessarily bisected by the first.

This follows both from the congruence of $\triangle$s $\frac{AEB}{AED}$, and from the congruence of $\triangle$s $\frac{BEC}{DEC}$, mentioned in (*e*).

(*h*) The diagonals of a kite are perpendicular; since $\triangle BDA$ and $\triangle BDC$ are isosceles triangles, the axis of symmetry AC is an altitude, as well as a median and an angle bisector, of these triangles.

(*i*) The diagonal AC is the perpendicular bisector of the diagonal BD, as a consequence of (*g*) and (*h*).

16.7 Sufficient Conditions for a Quadrilateral to be a Kite

We can show that a quadrilateral is a kite if *one* of the properties (*a*), (*b*), (*c*), (*e*), (*f*) and (*i*) can be identified. Theorems 16.2 and 16.3 state this for properties (*a*), (*b*) and (*c*), and it can be demonstrated in the other cases by studying the congruent triangles of the kite.

Theorem 16.2: *If the four sides of a quadrilateral are two pairs of adjacent sides of equal length, the quadrilateral is a kite.*

All the other properties of a kite follow from this, in particular the symmetry about a diagonal.

Suppose that, in a quadrilateral $ABCD$, $|AB| = |AD|$ and $|CB| = |CD|$ (Fig. 16.14).

$|AB| = |AD| \Rightarrow \triangle ABD$ is isosceles and symmetrical about AE (E is the mid-point of BD);

also $\angle AEB = 90°$.

$|BC| = |CD| \Rightarrow \triangle BDC$ is isosceles and symmetrical about EC;

also $\angle BEC = 90°$.

Therefore, $\angle AEC = 90° + 90°$, or AC is a straight line and is the axis of symmetry of $ABCD$.

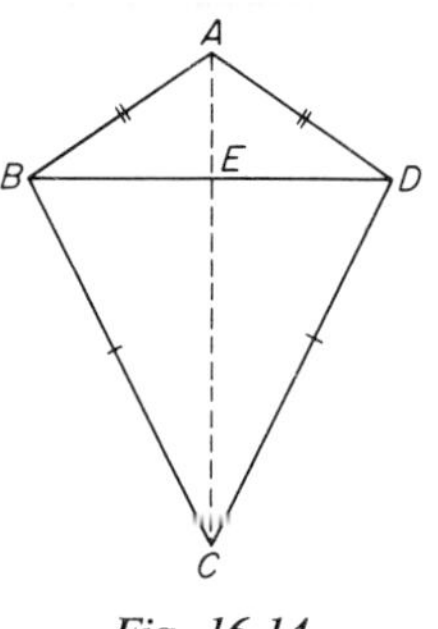

Fig. 16.14

Theorem 16.3: *If a quadrilateral is symmetrical about one of its diagonals, it is a kite and therefore two pairs of adjacent sides are of equal length.*

Suppose that the quadrilateral $ABCD$ is symmetrical about AC (Fig. 16.15).

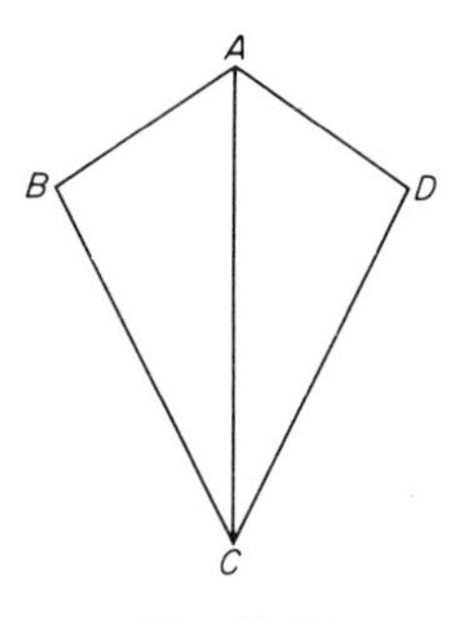

Fig. 16.15

When folded about the line AC, the line segment AB will therefore coincide with the line segment AD

$$\Rightarrow |AB| = |AD|, \text{ and } \triangle ABD \text{ is isosceles.}$$

Also the line segment BC will coincide with the line segment DC

$$\Rightarrow |BC| = |DC| \text{ and } \triangle BCD \text{ is isosceles.}$$

In other words, the line segments AB and AD (meeting at the point A and therefore adjacent) are of equal length and the line segments BC and DC (meeting at the point C and therefore adjacent) are of equal length.

16.8 Exercise

Show that each of the properties (e) and (i) in Section 16.6 is a sufficient condition for a quadrilateral to be a kite.

16.9 Constructing a Kite

Construction 16.1: *A kite.*

Properties (*a*), (*b*), (*c*) and (*i*) can each be used as the basis of a construction of a kite; the first three of these are the most useful.

A kite can be constructed with compasses and ruler in three ways:

(*a*) by constructing an isosceles triangle on either side of a line segment (Fig. 16.16);

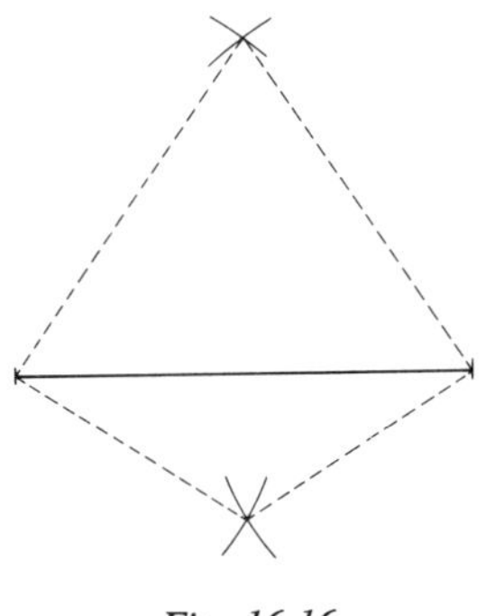

Fig. 16.16

(*b*) by constructing two congruent triangles on either side of a line segment (Fig. 16.17), making sure that the corresponding sides meet at the end points of the line segment, which is then the axis of symmetry of the kite;

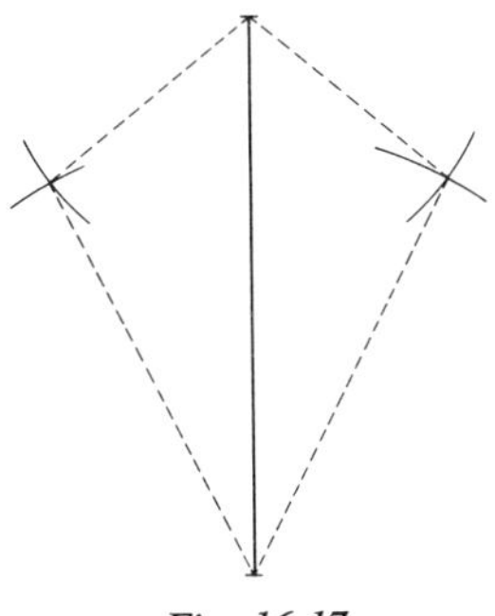

Fig. 16.17

(*c*) by using the construction of the isosceles triangle (Section 15.20) to draw an angle A, mark off equal lengths $|AB|$ and $|AD|$ on the arms of angle A, and construct another isosceles $\triangle CDB$ on the base BD by marking off arcs of circles using points B and D as centres (Fig. 16.18).

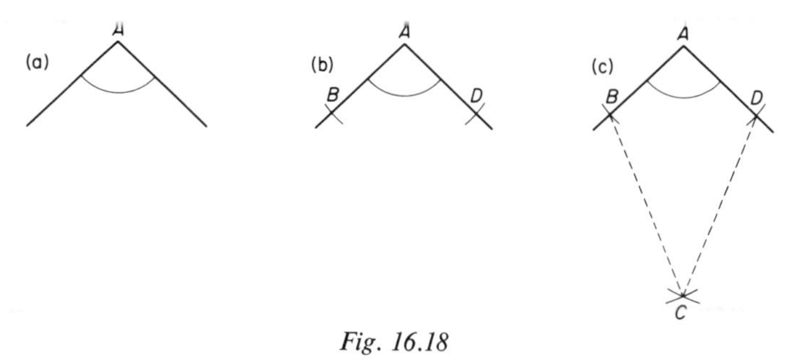

Fig. 16.18

16.10 Exercise

Construct a pair of intersecting perpendicular lines. Use either property (g) or property (i) to construct a kite with these lines as the diagonals.

16.11 Constructions Based on the Kite: Bisecting an Angle

The kite is particularly important because of the constructions that are based on it.

Construction 16.2: *To bisect a given angle using compasses and ruler.*

Construct a kite using the given angle A as one of the two vertices where two equal sides meet (see Construction 16.1(c), Section 16.9).

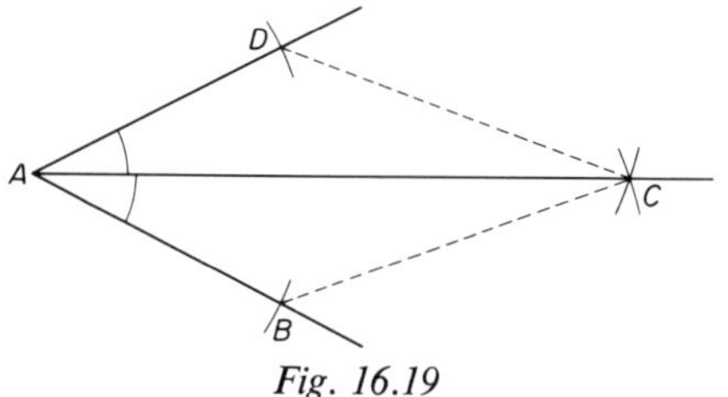

Fig. 16.19

Since the diagonal AC in Fig. 16.19 bisects the angles A and C (property (e) in Section 16.6), the line AC is the required line. (It is unnecessary to draw the line segments DC and BC.)

If the setting of the points of the compasses is left unaltered during the construction, so that

$$|AD| = |AB| = |BC| = |DC|,$$

the figure $ABCD$ is a special kite called the *rhombus*, which we will discuss further in Section 16.24.

16.12 Exercises

1. Construct an angle of 60° (see Construction 15.6 in Section 15.24); bisect this angle and so construct an angle of 30°.
2. Construct an angle of 15°.
3. Construct an angle of 150°. [Hint: $150° + 30° = 180°$, a straight line.]
4. Draw a line l and mark a point A on this line; we may speak of angle A as an angle of 180°. Bisect this angle A, using the procedure for bisecting any angle. Compare this construction with the construction of a right angle (Section 16.13).
5. Construct an angle of 45°.

16.13 More Constructions Based on the Kite

Construction 16.3: *A line perpendicular to a given line l through a point P not on this line.*

Again, construct a kite $PABC$ (Fig. 16.20) so that the point P is one of the vertices where two equal sides meet, and the points A and C are points on the line l. The diagonal PB is perpendicular to the diagonal AC and is therefore the required line.

(*a*) Using compasses, mark two points A and C on the line l so that $|AP| = |CP|$ ($\triangle ACP$ is isosceles);
(*b*) construct another isosceles triangle ACB on the base AC, marking equal distances from A and C;
(*c*) join the points P and B by a straight line.

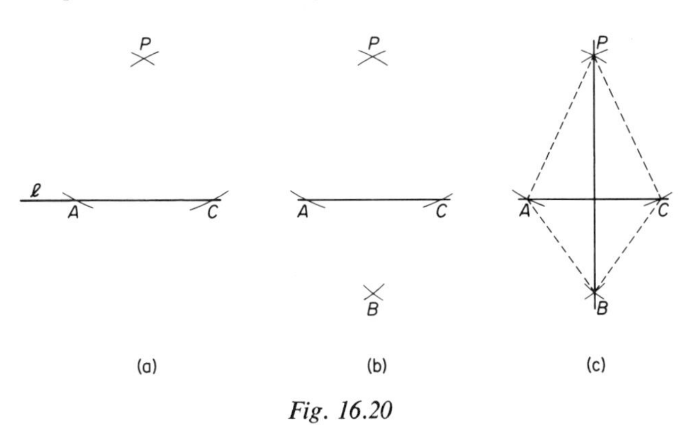

Fig. 16.20

Construction 16.4: *The perpendicular bisector of a given line segment.*

Construct two isosceles triangles ABC and ABD, both having the given line segment AB as their base. In the kite $CADB$ (Fig. 16.21), the diagonal CD bisects the diagonal AB at right angles, and CD is the required line.

The simplest procedure is to construct two congruent isosceles triangles ABC, ABD, in which $|AB| = |BC| = |AD| = |BD|$, so forming a rhombus (Section 16.24); the setting of the points of the compasses need not then be altered during the construction.

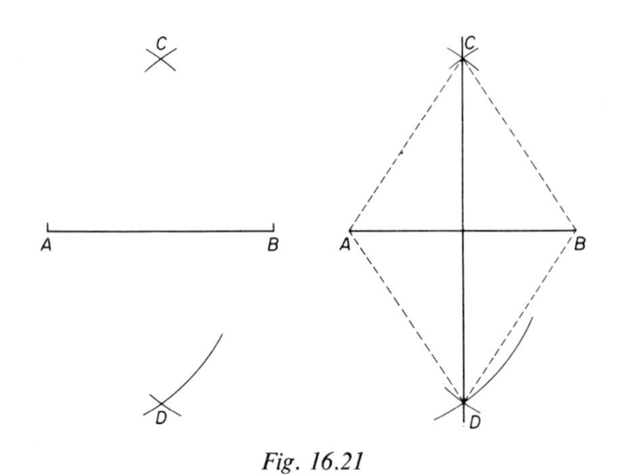

Fig. 16.21

Construction 16.5: *To find the mid-point of a line segment by construction.*

Since the perpendicular bisector of a line segment passes through its mid-point, Construction 16.4 provides a method of finding this point. There is no need to draw the perpendicular bisector; it is sufficient to mark neatly the point where it intersects the line segment (Fig. 16.22).

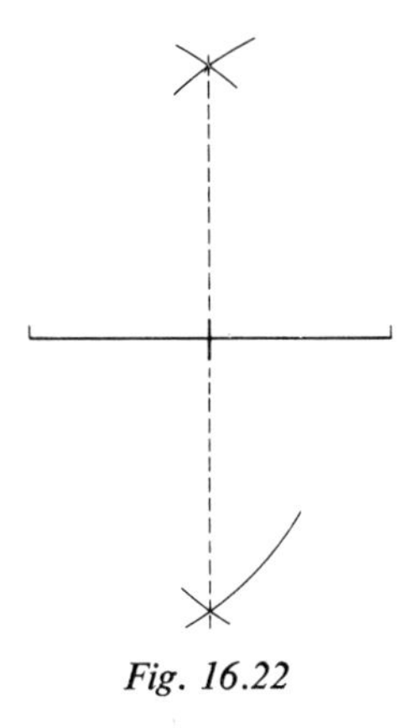

Fig. 16.22

Practise all these constructions until you are thoroughly familiar with them.

16.14 The Parallelogram

Another quadrilateral with interesting properties is the parallelogram (Fig. 16.23). A parallelogram is a quadrilateral in which the opposite sides are parallel and of equal length; the opposite angles are equal and each of the diagonals divides the parallelogram into two congruent triangles.

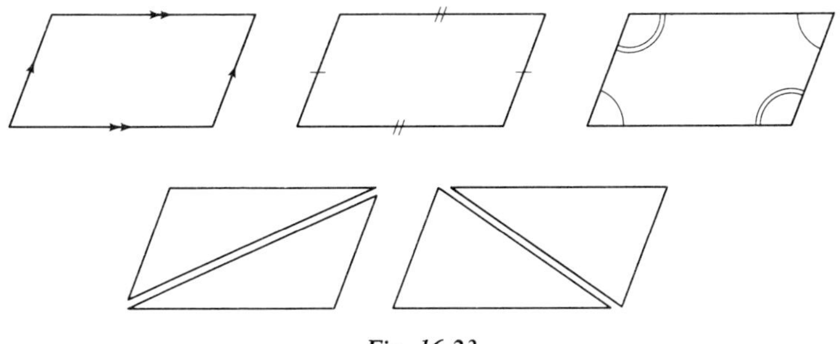

Fig. 16.23

The two congruent halves of the parallelogram are not mirror images of each other in the diagonal, unlike the component halves of the kite: the parallelogram is *not* symmetrical about the diagonal. You can demonstrate this by drawing a parallelogram on tracing paper and folding the paper along the line of the diagonal (Fig. 16.24): the corresponding sides of the two congruent triangles do not coincide (unless the parallelogram happens to be a rhombus, which is a kite as well as a parallelogram).

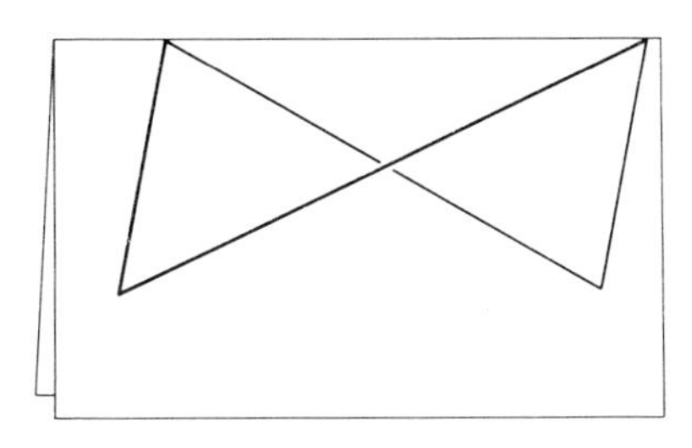

Fig. 16.24

If we want to place the two triangles formed by one diagonal of a parallelogram one over the other to demonstrate their congruence, we have to cut the parallelogram along the diagonal and rotate one of the sheets of paper. In fact, a parallelogram can be constructed by drawing a triangle ABC and rotating it through an angle of 180°:

On a sheet of stiff paper draw a triangle PQR (Fig. 16.25(*a*)) and mark the mid-point of one of the sides, say point M, the mid-point of PR. Cut this triangle out, leaving a little of the paper around point M.

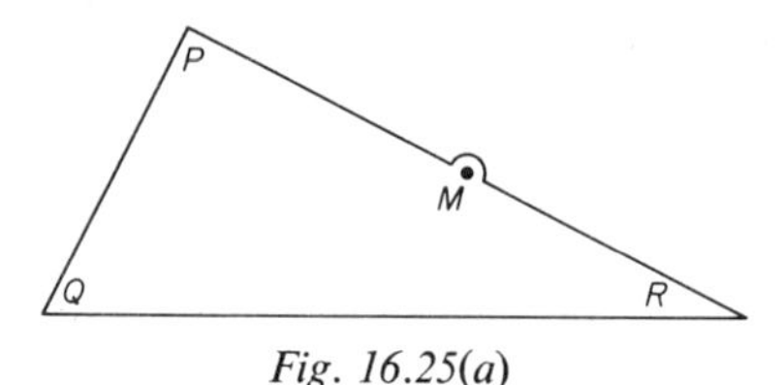

Fig. 16.25(a)

Use this triangle as a template to draw a triangle ABC, congruent with $\triangle PQR$.

Hold the point M in the same position (use a pin) and rotate the template through an angle of 180°, and again draw lines along the sides of the template, giving triangle ACD (Fig. 16.25(*b*)).

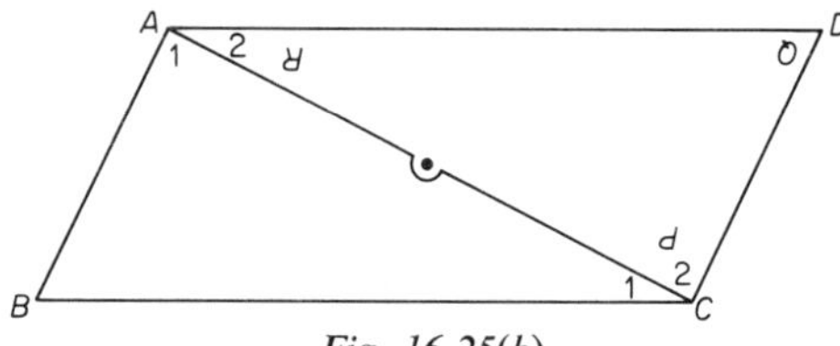

Fig. 16.25(b)

It is obvious from the construction that the quadrilateral $ABCD$ consists of two congruent triangles $\begin{matrix} ABC \\ CDA \end{matrix}$

(*a*) $\Rightarrow |AB| = |CD|$ and $|BC| = |DA|$
$\Rightarrow$ the opposite sides of $ABCD$ are of equal length.

(*b*) $\Rightarrow \angle B = \angle D$

and $\angle A_1 = \angle C_2$
$\angle A_2 = \angle C_1$

Adding, $\angle A_1 + \angle A_2 = \angle C_1 + \angle C_2$
$\Rightarrow \angle A = \angle C$
$\Rightarrow$ the opposite angles are equal.

(*c*) $\Rightarrow \angle A_1 = \angle C_2 \Rightarrow AB$ and CD are parallel lines (alternate angles), and
$\Rightarrow \angle A_2 = \angle C_1 \Rightarrow AD$ and BC are parallel lines (alternate angles)
$\Rightarrow$ the opposite sides are parallel.

(*d*) If the diagonal BD is drawn, $\triangle$s $\begin{matrix} ABD \\ CDB \end{matrix}$ are congruent (S–S–S):

$|AB| = |CD|$
$|DA| = |BC|$
$|BD| = |DB|$ (common side of both triangles).

Each of the diagonals therefore divides the parallelogram into two congruent triangles.

We have 'generated' a parallelogram by rotating a triangle and have shown that a parallelogram has *rotational symmetry* (see Section 14.2).

Note. The symbol / / is used as a mathematical notation for 'is parallel to'; we may, for example, write AD / / BC, that is, AD is parallel to BC.

16.15 Exercises

1. Use the template of triangle PQR to draw a parallelogram $ABCD$ with diagonal AC; mark all equal angles and sides and indicate pairs of parallel lines by arrows.
2. Draw another parallelogram $ABCD$ with both diagonals AC and BD; mark all equal angles and line segments. Use this to confirm the properties of the parallelogram detailed in Section 16.16.

16.16 Properties of the Parallelogram

The properties of the parallelogram, including those mentioned already, are:

(*a*) Each diagonal divides a parallelogram into two congruent triangles.
(*b*) Each diagonal bisects the area of a parallelogram (see also Section 20.9).
(*c*) Both pairs of opposite sides of a parallelogram are of equal length.
(*d*) Both pairs of opposite sides of a parallelogram are parallel.
(*e*) Both pairs of opposite angles are equal.
(*f*) The two diagonals together divide a parallelogram into two pairs of congruent triangles.

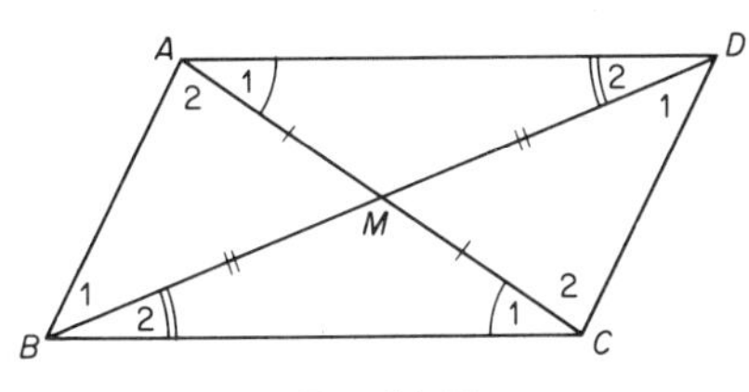

Fig. 16.26

Using parallelogram $ABCD$ of Exercise 2 of Section 16.15,
$\angle A_1 = \angle C_1$ (alternate angles formed by the parallel sides AD and BC and diagonal AC);
$|AD| = |BC|$ (property (*c*));
$\angle D_2 = \angle B_2$ (alternate angles).
$\triangle\text{s}\,\genfrac{}{}{0pt}{}{ADM}{CBM}$ are therefore congruent (*A–S–A*).

Similarly, $\triangle\text{s}\,\genfrac{}{}{0pt}{}{ABM}{CDM}$ are congruent (*A–S–A*). (Show why this is so, as an exercise.)

(*g*) The two diagonals of a parallelogram bisect each other, that is, they divide each other into two line segments of equal length; this follows immediately from the congruence of $\triangle\text{s}\,\genfrac{}{}{0pt}{}{ADM}{CBM}$ (property (*f*)). BM and DM are corresponding sides, and CM and AM are corresponding sides; therefore, $|BM| = |DM|$ and $|CM| = |AM| \Rightarrow$ the two diagonals bisect each other.

16.17 Sufficient Conditions for a Quadrilateral to be a Parallelogram

Each of these seven properties is a sufficient condition for a quadrilateral to be a parallelogram; that is, if a quadrilateral has any one of these seven properties, it must be a parallelogram.

Section 16.14 showed that all the properties of the parallelogram follow from the congruence of the triangles formed by each of the diagonals (property (*a*)). Property (*a*) is therefore a sufficient condition for a quadrilateral to be a parallelogram.

To demonstrate the sufficiency of most of the other properties, we show in each case that the property concerned implies the congruence of these triangles.

Theorem 16.4: *If the opposite sides of a quadrilateral are of equal length, the quadrilateral is a parallelogram.*

In a quadrilateral $ABCD$ (Fig. 16.27),

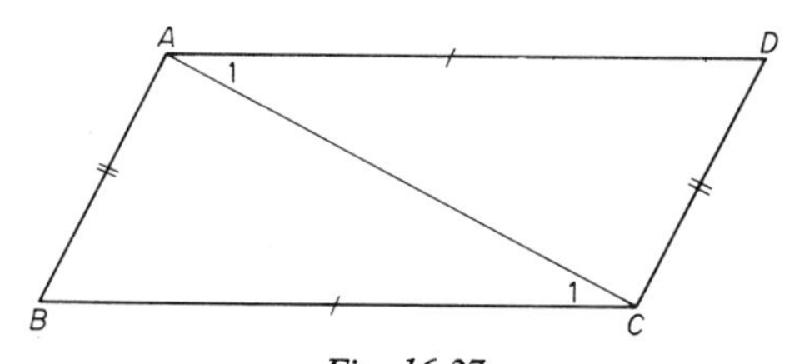

Fig. 16.27

$$|AB| = |CD| \text{ and}$$
$$|AD| = |CB|.$$

We draw the diagonal AC which is the common side of $\triangle ABC$ and $\triangle CDA$

$$\text{and } |AC| = |CA|.$$

Then $\triangle$s $\begin{matrix} ABC \\ CDA \end{matrix}$ are congruent (S–S–S).

Similarly, $\triangle$s $\begin{matrix} ABD \\ CDB \end{matrix}$ are congruent (S–S–S).

All the other properties of the parallelogram follow from this congruence. For example, AD and BC are parallel (since the alternate angles $\angle A_1$ and $\angle C_1$ are equal), and similarly AB and CD are parallel.

Theorem 16.5: *If the opposite sides of a quadrilateral are parallel, the quadrilateral is a parallelogram.*

In a quadrilateral $ABCD$ (Fig. 16.28),

AB is parallel to CD and AD is parallel to BC

or

$AB \,//\, CD$ and $AD \,//\, BC$

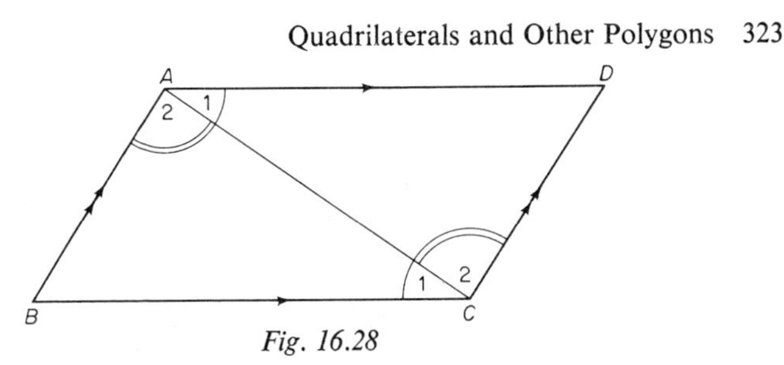

Fig. 16.28

Again, we draw the diagonal AC which divides $ABCD$ into two triangles BAC and DCA.

$$\angle A_1 = \angle C_1 \text{ (alternate angles of parallel sides } AD \text{ and } BC);$$
$$|AC| = |CA| \text{ (common side of } \triangle BAC \text{ and } \triangle DCA);$$
$$\angle A_2 = \angle C_2 \text{ (alternate angles)}$$
$$\Rightarrow \triangle\text{s } \begin{matrix} BAC \\ DCA \end{matrix} \text{ are congruent } (A\text{–}S\text{–}A).$$

Similarly, $\triangle\text{s } \begin{matrix} ABD \\ CDB \end{matrix}$ are congruent.

All the other properties of the parallelogram follow from this congruence; for example,

$|AB| = |CD|$ and $|AD| = |CB| \Rightarrow$ the opposite sides of $ABCD$ are of equal length;

$\angle B = \angle D$ and $\angle A_1 + \angle A_2 = \angle C_1 + \angle C_2 \Rightarrow \angle A = \angle C \Rightarrow$ opposite angles are equal.

Theorem 16.6: *If the diagonals of a quadrilateral bisect each other, the quadrilateral is a parallelogram.*

Let M be the point of intersection of the diagonals of a quadrilateral $ABCD$ (Fig. 16.29).

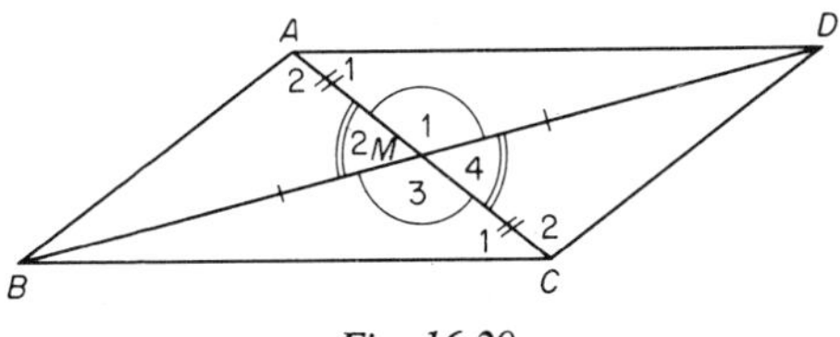

Fig. 16.29

We know that $|AM| = |MC|$

and $|DM| = |MB|$;

also $\angle M_1 = \angle M_3$ (vertically opposite angles);

$$\Rightarrow \triangle\text{s } \begin{matrix} AMD \\ CMB \end{matrix} \text{ are congruent } (S\text{–}A\text{–}S).$$

Similarly, $\triangle$s $\begin{matrix}AMB\\CMD\end{matrix}$ are congruent (S–A–S).

From these congruences follow the congruences of $\triangle$s $\begin{matrix}ABD\\CDB\end{matrix}$ and $\triangle$s $\begin{matrix}BAC\\DCA\end{matrix}$ and all the other properties of the parallelogram; for example,

$$\angle A_1 = \angle C_1 \Rightarrow AD \text{ and } BC \text{ are parallel, and}$$
$$\angle A_2 = \angle C_2 \Rightarrow AB \text{ and } CD \text{ are parallel.}$$

Theorem 16.7: *If one pair of opposite sides of a quadrilateral are parallel and of equal length, the quadrilateral is a parallelogram.*

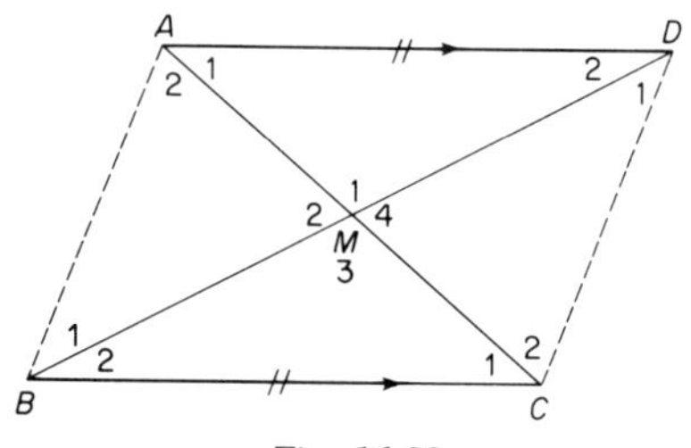

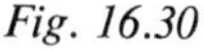

Fig. 16.30

In a parallelogram $ABCD$ (Fig. 16.30),

$$|AD| = |CB|, \text{ and } AD\,//\,CB;$$
$$\Rightarrow \angle A_1 = \angle C_1 \text{ (alternate angles);}$$
$$\angle D_2 = \angle B_2 \text{ (alternate angles);}$$
$$\Rightarrow \triangle\text{s } \begin{matrix}ADM\\CBM\end{matrix} \text{ are congruent } (A\text{–}S\text{–}A).$$

From this congruence, it follows that

$$|AM| = |CM|$$

and $$|BM| = |DM|.$$

Also $$\angle M_2 = \angle M_4 \text{ (vertically opposite angles)}$$

$$\Rightarrow \triangle\text{s } \begin{matrix}AMB\\CMD\end{matrix} \text{ are congruent } (S\text{–}A\text{–}S).$$

As in Theorem 16.6, the other properties of the parallelogram (in particular that AB and DC are parallel) follow.

Theorem 16.8: *If the opposite angles of a quadrilateral are equal, the quadrilateral is a parallelogram.*

In the parallelogram $ABCD$ (Fig. 16.31),

$$\angle A_1 = \angle C_1$$

and $$\angle B = \angle D.$$

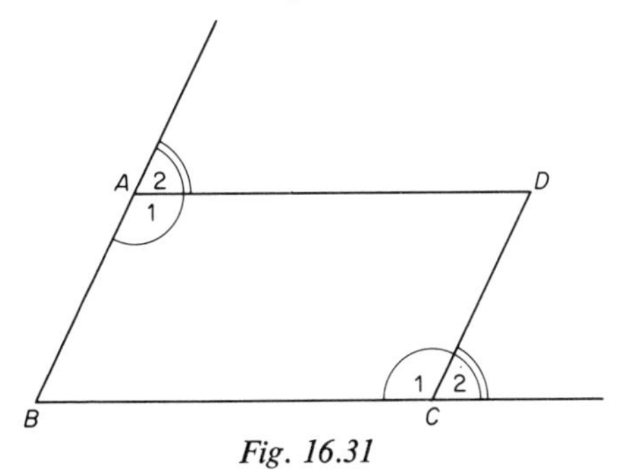

Fig. 16.31

$\angle A_1 + \angle B + \angle C_1 + \angle D = 360°$ (the angle sum of a quadrilateral); substituting $\angle A_1$ for $\angle C_1$ and $\angle B$ for $\angle D$,

$$\angle A_1 + \angle B + \angle A_1 + \angle B = 360°$$
$$\Rightarrow 2\angle A_1 + 2\angle B = 360°$$
$$\Rightarrow 2(\angle A_1 + \angle B) = 360°$$
$$\Rightarrow \angle A_1 + \angle B = 180°$$
$$\Rightarrow \angle B = 180° - \angle A_1.$$

Also, $\angle A_2 = 180° - \angle A_1$ ($\angle A_1$ and $\angle A_2$ form a straight line)
$\Rightarrow \angle B = \angle A_2 \Rightarrow AD$ and BC are parallel (corresponding angles are equal).
$\angle C_2 = 180° - \angle C_1 = 180° - \angle A_1$ (since $\angle C_1 = \angle A_1$);
$\angle B = 180° - \angle A_1$ (see above)
$\Rightarrow \angle C_2 = \angle B \Rightarrow AB$ and DC are parallel.
Therefore $ABCD$ is a parallelogram, and all the other properties follow by Theorem 16.5.

16.18 Exercises

1. Use each of the sufficient conditions (Section 16.16) as a basis for constructing a parallelogram.
   [For example, use Theorem 16.5 to construct a parallelogram by constructing two pairs of parallel lines; alternatively, draw a triangle ABC and through two of its vertices draw lines parallel to the opposite sides.]
2. In the parallelogram $PQRS$ (Fig. 16.32), the lines PT and SV are perpendicular to the line QR.

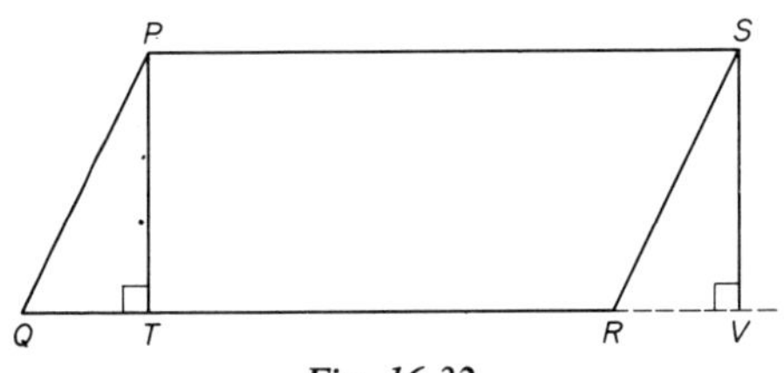

Fig. 16.32

Show (by drawing and cutting out) that $\Delta s \begin{matrix} PQT \\ SRV \end{matrix}$ are congruent and that therefore

(*a*) $|PT| = |SV|$, and
(*b*) the areas of triangles PQT and SRV are equal.

3. In a triangle ABC mark the mid-point M of AC and draw lines through the point M parallel to AB and BC. If these lines meet AB and BC respectively at N and L, show that

   (*a*) $BLMN$ is a parallelogram, and
   (*b*) $|BN| = |ML|$.

16.19 Intercept and Mid-point Theorems

There is an interesting way to divide a triangle into four congruent triangles.

Draw and cut out any triangle ABC (Fig. 16.33); by folding, find L, M and N, the mid-points of the sides (fold the triangle so that the point A covers the point C to find M, etc.)

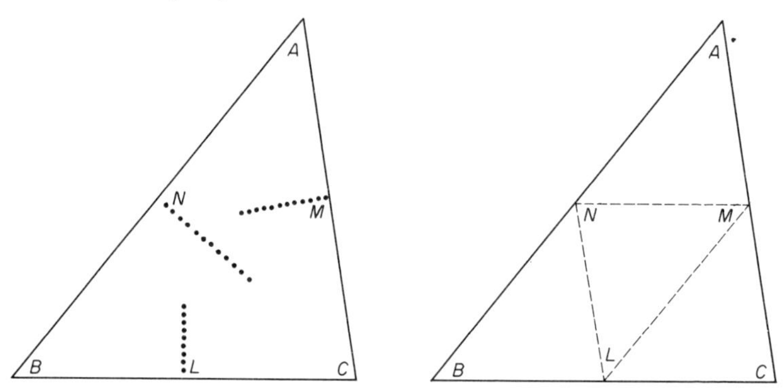

Fig. 16.33

Fold, and then cut, the triangle along the lines LM, MN and NL; by laying the pieces over each other, confirm that the four triangles $\begin{matrix} AMN \\ LNM \\ NLB \\ MCL \end{matrix}$ are congruent.

Put the fragments back into their original positions relative to each other, and confirm that:

the line LM is parallel to BA,
the line MN is parallel to CB,
the line NL is parallel to CA;
and

$$|LM| = \tfrac{1}{2}|BA|,$$
$$|MN| = \tfrac{1}{2}|CB|,$$
$$|NL| = \tfrac{1}{2}|CA|;$$

and

AMLN is a parallelogram,
BLMN is a parallelogram,
CMNL is a parallelogram.

These properties follow necessarily if the mid-points of the sides of a triangle are joined (see Section 16.21, The Mid-point Theorem).

Theorem 16.9: *The line drawn through the mid-point of one of the sides of a triangle, parallel to another side, is half the length of that side and bisects the third side.*

(This is a special case of the more general Intercept Theorem, which will be discussed in Section 18.1.)

In a triangle ABC (Fig. 16.34), we mark the mid-point M of the side AC and draw lines through the point M parallel to AB and BC, meeting AB and BC at N and L respectively.

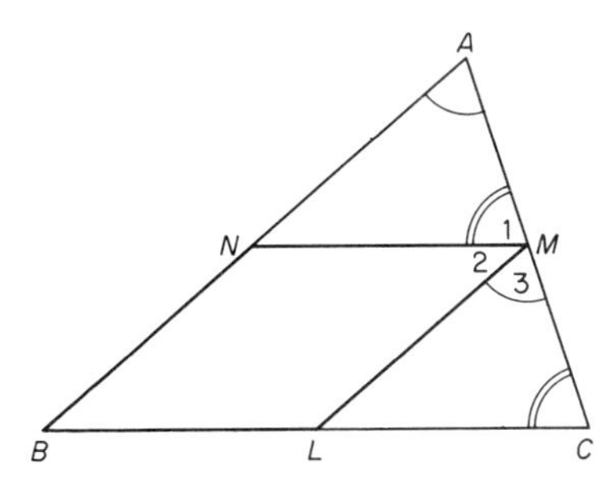

Fig. 16.34

By Theorem 16.5, $NBLM$ is a parallelogram.
We show that $|NM| = \frac{1}{2}|BC|$ and that $|AN| = |NB|$, i.e. that N is the mid-point of AB.

$$\angle A = \angle M_3 \text{ (corresponding angles of parallel lines);}$$
$$|AM| = |MC| \text{ (}M\text{ is the mid-point of }AC\text{);}$$
$$\angle M_1 = \angle C \text{ (corresponding angles of parallel lines)}$$
$$\Rightarrow \triangle\text{s } \begin{matrix} ANM \\ MLC \end{matrix} \text{ are congruent } (A\text{–}S\text{–}A),$$
$$\Rightarrow |NM| = |LC|;$$

also
$$|NM| = |BL| \text{ (opposite sides of parallelogram } NBLM)$$
$$\Rightarrow |BL| = |LC| \Rightarrow L \text{ is the mid-point of } BC, \text{ or}$$
$$|BL| = |LC| = \tfrac{1}{2}|BC|$$
$$\Rightarrow |NM| = \tfrac{1}{2}|BC|.$$

Similarly,
$$|AN| = |ML| \left(\text{corresponding sides of congruent } \triangle\text{s } \begin{matrix} ANM \\ MLC \end{matrix}\right)$$

and
$$|NB| = |ML| \text{ (opposite sides of the parallelogram } NBLM)$$
$$\Rightarrow |AN| = |NB|.$$

16.20 Exercise

Join the points N and L of the diagram above; use Theorem 16.9, the special case of the Intercept Theorem, to show the congruence of the $\triangle$s $\begin{matrix} AMN \\ LNM \\ NLB \\ MCL \end{matrix}$.

16.21 The Mid-point Theorem

Theorem 16.10: *The line joining the mid-points of two sides of a triangle is parallel to the third side and half the length of the third side.*

This theorem, the Mid-point Theorem, is the converse of Theorem 16.9.

In a triangle ABC (Fig. 16.35), M is the mid-point of AC ($|AM| = |MC|$) and N is the mid-point of AB ($|AN| = |NB|$).

We will show that NM is parallel to BC, and that $|NM| = \frac{1}{2}|BC|$. We extend (or produce) the line NM, and draw a line through the point C parallel to BA; these two lines meet at the point P.

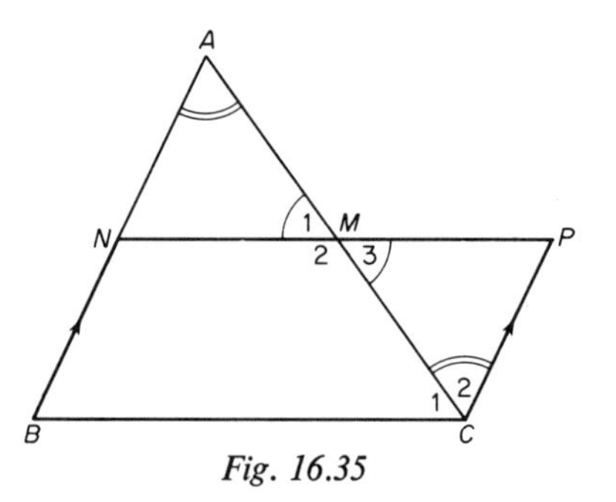

Fig. 16.35

First, we show that $\triangle$s $\begin{matrix} ANM \\ CPM \end{matrix}$ are congruent.

$$\angle A = \angle C_2 \text{ (alternate angles of parallel lines),}$$
$$|AM| = |MC| \text{ (} M \text{ is the mid-point),}$$
$$\angle M_1 = \angle M_3 \text{ (vertically opposite angles)}$$
$$\Rightarrow \triangle\text{s } \begin{matrix} ANM \\ CPM \end{matrix} \text{ are congruent } (A\text{–}S\text{–}A)$$
$$\Rightarrow |PC| = |AN|$$
$$\Rightarrow |PC| = |NB| \text{ (because } |AN| = |NB|)$$

and PC was drawn parallel to NB

$\Rightarrow$ $BCPN$ is a parallelogram (Theorem 16.7),

$\Rightarrow$ NP and BC are parallel

$\Rightarrow$ NM and BC are parallel.

$$|NM| = |MP| \left(\text{corresponding sides of congruent triangles } \begin{matrix} ANM \\ CPM \end{matrix} \right)$$

$\Rightarrow |NM| = \frac{1}{2}|NP|$;

but $|NP| = |BC|$ (opposite sides of the parallelogram $BCPN$),

$\Rightarrow |NM| = \frac{1}{2}|BC|$.

16.22 Exercise

In any triangle ABC, mark the mid-points N, L and M of the sides AB, BC and CA. Use the Mid-point Theorem to prove that the $\triangle$s AMN, LNM, NLB, MCL are congruent.

16.23 A Surprising Corollary of the Mid-point Theorem

Theorem 16.11: *The mid-points of the sides of any quadrilateral joined in order form a parallelogram.*

Draw a quadrilateral $ABCD$ of any shape and mark (by construction) the mid-points P, Q, R and S of the sides AB, BC, CD and DA respectively (Fig. 16.36).

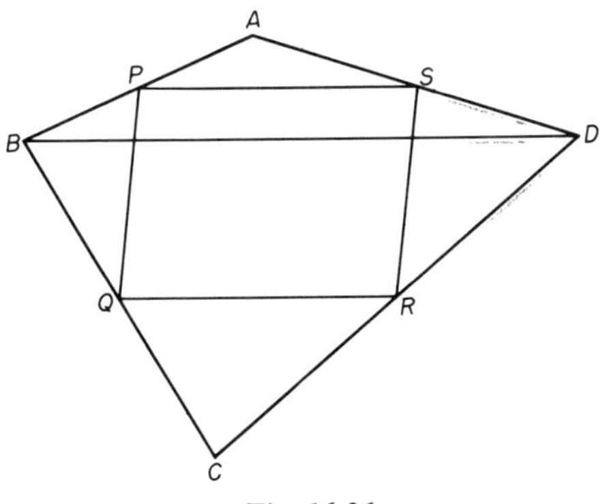

Fig. 16.36

By the Mid-point Theorem, $|PS| = \frac{1}{2}|BD|$ and $PS // BD$; and, also by the Mid-point Theorem, $|QR| = \frac{1}{2}|BD|$ and $QR // BD$

$\Rightarrow |PS| = |QR|$ and $PS // QR$

$\Rightarrow PQRS$ is a parallelogram (by Theorem 16.7).

16.24 The Rhombus

A rhombus is a quadrilateral whose sides are all of the same length. It is a special parallelogram and a special kite, and therefore displays all the properties of both the kite and the parallelogram; bearing in mind the proviso that all sides are of equal length, it can be constructed using any of the constructions of the kite or the parallelogram.

If no particular angle or side is specified, the simplest construction of a rhombus is to draw any arbitrary angle and use Construction 16.1(*c*) in Section 16.9.

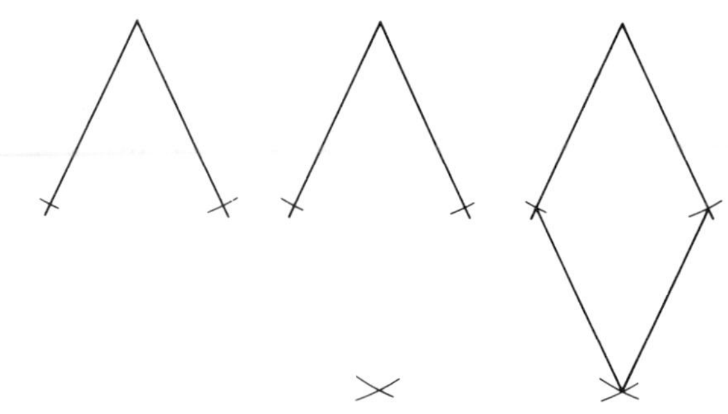

Fig. 16.37

16.25 Exercises

1. Use in turn the various constructions of the kite and the parallelogram (Sections 16.9 and 16.18) to construct a rhombus.
2. Construct a rhombus and cut it out. Fold it along the lines of the diagonals, and investigate the bilateral symmetry of the rhombus.
3. Draw an isosceles triangle PQR ($|PQ| = |QR|$) on stiff paper and cut it out; use it as a template to construct a rhombus by rotating the template about the mid-point of PR and so investigate the rotational symmetry of the rhombus.
4. Construct a rhombus $ABCD$ and draw the diagonals; mark all equal angles, sides and line segments and all right angles.

   Name the congruent triangles formed by the sides and diagonals and confirm the properties of the rhombus detailed in Section 16.26.

16.26 Properties of the Rhombus

(*a*) All sides of a rhombus are of equal length (Fig. 16.38).

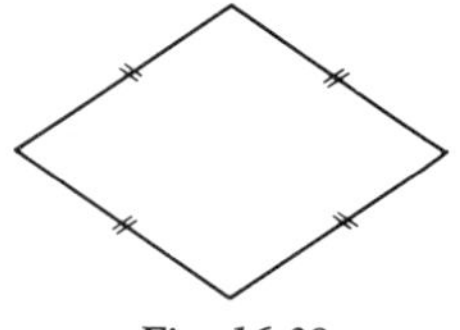

Fig. 16.38

(*b*) Every pair of adjacent sides of a rhombus with a diagonal forms an isosceles triangle (Fig. 16.39).

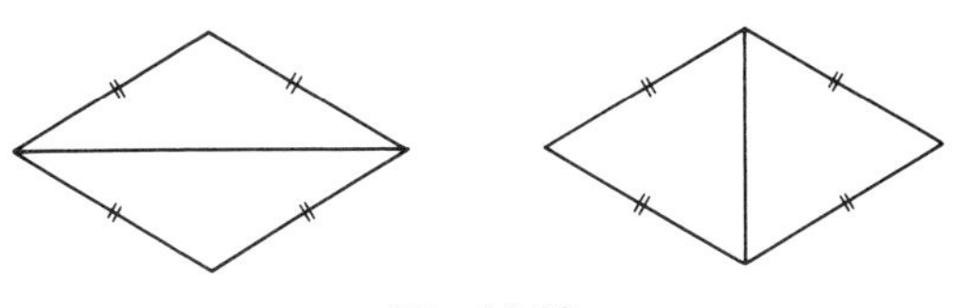

Fig. 16.39

(*c*) A rhombus is symmetrical about each of the diagonals (bilateral symmetry) (Fig. 16.40).

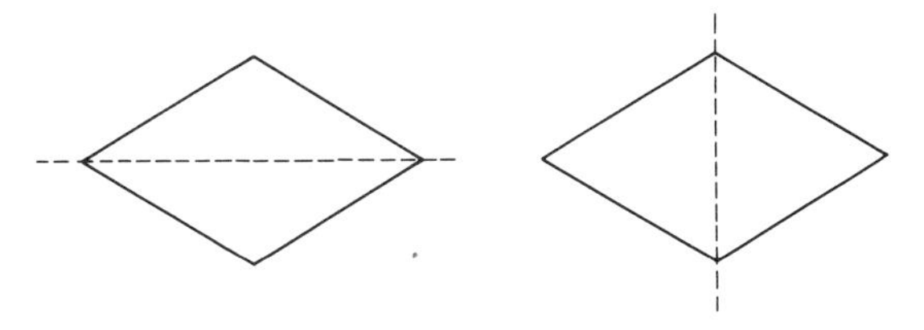

Fig. 16.40

(*d*) A rhombus is symmetrical about the point of intersection of the diagonals (rotational symmetry) (Fig. 16.41).

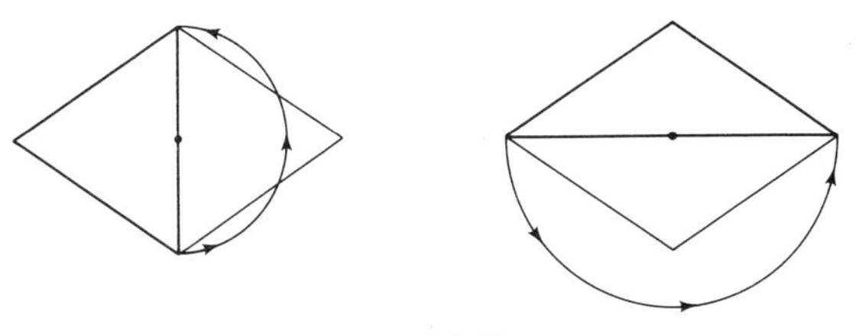

Fig. 16.41

(*e*) The diagonals of a rhombus bisect its angles (Fig. 16.42).

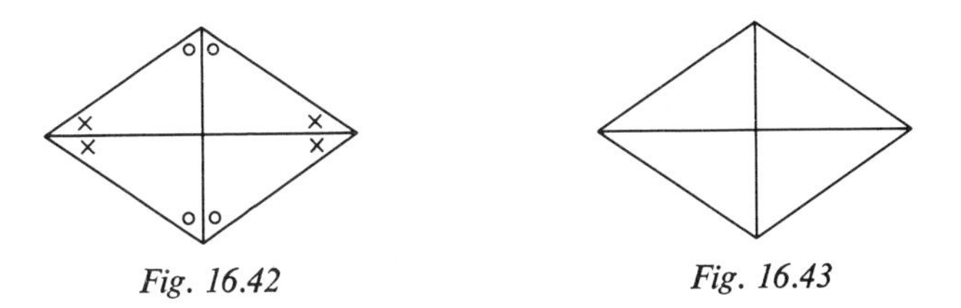

Fig. 16.42 *Fig. 16.43*

(*f*) The diagonals divide a rhombus into four congruent triangles (Fig. 16.43).

(*g*) The opposite sides of a rhombus are parallel (Fig. 16.44).

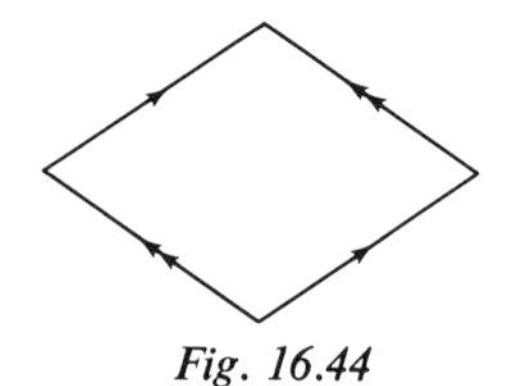

Fig. 16.44

(*h*) The diagonals of a rhombus bisect each other at right angles (Fig. 16.45).

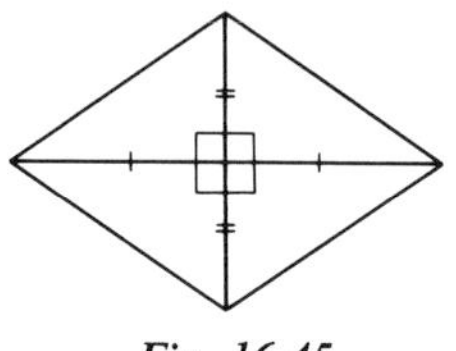

Fig. 16.45

Sufficient conditions for a quadrilateral to be a rhombus

If a quadrilateral has any one of the properties (*a*), (*b*), (*c*), (*e*), (*f*) or (*h*), it must be a rhombus; in other words, each of these properties is a sufficient condition for a quadrilateral to be a rhombus.

Theorem 16.12: *If the four sides of a quadrilateral are of equal length, the quadrilateral has all the other properties of the rhombus.*

In the quadrilateral $ABCD$ (Fig. 16.46) $|AB| = |BC| = |CD| = |DA|$.

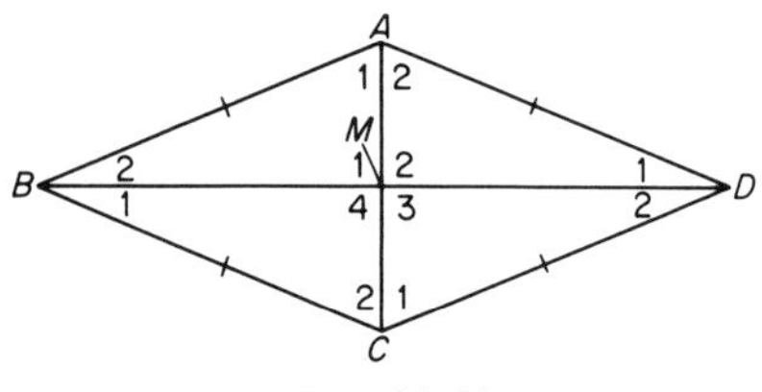

Fig. 16.46

Clearly, the triangles BAD, DCB, ABC and CDA are isosceles triangles (property (*b*)).

Moreover, $\triangle$s $\begin{matrix} BAD \\ BCD \end{matrix}$ are congruent (S–S–S) (BD is a common side),

and $\triangle$s $\begin{matrix} ABC \\ ADC \end{matrix}$ are congruent (S–S–S) (AC is a common side),

$\Rightarrow$ $ABCD$ is symmetrical about BD and AC (property (*c*)).

Since $\triangle$s $\begin{matrix} ABC \\ CDA \end{matrix}$ are congruent, $ABCD$ has rotational symmetry about the mid-point of AC (property (*d*)).

$$\angle A_1 = \angle A_2 \text{ and } \angle C_1 = \angle C_2 \quad \left(\text{from the congruence of } \triangle\text{s } \begin{matrix} ABC \\ ADC \end{matrix}\right)$$

$$\angle B_1 = \angle B_2 \text{ and } \angle D_1 = \angle D_2 \quad \left(\text{from the congruence of } \triangle\text{s } \begin{matrix} BAD \\ BCD \end{matrix}\right)$$

$\Rightarrow$ the diagonals bisect the angles of $ABCD$ (property (e)).

Also, $\angle A_1 = \angle C_2$ ($\triangle ABC$ is isosceles),
$\angle B_1 = \angle D_2$ ($\triangle BDC$ is isosceles),
$\Rightarrow \angle A_1 = \angle A_2 = \angle C_1 = \angle C_2$

and $\angle B_1 = \angle B_2 = \angle D_1 = \angle D_2$

$$\Rightarrow \triangle\text{s } \begin{matrix} ABM \\ CBM \\ CDM \\ ADM \end{matrix} \text{ are congruent } (A\text{–}S\text{–}A) \text{ (property } (f)).$$

Also, BA and CD are parallel ($\angle A_1 = \angle C_1$, alternate angles),
and BC and AD are parallel ($\angle A_2 = \angle C_2$, alternate angles) (property (g)).

From the congruence of $\triangle$s $\begin{matrix} ABM \\ CBM \\ CDM \\ ADM \end{matrix}$, it follows that

$$\angle M_1 = \angle M_2 = \angle M_3 = \angle M_4.$$

Since $\angle M_1 + \angle M_2 + \angle M_3 + \angle M_4 = 360°$,
$\angle M_1 = \angle M_2 = \angle M_3 = \angle M_4 = 90°$.

Also $|BM| = |MD|$
$|AM| = |MC|$

(from the same congruences) and therefore

the diagonals bisect each other at right angles (property (h)).

16.27 Exercises

1. Show that property (c) is a sufficient condition for a quadrilateral to be a rhombus. [Hint: show the equality of lengths of the sides of the rhombus by the congruence of triangles.]
2. In the same way, show that each of the properties (h) and (e) is a sufficient condition for a quadrilateral to be a rhombus.
3. Show that if two adjacent sides of a *parallelogram* are of equal length, the parallelogram is a rhombus.
4. On stiff paper draw an isosceles triangle PQR ($|PQ| = |PR|$); cut it out, leaving a little of the paper around point M, the mid-point of QR.

Use it as a template to draw the isosceles triangle ABC. Hold M in the same position (use a pin), and rotate the template through 180° and draw lines again along the template.

The completed figure is a rhombus, demonstrating its rotational symmetry.

16.28 The Rectangle

The most familiar quadrilateral is probably the rectangle, named after its most characteristic property: each of the four angles is a right angle. It is a special kind of parallelogram, and has therefore all the properties of the parallelogram.

By folding a rectangular sheet of paper, we can show another of its properties (Fig. 16.47): the rectangle is symmetrical about the two lines which join the mid-points of each pair of opposite sides.

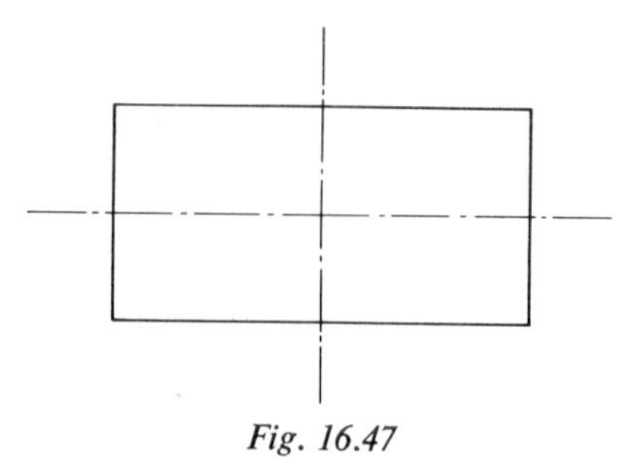

Fig. 16.47

These two lines of symmetry are perpendicular and bisect each other; they divide the rectangle into four congruent rectangles. Each of the diagonals of a rectangle $ABCD$ divides the rectangle into two congruent right-angled triangles (Fig. 16.48).

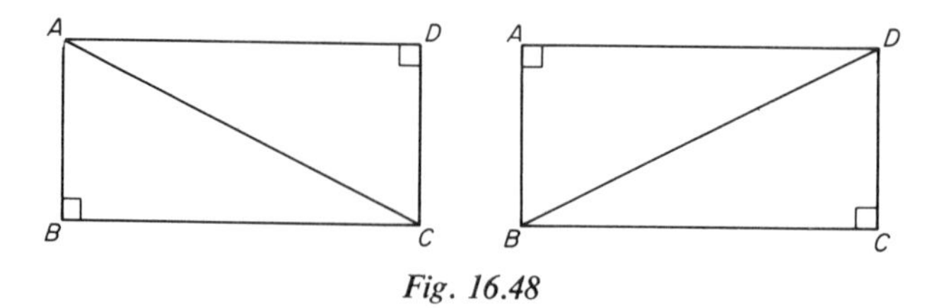

Fig. 16.48

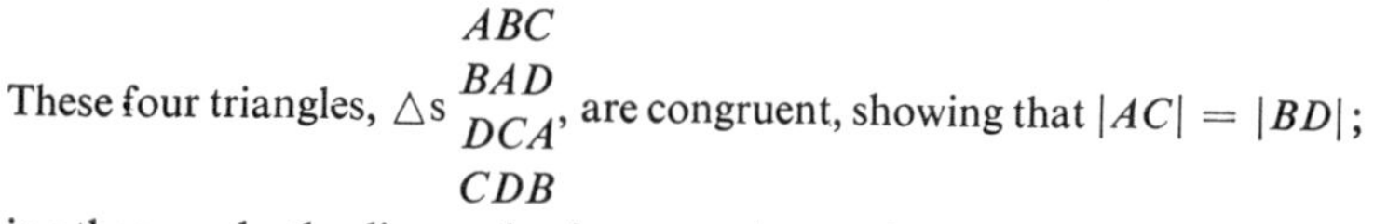

These four triangles, $\triangle$s $\begin{matrix} ABC \\ BAD \\ DCA \\ CDB \end{matrix}$, are congruent, showing that $|AC| = |BD|$;
in other words, the diagonals of a rectangle are of equal length.

The properties of the rectangle

(*a*) All the angles of a rectangle are right angles.
(*b*) The opposite sides of a rectangle are parallel and of equal length.
(*c*) A rectangle is symmetrical about the lines joining the mid-points of the opposite sides.
(*d*) The diagonals of a rectangle bisect each other.
(*e*) The diagonals of a rectangle are of equal length.

Sufficient conditions for a quadrilateral to be a rectangle

Theorem 16.13: *If each of the angles of a quadrilateral is a right angle, the quadrilateral has all the properties of the rectangle, for example, the opposite sides are equal and parallel, and the diagonals are equal.*

In the quadrilateral $ABCD$ (Fig. 16.49), $\angle A = \angle B = \angle C = \angle D = 90°$.

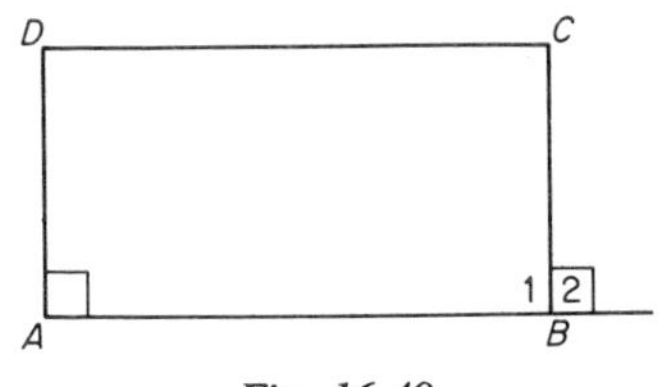

Fig. 16.49

$\angle A = \angle B_1 = \angle B_2\ (= 90°) \Rightarrow$ DA and CB are parallel (corresponding angles equal);
similarly, $\angle A = \angle D \Rightarrow AB$ and DC are parallel,
$\Rightarrow ABCD$ is a parallelogram
$\Rightarrow |AB| = |DC|$ and $|BC| = |AD|$.
$\angle A = \angle D\ (= 90°)$,
$|AD| = |AD|$ (common side of $\triangle$s $\begin{matrix} DAB \\ ADC \end{matrix}$).

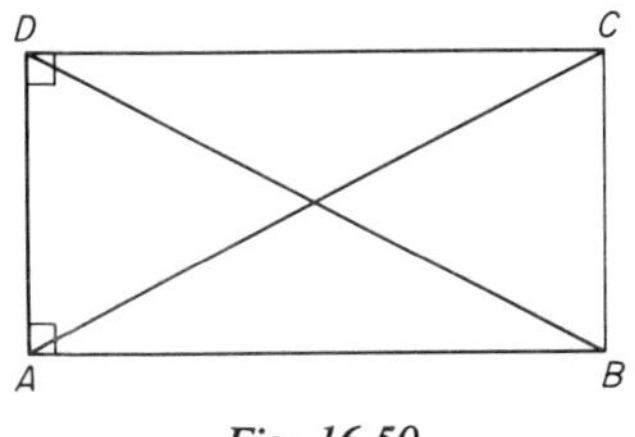

Fig. 16.50

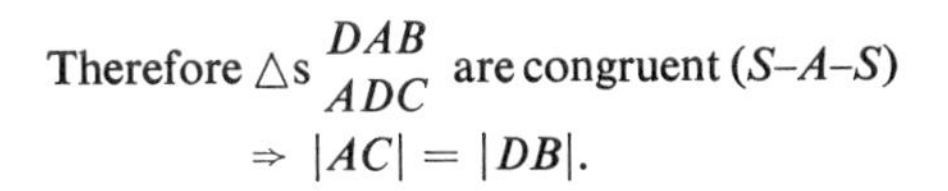
Therefore $\triangle$s $\begin{matrix} DAB \\ ADC \end{matrix}$ are congruent (S–A–S)
$\Rightarrow |AC| = |DB|$.

Theorem 16.14: *If the diagonals of a parallelogram are of equal length, the parallelogram is a rectangle.*

If the quadrilateral $ABCD$ (Fig. 16.51) is a parallelogram, and $|AC| = |BD|$,

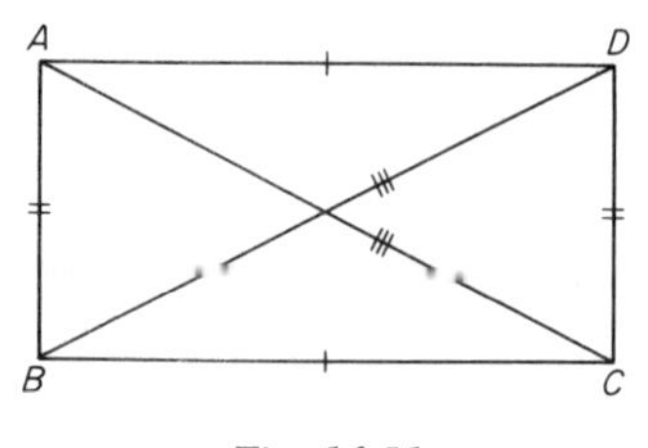

Fig. 16.51

we will show that each of the angles A, B, C and D is a right angle.

$ABCD$ is a parallelogram $\Rightarrow$ $|AD| = |BC|$
and $|AB| = |DC|$.

We know too that $|AC| = |BD|$.

$$\Rightarrow \triangle\text{s}\ \begin{matrix} ABD \\ BAC \end{matrix}\ \text{are congruent } (S\text{–}S\text{–}S)$$

$$\Rightarrow \angle A = \angle B.$$

Moreover, $\angle A + \angle B = 180°$ (since AD and BC are parallel)

$$\Rightarrow \angle A = \angle B = 90°.$$

We can show similarly that $\angle C = \angle D = 90°$.

16.29 Exercises

1. Show that, if one of the angles of a parallelogram is a right angle, each of the other three angles is a right angle and the parallelogram is therefore a rectangle. (This is another sufficient condition for a quadrilateral to be a rectangle.)
2. The diagonals of a quadrilateral $PQRS$ bisect each other and are of equal length. Show that $PQRS$ must be a rectangle.
3. If one of the sides of a rectangle is 6 cm and another is 8 cm, what is the length of the diagonal? [Hint: use Pythagoras' Theorem, Section 15.28.]
4. Construct a right angle, and use it to construct a rectangle whose sides are 9 cm and 3 cm.
5. Draw two line segments of equal length which intersect but do not bisect each other. Join the end-points of these segments to form a quadrilateral and show that property (*e*) (Section 16.28) alone is not a sufficient condition for a quadrilateral to be a rectangle.

16.30 The Square

The square is a quadrilateral in which all four sides are of equal length and all four angles are right angles. It is a parallelogram, a rhombus, a kite and a rectangle all in one and combines the properties of all these shapes.

Properties of the square

(*a*) A square is symmetrical about each of its diagonals and about the lines joining the mid-points of the opposite sides (Fig. 16.52).

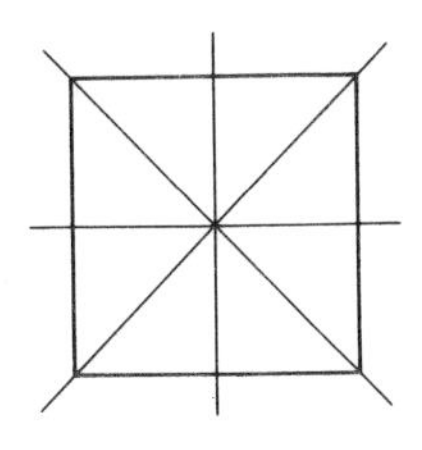

Fig. 16.52

Check this by cutting out a square and folding it in different ways.

(*b*) All the sides of a square are of equal length (Fig. 16.53).

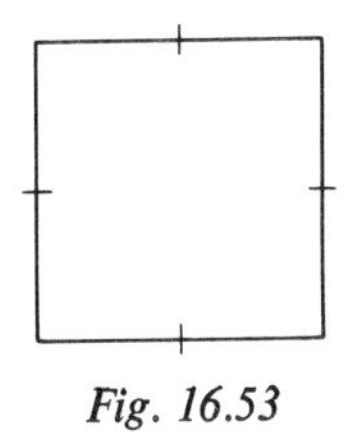

Fig. 16.53

(*c*) Each of the four angles of a square is a right angle (Fig. 16.54).

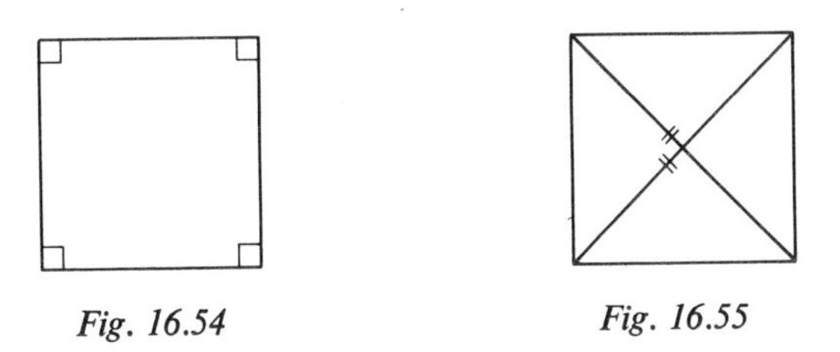

Fig. 16.54

Fig. 16.55

(*d*) The diagonals of a square are of equal length (Fig. 16.55).

(e) The diagonals of a square bisect each other at right angles (Fig. 16.56).

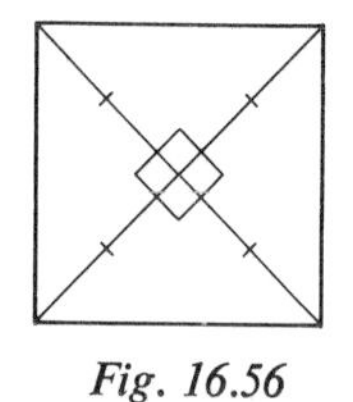

Fig. 16.56

Sufficient conditions for a quadrilateral to be a square

Theorem 16.15: *If one of the angles of a rhombus is a right angle, the rhombus is a square.*

If $ABCD$ is a rhombus, and if $\angle A = 90°$, we will show that all four angles are right angles.

$ABCD$ is a rhombus $\Rightarrow \angle C = \angle A = 90°$ (opposite angles of a rhombus)
$\Rightarrow \angle A + \angle C = 180°$
$\Rightarrow \angle B + \angle D = 180°$ (angle sum of a quadrilateral,

$$\angle A + \angle B + \angle C + \angle D = 360°)$$

but $\angle B = \angle D$ (opposite angles of a rhombus are equal)
$\Rightarrow \angle B = \angle D = 90°$.

Therefore all four angles of $ABCD$ are right angles and $ABCD$ has all the properties of the rectangle as well as the properties of the rhombus: it is therefore a square.

Theorem 16.16: *If the diagonals of a quadrilateral are of equal length and bisect each other at right angles, the quadrilateral is a square.*

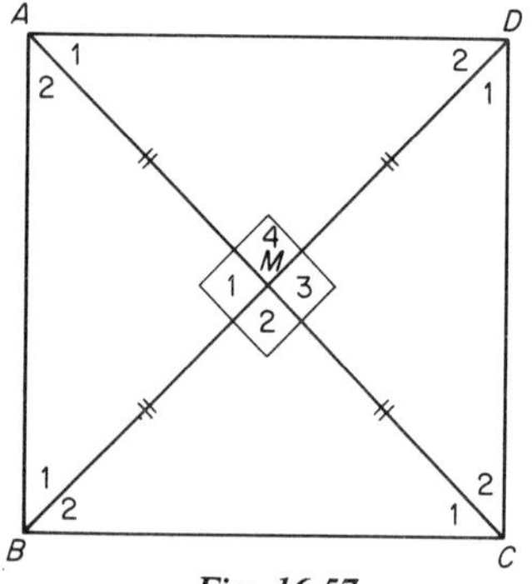

Fig. 16.57

In $ABCD$ (Fig. 16.57), $|AM| = |CM| = |DM| = |BM|$ and $\angle M_1 = \angle M_2 = \angle M_3 = \angle M_4 = 90°$.

Then $\triangle$s $\begin{matrix} AMB \\ BMC \\ CMD \\ DMA \end{matrix}$ are congruent (S–A–S) and isosceles

$\Rightarrow$ (*a*) $|AB| = |BC| = |CD| = |DA|$, or *all sides of ABCD are of equal length.*

(*b*) $\angle A_1 = \angle D_2 = 45°$ ($\triangle AMD$ is isosceles and $\angle M_4 = 90°$)

$\Rightarrow \angle A_1 = \angle A_2 = \angle B_1 = \angle B_2 = \angle C_1 = \angle C_2 = \angle D_1 = \angle D_2 = 45°$

$\Rightarrow \angle A_1 + \angle A_2 = 90°, \angle B_1 + \angle B_2 = 90°$, and so on.

$\Rightarrow$ *all angles of ABCD are right angles.*

Various combinations of properties can be a sufficient condition for a quadrilateral to be a square. In each case, it is sufficient to show they necessarily imply equality of length of all four sides, and that at least one angle is a right angle.

16.31 Exercises

1. Show that, if the diagonals of a parallelogram are of equal length and intersect at right angles, the parallelogram is a square.
2. Show that if two adjacent sides of a rectangle are of equal length, the rectangle is a square.
3. Use Pythagoras' Theorem to calculate the length of the diagonals of a square whose sides are each 5 cm.
4. Construct an isosceles triangle ABC, right-angled at B ($|AB| = |BC|$). Cut it out, and use it as a template to generate a square by rotating it around B through an angle of 90°, and then repeating this rotation until the triangle has made one complete turn. You can demonstrate in this way the rotational symmetry of the square.

16.32 Constructing a Square

We may either use Theorem 16.15 and start from the right-angle property of the angles of the square, or use Theorem 16.16, starting from the right-angle property of the intersection of the diagonals.

Construction 16.6: *A square whose side is of a given length.*

We use the construction of a right angle on a given line segment (Construction 16.3, Section 16.13) and the construction of the rhombus (Section 16.24):

(*a*) Draw a line segment of the required length (Fig. 16.58).
(*b*) Construct a right angle at A (or B), using any one of the constructions in Section 16.13.
(*c*) Using compasses, mark the point D so that $|AD| = |AB|$.

(*d*) Find point C by marking arcs of circles of radius AB using point D and point B as centres. Then join the points B and C, C and D.
Alternatively, having found point D as before, construct a line through D parallel to AB and a line through B parallel to AD.

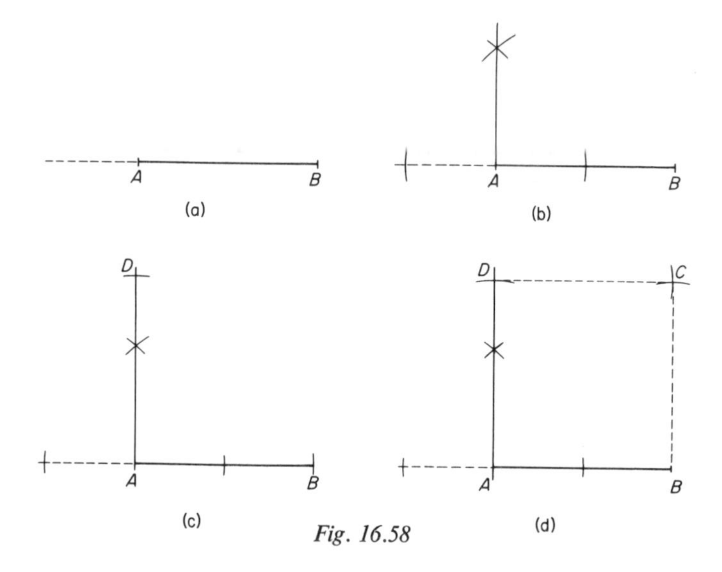

Fig. 16.58

Construction 16.7: *A square whose diagonal is of a given length.*

We use the property that the diagonals of the square bisect each other at right angles (Fig. 16.59).

(*a*) Construct a right angle, extending the line segments on both sides of the point of intersection.
(*b*) Place the point of the compasses on the point of intersection, T, of the perpendicular lines and mark the points P, Q, R and S at equal distances (half the given length of the diagonal) from T.
(*c*) Join the points P, Q, R and S.

16.33 Exercises

1. By folding paper, 'construct' a right angle. [Hint: folding paper once will produce a straight edge; by folding it again, you can 'bisect' an angle of 180°.]
2. Construct a square by folding a rectangular sheet of paper.
3. Use a ruler to draw a straight line. Hold the ruler in this position and draw two parallel lines, both perpendicular to the first line, using a set-square. Then use compasses and ruler to complete this shape to form a square.

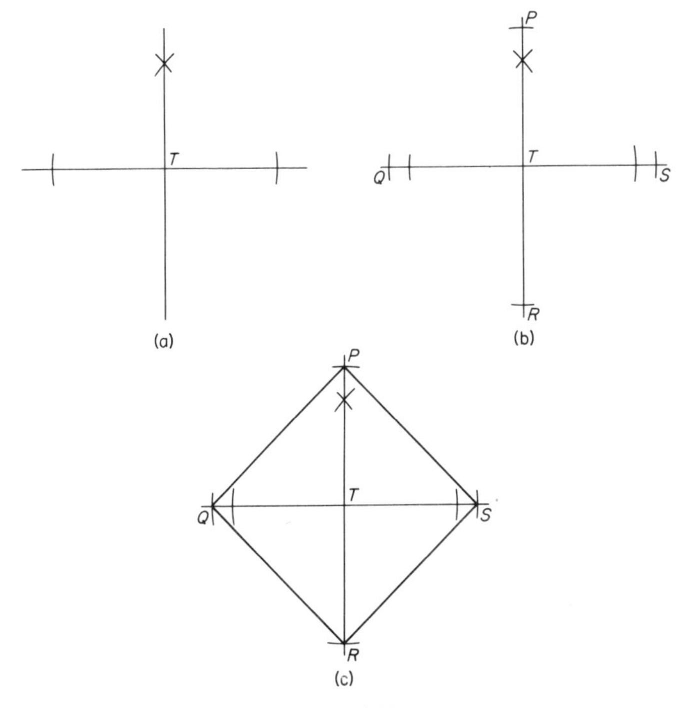

Fig. 16.59

16.34 Regular Polygons

The term 'polygon' (Greek: 'many-angled') is used to denote any plane shape bounded by straight line segments. Triangles and quadrilaterals are polygons: we could call them 'three-sided polygons' and 'four-sided polygons' respectively. Similarly, a plane shape, enclosed by five straight line segments, is a 'five-sided polygon'. For the more familiar polygons, the Greek numbers are substituted for the prefix 'poly' ('many'):

a five-sided polygon is called a *pentagon*;
a six-sided polygon is called a *hexagon*;
a seven-sided polygon is called a *heptagon*;
an eight-sided polygon is called an *octagon*;
a nine-sided polygon is called a *nonagon*;
a ten-sided polygon is called a *decagon*;
a twelve-sided polygon is called a *duodecagon*.

Remember that both the equilateral triangle and the square have the following properties:

(*a*) Their angles are equal.
(*b*) The lengths of their sides are equal.
(*c*) They have rotational symmetry about a central point.

A polygon with these characteristics is called a *regular polygon*; the equilateral triangle is a regular three-sided polygon, and the square is a regular four-sided polygon.

16.35 The Angles of a Regular Polygon

We have already established that the sum of the angles of a polygon of n sides, is $(n-2) \times 180°$ (see Section 16.2).

The sum of the angles of a pentagon—regular or not—is therefore $(5-2) \times 180° = 540°$. In a *regular* pentagon, all five angles are of equal size and each angle is $\frac{540°}{5} = 108°$.

Similarly, the sum of the angles of a hexagon is $(6-2) \times 180° = 720°$, and each angle of a *regular* hexagon is $\frac{720°}{6} = 120°$.

In general, if a regular polygon has n sides, the sum of its n angles is $(n-2) \times 180°$ and

$$\boxed{\text{each angle is } \frac{(n-2) \times 180°}{n}}$$

16.36 Exercise

Calculate the size of each angle of a regular heptagon, a regular nonagon, a regular decagon and a regular duodecagon.

16.37 The Rotational Symmetry of Regular Polygons

We have studied the rotational symmetry of the equilateral triangle and the square, and seen how this meant that we could generate these regular shapes by rotating a particular isosceles triangle about a central point (see Section 15.26 and Exercise 4 in Section 16.31).

We found that three particular congruent isosceles triangles together formed the equilateral triangle, and four particular isosceles triangles formed the square. The vertical angle of these isosceles triangles—or each of the angles at

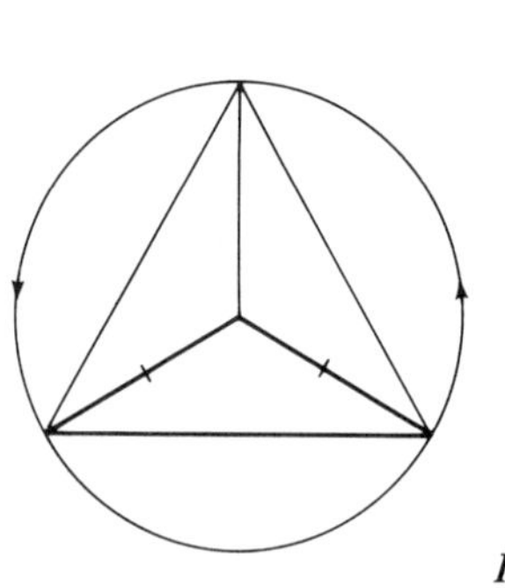

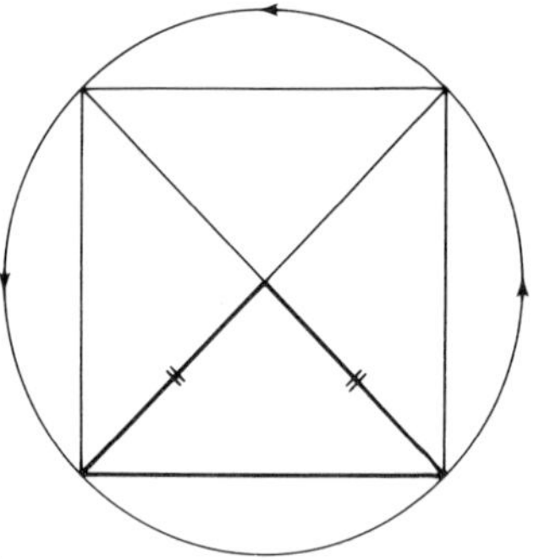

Fig. 16.60

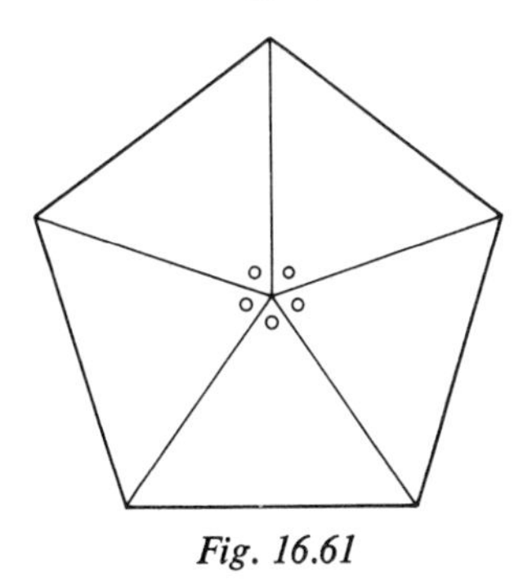

Fig. 16.61

the centre—had to be 120° in the case of the equilateral triangle (three angles of equal size totalling 360°), and 90° $\left(=\frac{360°}{4}\right)$ in the case of the square.

The regular pentagon (five-sided) can be divided into five congruent isosceles triangles (Fig. 16.61), the vertical angle of each of these isosceles triangles, or each of the five angles at the centre of the regular pentagon, being $\frac{360°}{5} = 72°$.

Similarly, each angle at the centre of a regular hexagon (six-sided) is $\frac{360°}{6} = 60°$.

In general, in a regular polygon of n sides:

> each of the n angles at the centre is $\frac{360°}{n}$

16.38 Exercises

1. Complete the following table:

number of sides	*name*	*size of the angle at each vertex*	*size of each angle at the centre*
3	equilateral triangle	60°	120°
4	square	90°	90°
5	regular pentagon	108°	72°
6	regular hexagon	120°	60°
7	regular heptagon		
8	regular octagon		
9	regular nonagon		
10	regular decagon		
12	regular duodecagon		

2. Compare the size of each angle at the vertex and each angle at the centre of each of the regular polygons. Can you account for their supplementary nature?

16.39 The Exterior Angles of a Regular Polygon

In the last exercise you found that, in every regular polygon, the angle at each vertex and one of the angles at the centre are supplementary. The reason for this interesting phenomenon becomes clear if we look at one of the congruent isosceles triangles formed by the central point and the sides of the regular polygon, for example $\triangle SAB$ in Fig. 16.62. (The diagram shows only three sides of a regular polygon; the total number of its sides does not matter in this context.)

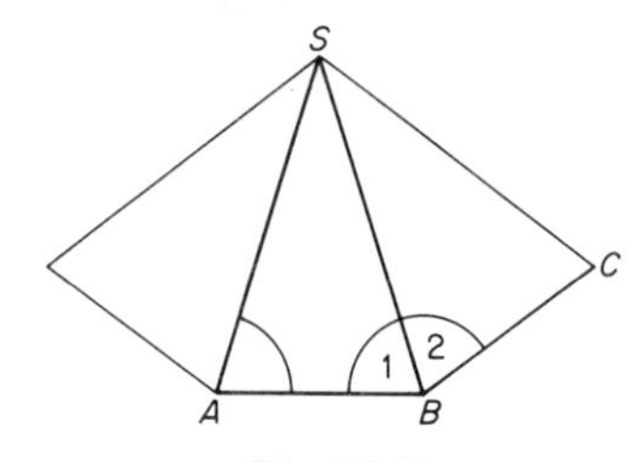

Fig. 16.62

$$\angle A + \angle B_1 + \angle S = 180^\circ.$$
$$\text{Since } \angle A = \angle B_2, \quad \angle B_2 + \angle B_1 + \angle S = 180^\circ;$$
$$\text{therefore} \quad \angle ABC + \angle S = 180^\circ.$$

In other words, $\angle ABC$, an angle at a vertex of the regular polygon, and $\angle S$, one of the angles at its centre, are supplementary.

The *exterior angle* of the regular polygon is also the supplementary angle of the angle at a vertex.

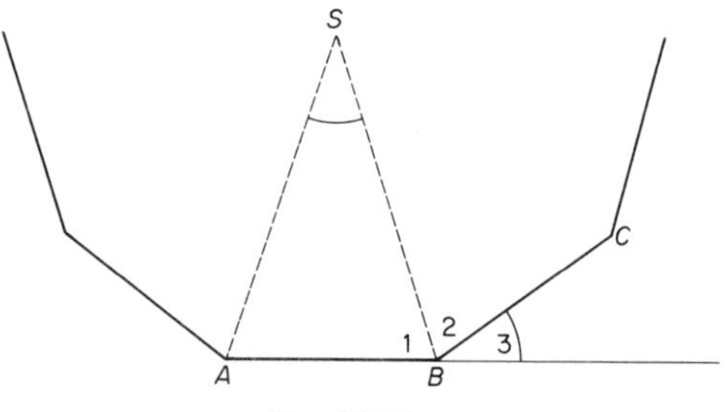

Fig. 16.63

In Fig. 16.63, for example, $\angle ABC + \angle B_3 = 180^\circ$ (they form a straight line).
Since both $\angle S$ and $\angle B_3$ are supplementary to $\angle ABC$, they are equal, or

$$\left.\begin{aligned} \angle ABC + \angle S &= 180^\circ \\ \angle ABC + \angle B_3 &= 180^\circ \end{aligned}\right\} \Rightarrow \angle S = \angle B_3.$$

The exterior angle of a regular polygon is of the same size as each of the angles at its centre.

16.40 Exercises

1. Confirm the last result by calculating the exterior angles of the regular polygons named in the table of the previous exercise.
2. Calculate
   (*a*) the sum of the five exterior angles of the regular pentagon;
   (*b*) the sum of the exterior angles of the regular hexagon;
   (*c*) the sum of the exterior angles of the regular octagon.
3. Show that the sum of the exterior angles of a regular polygon is always 360°. [Hint: the size of each of the n exterior angles of a regular n-sided polygon equals that of each of the n angles at its centre.]

16.41 Constructing Regular Polygons

Construction 16.8: *Regular polygons.*

Any regular polygon of n sides can be constructed by drawing n equal angles meeting at one point (S), each being of size $\frac{360^\circ}{n}$, and marking off equal distances from S on the sides of the angles. (Compare the construction of the equilateral triangle in Section 15.27, and Construction 16.6, Section 16.32, of the square.)

Let us take the *regular pentagon* as an example.

(*a*) Calculate the size of the angle at the centre: the pentagon has five sides, therefore the angle at the centre is $\frac{360^\circ}{5} = 72^\circ$.

(*b*) Use a protractor to mark five angles of 72° each, meeting at a point (mark a point S on your paper and place the centre of the protractor on it; then mark points along the protractor where the scale reads 0°, 72°, 144°, 216° and 288° and join these points to the centre S).

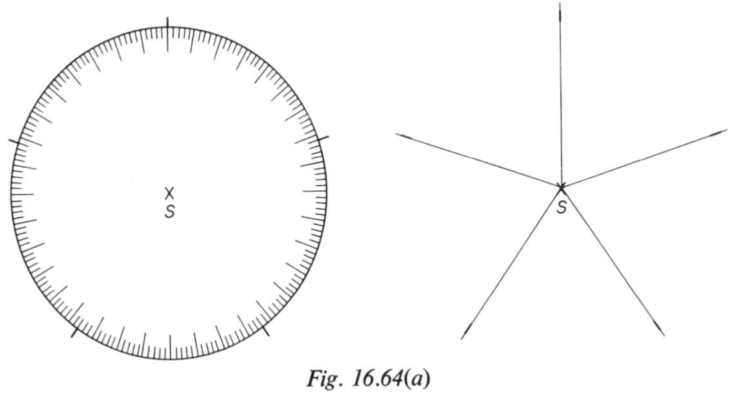

Fig. 16.64(a)

(c) Use compasses to mark off equal distances from S on the sides of the angles (mark arcs of a circle with S as centre, or draw a complete circle with S as centre).

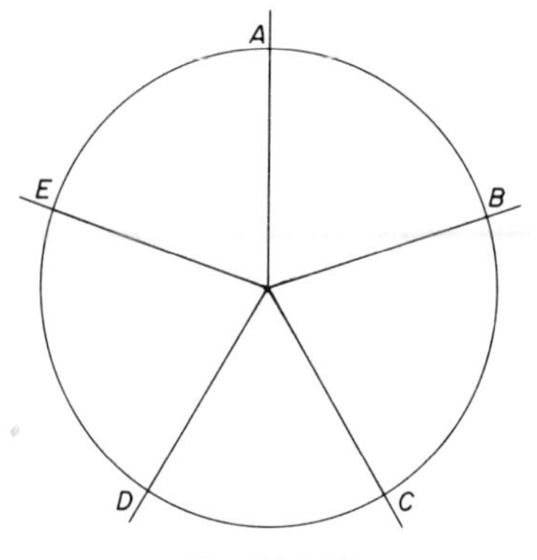

Fig. 16.64(b)

(d) Complete the regular pentagon by joining the points A, B, C, D, E and A in order.

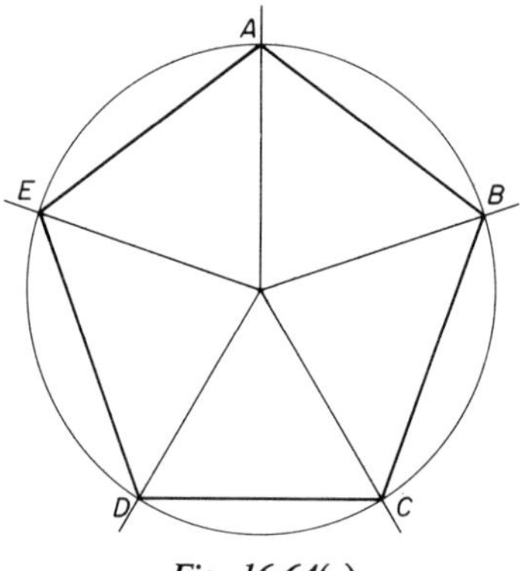

Fig. 16.64(c)

Alternatively (Fig. 16.65):

(a) Draw a circle, centre S.
(b) Mark an arbitrary point A on the circumference of this circle and draw the line segment SA.
(c) Use a protractor to draw a line forming an angle of 72° with SA, intersecting the circumference of the circle at B.
(d) Measure the distance AB with your compasses.
(e) Mark off this distance all round the circumference of the circle, so that $|AB| = |BC| = |CD| = |DE| = |EA|$.
(f) Join the points A, B, C, D, E and A in order.

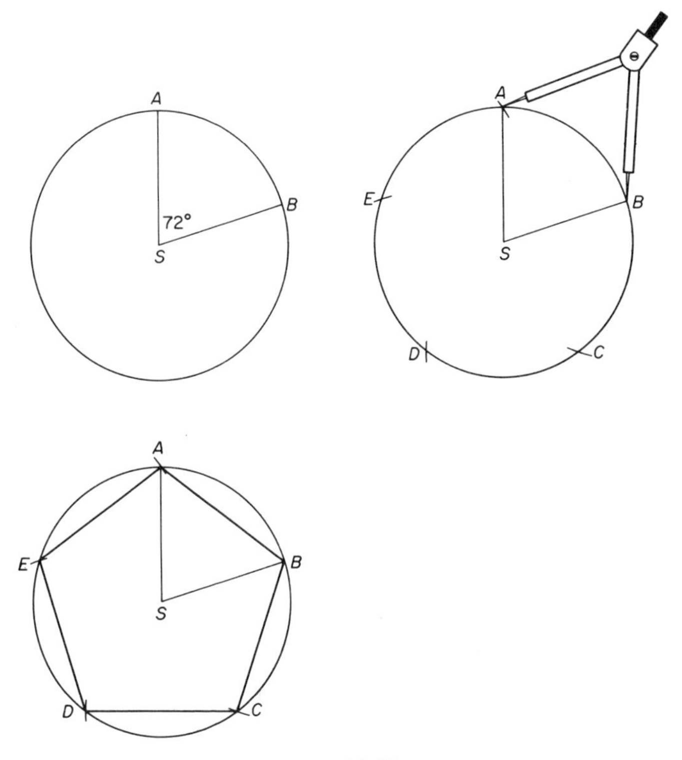

Fig. 16.65

16.42 Exercise

Using either of the methods in Section 16.41, construct:

a regular hexagon;
a regular octagon;
a regular nonagon;
a regular duodecagon.

16.43 Sufficient Conditions for a Polygon to be Regular

A simple experiment with meccano strips (Fig. 16.66) of equal lengths will confirm that equality of lengths of sides does not necessarily imply that the angles are equal: except in the case of the equilateral triangle, the strips will not make a rigid framework.

Neither is the equality of the size of all angles of a polygon a sufficient condition for the polygon to be regular, as Fig. 16.67 makes clear.

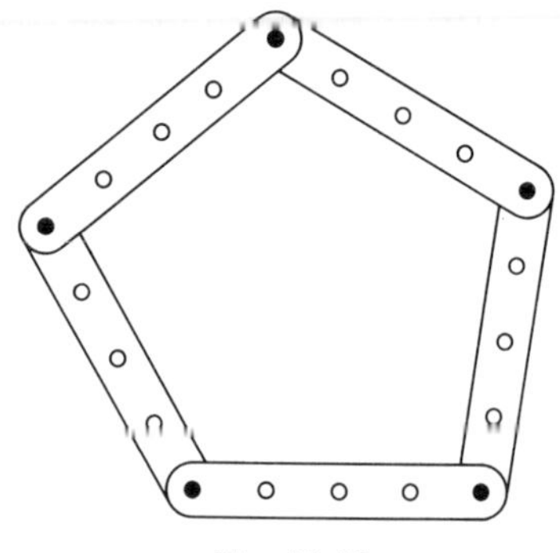

Fig. 16.66

However, a combination of *both* these properties is a sufficient condition for a polygon to be regular;. its rotational symmetry—property (*c*) of a regular polygon (Section 16.34)—follows necessarily.

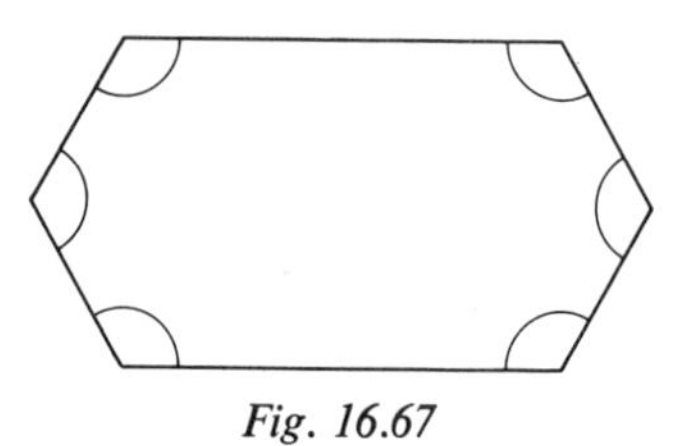

Fig. 16.67

16.44 Exercises

1. Show that if all the sides of a polygon are of equal length and all its angles are of equal size, the polygon has rotational symmetry.
2. Investigate, by folding, the *bilateral symmetry* of the regular pentagon, the regular hexagon, the regular nonagon and the regular decagon.
3. Show that regular polygons with an even number of sides are symmetrical about their diagonals which pass through the central point.

16.45 The Regular Hexagon

We found (Section 16.37) that each of the six angles at the centre of a regular hexagon is $\frac{360°}{6} = 60°$ (Fig. 16.68).

The isosceles triangle SAB which can generate the regular hexagon $ABCDEF$ is therefore an *equilateral* triangle (its three angles are each 60°), and $|AB| = |AS| = |BS|$.

Since $ABCDEF$ is a *regular* hexagon, all its sides are of equal length, each of them the same length as the line segment joining the central point to any of the vertices.

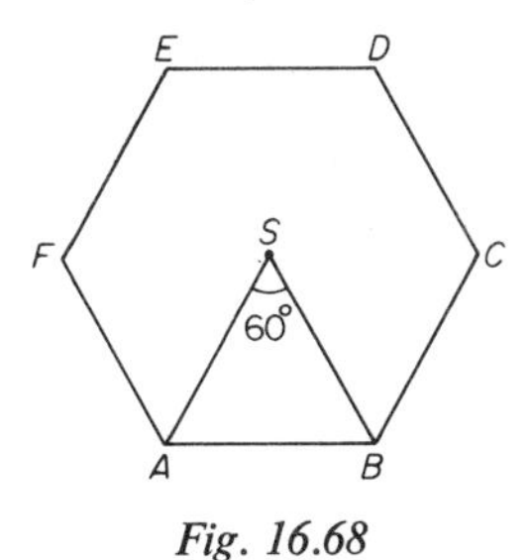

Fig. 16.68

This special property of the regular hexagon simplifies the alternative procedure (Section 16.41) for its construction: the radius of the circle can be used as the distance which is 'stepped off' all round the circumference of the circle.

Construction 16.9: *The regular hexagon.*

(*a*) Draw a circle of arbitrary radius, say 5 cm, and mark a point A anywhere on the circumference of this circle.

(*b*) Leaving the distance between the points of the compasses unchanged (5 cm), 'step off' equal distances of 5 cm on the circumference of the circle starting from point A.

(*c*) Join the marked points in order.

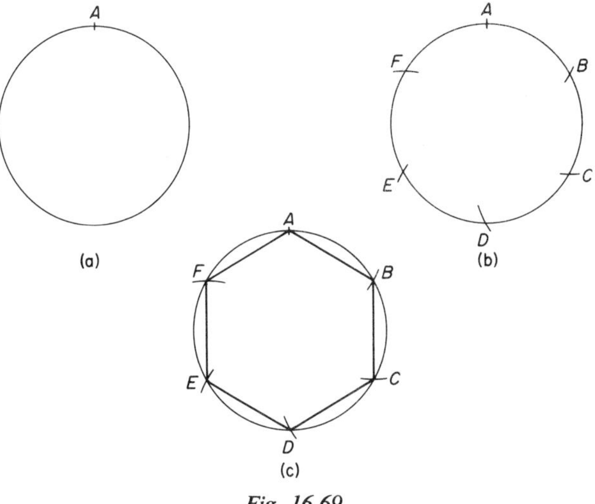

Fig. 16.69

16.46 Exercises

1. Use the method in Section 16.45 to construct a regular hexagon $ABCDEF$. Draw the lines AC, CE and EA. Show that $\triangle ACE$ is an equilateral triangle.

   [Hint: show the congruence of $\triangle$s $\begin{matrix} BCA \\ DCE. \\ FEA \end{matrix}$]

2. Construct a regular hexagon $KLMNOP$ and draw the diagonals which pass through the central point. By bisecting the angles at the centre, construct a regular twelve-sided polygon, a regular duodecagon.
3. Use a square in the same way to construct a regular octagon.

Unit Seventeen

Loci and Circles

17.1 The Mathematical Locus

Whenever we have used compasses in our constructions, we have used the most characteristic property of the circle: all the points of its circumference are at the same distance from one central point. If you draw a circle with compasses, closely watching your actions as you do so, you will see that the circumference of the circle is simply an innumerable set of points traced out by the pencil which is kept at a fixed distance from the other compass point. Any simple instrument, such as a piece of card or string, which keeps a marker at a fixed distance from a central point can be used to draw circles (Fig. 17.1).

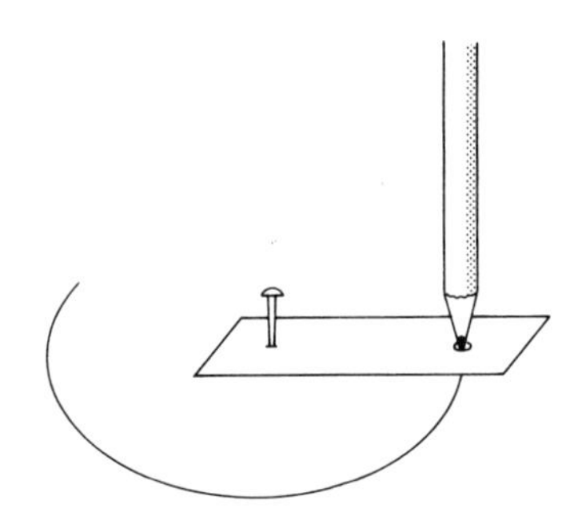

Fig. 17.1

The *circumference* of the circle—that is, the bounding curve—represents *all* the points on the paper at a given distance from the centre. Take, for example, the circle with centre P and radius 5 cm: any point inside the circumference (and not on the circumference) is less than 5 cm from the centre, and any point outside the circle is more than 5 cm from the centre. The *circumference* of this circle is the set of *all* points at a distance of 5 cm from the point P.

In mathematics, such a set of points is called a *locus* (Latin for 'place'; plural, loci). For a set of points to be a locus,

(*a*) the points must have a *common property*, and
(*b*) the set must contain *all* the points which have this common property.

These points may together form either a straight line, a line segment, a curve or a region.

The graphs discussed in Units 12 and 13 are examples of loci: for example, the graph of the equation $y + x = 5$ is the set, or locus, of all points the sum of whose x and y co-ordinates is 5. Similarly, the graph of equation $y = x - 2$ is the set, or locus, of all points whose x-co-ordinate is 2 more than its y-co-ordinate, and the graph of the inequality $y + x < 6$ is the locus of all points the sum of whose co-ordinates is less than 6.

The common property, the locus property, is often equality of distance from a point or a line; we then speak of the points as being *equidistant* from the point or the line.

The *distance* between two points is the length of the straight line segment with these two points as end-points. The *distance of a point from a line* always implies the *perpendicular distance* of the point from the line. In Fig. 17.2, for example, the distance of the point P from the line l is the length of the line segment PQ, perpendicular to l (the shortest distance between the point P and the line l).

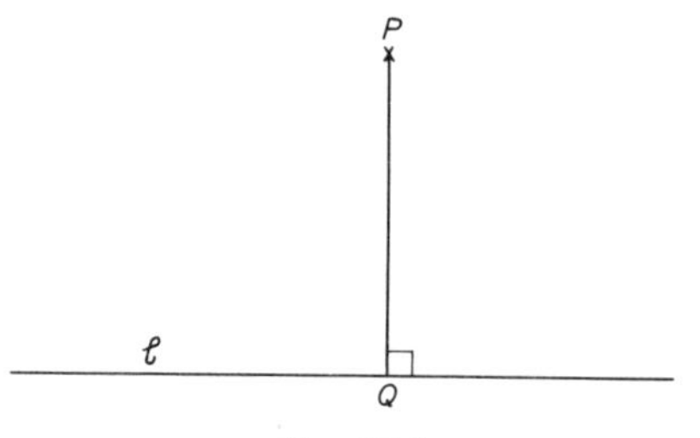

Fig. 17.2

Point Q is called the *foot* of the perpendicular from P to the line l.

We can then describe the circumference of a circle as a locus as follows:

Locus 17.1: *The locus of all points equidistant from one point O is the circumference of a circle with centre O.*

Locus 17.2: *The locus of all points at less than a given distance r from a given point O is the whole region inside the circumference of the circle of centre O and radius of length r.*

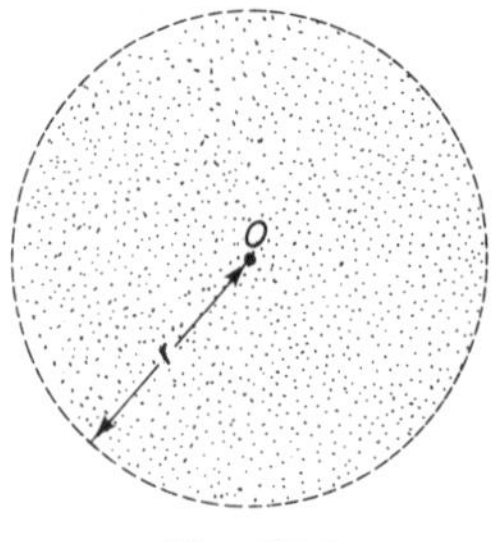

Fig. 17.3

The circle itself, the whole of the plane shape, includes both the curved boundary and the space inside it, and is in fact the union of the two sets described by Locus 17.1 and Locus 17.2.

Locus 17.3: *The locus of all points at a distance r or less than r from a given point O is the whole of the circle of centre O and radius of length r.*

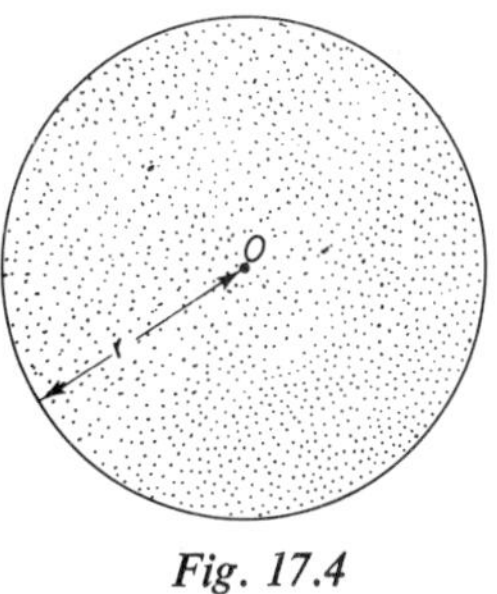

Fig. 17.4

To find a locus, we may start by marking a number of points which have the locus property. These points give us an idea of the general appearance of the locus.

If these points suggest a curve or straight line, we can join them, but we must check whether the intermediate points also have the locus property.

Finally, we must make sure that this line (or lines, or region), includes *all* the points which have the locus property.

Example

Find the locus of all points equidistant from a given line.

We will use the procedure we have just described to find the locus of all points at a distance of 4 cm from a given line l.

We measure and mark various points at a distance—that is, a perpendicular distance—of 4 cm from the line l, on either side of the line (Fig. 17.5).

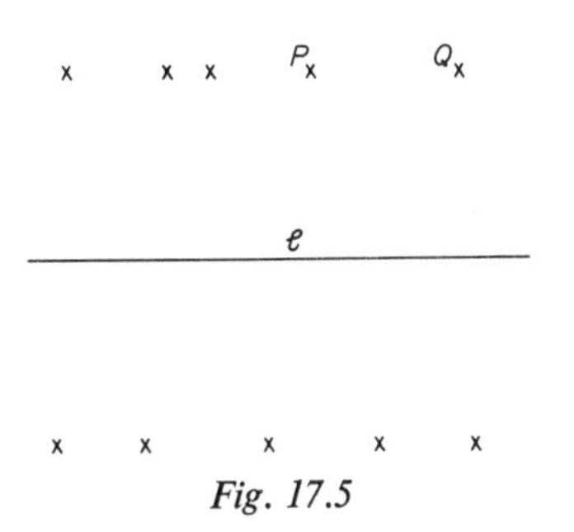

Fig. 17.5

The points suggest two straight lines, one on either side of l, and parallel to it. We show that, if P and Q are any two points on one side of l at a (perpendicular) distance of 4 cm from l, any point S on the line segment joining P and Q is also 4 cm from l (Fig. 17.6).

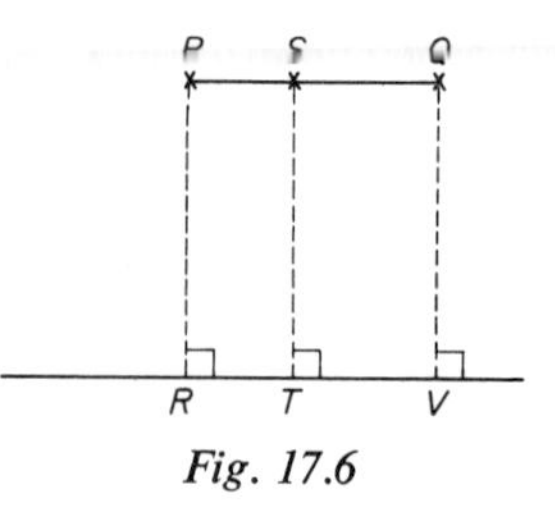

Fig. 17.6

PR and QV are parallel ($\angle R = \angle V = 90°$, corresponding angles), and
$|PR| = |QV| = 4$ cm
$\Rightarrow$ $PRVQ$ is a parallelogram (see Section 16.17) and a rectangle (see Section 16.28)
$\Rightarrow$ PS and RT are parallel and PR and ST are parallel ($\angle R = \angle T = 90°$)
$\Rightarrow$ $PRTS$ is a rectangle
$\Rightarrow$ $|ST| = |PR| = 4$ cm.

Now we must check whether these two lines on either side of l include *all* the points in the plane at 4 cm from l.

All points between these two parallel lines are at a distance less than 4 cm from l; all points 'outside' the parallel lines are more than 4 cm from l.

We can therefore say that the locus of all points in the plane at a distance of 4 cm from a given line l is a pair of lines parallel to l, one on either side of l, at a distance of 4 cm from l. (Remember that 'the distance between two parallel lines' means the *perpendicular* distance of any point on one line from the other line.)

In general terms,

Locus 17.4: *The locus of all points in the plane equidistant from a given line l is a pair of lines parallel to l, one on either side of l at the given distance from l.*

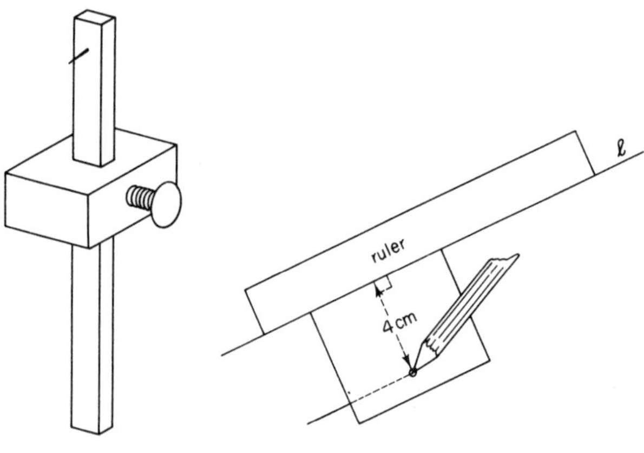

Fig. 17.7

The carpenter's *scriber* (Fig. 17.7) is a tool or instrument for marking lines parallel to a straight edge.

You can make a simple 'scriber' to draw lines—sets of points, that is—at a given distance, say 4 cm, from a given line.

Take a small piece of cardboard, plastic or plywood with at least one straight edge, and make a small hole at a distance of 4 cm from the straight edge. Pass the point of a pencil or marker through the hole; then, if you slide the 'scriber' with its straight edge along a ruler, it will mark a line parallel to the line of the ruler and 4 cm from it (Fig. 17.17).

17.2 Exercises

1. Use the home-made scriber, with compasses, to show that the locus of all points at a distance of 4 cm from a line *segment* has the 'running-track' appearance (Fig. 17.8).

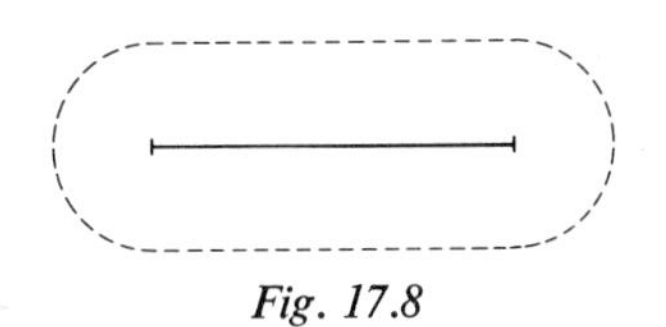

Fig. 17.8

2. Draw a large triangle and find the locus of all points at a distance of 4 cm from its sides. [Hint: there are such points both inside and outside the triangle.]
   Similarly, find the locus of points at a distance of 4 cm from the sides of a large square.
3. Show that the locus of all points in the plane equidistant from two parallel lines, is a line parallel to these two lines, half-way between them.

17.3 Two Important Loci

Theorem 17.1: *The locus of all points equidistant from two given points A and B is the perpendicular bisector of the line segment AB.*

We have already noticed (Section 15.16) that one of the properties of an isosceles triangle is the coincidence of the median, angle bisector, altitude and perpendicular bisector which cut the side between the equal angles. If, in a triangle ABC, the perpendicular bisector of AB passes through the vertex C, $\triangle ABC$ is isosceles and $|AC| = |BC|$ because $\triangle$s $\begin{matrix} AMC \\ BMC \end{matrix}$ are congruent (S–A–S) (Fig. 17.9).

By the same argument, any point P on the perpendicular bisector of AB forms an isosceles $\triangle ABP$; that is, any point on this line is equidistant from the points A and B.

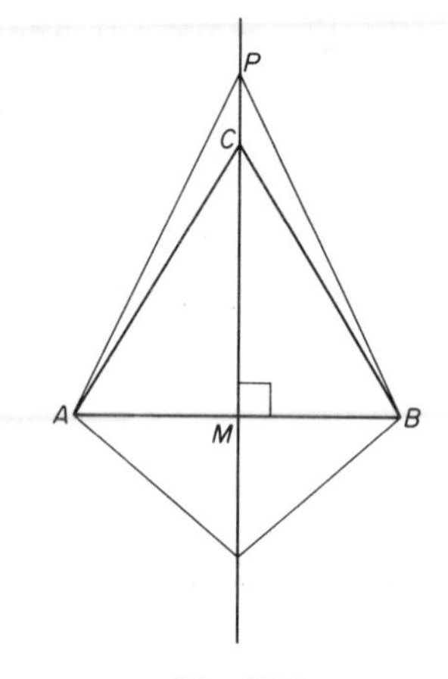

Fig. 17.9

To prove that the perpendicular bisector of AB is the *locus* of *all* points equidistant from the points A and B, we must also show that any point which is equidistant from A and B must lie on the perpendicular bisector of AB.

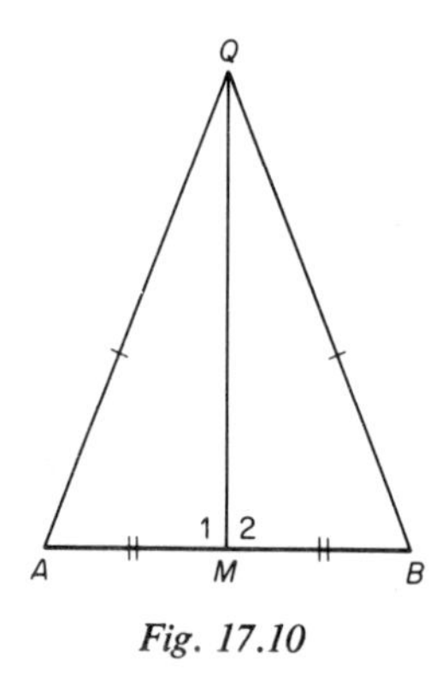

Fig. 17.10

Let Q (Fig. 17.10) be a point equidistant from A and B, or $|AQ| = |BQ|$, and let M be the mid-point of AB.

We show that QM is perpendicular to AB:

$|AM| = |BM|$ (M is the mid-point of AB)
$|AQ| = |BQ|$ (given)
$|QM| = |QM|$ (a common side of $\triangle AMQ$ and $\triangle BMQ$)

$\Rightarrow \triangle\text{s}\ \begin{matrix} AMQ \\ BMQ \end{matrix}$ are congruent (S–S–S)

$\Rightarrow \angle M_1 = \angle M_2$.

Since $\angle M_1 + \angle M_2 = 180°$, $\angle M_1 = \angle M_2 = 90°$.

QM is therefore the perpendicular bisector of AB, or Q is a point on the perpendicular bisector of AB.

Theorem 17.2: *The locus of all points equidistant from two intersecting lines is the pair of bisectors of the angles formed by the two lines.*

We show first that any point on the angle bisector is equidistant from the two lines forming the angle.

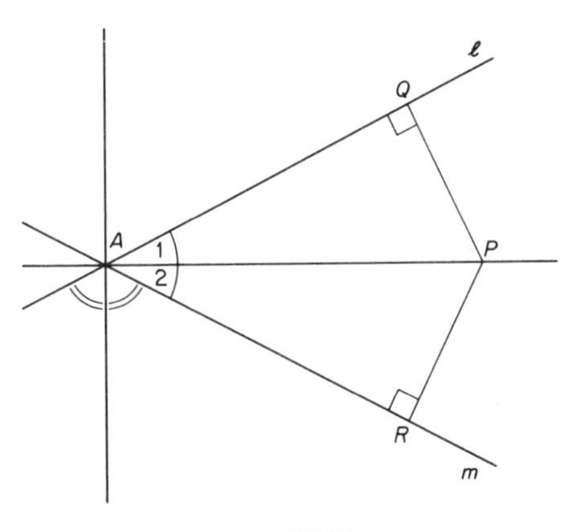

Fig. 17.11

Let P be any point on one of the two bisectors of the angles between the lines l and m (Fig. 17.11), and Q and R the feet of the perpendiculars from point P to the two lines.
Then

$\angle A_1 = \angle A_2$
(AP is the angle bisector)
$|AP| = |AP|$
(AP is a common side of $\triangle APQ$ and $\triangle APR$)
$\angle Q = \angle R\,(= 90°)$
$\Rightarrow \triangle\text{s}\ \begin{matrix} APQ \\ APR \end{matrix}$ are congruent (A–S–A)
$\Rightarrow |PQ| = |PR|$

that is, point P is equidistant from the lines AQ and AR.

If the angle bisectors are the locus of *all* points equidistant from two intersecting lines, it must also be true that any point which is equidistant from these two lines is a point on one of the angle bisectors.

Two lines l and m intersect at A (Fig. 17.12). A point P is equidistant from l and m; in other words, if S and T are the feet of the perpendiculars from point P to l and m, $|PS| = |PT|$.

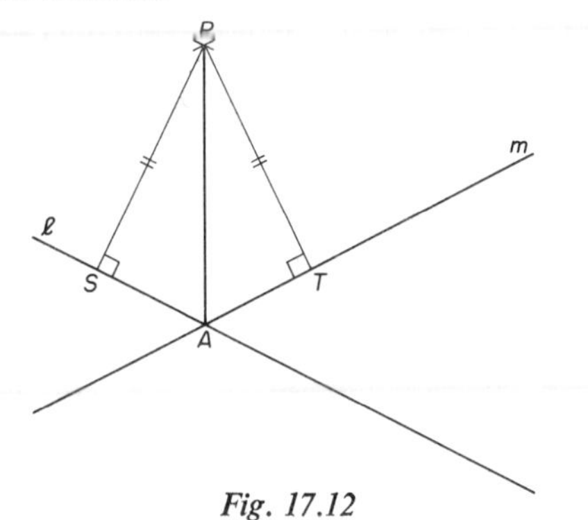

Fig. 17.12

We show that $\triangle$s $\begin{matrix} APS \\ APT \end{matrix}$ are congruent:

$$\angle S = \angle T \,(= 90°),$$
$$|AP| = |AP|$$
(*AP* is a common side of $\triangle APS$ and $\triangle APT$)
$$|PS| = |PT| \text{ (given)}$$
$$\Rightarrow \triangle s \begin{matrix} APS \\ APT \end{matrix} \text{ are congruent } (R\text{–}H\text{–}S)$$
$$\Rightarrow \angle SAP = \angle TAP;$$

that is, *AP* is the angle bisector of $\angle SAT$.

17.4 Applications of Loci

The graphical method of solving simultaneous equations and simultaneous inequalities (Sections 12.23 and 12.27) was based on the principle of intersection of two sets. We said that if one set contains all the points having one property and another set contains all the points having another property, all the points in the intersection, or the overlap of the two sets, have *both* the first and the second properties. In terms of loci, the point(s) of intersection of two loci have the properties of both loci. We have already used this principle many times in our constructions, for example in constructing $\triangle ABC$, when we knew the lengths of the sides to be *a* cm, *b* cm and *c* cm (Construction 15.1 in Section 15.8).

We began by drawing one line segment, say the line segment *BC* which is *a* cm in length; the position of the vertex *A* is determined by the two distances, *b* cm and *c* cm (Fig. 17.13). Now all the points at *c* cm from *B* are on the circumference of a circle of centre *B* and radius *c* cm, and all the points at a distance of *b* cm from point *C* are on the circumference of a circle of centre *C* and radius *b* cm. The two points of intersection of these two loci are the two alternative positions of the vertex *A*. If the two circles do not intersect,

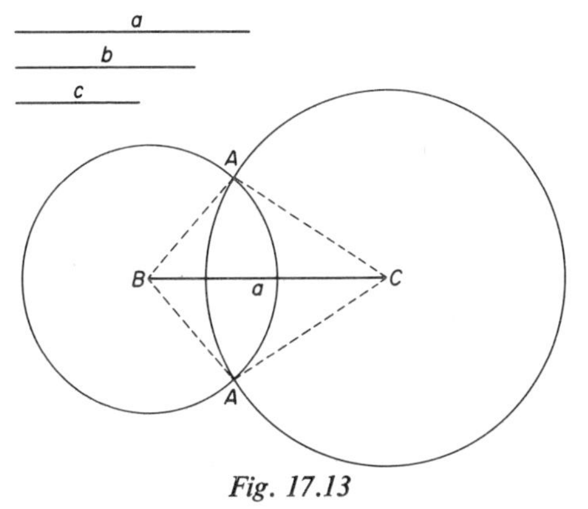

Fig. 17.13

$b + c < a$; if they intersect at only one point (that is, if they just touch) $b + c = a$. This confirms the triangle inequality $b + c > a$ (see Section 15.2).

Here we have been using the principle of *intersection* of two loci to find the exact location of a point (or points) which satisfies (satisfy) two conditions.

If each of the two loci is a *straight* line, there is only *one point* which satisfies these conditions, the point of intersection of the two lines (if they do in fact intersect). As in the solution of two simultaneous linear equations, we can therefore use this principle to prove that sometimes there is only *one* point which satisfies certain conditions.

Theorems 17.3 and 17.4 state two such cases.

Theorem 17.3: *There is only one point in the plane equidistant from the three vertices of a triangle.*

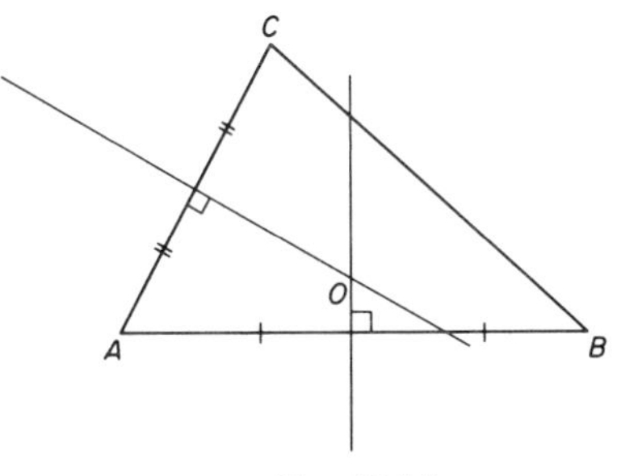

Fig. 17.14

All the points equidistant from the vertices A and B of $\triangle ABC$ are on a straight line, the perpendicular bisector of AB (Theorem 17.1). Similarly, all the points equidistant from the vertices A and C are on the perpendicular bisector of AC.

These two straight lines are *not* parallel but intersect at an angle of $180° - \angle A$; this follows from the property of the angle sum of quadrilaterals.

We name this *unique* point of intersection O.

Then $|OA| = |OB|$, as O is a point on the perpendicular bisector of AB, and also $|OA| = |OC|$, as O is a point on the perpendicular bisector of AC. Therefore, $|OA| = |OB| = |OC|$, and O is equidistant from the three points A, B and C.

Theorem 17.3 can be worded alternatively as:

There is one, and only one, point in the plane equidistant from three non-collinear points (i.e. points which are not all on one straight line).

Since $|OB| = |OC|$, point O also lies on the perpendicular bisector of BC (all the points equidistant from B and C are on the perpendicular bisector of BC, by Theorem 17.1); in other words,

the three perpendicular bisectors of the sides of the triangle are concurrent (that is, they pass through the same point).

Since $|OA| = |OB| = |OC|$, we can draw a circle with O as the centre and OA as radius, whose circumference will pass through the three vertices of the triangle, the points A, B and C (Fig. 17.15).

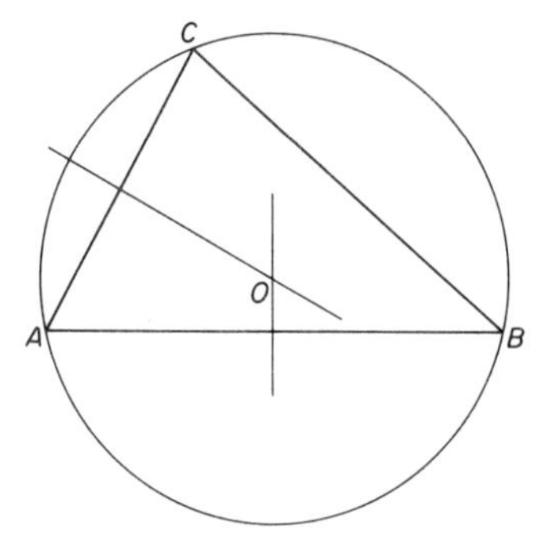

Fig. 17.15

This circle is called the *circumscribed* (Latin, 'drawn around') *circle* or the *circumcircle* of $\triangle ABC$ and the unique point O, where the three perpendicular bisectors meet, is called the *circumcentre* of the triangle.

Construction 17.1: *The circumscribed circle of a triangle.*

(*a*) Find the circumcentre O of the triangle ABC, the intersection of the perpendicular bisectors, by construction; it is, of course, only necessary to draw two perpendicular bisectors (Fig. 17.16).

(*b*) Draw the circle with O as centre and radius of length $|OA|$.

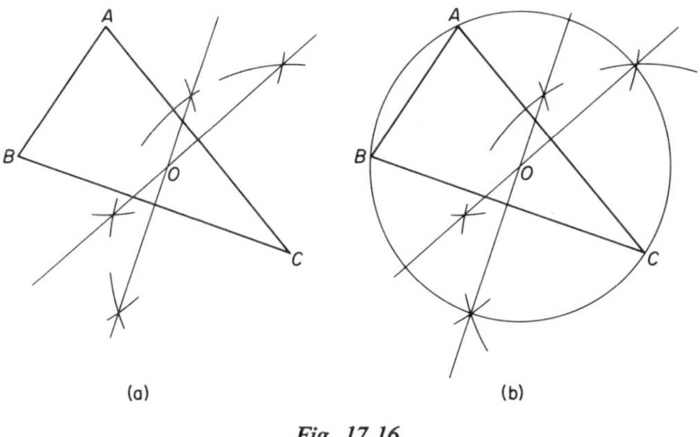

Fig. 17.16

Theorem 17.4: *There is only one point in the plane equidistant from the three sides of a triangle.*

All the points equidistant from the two sides PQ and PR of a $\triangle PQR$ are on the bisector of $\angle P$ (Fig. 17.17).

All the points equidistant from the sides RP and RQ are on the bisector of $\angle R$.

These two straight lines are *not* parallel but intersect at a point I. (The angle at which they intersect is $180° - (\frac{1}{2}\angle P + \frac{1}{2}\angle R)$; work this out for yourself as an exercise.)

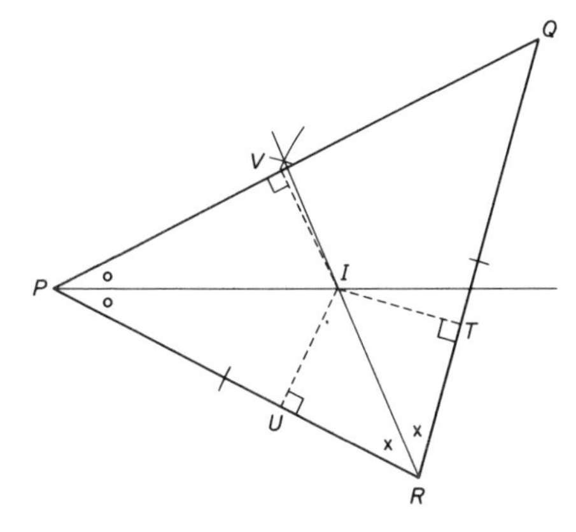

Fig. 17.17

If the points T, U and V are the feet of the perpendiculars from the point I to the sides of $\triangle PQR$, then

$|IU| = |IV|$, since I is a point on the bisector of $\angle P$,
$|IU| = |IT|$, since I is a point on the bisector of $\angle R$.

Therefore, $|IT| = |IV| = |IU|$, and the point I is equidistant from the three sides of $\triangle PQR$.

Since $|IV| = |IT|$, point I is also a point on the locus of all points equidistant from the sides QP and QR, the bisector of $\angle Q$. In other words, *the three bisectors of the angles of a triangle pass through one point; they are concurrent.*

Since $|IT| = |IV| = |IU|$, we can draw a circle of centre I which will pass through the points T, V and U, the *inscribed circle* or *incircle of* $\triangle PQR$, and the point I in Fig. 17.17, the unique point at which the three bisectors of the angles of $\triangle PQR$ meet, is called the *in-centre* of the triangle.

Construction 17.2: *The inscribed circle of a triangle.*

(*a*) Construct the bisectors of two angles of the triangle PQR, which intersect at a point I.

(*b*) Draw a perpendicular through I to one of the sides, intersecting it at a point F.

(*c*) Draw the circle with I as centre and radius of length $|IF|$.

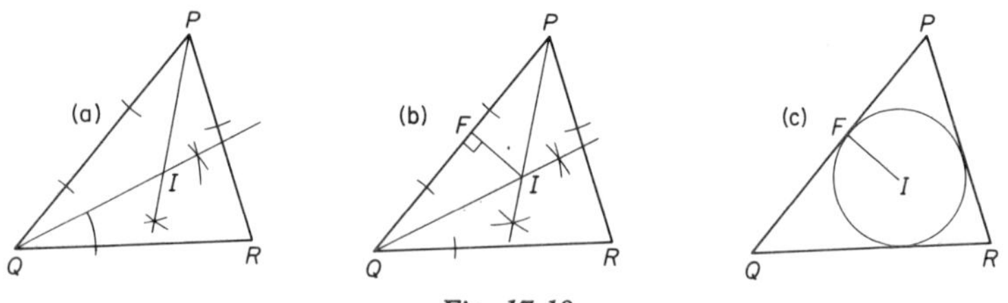

Fig. 17.18

17.5 Exercises

1. Draw a triangle PQR and construct its circumscribed circle, centre O. At the points P, Q and R construct lines perpendicular to OP, OQ and OR, respectively (Fig. 17.19). If these lines intersect at the points A, B and C, the circumscribed circle of $\triangle PQR$ is the inscribed circle of $\triangle ABC$.
   Show that $|AQ| = |AR|$, $|BP| = |BR|$ and $|CP| = |CQ|$.
   [Hint: show that $AQOR$, $BPOR$ and $CPOQ$ are kites; alternatively, show the congruence of $\triangle$s $\begin{matrix} AQO \\ ARO \end{matrix}$ (*R–H–S*), and so on.]
2. Practise the constructions of inscribed and circumscribed circles of various types of triangles.

17.6 The Circle

The most natural shapes and the ones most frequently found on this Earth and elsewhere in the universe are probably the 'rounded' shapes of the sphere

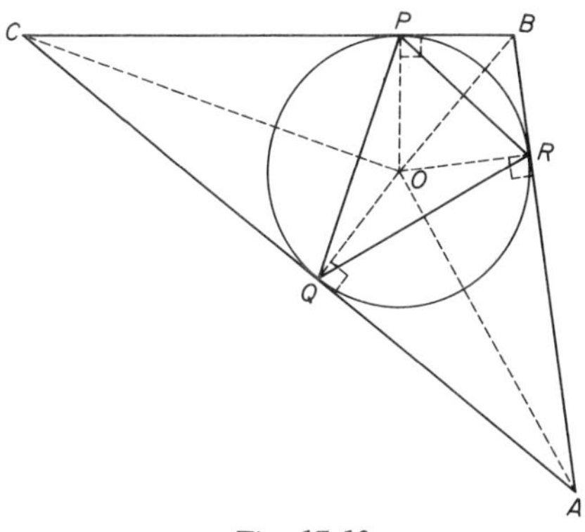

Fig. 17.19

and circle. Often these shapes are created by a force which acts from one central point in various directions, like the circular wave on the water surface made by a falling stone. Sometimes, too, the circle is the most suitable shape for a particular purpose because of its perfect symmetry.

Man has always been fascinated by the circle. Magic properties were attributed to it by many ancient religions, demonstrated by the circles of stones which still stand at places like Stonehenge and Avebury. The development of his understanding of the special nature of the circle led to his invention of the wheel. If the rigidity of the triangle can be called the basis of all structural engineering, the perfect symmetry of the circle is the basis of all mobile engineering.

Terms. The *circle*, or circular disc, is a closed plane shape, bounded by a curved line, its *circumference*, which is a set of points all at the same distance from one point, the *centre of the circle.*

If A and B are two points on the circumference of a circle (Fig. 17.20), the straight line segment AB is a *chord of the circle.*

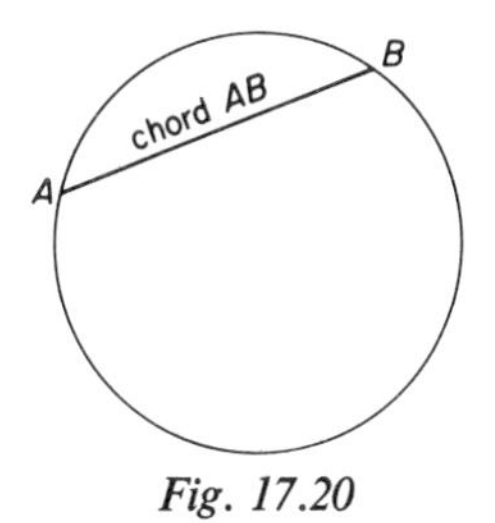

Fig. 17.20

The (curved) part of the circumference between the points A and B is called an *arc of the circle*; all its points are at the same distance from the centre of the circle. Every pair of points on the circumference of a circle divides the circumference into two arcs, and to avoid confusion, we distinguish between them by mentioning an intermediate point on the arc. Thus in Fig. 17.21, 'the arc AB' can refer either to the *minor arc AB* (the arc APB) or the *major arc AB* (the arc AQB).

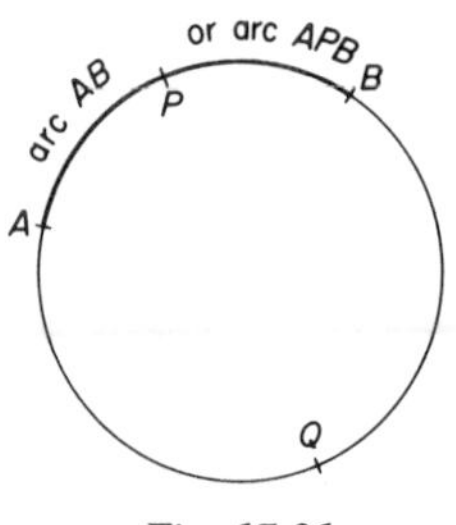

Fig. 17.21

A chord PQ of a circle which passes through the centre of the circle is called a *diameter of the circle*. A line segment with end-points lying respectively on the circumference and at the centre of the circle is called a *radius of the circle*. Clearly, the length of the diameter is twice the length of the radius; in Fig. 17.22, $|PQ| = 2|PO| = 2|OQ|$.

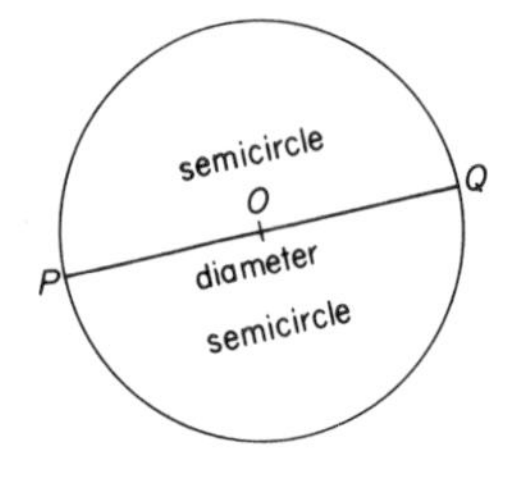

Fig. 17.22

Each diameter divides the circle into two congruent halves, each known as a *semicircle*. Other chords divide the circle into two *segments*, called the *major segment* and the *minor segment* respectively (Fig. 17.23).

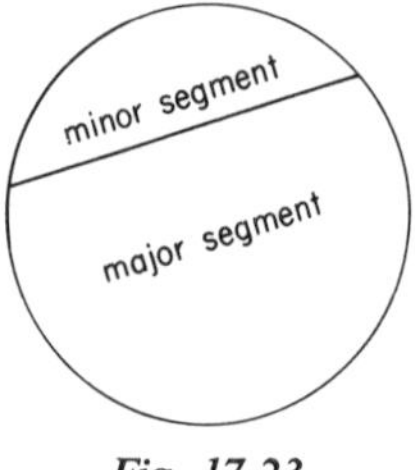

Fig. 17.23

A part of the circle bounded by two radii OA and OB and the arc AB is a *sector of the circle*, and $\angle AOB$ (Fig. 17.24), the angle in the sector formed by the two bounding radii, is called the angle of the sector. A semicircle is both a segment and a sector of the circle.

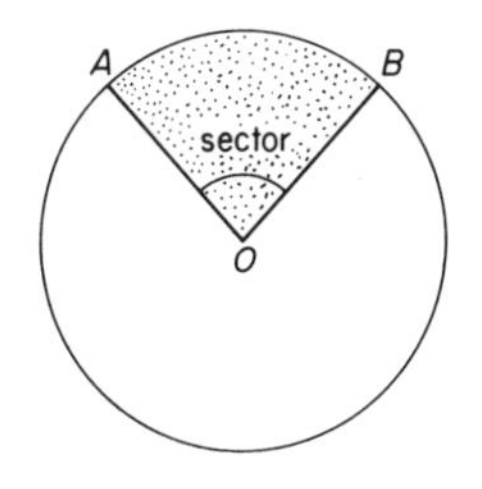

Fig. 17.24

A line which meets the circle at *one* point only (Fig. 17.25) is said to be a *tangent* of (or to) the circle (Latin, 'touching').

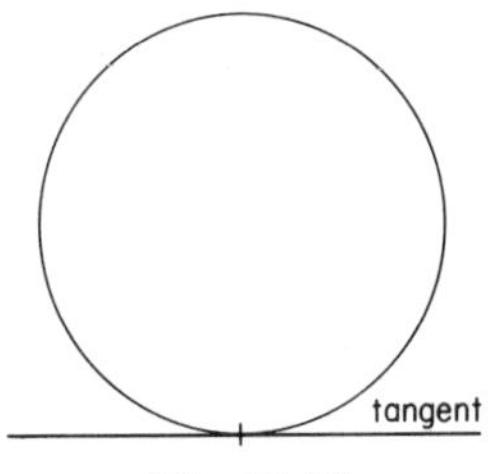

Fig. 17.25

17.7 Exercises

1. Draw a circle and cut it out. Fold it in half, and check whether the line of the fold passes through the centre of the circle. Fold it again and again, and show that the circle has *bilateral symmetry* about every diameter and therefore has an infinite number of axes of symmetry.
2. The *rotational symmetry* of the circle is clearly shown by the way that it is drawn using compasses, tracing out the points of the circumference by the end-point of a rotating line segment, the other end of which is fixed.

   Similarly, the *whole* of the circle, the interior as well as the circumference, is generated by a line segment rotating about one of its end-points. Try to demonstrate this experimentally using a pencil lead or chalk, or with a stick rotating in sand.

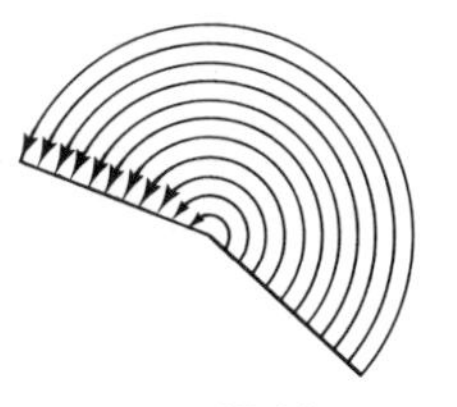

Fig. 17.26

3. Use the triangle inequality (Section 13.2) to show that the diameter is the largest chord in the circle. [Hint: let AB (Fig. 17.27) be a chord which is not a diameter of the circle of which the centre is O.

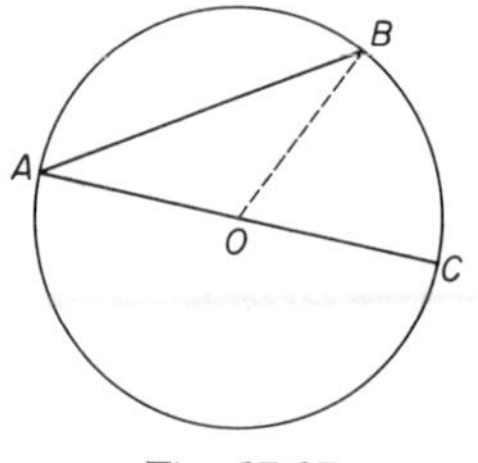

Fig. 17.27

Draw the diameter AOC and show that $|AB| < |AC|$.]
The principle of slide calipers (Fig. 17.28) is based on this geometrical fact.

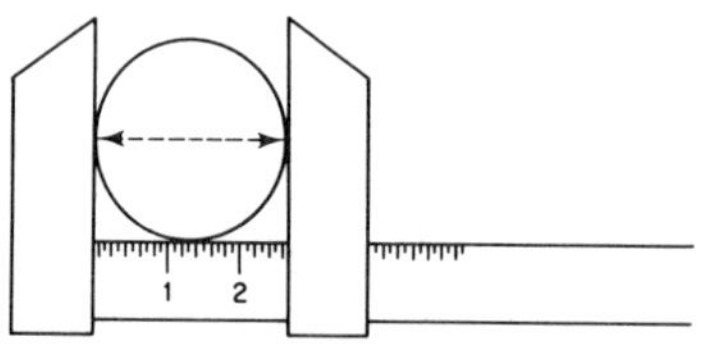

Fig. 17.28

4. AB and CD are two chords in a circle of centre O, and $|AB| = |CD|$.
   Show that $\triangle$s $\begin{matrix} AOB \\ COD \end{matrix}$ are congruent.

17.8 The 'Length' of the Circumference of a Circle

We defined the length of a line segment as the shortest distance between the two end-points (Section 14.4). We can only use the notion of length in relation to a segment of a curve if we imagine the curve to be 'straightened out'.

In Exercise 2 of Section 1.10, we made a rough measurement of the 'length' of the circumference of a circle by placing a piece of string or tape round its rim.

Another method is to roll a disc or wheel along a straight line (Fig. 17.29).

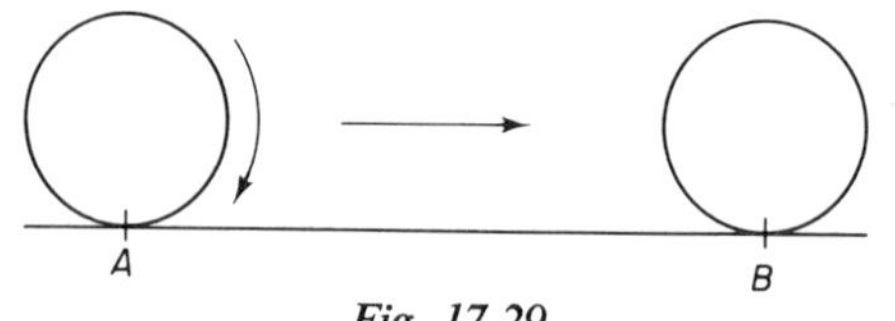

Fig. 17.29

A marked point on the rim of the wheel is placed on a point, marked A, of a straight line and the wheel carefully rolled along this line. After one revolution the marked point on the wheel will meet the line again at a point B; we can then measure AB, the length of the circumference of the wheel.

Mathematicians noticed even in ancient times that the ratio of the length of the circumference of a circle to its diameter is always the same number, whatever the size of the circle, and the search went on for centuries for the precise value of this number.

The ancient Chinese and Hebrews took it to be the number 3, but it is not a whole number, nor even a rational number. It is now called the number π (the Greek letter *pi*), a name given to it by William Jones in 1707.

At the time of Archimedes (born 287 B.C.) an approximate value for π of $3\frac{1}{7}$ or $\sqrt{10}$ was used; a closer approximation, which the Greeks knew by around 200 B.C., is 3·1416 (or $3{\cdot}14 < \pi < 3\frac{1}{7}$), and this figure is precise enough for most practical purposes.

On most slide rules the value of π is specially marked (Section 6.17). The relation between π, the circumference of a circle and its diameter is simply

$$\pi = \frac{\text{length of the circumference of a circle}}{\text{length of the diameter of the circle}}.$$

If r is the length of the radius, and d the length of the diameter, of the circle, we may write down the formulae:

$$\boxed{\begin{aligned}\text{length of the circumference of the circle} &= 2\pi r\\ &= \pi d\end{aligned}}$$

17.9 Exercises

1. Cut out a circular cardboard disc, 10 cm in diameter. Find the length of its circumference by rolling it along a straight line.
   Alternatively, use any wheel and measure its diameter and circumference.
2. Wind a thread 10 times round a small circular object, such as a pipe or a bottle. Measure the total length of the thread used and the diameter of the circular object. Calculate a value for π.
3. The diameter of a car wheel is 60 cm. Calculate the distance the car has travelled after 1000 rotations of the wheels, using your slide rule.
   How many times do the wheels rotate while this car travels 1 km?
4. Show that the ratio of the lengths of the circumferences of two circles is the same as the ratio of the lengths of their diameters.
   A combination of two wheels or pulleys provides a gear of 5:2, that is, the ratio of the lengths of the circumferences of the two wheels is 5:2. Calculate the diameter of the smaller wheel, if the diameter of the larger wheel is 24 cm (Fig. 17.30).

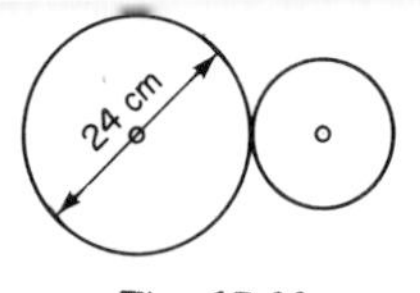

Fig. 17.30

17.10 Congruence of Sectors of Circles, and Lengths of Arcs

Clearly, two circles are congruent if their radii are of the same lengths; if their centres are made to coincide, their circumferences too will coincide, because of the locus property of the circle (Section 17.1).

We can draw a sector of a circle if we know its angle and the radius of the circle: there is only one sector that can be drawn for a given angle and a given radius. Thus, to construct a sector of angle 30° and radius 5 cm (Fig. 17.31):

(*a*) Construct an angle of 30°.
(*b*) Draw a circle with the vertex of the angle as its centre and with radius 5 cm.

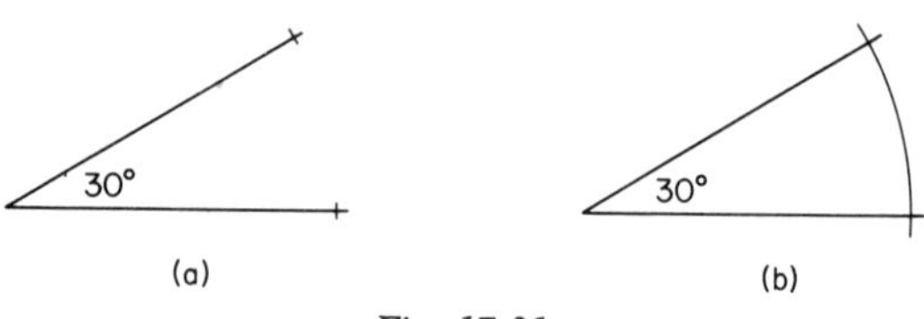

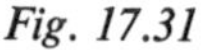

Fig. 17.31

Two sectors of the same circle are therefore congruent if their angles are of equal size. More generally,

> *two sectors of circles are congruent if their angles are of the same size and the radii of their circles are of the same length.*

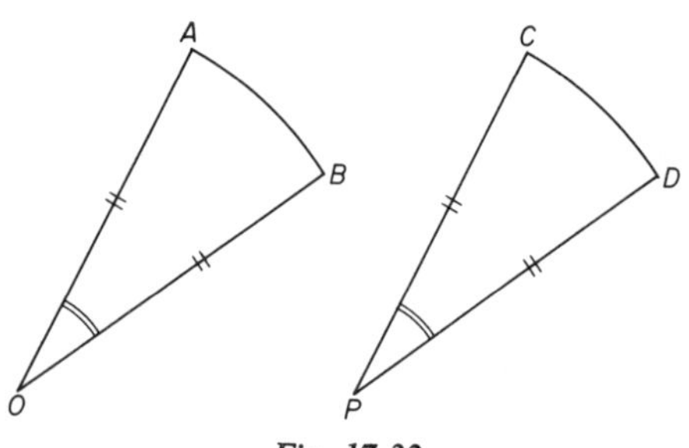

Fig. 17.32

In the two sectors OAB and PCD in Fig. 17.32,

$$|OA|\ (=|OB|) = |PC|\ (=|PD|) = 3\text{ cm}$$
$$\text{and } \angle AOB = \angle CPD.$$

If the two angles are made to coincide, point O with point P, A with C and B with D, the arc AB (or CD) is the locus of all points at 3 cm from O (or P) between the half lines forming the angle; therefore the two arcs AB and CD coincide and *are of the same length.*

The length of an arc of a circle

If the points A, B, C, D, E and F are points on the circumference of a circle centre O so that $ABCDEF$ is a regular hexagon, the sectors OAB, OBC, OCD, ODE, OEF and OFA are congruent and the arcs AB, BC, CD, DE, EF and FA are of equal length (Fig. 17.33).

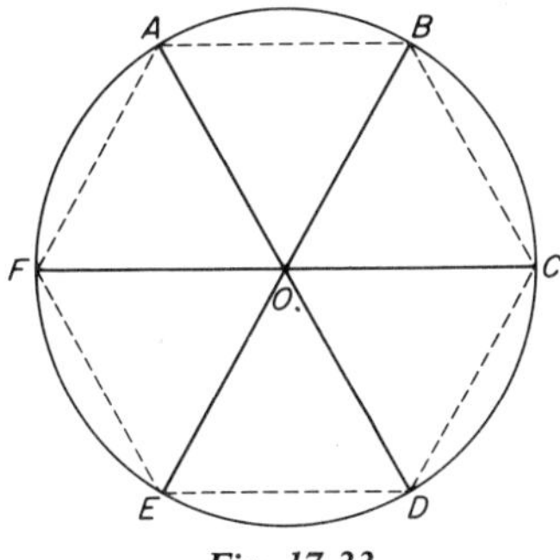

Fig. 17.33

The sum of their lengths is clearly the length of the circumference of the circle; the length of each of these arcs is therefore one-sixth of the length of the circumference. For example, if the diameter of the circle in Fig. 17.33 is 10 cm, and π is taken to be 3·1416, the circumference of the circle is $3{\cdot}1416 \times 10$ cm $= 31{\cdot}416$ cm. The length of the arc AB (or any of the other five arcs) is $\frac{31{\cdot}416}{6}$ cm $= 5{\cdot}24$ cm (correct to 3 SF).

The regular pentagon $PQRST$ and its circumscribed circle (Fig. 17.34) can be considered in the same way.

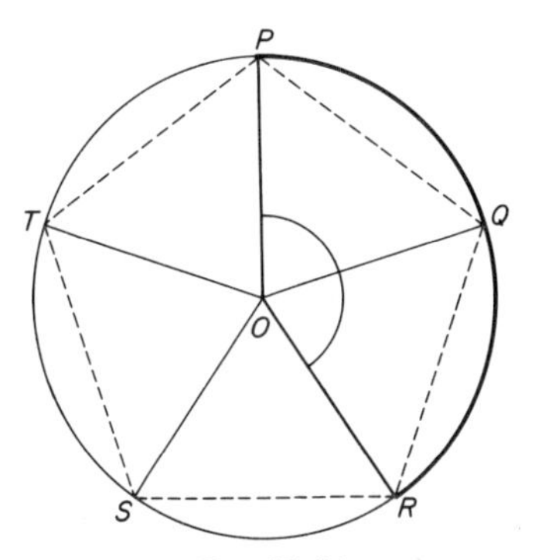

Fig. 17.34

Each of the five sectors OPQ, OQR, ORS, OST and OTP, is one-fifth of the complete circle, each of the angles at the centre is $\frac{360°}{5}$, and each of the five arcs PQ, QR, RS, ST and TP is one-fifth of the circumference.

Also $\angle POR$ is obviously $\frac{2}{5}$ of 360° and the arc PQR is $\frac{2}{5}$ of the circumference of the circle.

In general, we can calculate the length of any arc AB of a circle with centre O as a fraction of the length of the circumference of the circle; this fraction is determined by the size of $\angle AOB$.

Example

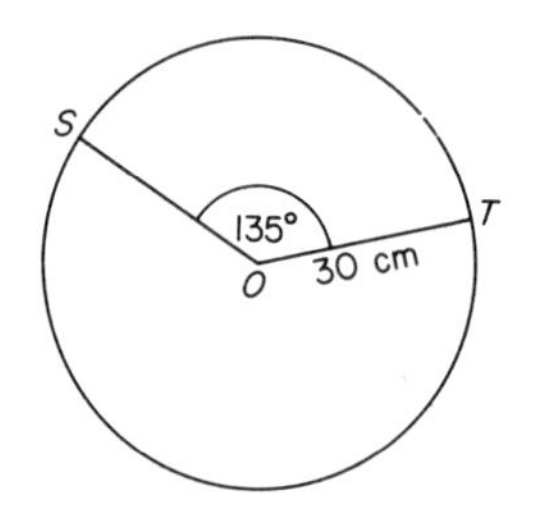

Fig. 17.35

The angle of a sector SOT of a circle is 135°. The radius of the circle is 30 cm. Calculate the length of the minor arc ST.

The ratio of the size of the angle of the sector to the size of the complete turn is $\frac{135}{360} = \frac{3}{8}$.

The length of the arc ST is therefore $\frac{3}{8} \times (\pi \times 60)$ cm $= 70{\cdot}7$ cm (using the slide rule).

Therefore, for any sector of a circle with an angle of $\alpha°$, the length of the arc of the sector is $\frac{\alpha}{360} \times$ length of the circumference, or, if the radius of the circle is r cm,

the length of the arc is $\frac{\alpha}{360} \times 2\pi r$ cm

17.11 Exercises

1. Calculate the total length of the circumference of a circle if the angle of a sector is 7·2° and the length of its arc is 800 km. (Eratosthenes used this method in 250 B.C. to calculate the length of the circumference of the Earth. He used the distance between Syene and Alexandria (800 km) as the length of the arc, and found the 'angle of the sector' from the shadows cast by the sun at these two places at the same time of day.)

Methods of calculating distances on the surface of the Earth will be discussed further in Section 21.18.

2. Use your slide rule to calculate the radius of the Earth if the length of the equator is given as 40 000 km.

17.12 Angles Formed by Lines in a Circle

Every arc of a circle determines one, and only one, sector of the circle, that is the sector formed by the arc and the two radii through the end-points of the arc (Fig. 17.36). (There is only one centre for each circle.) The angle of the sector is sometimes called the *angle at the centre standing on the arc*, or the *angle at the centre subtended by the arc.*

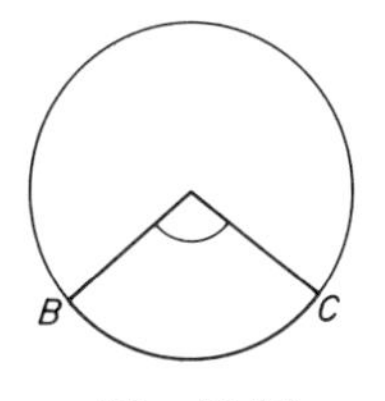

Fig. 17.36

If two *chords* are drawn, each joining an end-point of the arc BC to one point A on the remainder of the circumference of the circle (Fig. 17.37), an angle is formed at A by the two chords, known as an *angle at the circumference standing on the arc BC*, or an *angle at the circumference subtended by the arc BC.*

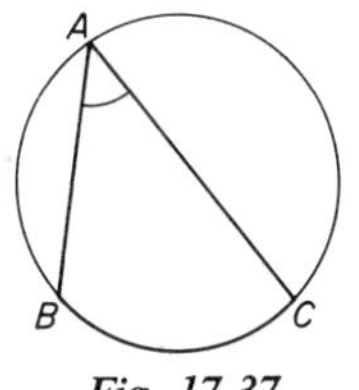

Fig. 17.37

The points B and C can be joined in this way to an infinitely large number of points on the circumference, so there are many such angles (Fig. 17.38).

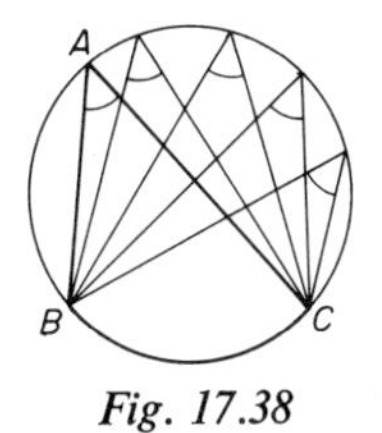

Fig. 17.38

If you measure these angles you will find that they are all of equal size.

17.13 Exercises

1. Draw a circle with centre O, and mark two points B and C on the circumference. Also mark any points P, Q, R and S on the circumference of the circle but not on the minor arc BC. Join each of these points to the points B and C.
   Measure the angles BPC, BQC, BRC, etc.
2. On the same diagram, draw the radii OB and OC and measure the angle BOC.

   You will find that the angle BOC (the angle at the centre) is twice the size of each of the angles BPC, BQC, BRC (the angles at the circumference standing on the arc BC).

17.14 The Angle at the Centre of a Circle

The size of the angle at the centre is always twice the size of the angle at the circumference which stands on the same arc.

We draw a circle with centre O, and mark an arc BC (Fig. 17.39). If one of the radii is extended, it will intersect the circle again at a point; in Fig. 17.39, CO is extended to the point A.

$\angle BOC$ is an exterior angle of $\triangle OAB$, formed by the two radii OA and OB and the chord $AB \Rightarrow \angle BOC = \angle B + \angle A$.

Since $\triangle OAB$ is isosceles because $|OA| = |OB|$ (radii), $\angle B = \angle A$ and $\angle BOC = 2\angle A$.

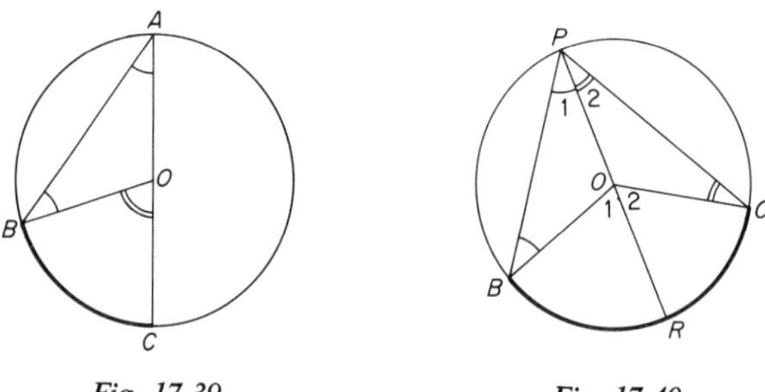

Fig. 17.39 Fig. 17.40

Now we will use the same proof to show:

Theorem 17.5: *The angle at the centre of a circle is twice the size of any angle at the circumference standing on the same arc.*

In a circle with centre O, BC is an arc and P is any point on the circumference of the circle, not on the arc BC (Fig. 17.40).

We draw the chords PB and PC, the radii OB, OC and OP and the diameter PR.

Then $\angle O_1 = \angle B + \angle P_1$

($\angle O$ is an exterior angle of $\triangle OBP$)

$|OB| = |OP|$ (radii)

$\Rightarrow \angle B = \angle P_1$.

Therefore, $\angle O_1 = \angle P_1 + \angle P_1 = 2\,\angle P_1$

In the same way,

$$\angle O_2 = \angle C + \angle P_2,$$
$$|OC| = |OP| \Rightarrow \angle C = \angle P_2$$

and $\angle O_2 = 2\angle P_2$.

Therefore,

$$\angle BOC = \angle O_1 + \angle O_2 = 2\angle P_1 + 2\angle P_2 = 2(\angle P_1 + \angle P_2) = 2\angle BPC.$$

This theorem is true whatever the size of $\angle BOC$; it can be an acute, obtuse or reflex angle.

Theorem 17.6: *All angles at the circumference standing on the same arc are equal.*

This follows immediately from Theorem 17.5. If the size of every angle at the circumference of a circle standing on the arc BC is half the size of the angle at the centre standing on the same arc BC, all the angles at the circumference standing on the arc BC are of the same size.

The angle at the point P of the circumference of a circle standing on the arc BC is sometimes called the *angle in the segment* of the circle formed by the chord BC and the arc BPC.

Theorem 17.6 can then also be expressed as:

Angles in the same segment are of equal size (Fig. 17.41).

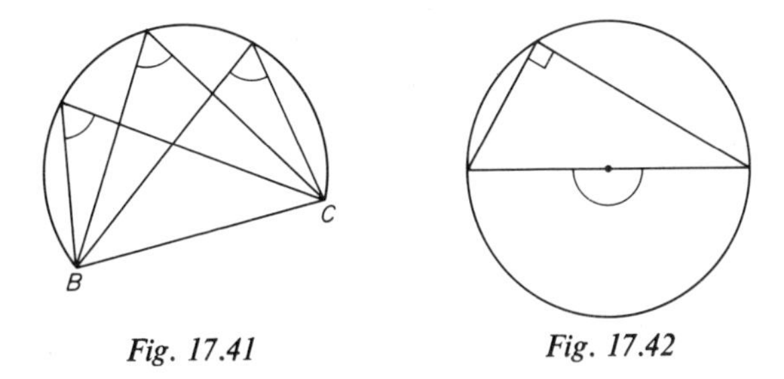

Fig. 17.41 *Fig. 17.42*

A specially important case of Theorem 17.6 concerns the angle in a semicircle. In this case, the angle at the centre is 180°, and so the angle at the circumference standing on the same arc is $\frac{180°}{2} = 90°$, a right angle (Fig. 17.42).

In terms of segments,

The angles in a semicircle are right angles.

Locus 17.5: Since every point on the arcs of either of the two semicircles forms a right angle with the two end-points of the diameter, *the two semicircular arcs of which the line segment AB is the diameter are the locus of all vertices of right-angled triangles which have AB as hypotenuse* (Fig. 17.43).

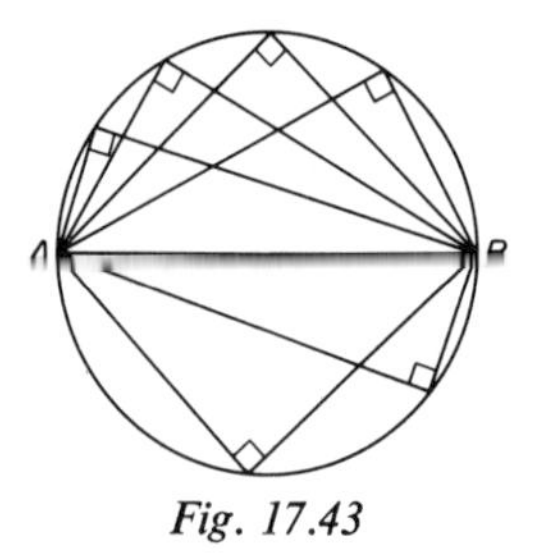

Fig. 17.43

17.15 Exercises

1. Draw a line segment PQ. Place a set-square over this line segment so that each of the two sides of the set-square forming the right angle passes through one of the two points P and Q. (You can keep the set-square in the right position by putting pins through the points P and Q.) Mark the point at the vertex of the right angle.

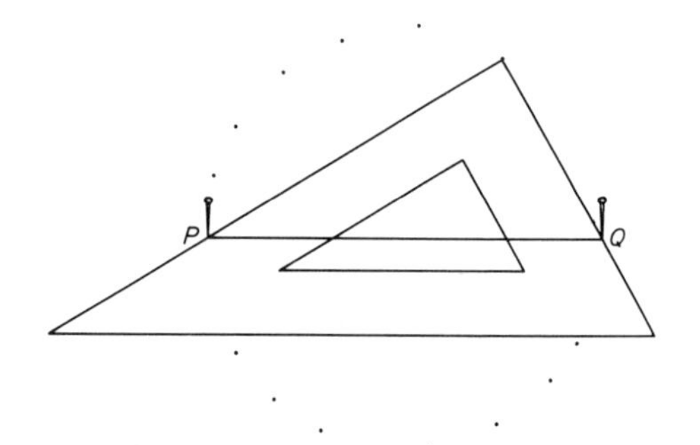

Fig. 17.44

Repeat this several times, and so confirm that any point forming a right-angled triangle with PQ as the hypotenuse lies on the circumference of a circle whose centre is the mid-point of PQ and whose radius is $\frac{1}{2}|PQ|$.

2. A triangle is right-angled at A, $|BC| = 8$ cm and $|BA| = 4$ cm. Construct $\triangle ABC$. [Hint: draw the hypotenuse BC; the point A is then the intersection of two loci.]
3. Show that the circumcentre of a right-angled triangle is the mid-point of its hypotenuse.

4. Calculate the size of the angles indicated by the letters a, b, c and d in Fig. 17.45.

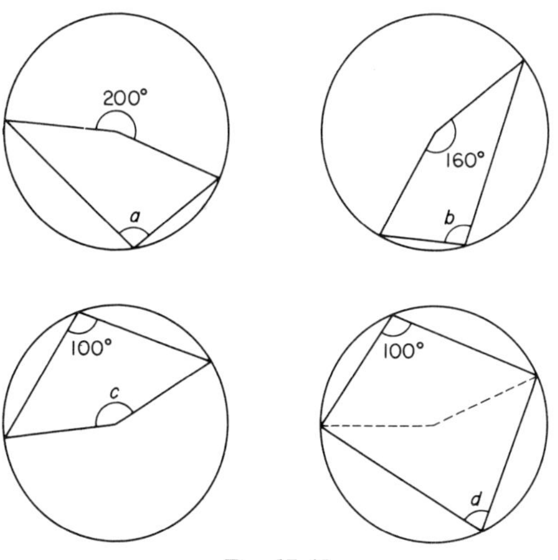

Fig. 17.45

5. The locus property of the right angle in a pair of semicircles described at the end of Section 17.14 is a special and important case, but it can be applied more generally to the angle contained in any pair of congruent segments which share a common chord. Of course, these two congruent segments do not together form one circle, except when they are semicircles.

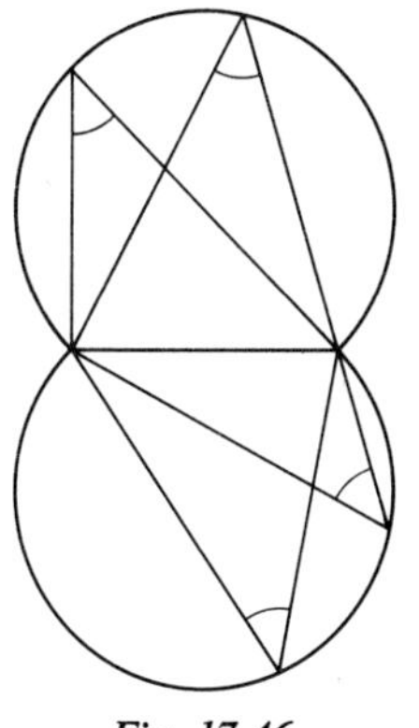

Fig. 17.46

Draw a line segment TS which is 5 cm in length.
Use a set-square with angles of 90°, 60° and 30° to draw the locus of all vertices P, such that $\angle TPS = 30°$.

17.16 Concyclic Points and Cyclic Quadrilaterals

We found (Theorem 17.3) that we can always draw one circle, and only one, whose circumference passes through the three vertices of a triangle. If the positions of four points P, Q, R and S are given, only one circle can be drawn whose circumference will pass through any three of these points, for example, through the points P, Q and R (Fig. 17.47). Point S *may* also lie on the circumference of this circle, but if the four points are arbitrarily chosen, this is very unlikely; you can confirm this by experiment.

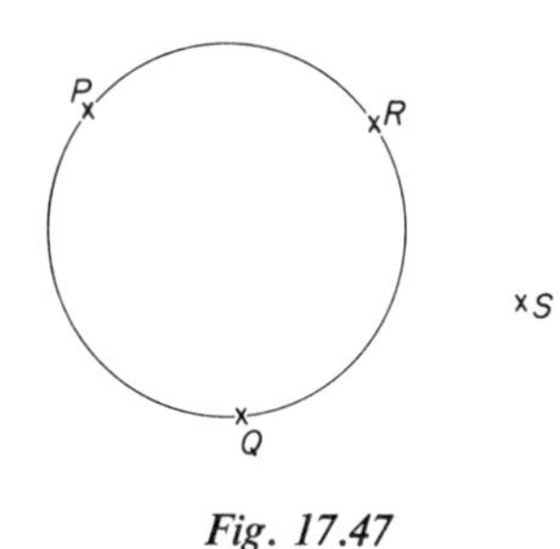

Fig. 17.47

In the exceptional case in which the four points P, Q, R and S all lie on the circumference of one circle, we call them *concyclic points*, and the quadrilateral formed by joining these points a *cyclic quadrilateral.*

(The simplest way of drawing a cyclic quadrilateral is by drawing a circle first and marking four points on its circumference.)

In Exercise 4 of Section 17.15 you may have noticed a special characteristic of the cyclic quadrilateral: the sum of its opposite angles is 180°, that is, the opposite angles of a cyclic quadrilateral are supplementary. We can confirm this by drawing an arbitrary cyclic quadrilateral, cutting it out, tearing off pairs of opposite corners and placing them side by side (Fig. 17.48).

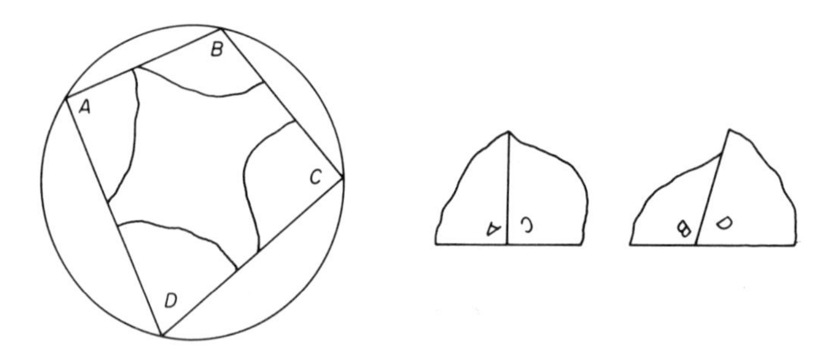

Fig. 17.48

Theorem 17.7: *The opposite angles of a cyclic quadrilateral are supplementary.*

If $ABCD$ is a cyclic quadrilateral, the radii OB and OD of the circumscribed circle form two angles at the centre, marked in Fig. 17.49 as $\angle O_1$ and $\angle O_2$, whose sum is 360°: $\angle O_1 + \angle O_2 = 360°$.

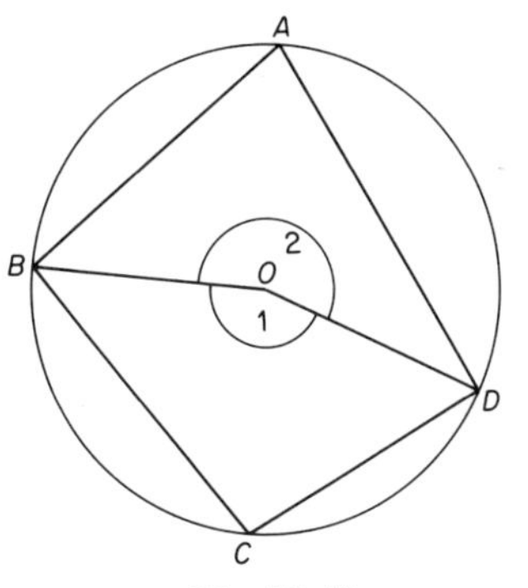

Fig. 17.49

$$\angle A = \tfrac{1}{2}\angle O_1, \text{ by Theorem 17.5,}$$
$$\angle C = \tfrac{1}{2}\angle O_2, \text{ similarly.}$$

Adding, $\angle A + \angle C = \frac{1}{2}\angle O_1 + \frac{1}{2}\angle O_2 = \frac{1}{2}(\angle O_1 + \angle O_2) = \frac{1}{2} \times 360° = 180°$. You can show in the same way that $\angle B$ and $\angle D$ are also supplementary. It also follows from the angle sum of a quadrilateral and the first part of the proof:

$$\angle A + \angle B + \angle C + \angle D = 360°,$$
$$\angle A + \angle C = 180°.$$

Subtracting, $\angle B + \angle D = 180°$.

17.17 Exercises

1. Calculate the size of the angles marked by the letters a, b, c, d, e and f in Fig. 17.50.

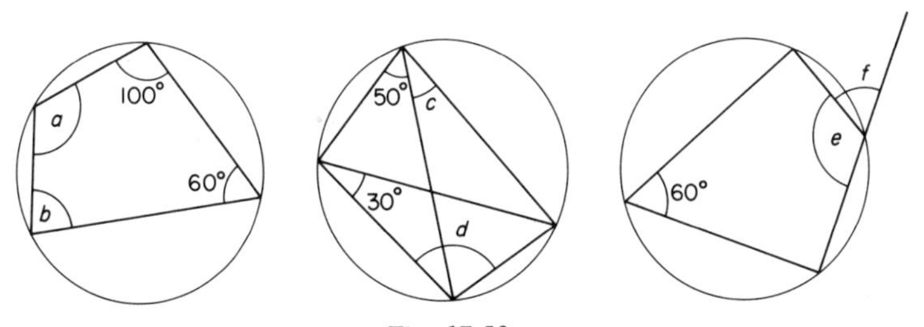

Fig. 17.50

2. Use Theorem 17.7 to show that *the exterior angle of a cyclic quadrilateral is equal in size to the opposite interior angle* (Fig. 17.51).

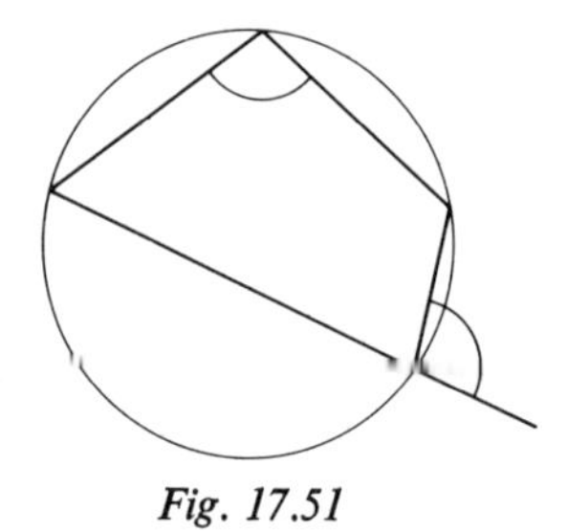

Fig. 17.51

17.18 Tests for Concyclic Points

Both Theorem 17.7 and the locus mentioned in Exercise 5 of Section 17.15 can be used to test whether or not a quadrilateral is cyclic, or if four points are concyclic.

(*a*) If the four points A, B, C and D, joined in order, form a quadrilateral and the sum of one pair of opposite angles is 180°, then the points A, B, C and D are concyclic and the quadrilateral $ABCD$ is a cyclic quadrilateral. (We only need to show that *one* pair of opposite angles is supplementary; it follows that the second pair is also supplementary, as we showed in Section 17.16.)

(*b*) If the points P, Q, R and S, joined in order, do not form a quadrilateral (the line segments PQ and RS intersect), and it is known that either $\angle P = \angle R$, or $\angle S = \angle Q$, then the four points P, Q, R and S are concyclic (Fig. 17.52).

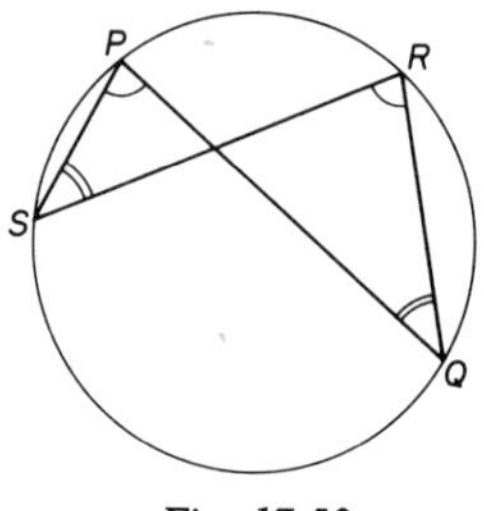

Fig. 17.52

17.19 Exercise

Use either of the two tests in Section 17.18 to decide whether the sets of four points in Fig. 17.53 are necessarily sets of concyclic points.

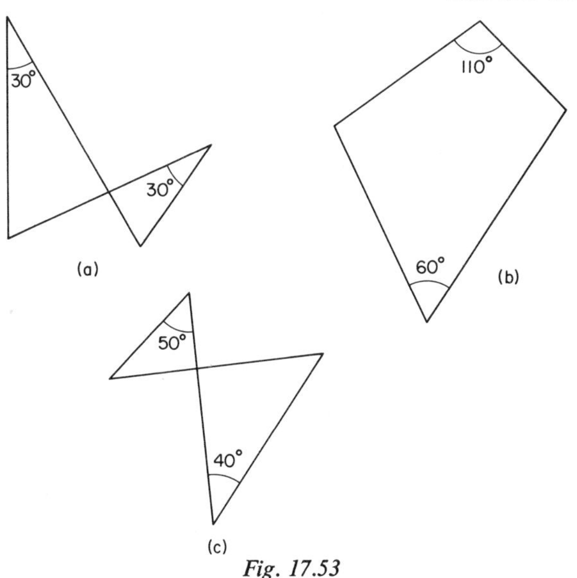

Fig. 17.53

17.20 Tangents to a Circle

A tangent to a circle is a line which intersects the circle at only one point, called the *point of contact*. A tangent can be drawn at any point on the circumference of a circle. If the point of contact P is joined to the centre O of a circle, the angle between the tangent and the radius OP is a right angle (Fig. 17.54).

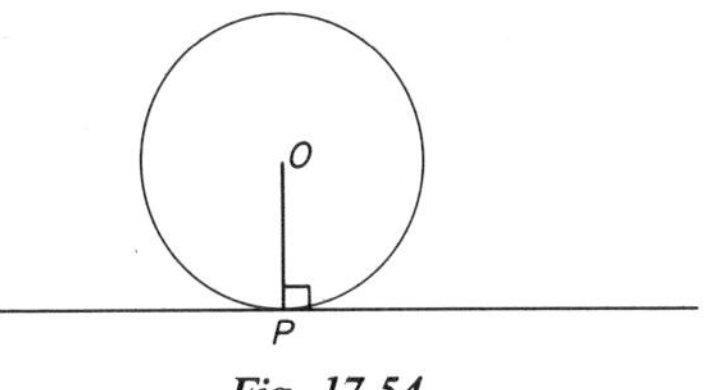

Fig. 17.54

Theorem 17.8: *The tangent to a circle is perpendicular to the radius through the point of contact.*

If point R is a point on a tangent to a circle (but not the same point as the point of contact P), the line segment RO will intersect the circumference of the circle at the point marked Q in Fig. 17.55.

If the length of the radius of the circle is r,

$$|OR| = |OQ| + |QR| = r + |QR|;$$

but $$|OP| = r,$$

therefore $$|OP| < r + |QR| = |OR|.$$

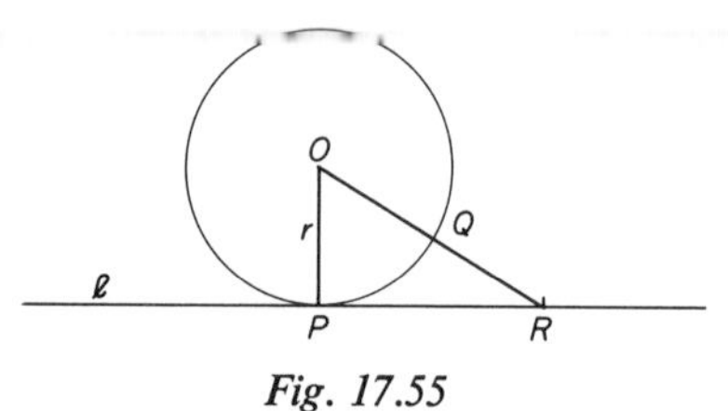

Fig. 17.55

Since this is true whatever the position of the point R on the tangent l, OP is the shortest distance from O to the line l and therefore perpendicular to l.

Construction 17.3: *Tangents from a given point to a circle.*

Obviously, no tangent can be drawn to a circle through a point *inside* the circle. Any line through such a point intersects the circle at *two* points, and is therefore not a tangent.

We can use Theorem 17.8 to construct a tangent at a point P on the circumference of a circle, centre O (Fig. 17.56).

(*a*) Draw the line OP.
(*b*) Construct a line at P at right angles with OP, using any of the constructions of a right angle, Section 16.13.

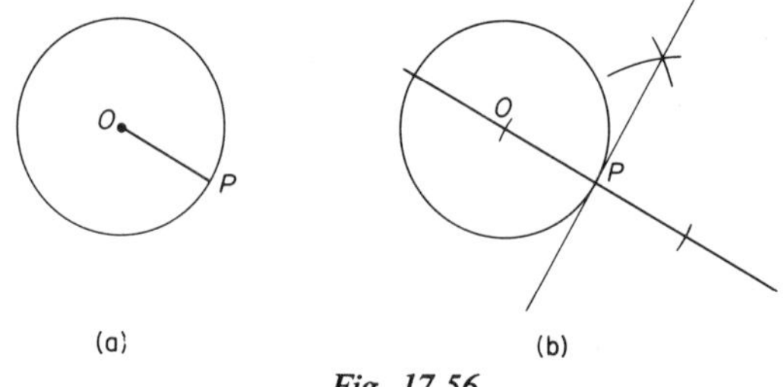

Fig. 17.56

To construct a tangent to a circle, centre O, through a point P *outside the the circle*, we have to find a point Q on the circumference of the circle such that $\angle OQP$ is a right angle; in other words, such that $\triangle OQP$ is right-angled at Q and OP is the hypotenuse (Fig. 17.57).

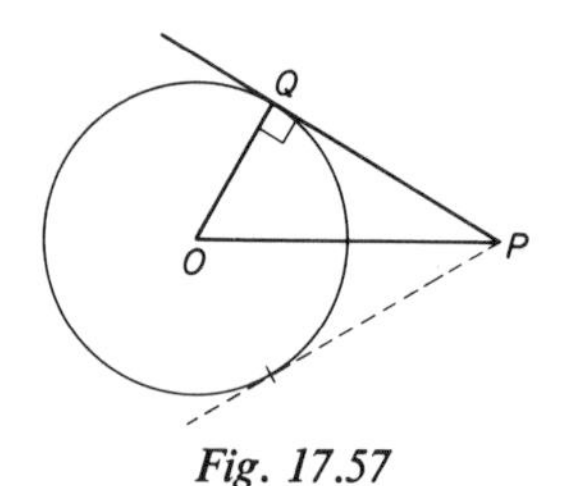

Fig. 17.57

Now the locus of all such points is the circumference of a circle on the diameter OP (Section 17.14). So we can find the position of Q simply by constructing a circle with OP as diameter; all we need to find is the mid-point of OP, the centre of this circle. The *two* points of intersection of the two circles are the two alternatives for the required point Q.

The procedure for the construction is as follows (Fig. 17.58):

(*a*) Join the points O and P.
(*b*) Find, by construction, the mid-point M of the line segment OP.
(*c*) Draw the circle with centre M and radius MO. You need not draw the whole circumference, only the arcs marking the intersections of the two circles.
(*d*) Join the point P to the points where the circumferences of the two circles intersect (the points marked Q and R). Then PQ and PR are the required tangents.

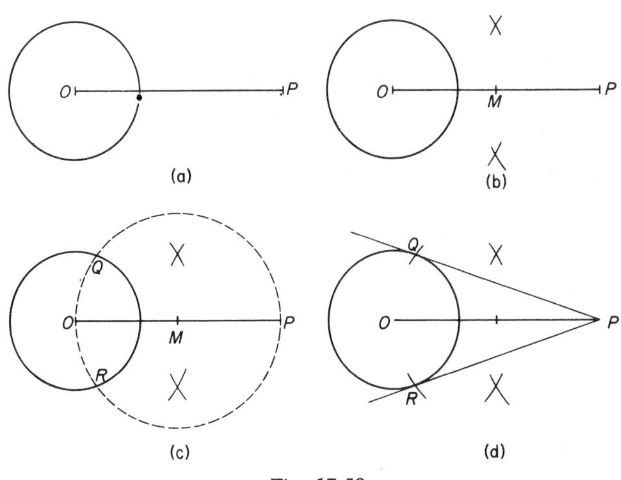

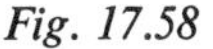
Fig. 17.58

17.21 Exercises

1. Draw a circle, centre O, and a point P outside the circle. Construct the two tangents from the point P to the circle. If the points of contact of these two tangents are Q and R, show that $\triangle$s $\begin{matrix}POQ\\POR\end{matrix}$ are congruent (R–H–S), and that therefore
   (*a*) the lengths of the line segments PQ and PR are equal;
   (*b*) the line PO is the bisector of $\angle QPR$.
2. Draw a large triangle ABC and construct its inscribed circle (the sides of the triangle are tangents to the circle). Mark the points of contact P, Q and

R on DC, CA and AD respectively and join P, Q and R by straight lines. The inscribed circle of $\triangle ABC$ is then the circumscribed circle of $\triangle PQR$.

If I is the centre of the inscribed circle of $\triangle ABC$, show that BI, CI and AI are the perpendicular bisectors of the sides of $\triangle PQR$. [Hint: show that $AQIR$, $BPIR$ and $CQIP$ are kites.]

3. $ABCD$ is a (special) quadrilateral which has an inscribed circle (Fig. 17.59).

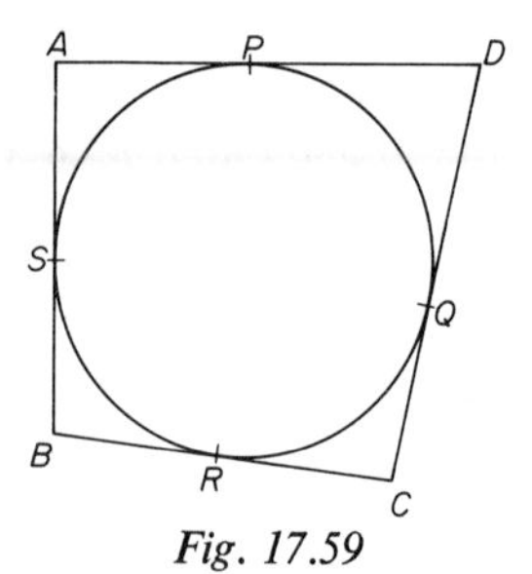

Fig. 17.59

Show that $|AB| + |CD| = |BC| + |DA|$. [Hint: use the result of the first example in this exercise; remember that AD, DC etc., are tangents to the circle, and, for example, $|AD| = |AP| + |PD|$.]

17.22 Common Tangents to Two Circles

A line l which is a tangent to two different circles is called the *common tangent* to these two circles.

The number of possible common tangents depends on the relative position of the two circles (Fig. 17.60).

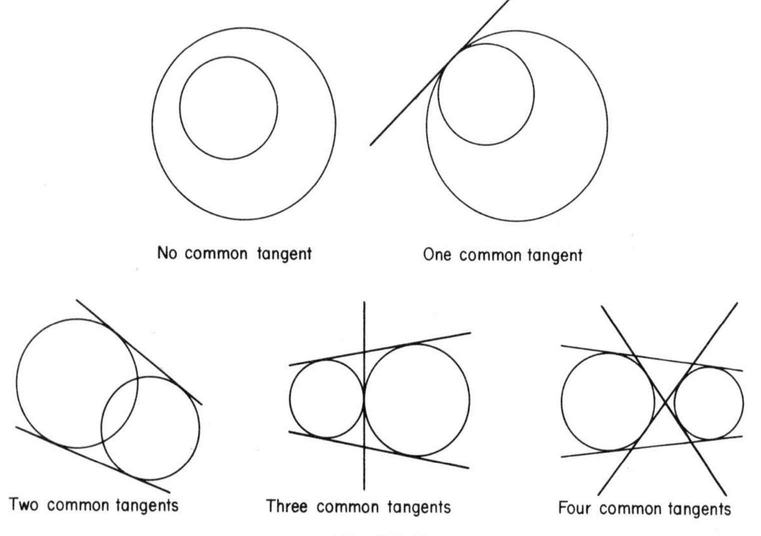

Fig. 17.60

17.23 The Alternate Segment Theorem

If l is a tangent to a circle at a point P, any chord drawn through the point P divides the circle into two segments, and the angle at P formed by the straight line l into two supplementary angles.

In relation to one of these angles, we call the segment 'on the other side' of the chord the *alternate segment*; in Fig. 17.61 the shaded segment is the alternate segment relative to the angle marked by the arc.

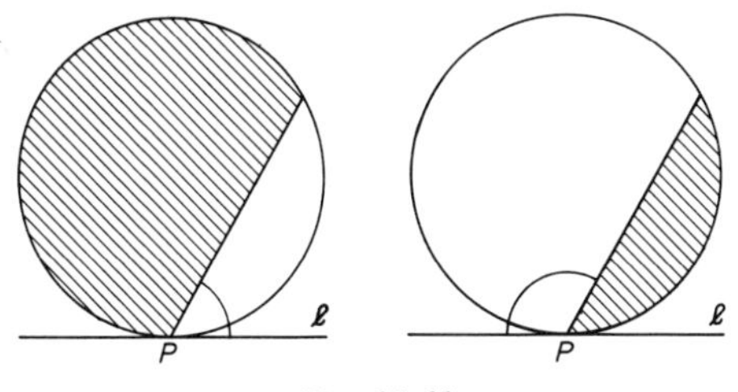

Fig. 17.61

Theorem 17.9: *The angle formed by a tangent and a chord through the point of contact is of the same size as the angle in the alternate segment.*

In Fig. 17.62, $\angle P_1 = \angle Q = \angle R = \angle S$, etc.

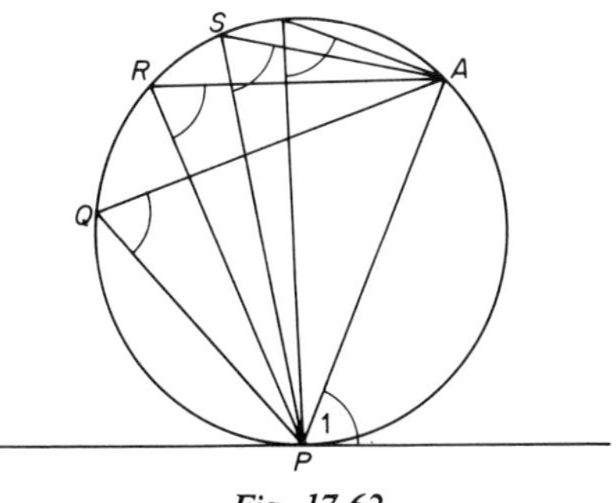

Fig. 17.62

Since all angles in the segment PQA are of equal size, we only need to show that one of them is of the same size as $\angle P_1$. We draw the diameter through point P, intersecting the circumference of the circle again at T, and we join the points T and A (Fig. 17.63).

Then $\angle TAP$ is a right angle (angles in a semicircle),
and $\angle P_3$ is a right angle (tangent and radius through the point of contact).
In the right-angled $\triangle TAP$,

$$\angle T + \angle P_2 = 90° \Rightarrow \angle T = 90° - \angle P_2;$$

also
$$\angle P_1 + \angle P_2 = 90° \Rightarrow \angle P_1 = 90° - \angle P_2.$$

Therefore
$$\angle T = \angle P_1.$$

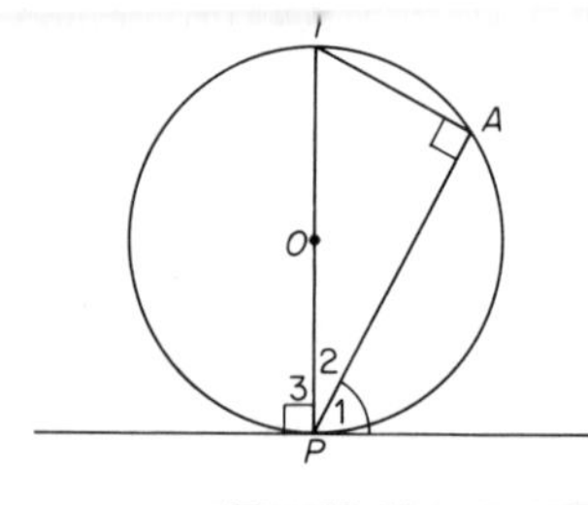

Fig. 17.63

Since all angles in the alternate segment PTA are of the same size, $\angle P_1$ is of the same size as every angle in the alternate segment.

17.24 Exercise

Use the Alternate Segment Theorem to calculate the size of the angles indicated in Fig. 17.64 by the letters a, b, c, d, e, f, g, h and i.

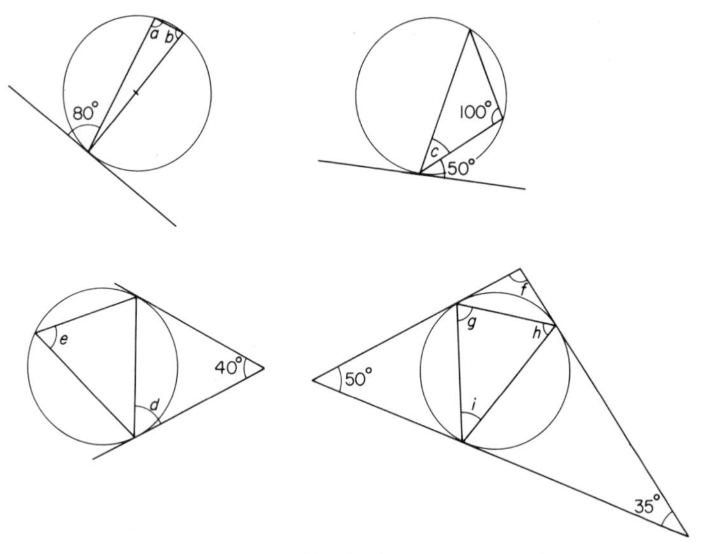

Fig. 17.64

Unit Eighteen

Similarity

18.1 Similar Figures and the General Intercept Theorem

At the beginning of Unit 14, we spoke of similar shapes rather vaguely as shapes which have the same general appearance but not necessarily the same size. Similarity was illustrated by the 'sameness' that can be found in two maps of England, or in a photograph and its enlargement. You may have noticed this similarity in shapes such as circles, equilateral triangles or squares, or in any regular polygons having the same number of sides. But it is obvious from our recognition of similarity in maps using different scales, or in photographs at different enlargements, that similarity does not only concern *regular* shapes. It is a relation high-lighted in the Mid-point and Intercept Theorems in Section 16.19, which studied the two triangles ABC and APQ, where P was the mid-point of AB and Q was the mid-point of AC (Fig. 18.1).

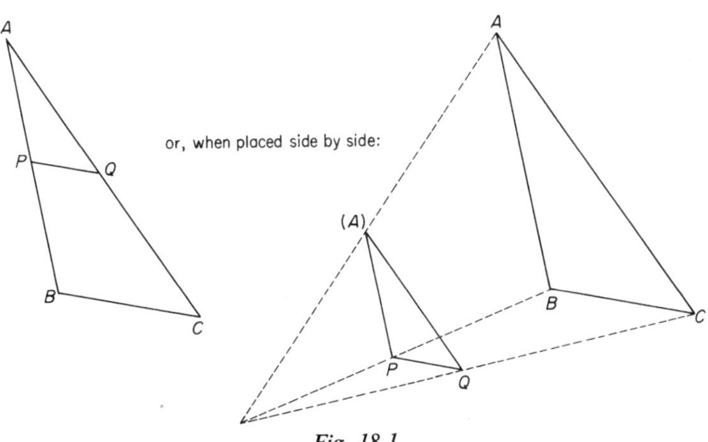

Fig. 18.1

Indeed, the Intercept and Mid-point Theorems reveal the most important aspects of similarity.

(*a*) Each angle in one figure corresponds to an angle of equal size in the other figure: we speak, therefore, of *corresponding angles.*

(*h*) If we call the sides occupying the same position relative to the corresponding equal angles the *corresponding sides*, the ratios of the lengths of the corresponding sides are the same, or constant; in Fig. 18.1,

$$\frac{|AB|}{|AP|} = \frac{|BC|}{|PQ|} = \frac{|CA|}{|QA|} = 2.$$

Either of these two properties is a sufficient condition for similarity of *triangles*. We have already shown this for one special case in the Intercept and Mid-point Theorems.

We will now examine the more general case of the Intercept Theorem and show that each of these properties is always sufficient for the similarity of triangles.

The General Intercept Theorem

We began the proof of the special case of the Intercept Theorem (Theorem 16.9 in Section 16.19), by drawing a line parallel to BC through the mid-point M of the side AC of $\triangle ABC$.

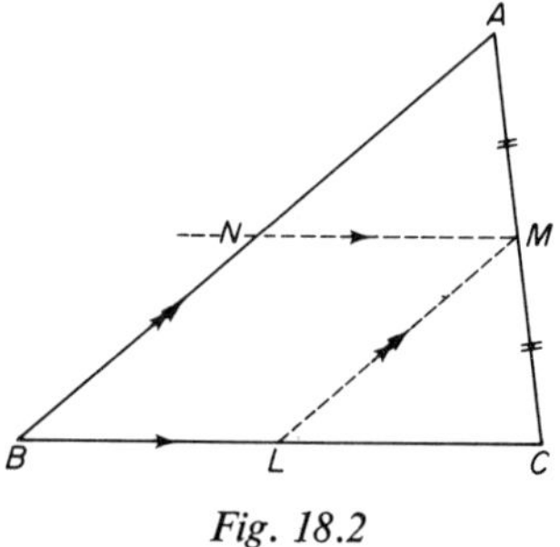

Fig. 18.2

We found that this line also bisects the side AB, or *makes equal intercepts* on the line segment AB, or $|AN| = |BN|$, using the congruence of $\triangle$s $\frac{ANM}{MLC}$.

The same argument can be used to prove the general case.

Theorem 18.1: *If parallel lines make equal intercepts on one line, they make equal intercepts on any other line that intersects them.*

In Fig. 18.3, $BK // CL // DM // EN$, and $|AK| = |KL| = |LM| = |MN|$, i.e. the lines BK, CL, DM and EN make equal intercepts on AN.

The line AE intersects the parallel lines at B, C, D and E respectively.

We show that $|AB| = |BC| = |CD| = |DE|$.

We draw lines through the points K, L and M, parallel to AE, intersecting the parallel lines at P, Q and R (Fig. 18.3); $BKPC$, $CLQD$ and $DMRE$ are therefore parallelograms.

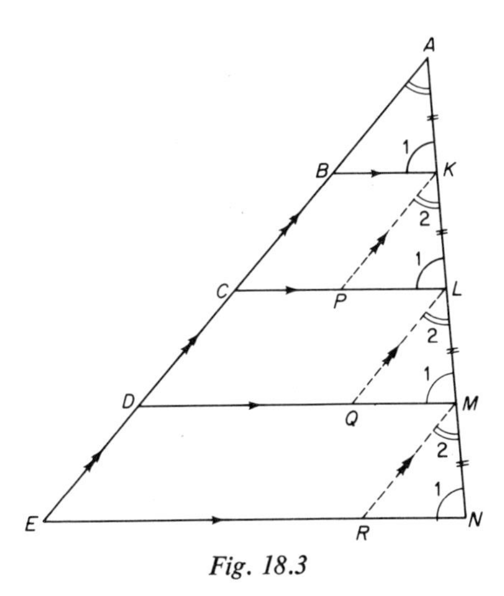

Fig. 18.3

Then $\angle A = \angle K_2 = \angle L_2 = \angle M_2$ (corresponding angles formed by AN and the parallel lines AE, KP, LQ and MR),
$|AK| = |KL| = |LM| = |MN|$ (given),
$\angle K_1 = \angle L_1 = \angle M_1 = \angle N_1$ (corresponding angles formed by AN and the parallel lines BK, CL, DM and EN).

Therefore $\triangle$s $\begin{matrix} ABK \\ KPL \\ LQM \\ MRN \end{matrix}$ are congruent (A–S–A)

$\Rightarrow |AB| = |KP| = |LQ| = |MR|$.

But $|KP| = |BC|$ (opposite sides of the parallelogram $BKPC$),
$|LQ| = |CD|$ (opposite sides of the parallelogram $CLQD$),
$|MR| = |DE|$ (opposite sides of the parallelogram $DMRE$).

Therefore

$$|AB| = |BC| = |CD| = |DE|;$$

that is, the 'intercepts' on AE are equal.

18.2 Exercises

1. If the line segments KP, LQ and MR in Fig. 18.3 are extended, and lines are drawn through the points B, C and D parallel to AN, the triangle AEN is divided into 16 small triangles (Fig. 18.4).
   Show that these 16 triangles are all congruent (compare the exercise in Section 16.22).
2. Use the congruence of these 16 triangles to show that
   $$|BK|:|CL|:|DM|:|EN| = 1:2:3:4.$$

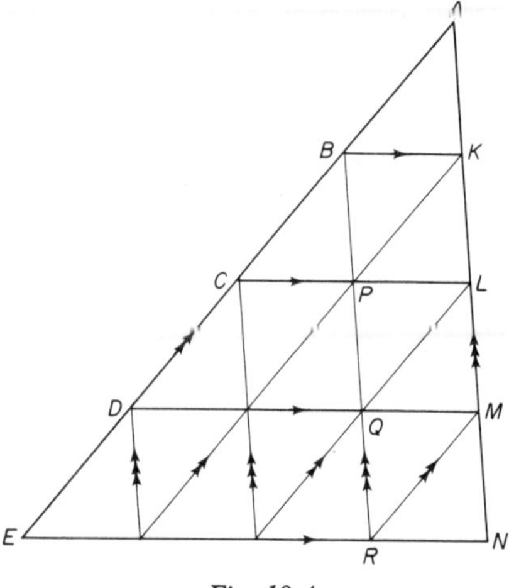

Fig. 18.4

Compare this with the ratios $|AB|:|AC|:|AD|:|AE|$ and $|AK|:|AL|:|AM|:|AN|$.

3. Write down the ratios of the lengths of the corresponding sides of

$$\triangle ACL \text{ and } \triangle ABK;$$
$$\triangle ADM \text{ and } \triangle ACL;$$
$$\triangle AEN \text{ and } \triangle ADM.$$

(If you find it confusing to have all these triangles drawn in one diagram, separate them, drawing them side by side.)

4. If $\triangle AEN$ consists of 16 triangles, each congruent with $\triangle ABK$ (and therefore each occupying the same amount of space), the ratio of the areas of the similar triangles $\triangle AEN$ and $\triangle ABK$ is $16:1$ or $4^2:1$.

   In the same way, investigate the ratio of the areas of the similar triangles,

$$\triangle ADM \text{ and } \triangle ABK;$$

also of

$$\triangle ACL \text{ and } \triangle ABK,$$
$$\triangle ADM \text{ and } \triangle ACL,$$
$$\triangle AEN \text{ and } \triangle ADM.$$

   Confirm that, if the ratios of the lengths of the corresponding sides of similar triangles is $n:k$, the ratio of their areas is $n^2:k^2$.
   (We will discuss this further in Unit 20.)

18.3 The Division of a Line Segment

Construction 18.1: *Division of a line segment into a given number of equal parts.*

We will divide a given line segment PQ into five segments of equal length, using the principle of the Intercept Theorem (Fig. 18.5).

(*a*) Through one of the end points of the segment PQ, say the point P, draw another line l.

(*b*) Use compasses to mark five points, A, B, C, D and E, on this line, stepping off equal distances, so that $|PA| = |AB| = |BC| = |CD| = |DE|$.

(*c*) Join the last of these points, point E, to the other end-point of the segment PQ, the point Q.

(*d*) Draw lines parallel to QE through the remaining points, the points A, B, C and D.

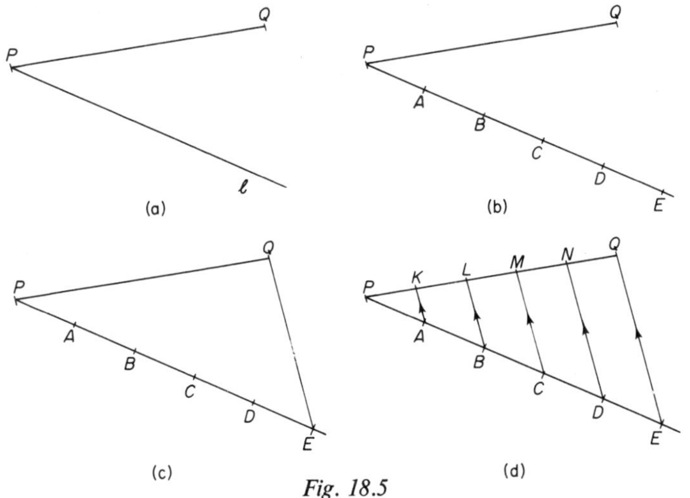

Fig. 18.5

The points where these parallel lines intersect the segment PQ, that is, *the points K, L, M and N divide PQ into five segments of equal length.*

Since the segments PK, KL, LM, MN and NQ are of the same length,

$$|PK|:|KQ| = 1:4,$$
$$|PL|:|LQ| = 2:3,$$
$$|PM|:|MQ| = 3:2,$$
$$|PN|:|NQ| = 4:1;$$

also, $$|PL|:|LN|:|NQ| = 2:2:1, \text{ and so on.}$$

This means that we can use this construction to divide $|PQ|$ in any of these ratios.

Construction 18.2: *Internal and external division of a line segment in a given ratio.*

If P is a point of the line segment AB—that is, a point on AB between the points A and B—so that $|AP|:|PB| = 5:3$, we say that the point P divides the segment *internally* in the ratio 5:3.

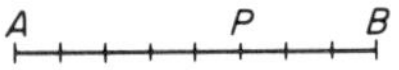

A P B

Fig. 18.6

Since $|AB| = |AP| + |PB|$, we may extend this ratio (see extension of ratios, Section 7.7):

$$|AP|:|PB|:|AB| = 5:3:8.$$

We can find the point P by construction (Fig. 18.7).

(*a*) Use Construction 18.1 to divide AB into eight equal parts, or units.
(*b*) The point P then lies at a distance of five units from A.

In general terms, we find the point P which divides the line segment AB *internally* in the ratio $p:q$ (p and q are whole numbers) as follows:

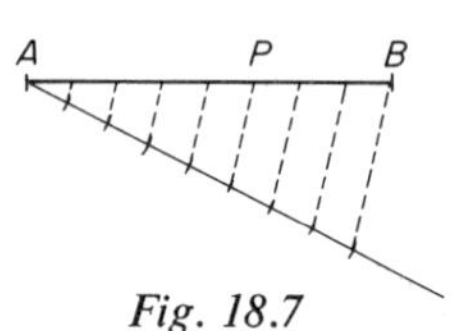

Fig. 18.7

(*a*) Divide AB by Construction 18.1 into $(p + q)$ equal parts.
(*b*) Point P is then at a distance of p of these units from A, or q units from B.

If the point Q lies on the line AB but *not* between the points A and B, so that $|AQ|:|BQ| = 5:3$ (Fig. 18.8), we say that the point Q divides the segment AB *externally* in the ratio 5:3.

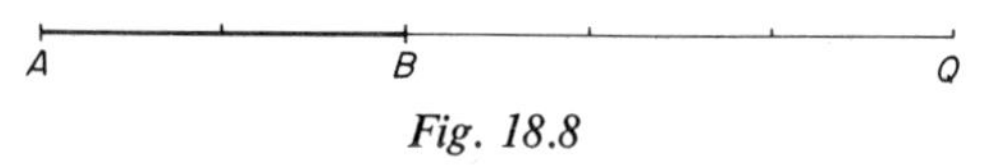

Fig. 18.8

Since $|AB| = |AQ| - |BQ|$, by extension of ratios,

$$|AQ|:|BQ|:|AB| = 5:3:2.$$

Therefore, we can find the point Q by construction (Fig. 18.9).

(*a*) Divide $|AB|$ into two equal parts, or units.
(*b*) Find the point Q by stepping off three units on the line AB starting from B (*five* units away from A).

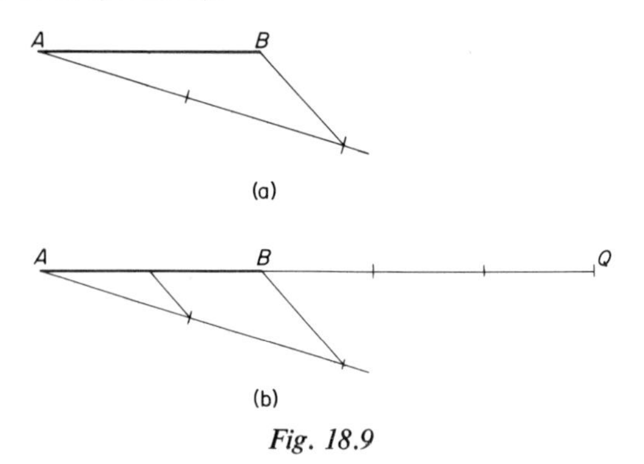

Fig. 18.9

It is clear from the ratio $AQ:BQ$ ($= 5:3$) that the point Q is nearer to B than to A ($5 > 3$); if Q is a point 'outside' the segment AB, nearer to B than to A, it must be a point 'on the same side' of A as the point B. But if the ratio $|AQ|:|BQ|$ were $3:5$, then the point Q is nearer to A than to B, and lies on the other side of A (Fig. 18.10).

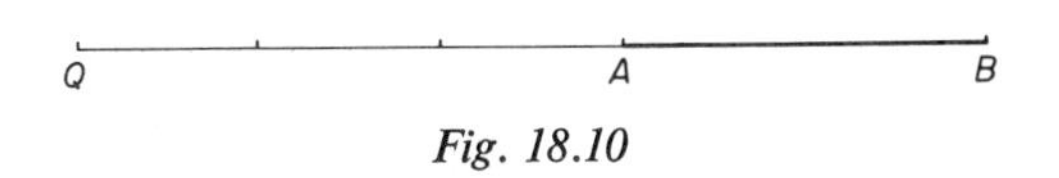

Fig. 18.10

In general, to find by construction the point Q which divides the segment AB externally in the ratio $p:q$,

(*a*) First establish which of p or q is the larger.

If $p > q$,
(*b*) Divide AB into $(p - q)$ equal parts, or units.
(*c*) Step off p units on the line AB starting from A in the direction of B and there mark the point Q, beyond B.

If $q > p$,
(*b*) Divide AB into $(q - p)$ equal parts.
(*c*) Find Q by stepping off q units on the line AB starting from B in the direction of A, but beyond A.

18.4 Exercises

1. Draw a line segment KL and divide it by construction into seven segments of equal length.
   Mark the point R, which divides the segment internally in the ratio $2:5$, that is, such that $|KR|:|RL| = 2:5$.
2. Draw a line segment AB, and divide it internally and externally in the ratio $7:3$.
   [Hint: Lined paper provides a ready-made set of parallel lines; they make equal intercepts on any line drawn across them.]

18.5 Another Application of the Intercept Theorem

In a more general form, the Intercept Theorem states that if P is a point on the side AB of $\triangle ABC$ and the line through P, parallel to BC, intersects AC at Q, then $|AP|:|PB| = |AQ|:|QC|$ (Fig. 18.11).

Since $|AB| = |AP| + |PB|$ and $|AC| = |AQ| + |QC|$,
by extension of ratios,

$$|AP|:|AB| = |AQ|:|AC|, \text{ or } \frac{|AP|}{|AB|} = \frac{|AQ|}{|AC|}.$$

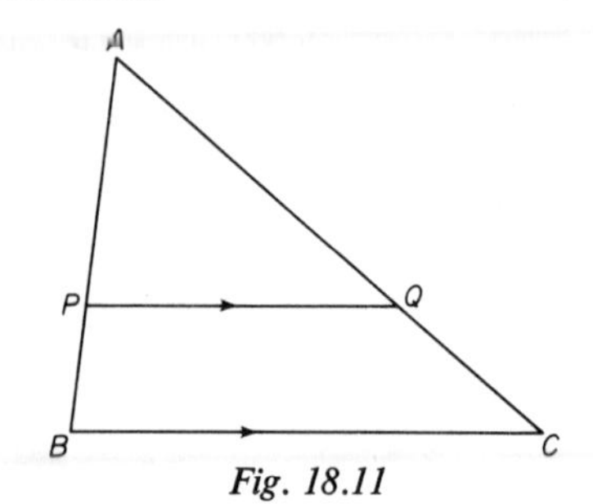

Fig. 18.11

Construction 18.3: *Division of a line segment in the ratio of the lengths of two given line segments*; this construction extends the use of the principles of Construction 18.2.

Suppose that, for example, three line segments are given, the line segment AB and two segments whose lengths are respectively p and q (Fig. 18.12). To divide the segment AB internally in the ratio $p:q$,

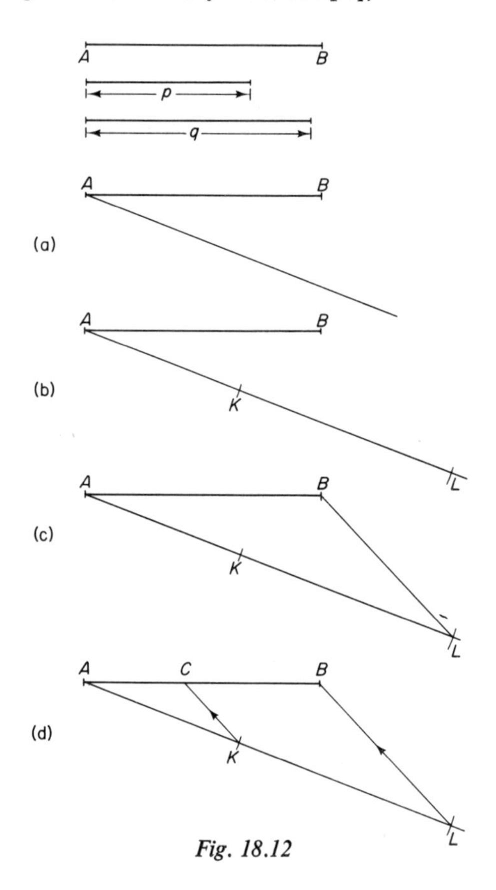

Fig. 18.12

(*a*) Draw a line through one of the end-points of AB, say the point A.
(*b*) Use compasses to mark two points K and L on this line, such that $|AK| = p$ and $|KL| = q$.
(*c*) Join the points B and L.
(*d*) Draw a line through K parallel to LB, intersecting AB at the point C. Then $|AC|:|CB| = p:q$.

Fig. 18.13 shows the steps in the exactly analogous construction of the *external* division of the segment AB in the ratio $p:q$.

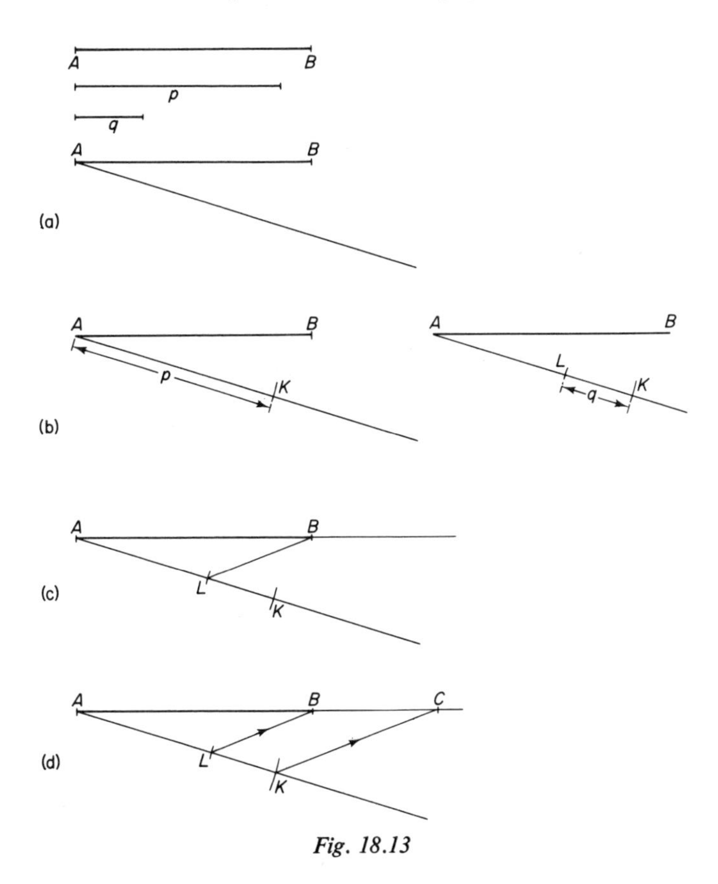

Fig. 18.13

18.6 Exercise

Draw a line segment PQ and two other segments of lengths a and b: choose any three arbitrary lengths.

Find by construction the points on the line PQ which divide it internally and externally in the ratio $a:b$.

18.7 The Angle Bisector Theorem

An alternative construction of the division of a line segment in a given ratio $p:q$ is based on Theorem 18.2, the *Angle Bisector Theorem.*

Theorem 18.2: *The bisectors of an angle of a triangle divide the side opposite internally and externally in the ratio of the lengths of the other two sides.*

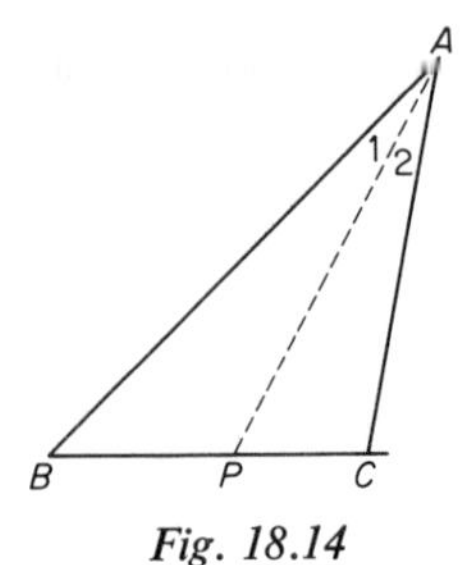

Fig. 18.14

(*a*) In Fig. 18.14, the *internal bisector* AP ($\angle A_1 = \angle A_2$) intersects BC at a point P such that $|BP|:|PC| = |AB|:|AC|$.

We draw a line through the vertex C parallel to AP, which intersects BA (produced) at point Q (Fig. 18.15).

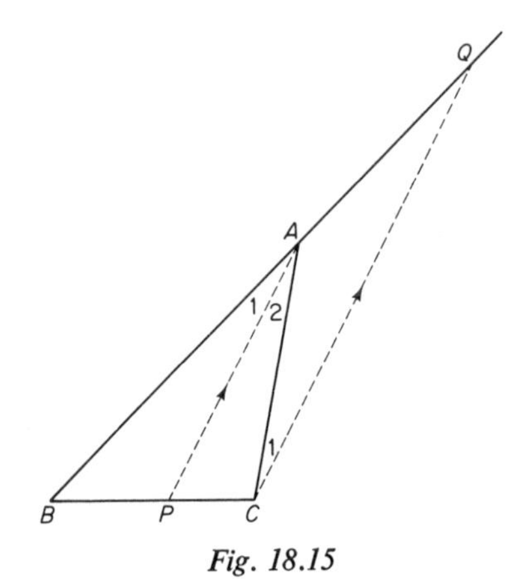

Fig. 18.15

$\triangle ACQ$ is isosceles, since

$$\begin{aligned} &\angle A_1 = \angle Q \text{ (corresponding angles of parallel lines),} \\ &\angle A_2 = \angle C_1 \text{ (corresponding angles of parallel lines),} \\ &\angle A_1 = \angle A_2 \text{ (} AP \text{ is the angle bisector)} \\ \Rightarrow\ &\angle Q = \angle C_1 \\ \Rightarrow\ &|AQ| = |AC|. \end{aligned}$$

By the Intercept Theorem,

$$|BP|:|PC| = |BA|:|AQ|.$$

Substituting $|AC|$ for $|AQ|$,

$$|BP|:|PC| = |AB|:|AC|.$$

(*b*) Similarly, we show that the *external bisector* AP' (that is, the bisector of the external angle at A) divides BC externally in the ratio $|AB|:|AC|$, by drawing a line through C parallel to AP' (Fig. 18.16).

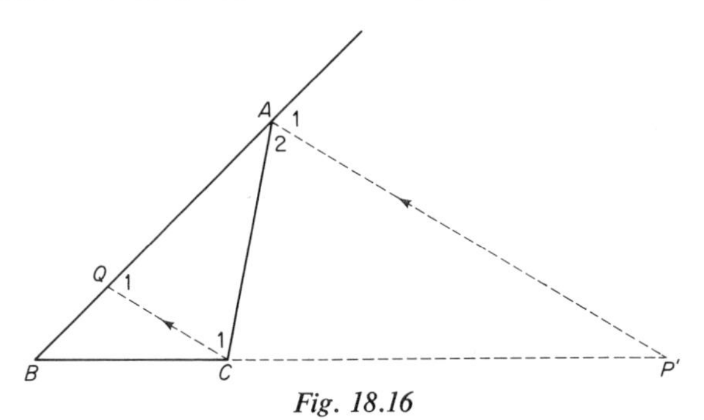

Fig. 18.16

$\triangle AQC$ is isosceles, since

$$\angle A_1 = \angle Q_1 \text{ (corresponding angles of parallel lines),}$$
$$\angle A_2 = \angle C_1 \text{ (corresponding angles of parallel lines),}$$
$$\angle A_1 = \angle A_2 \text{ (}AP\text{ is the angle bisector).}$$
$$\Rightarrow \angle Q_1 = \angle C_1$$
$$\Rightarrow |AQ| = |AC|.$$

By the Intercept Theorem,

$$|BP'|:|CP'| = |BA|:|QA|.$$

Substituting $|AC|$ for $|AQ|$,

$$|BP'|:|CP'| = |AB|:|AC|.$$

18.8 Exercises

1. (*a*) In $\triangle ABC$ (Fig. 18.17), $|AB| = 7$ cm, $|BC| = 11$ cm and $|CA| = 6$ cm. $\angle BAP = \angle PAC$ (AP bisects $\angle A$). Use the Angle Bisector Theorem to calculate $|BP|$ and $|PC|$ (to 3 SF).

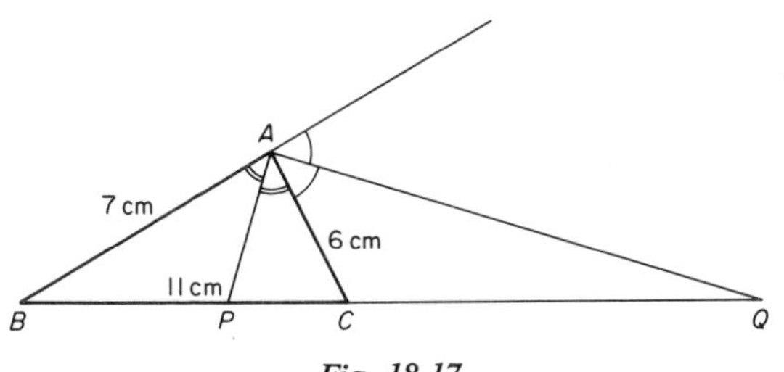

Fig. 18.17

(*b*) If the external bisector of $\angle A$ intersects BC produced at Q, calculate $|BQ|$ and $|CQ|$.

2. Construct $\triangle PQR$ whose sides measure 5 cm, 7 cm and 8 cm, respectively (Fig. 18.18).
   Use the Angle Bisector Theorem to find by construction the point K on PQ which divides the side PQ in the ratio 8:5.

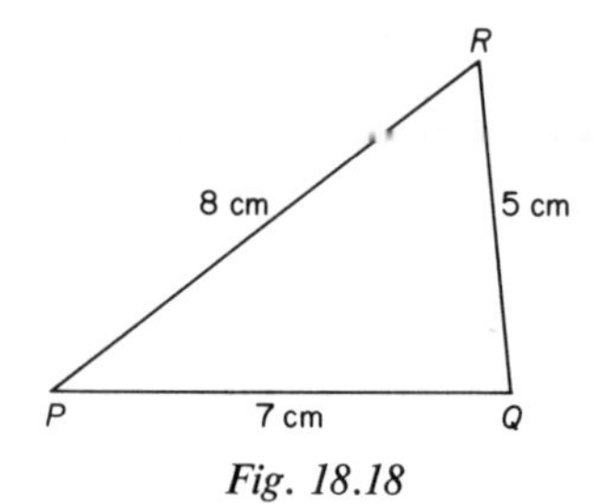

Fig. 18.18

3. Use the Angle Bisector Theorem to divide a line segment of 9 cm in the ratio 4:7.

18.9 Similar Triangles

If each angle of one triangle corresponds to an angle of equal size in another triangle, the triangles are said to be *equiangular*.

A simple experiment shows that the ratios of the lengths of the corresponding sides of two equiangular triangles are constant, and that therefore *equiangular triangles are similar.*

Draw two equiangular triangles ABC and KLM. Place the smaller triangle over the larger so that one pair of corresponding angles coincides, and the corresponding vertices are on the same line (Fig. 18.19).

The sides opposite the coincident angles are parallel, and by the Intercept Theorem

$$\frac{|KL|}{|AB|} = \frac{|LM|}{|BC|} = \frac{|MK|}{|CA|},$$

that is, the ratios of the lengths of the corresponding sides are constants.

Therefore *equiangular triangles* $\Leftrightarrow$ *similar triangles.*

Notes. (*a*) Equality of corresponding angles is *not* a sufficient condition for the similarity of polygons of more than three sides; for example, all rectangles are 'equiangular' but not necessarily similar. Compare also Section 16.43.

(*b*) For two triangles to be similar, it is sufficient that two pairs of corresponding angles are of equal size; from the angle sum of triangles it follows that the other two angles must also be of equal size.

Similarly, *if the ratios of the lengths of the corresponding sides of a triangle are constant, their corresponding angles are of equal size and the triangles are similar.*

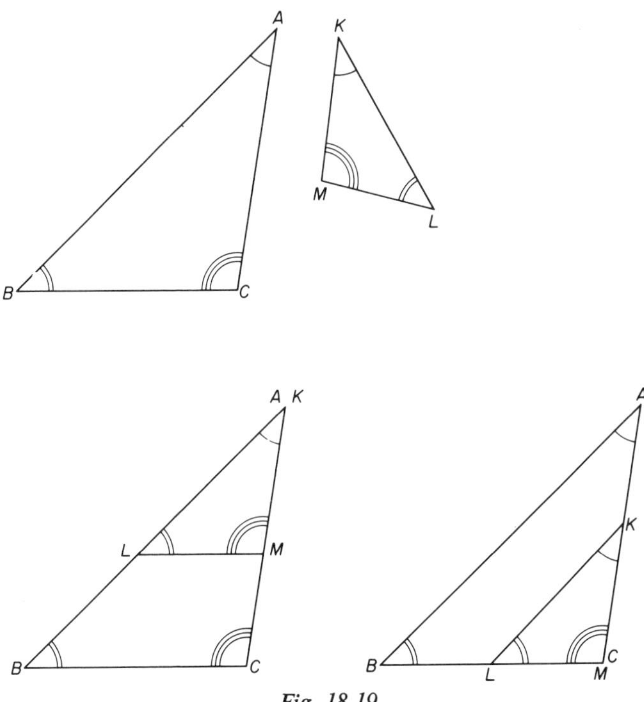

Fig. 18.19

This is obvious if the ratios are 1:1 (in other words, if the corresponding sides are of equal length), because the triangles are then congruent (*S–S–S*) and so of course similar as well.

If the length of each side of $\triangle PQR$ is k times the length of the corresponding side of $\triangle ABC$, each side of $\triangle ABC$ can be enlarged k times to form a triangle $\triangle A'B'C'$, which is then congruent with $\triangle PQR$ (*S–S–S*) (Fig. 18.20).

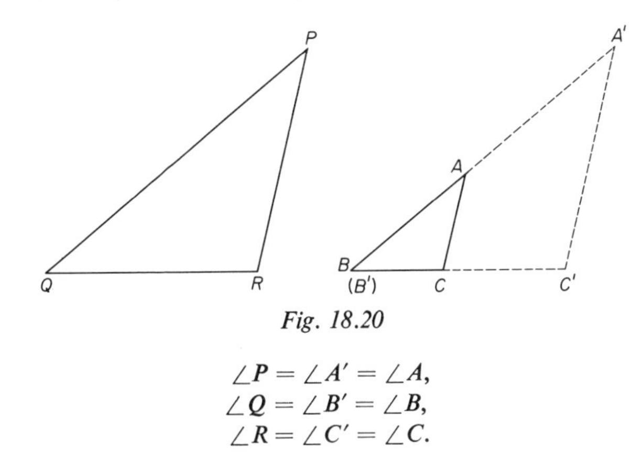

Fig. 18.20

Then

$$\angle P = \angle A' = \angle A,$$
$$\angle Q = \angle B' = \angle B,$$
$$\angle R = \angle C' = \angle C.$$

Polygons of more than three sides are *not* necessarily similar if the ratios of the lengths of their sides are constant. Compare, for example, a square whose sides are 4 cm with a rhombus whose sides are 2 cm and one angle of which is 30°: the ratio of the lengths of their sides is constant (2:1) but they are not similar quadrilaterals, for their corresponding angles are not of equal size.

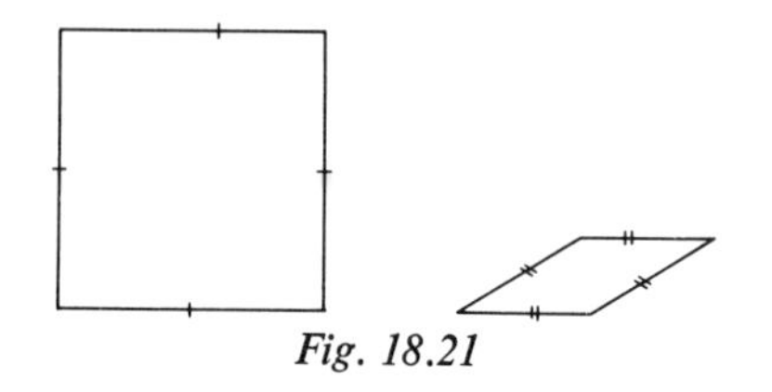

Fig. 18.21

To show that polygons of more than three sides are similar, we must show that their corresponding angles are equal *and* that the ratios of the lengths of their corresponding sides are the same.

18.10 Exercises

1. Test the pairs of triangles in Fig. 18.22 for similarity, by checking whether the ratios of corresponding sides are the same.

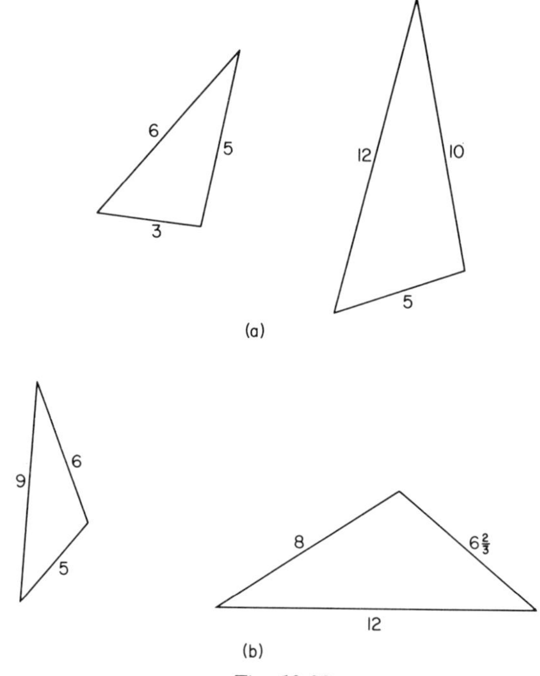

Fig. 18.22

2. Copy Fig. 18.23, and mark in the sizes of all the angles.

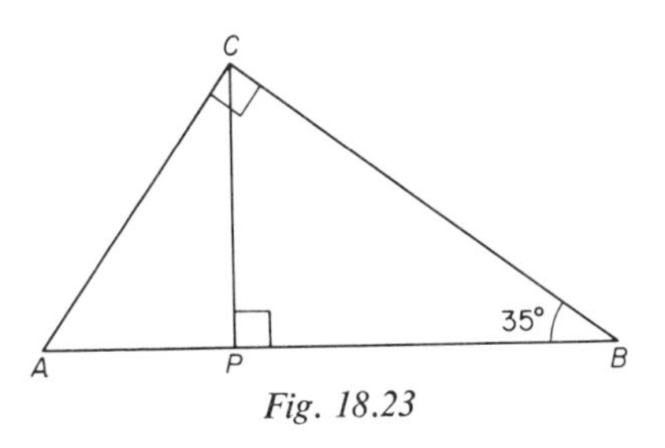

Fig. 18.23

Confirm that $\triangle$s ABC, ACP, CBP are similar.

18.11 Applications of Similarity

Both aspects of similarity, the equality of corresponding angles and the constant ratio of lengths of corresponding sides, have a wide range of useful applications in mathematics and everyday life. In navigation (see Section 14.8 on Bearings), we use the 'similarity' of regions of the world with their maps to find positions and directions. We will use the equality of corresponding angles of similar triangles in Unit 19 on Trigonometric Ratios to relate angles and lengths of sides. In scale drawing for architectural or engineering purposes, in model making and in calculating distances from maps the constant ratio, or scale (see also Section 7.5), is the key to finding 'unknown' lengths.

Finding unknown lengths in similar triangles

Once we have established that two triangles are similar (i.e. equiangular) we can:

(*a*) find the constant ratio if the lengths of two corresponding sides are known;
(*b*) if the length of another side of one of the triangles is also known, we can calculate the length of the side corresponding to it.

Examples

1. In $\triangle ABC$ and $\triangle KLM$ (Fig. 18.24),
$\angle A = \angle K$ and $\angle B = \angle L$,
$|AB| = 6$ cm, $|KL| = 2$ cm, $|LM| = 3$ cm and $|MK| = 4$ cm.

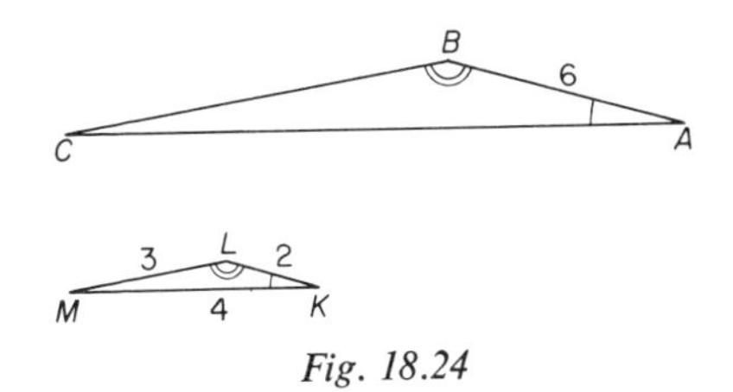

Fig. 18.24

Then $\triangle$s $\begin{matrix}ABC\\KLM\end{matrix}$ are similar (corresponding angles are equal).

The constant ratio is the ratio $|AB|:|KL|$ or $\frac{|AB|}{|KL|} = \frac{6}{2} = \frac{3}{1}$, that is the length of each side of $\triangle ABC$ is 3 times the length of the corresponding side of $\triangle KLM$.

Therefore $\qquad |BC| = 3\,|LM| = 3 \times 3 \text{ cm} = 9 \text{ cm};$
$\qquad\qquad\qquad |CA| = 3\,|MK| = 3 \times 4 \text{ cm} = 12 \text{ cm}.$

Alternatively, we can write down the equations of ratios and substitute the known lengths in them.

$\frac{|AB|}{|KL|} = \frac{|BC|}{|LM|} \Rightarrow \frac{6 \text{ cm}}{2 \text{ cm}} = \frac{|BC|}{3 \text{ cm}} \Rightarrow \frac{3}{1} = \frac{|BC|}{3 \text{ cm}} \Rightarrow |BC| = 9$ cm (by cross-multiplication);

similarly, $\frac{|AB|}{|KL|} = \frac{|AC|}{|KM|} \Rightarrow \frac{3}{1} = \frac{|AC|}{4 \text{ cm}} \Rightarrow |AC| = 12$ cm.

2. Special care is needed when the triangles partly coincide.

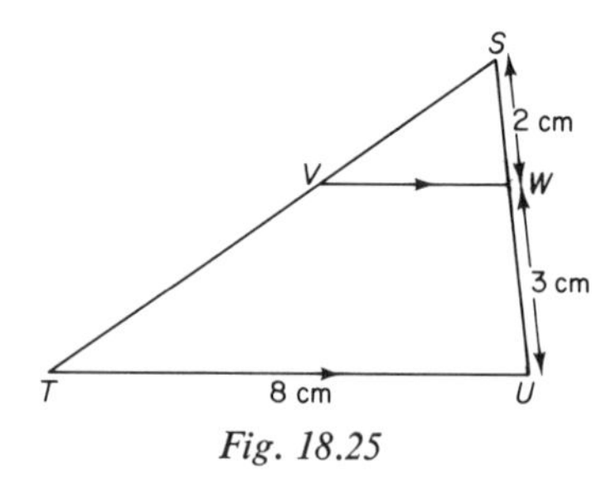

Fig. 18.25

In Fig. 18.25, $\triangle$s $\begin{matrix}STU\\SVW\end{matrix}$ are similar.

The 'known' corresponding sides of the triangles are SW and SU (*not* WU), and the ratio $|SW|:|SU| = 2:5$ (*not* $2:3$).
Therefore

$$\frac{|VW|}{|TU|} = \frac{2}{5} \Rightarrow \frac{|VW|}{8 \text{ cm}} = \frac{2}{5} \Rightarrow |VW| = \frac{2}{5} \times 8 \text{ cm} = \frac{16}{5} \text{ cm, or } 3{\cdot}2 \text{ cm}.$$

The comment in Section 15.10 about the corresponding sides of congruent triangles also applies to similar triangles: corresponding sides are not always parallel to each other or conveniently drawn in the same position. It is a good habit to identify the pairs of corresponding sides by their position opposite the equal corresponding angles.

$\triangle$s $\begin{matrix}PQR\\PKL\end{matrix}$ in Fig. 18.26 are similar.

The pairs of corresponding sides are: the sides opposite $\angle K$ in $\triangle PKL$ and opposite $\angle Q$ in $\triangle PQR$, that is PL and PR; the sides opposite $\angle L$ in $\triangle PKL$ and opposite $\angle R$ in $\triangle PQR$, PK and PQ; and so on.

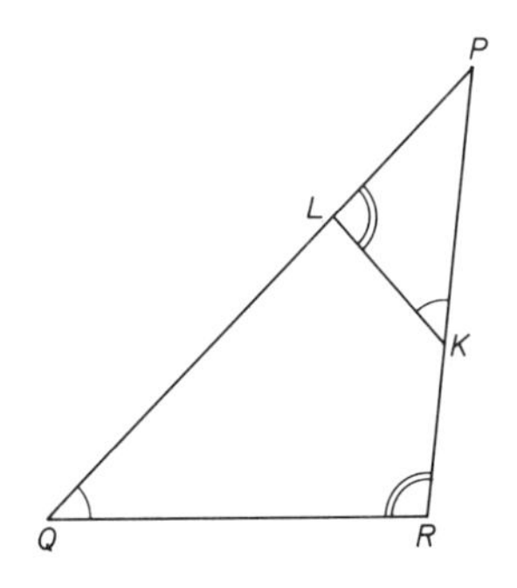

Fig. 18.26

If we follow the convention we have used for congruent triangles, and name the letters indicating corresponding angles and vertices in the same order, for example,

$$\begin{matrix} P & Q & R \\ \updownarrow & \updownarrow & \updownarrow \\ P & K & L \end{matrix}$$

the corresponding sides can be read off immediately:

PQ corresponds to PK,
QR corresponds to KL,
PR corresponds to PL.

18.12 Exercises

1. Check whether the pairs of triangles shown in Fig. 18.27(*a*) and (*b*) are similar. If so, write down the pairs of corresponding angles and sides.

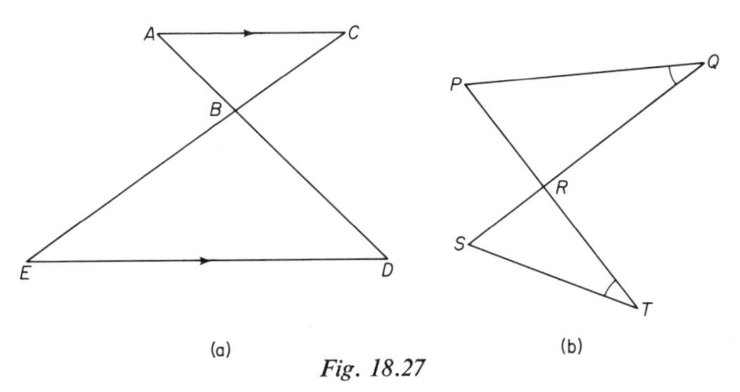

Fig. 18.27

2. Use the information given in Fig. 18.28 to calculate the following lengths:

$$|AC|; |BC|; |HI|; |HJ|; |RT|; |TS|.$$

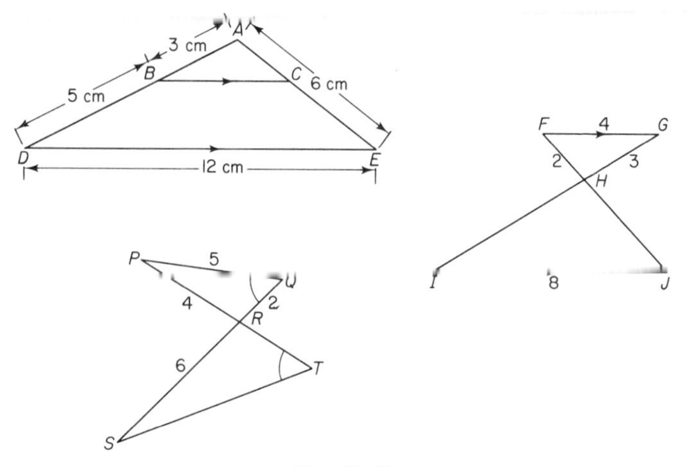

Fig. 18.28

18.13 Scale Drawing as a Method of Calculating Heights and Distances

'Real' distances and the actual dimensions of machinery can be calculated from maps or engineering drawings respectively by multiplying the corresponding distance on the map or plan by the appropriate scale number (see Section 7.5).

Heights of features like buildings, trees or mountains and distances from inaccessible objects can be calculated in the same way from a scale drawing.

Example

A television mast is held in position by two cables which are attached to the top; both cables are secured to the ground at points which are each at a distance of 50 m from the foot of the mast. Each rope makes an angle of 60° with the ground (Fig. 18.29).

The mast, one of the ropes and the level ground form a triangle RST; this is a right-angled triangle, since the mast forms a right angle with the line on the ground joining the foot of the mast to the point where the cable is secured. $\triangle RST$ is similar to any triangle whose angles are 60°, 90° and 30°.

We construct a $\triangle ABC$ (Fig. 18.30), such that $\angle A = 60°$ and $\angle B = 90°$. $\triangle ABC$ is a scale drawing of $\triangle RST$.

In Fig. 18.30, $|AB| = 5$ cm; but the distance $|RS| = 50$ m and the scale chosen is therefore $\dfrac{|AB|}{|RS|} = \dfrac{5 \text{ cm}}{5000 \text{ cm}} = \dfrac{1}{1000}$.

We find the height of the mast by measuring the corresponding side in the scale drawing; we find $|BC|$ to be 8·7 cm.

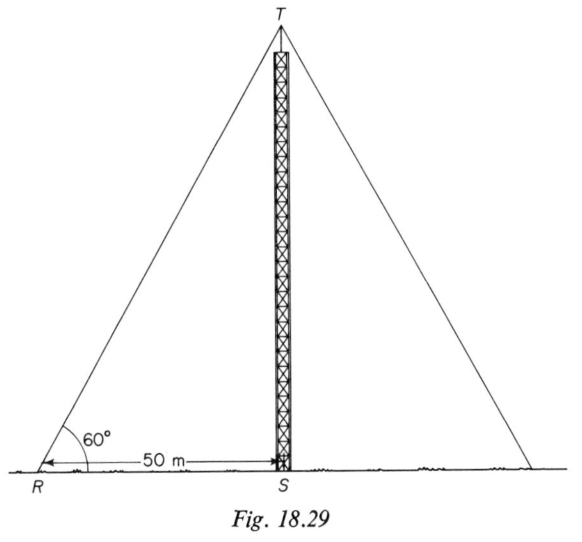

Fig. 18.29

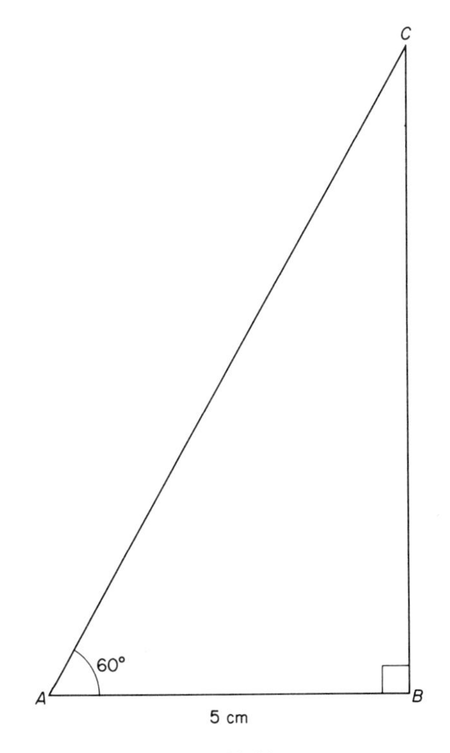

Fig. 18.30

The height of the mast is therefore 1000×8.7 cm $= 8700$ cm $= 87$ m (to 2 SF).

Usually, however, there is no convenient cable attached to the top of the building, or mast, whose height is to be found. Instead we consider the triangle whose sides are

> the face line of the building,
> the horizontal line through the point of observation, and
> the line between the point of observation and the top of the building.

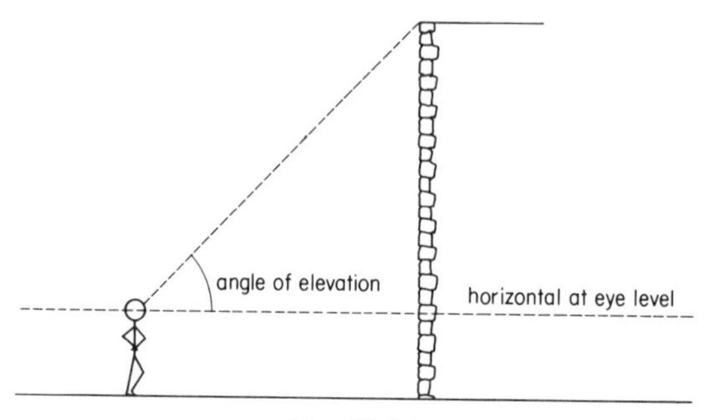

Fig. 18.31

The angle between the horizontal and the line joining the point of observation to the point observed is called the *angle of elevation* (Latin, 'lifting up', i.e. of eyes).

Similarly, if a point at a lower level is being observed, we speak of an *angle of depression* (Fig. 18.32).

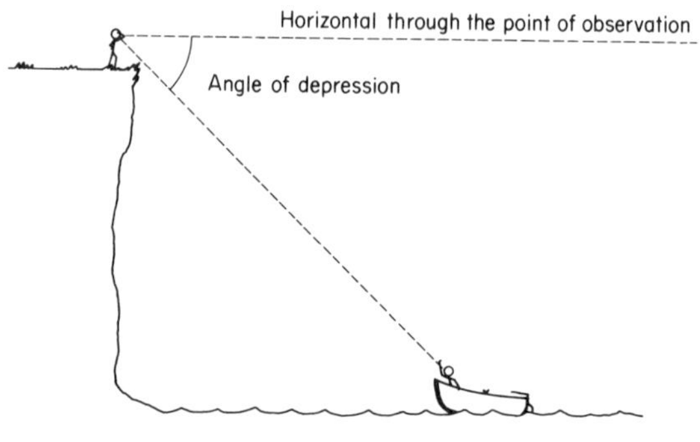

Fig. 18.32

Surveyors use a *theodolite* for accurate measurement of angles of elevation or depression. A simpler instrument for the same purpose is called a *clinometer*. You can make a clinometer by attaching a plumb line to the centre of a protractor, and then gluing a drinking straw along the diameter joining the 90° mark and the 270° mark (Fig. 18.33). The straw is used as a telescopic sighter, and the plumb line then marks the angle of elevation (or depression).

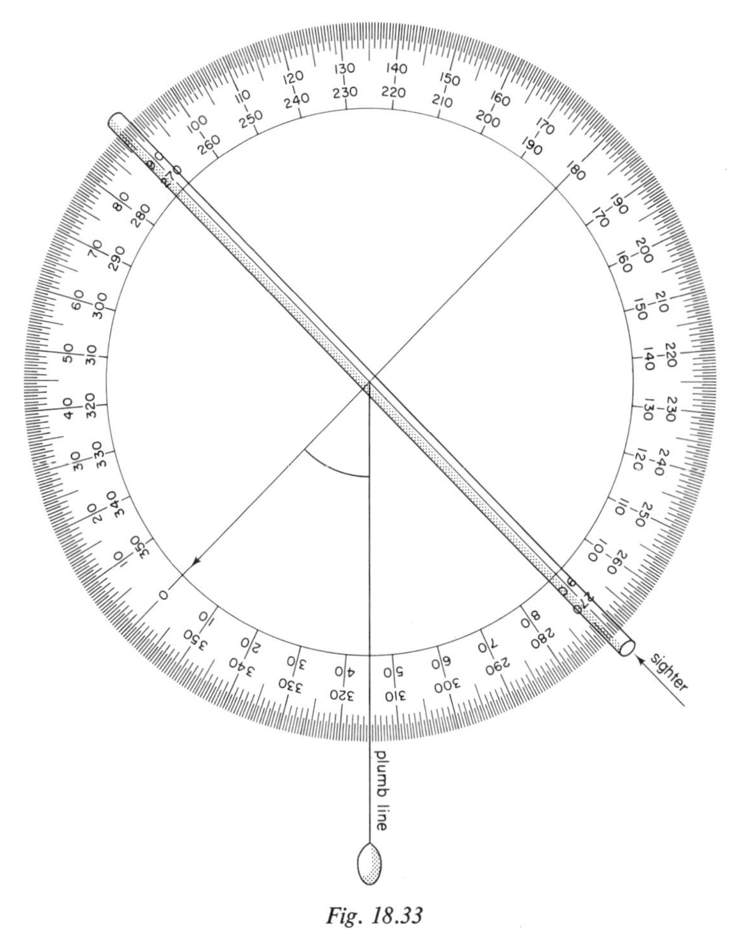

Fig. 18.33

We can assume that an object such as a tree, a mast or a factory chimney, makes a right angle with the ground; the only angle to be measured is then the angle of elevation from one point of observation. But scale drawings or the principle of similar triangles can also be used to find heights of objects which do not form a right angle with the ground.

Example

The angle of elevation of the top of a pyramid observed from a point A is $22°$; when it is observed from a point B the angle of elevation is $50°$. Point B is on the straight line between the point A and the foot of the pyramid, at a distance of 150 m from A.

We can find the height of the pyramid from a scale drawing. We draw the line segment AB, choosing a convenient length, such as $|AB| = 15$ cm. The scale is then $\frac{15 \text{ cm}}{15\ 000 \text{ cm}} = \frac{1}{1000}$.

At the end-points of the segment AB we construct angles of 22° and 130°, completing the triangle ABC (Fig. 18.34).

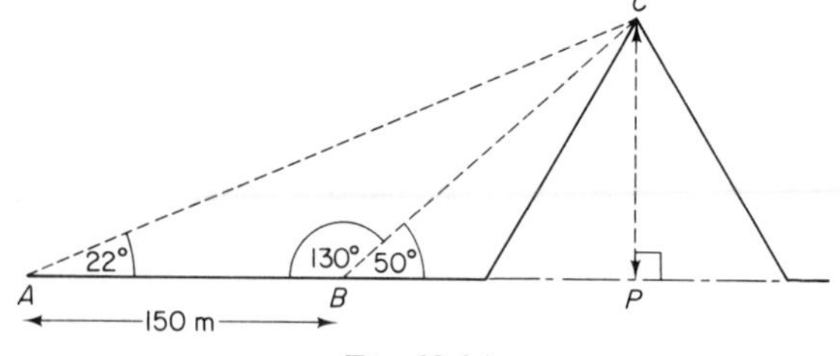

Fig. 18.34

The perpendicular through C to the line AB represents the height of the pyramid; by measurement on the scale drawing, we find $|CP| = 9{\cdot}2$ cm. The real height is therefore approximately 92 m (to 2 SF). (The accuracy of the approximation depends both on the accuracy of the instruments with which we measure angles and distances, and also on the accuracy and size of the scale drawing.)

Thales (640–546 B.C.) calculated the height of the Great Pyramid of Cheops by this method. More modern surveying methods using trigonometric ratios (Section 19.18) are based on the same principle of similarity.

Note. If the point of observation is not at the same horizontal level as the foot of the object whose height is to be found, an adjustment must be made (Fig. 18.35).

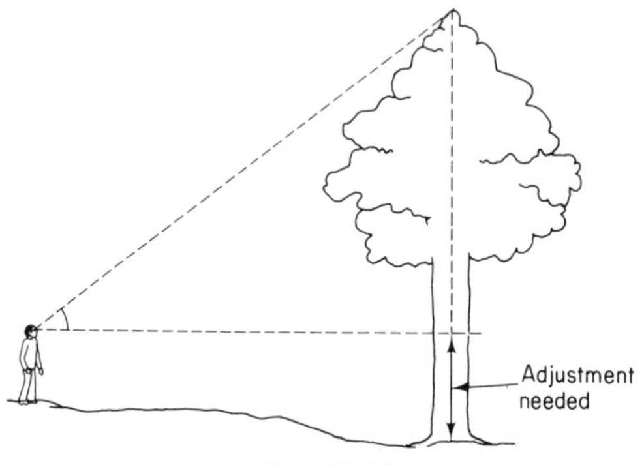

Fig. 18.35

18.14 Exercises

1. Choose a suitable object, such as a tree. From some observation point at ground level, measure the angle of elevation of the top of this tree and the distance of the observation point from the foot of the tree. Find the height of the tree by scale drawing.

2. The method of scale drawing can be used to calculate distances from inaccessible objects in a horizontal plane. From a point A on the bank of a river (which runs due south), a tree is observed on the opposite bank of the river on a bearing of 110°. When observations are taken from a point B on the river bank 300 m south of A, the bearing of the tree is 050°.
   Find the width of the river by scale drawing.

18.15 Two Special Cases of Similar Triangles

In any right-angled triangle ABC, in which $\angle C$ is the right angle, the perpendicular through C divides the triangle into two triangles, both similar to $\triangle ABC$ and to one another.

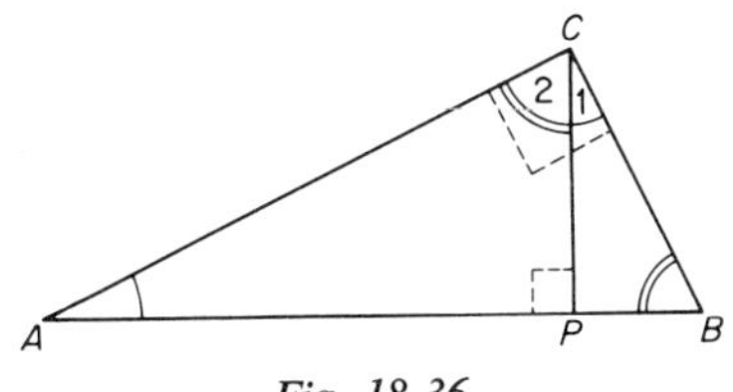

Fig. 18.36

In Fig. 18.36, $\triangle$s $\begin{matrix}ABC\\ ACP\\ CBP\end{matrix}$ are similar. We have already noticed this in a particular case in Exercise 2 in Section 18.10; but we can easily show that $\triangle$s $\begin{matrix}ABC\\ ACP\\ CBP\end{matrix}$ are similar, whatever the size of $\angle B$.

In each triangle one angle is a right angle.

Moreover, $\triangle ABC$ and $\triangle ACP$ share a common angle, $\angle A$.

Having two pairs of angles of equal size, $\triangle$s $\begin{matrix}ABC\\ ACP\end{matrix}$ are similar.

$\triangle ABC$ and $\triangle CBP$ also have a second angle of equal size since they share $\angle B$; therefore $\triangle$s $\begin{matrix}ABC\\ CBP\end{matrix}$ are also similar.

We can also show that $\angle A = \angle C_1$ and $\angle B = \angle C_2$:

$$\angle A + \angle B = 90° \Rightarrow \angle A = 90° - \angle B,$$
$$\angle C_1 + \angle B = 90° \Rightarrow \angle C_1 = 90° - \angle B$$
$$\Rightarrow \angle A = \angle C_1.$$

Also,
$$\angle C_2 + \angle A = 90° \Rightarrow \angle C_2 = 90° - \angle A,$$
$$\angle B + \angle A = 90° \Rightarrow \angle B = 90° - \angle A$$
$$\Rightarrow \angle B = \angle C_2.$$

The similarity of these triangles reveals an interesting identity:

$$\frac{|PA|}{|PC|} = \frac{|PC|}{|PB|} \Rightarrow |PA| \,.\, |PB| = |PC|^2 \text{ (by cross-multiplication).}$$

We can use this similarity to prove Pythagoras' Theorem.

An alternative proof of Pythagoras' Theorem

If the lengths of the sides of a $\triangle ABC$ right-angled at C, are respectively a, b and c units, Pythagoras' Theorem states that $c^2 = a^2 + b^2$ (Theorem 15.8 in Section 15.28).

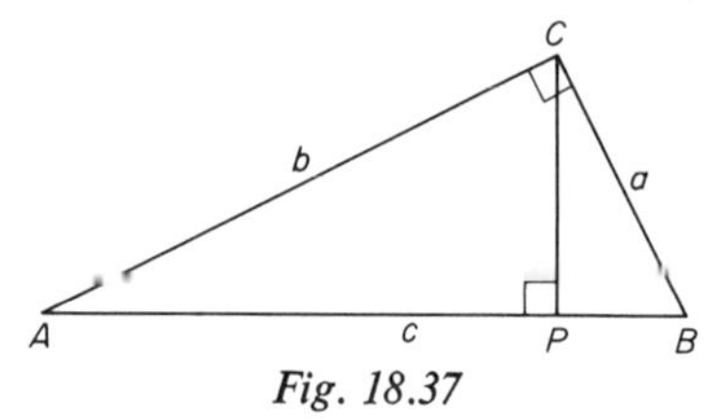

Fig. 18.37

$$\triangle\text{s}\ \begin{matrix}ACP\\ABC\end{matrix}\ \text{are similar} \Rightarrow \frac{|AP|}{|AC|} = \frac{|AC|}{|AB|}\ \text{or}\ \frac{|AP|}{b} = \frac{b}{c}$$

$$\Rightarrow |AP| \,.\, c = b^2\ \text{(by cross-multiplication)} \Rightarrow |AP| = \frac{b^2}{c}.$$

$$\triangle\text{s}\ \begin{matrix}CBP\\ABC\end{matrix}\ \text{are similar} \Rightarrow \frac{|BP|}{|BC|} = \frac{|BC|}{|AB|}\ \text{or}\ \frac{|BP|}{a} = \frac{a}{c}$$

$$\Rightarrow |BP| \,.\, c = a^2 \Rightarrow |BP| = \frac{a^2}{c}.$$

But $|AP| + |BP| = c$;

substituting $\frac{b^2}{c}$ for $|AP|$ and $\frac{a^2}{c}$ for $|BP|$,

$$\frac{b^2}{c} + \frac{a^2}{c} = c$$

$$\Rightarrow b^2 + a^2 = c^2\ \text{(multiplying both sides by } c\text{)}.$$

The similar triangles resulting from the equality of angles formed by lines and chords of circles are also of particular interest. They are similar triangles with a special property which concerns the products of lengths of sides.

Theorem 18.3: *The Intersecting Chord Theorem.*

(*a*) From Theorem 17.6 (Section 17.14), we know that if two chords PQ and RS of a circle intersect at T (inside the circle) (Fig. 18.38), then

$$\angle P = \angle R\ \text{(both standing on the arc } SQ\text{)},$$
$$\angle Q = \angle S\ \text{(both standing on the arc } PR\text{)}.$$

Therefore $\triangle\text{s}\ \begin{matrix}PST\\RQT\end{matrix}$ are similar (compare Exercise 1 in Section 18.12).

Equating ratios of lengths of corresponding sides,

$\frac{|TP|}{|TR|} = \frac{|TS|}{|TQ|} \Rightarrow |TP| \,.\, |TQ| = |TR| \,.\, |TS|$ (by cross-multiplication); or,

(distance of P from T) $\times$ (distance of Q from T)

= (distance of R from T) $\times$ (distance of S from T).

This is called the *Intersecting Chord Theorem for an internal point.*

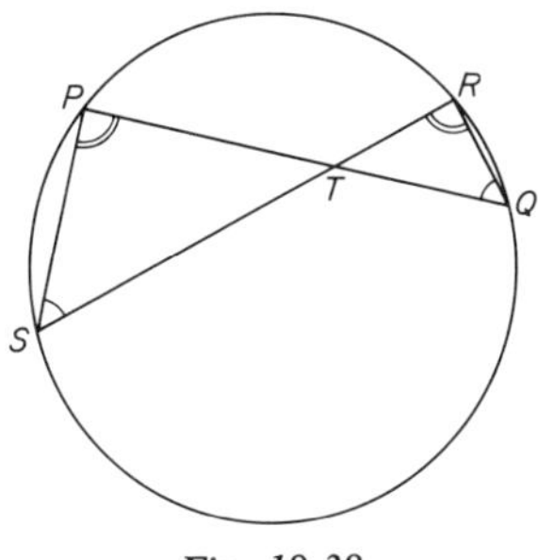

Fig. 18.38

(*b*) If two opposite sides of a cyclic quadrilateral are extended to intersect outside the circle (Fig. 18.39), the triangles so formed are similar.

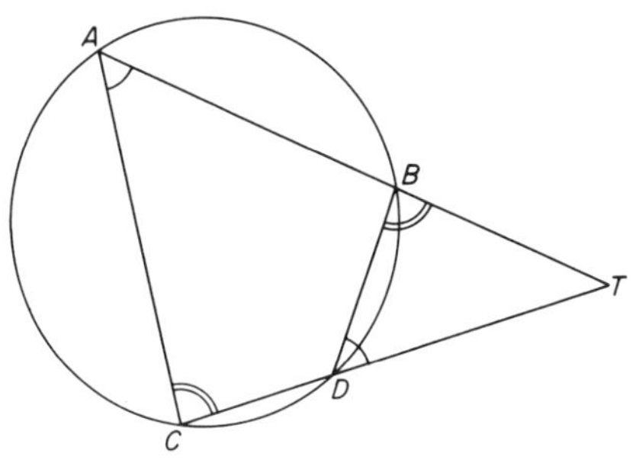

Fig. 18.39

The triangles $\begin{matrix} TAC \\ TDB \end{matrix}$ in the diagram are similar, since

$\angle A = \angle D$ (exterior angle of a cyclic quadrilateral), and
$\angle C = \angle B$ (exterior angle of a cyclic quadrilateral).

Equating ratios of lengths of corresponding sides,

$\frac{|TA|}{|TD|} = \frac{|TC|}{|TB|} \Rightarrow |TA| \, . \, |TB| = |TC| \, . \, |TD|$, where T is the point of intersection of the extended chords AB and CD.

This result is therefore called the *Intersecting Chord Theorem for an external point.*

Again, if T is the point of intersection of the chords AB and CD,
(distance of A from T) $\times$ (distance of B from T)
$=$ (distance of C from T) $\times$ (distance of D from T).

(*c*) TP is a tangent to a circle, and another line through T intersects the circumference of the circle at the points A and B (Fig. 18.40).

We can show that $\triangle TPA$ and $\triangle TBP$ are similar, by showing that two pairs of angles are of equal size.

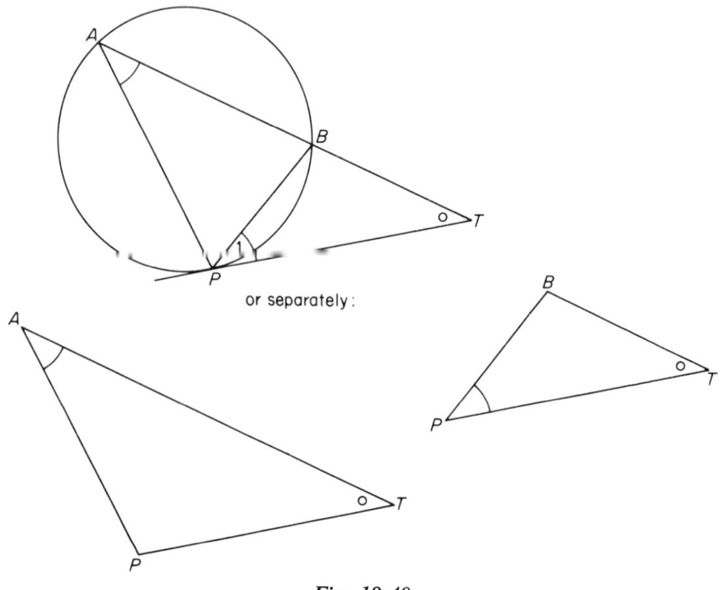

Fig. 18.40

By the Alternate Segment Theorem, $\angle P_1 = \angle A$;
also, $\angle T$ is a common angle of both $\triangle TPA$ and $\triangle TBP$.

Therefore $\triangle$s $\dfrac{TPA}{TBP}$ are similar

$$\Rightarrow \frac{|TA|}{|TP|} = \frac{|TP|}{|TB|} \Rightarrow |TA| \,.\, |TB| = |TP| \,.\, |TP| = |TP|^2.$$

We could consider this too as a special case of the result in (*b*), $|TA| \,.\, |TB| = |TC| \,.\, |TD|$, in which the two points C and D coincide at P; that is, $|TA| \,.\, |TB| = |TP| \,.\, |TP| = |TP|^2$.

Note. Special care is needed when these results are used in calculations. Suppose, for example, $|AB| = 3$ cm and $|BT| = 5$ cm;
then $\qquad |AT| = 8$ cm $\quad (|AB| + |BT'| = |AT|)$.
In other words, avoid the common mistake of writing down

$$|TB| \times |BA| = |TD| \times |DC|.$$

18.16 Exercise

Use the Intersecting Chord Theorem to calculate the lengths of the segments indicated by a, b, c and d in Fig. 18.41. [Hint: in the last diagram, $|TB| = d$; $|TA| = |TB| + |BA| = d + 8 + 8$; and so on.]

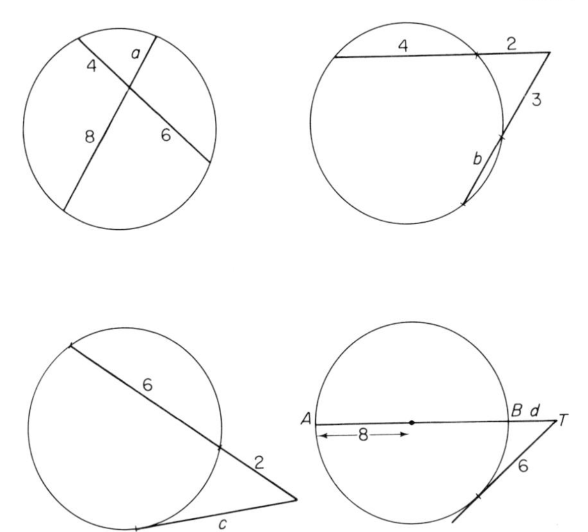

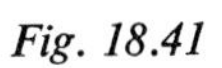

Fig. 18.41

Unit Nineteen

Trigonometric Ratios

19.1 Introduction to Trigonometry

So far, whenever we have compared the lengths of sides of similar triangles, we have been concerned with pairs of corresponding sides. We found that if, for example, $\triangle\text{s}\ \substack{ABC \\ PQR}$ are similar and $|AB|:|PQ| = 2:1$, then also $|BC|:|QR| = 2:1$ and $|CA|:|RP| = 2:1$.

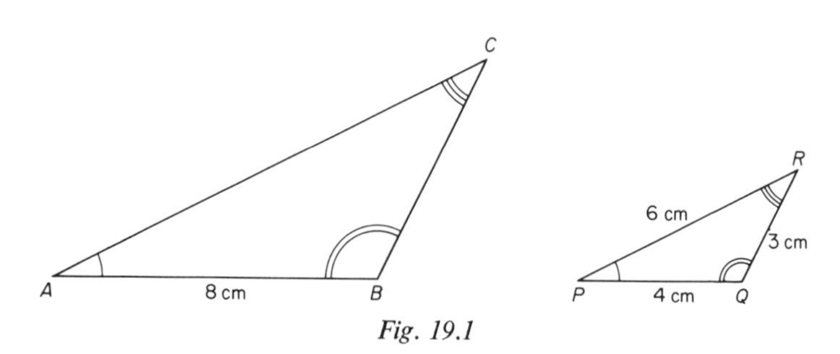

Fig. 19.1

If, as in Fig. 19.1, we know the lengths of the sides of $\triangle PQR$, we can calculate the lengths of the sides of $\triangle ABC$.

$$|BC| = 2 \times |QR| = 2 \times 3\,\text{cm} = 6\,\text{cm, and}$$
$$|CA| = 2 \times |RP| = 2 \times 6\,\text{cm} = 12\,\text{cm.}$$

If we now look at the ratios of the lengths of the sides of $\triangle PQR$, we see that

$$\frac{|PQ|}{|QR|} = \frac{4}{3}, \text{ or } |PQ| = \frac{4}{3}\,|RQ|.$$

Turning then to $\triangle ABC$ we find a similar relationship:

$$\frac{|AB|}{|BC|} = \frac{8}{6} = \frac{4}{3}, \text{ or } |AB| = \frac{4}{3}\,|BC|.$$

Also,
$$\frac{|QR|}{|RP|} = \frac{3}{6} = \frac{1}{2} \text{ and } \frac{|BC|}{|CA|} = \frac{6}{12} = \frac{1}{2}.$$

In other words, the ratio of the lengths of two sides of $\triangle ABC$ is the same as the ratio of the lengths of the corresponding sides of $\triangle PQR$, or $|PQ|:|QR|:|RP| = 4:3:6$ and $|AB|:|BC|:|CA| = 4:3:6$.

If we draw *any* other triangle KLM which is equiangular with $\triangle PQR$, and therefore similar with $\triangle PQR$, the ratios of the lengths of its sides are always $|KL|:|LM|:|MK| = 4:3:6$.

In general, *the ratio of the lengths of the sides of any triangle ABC is the same as the ratio of the lengths of the sides of a similar* $\triangle PQR$.

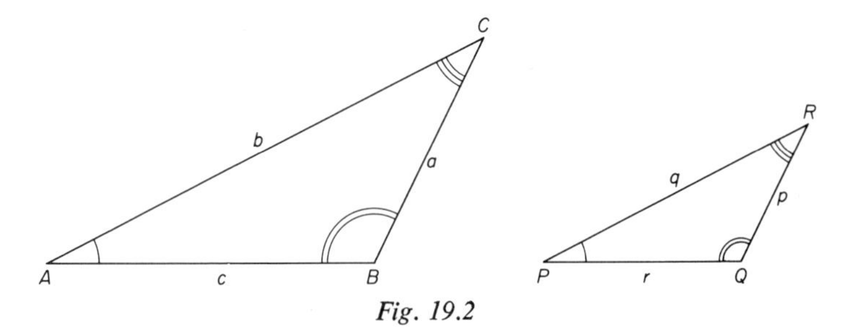

Fig. 19.2

If a, b and c are the lengths of the sides of $\triangle ABC$ and p, q and r are respectively the lengths of the corresponding sides of $\triangle PQR$ (Fig. 19.2), then

$$\frac{|AB|}{|PQ|} = \frac{|BC|}{|QR|} \quad \text{or} \quad \frac{c}{r} = \frac{a}{p};$$

multiplying both sides of this equation by $\frac{r}{a}$,

$$\frac{r}{a} \times \frac{c}{r} = \frac{r}{a} \times \frac{a}{p} \Rightarrow \frac{c}{a} = \frac{r}{p}.$$

In the same way we can show

$$\frac{a}{b} = \frac{p}{q} \text{ and}$$
$$\frac{b}{c} = \frac{q}{r}.$$

The angles of the particular $\triangle PQR$ in Fig. 19.1 are measured as 27°, 115° and 38° respectively. For future reference we *might* be able to remember, whenever we meet a triangle with these angles, that the ratios of their sides are 4:3:6. We might even consider drawing up a table of triangles listing every possible combination of three angles, and the corresponding ratio of lengths of sides for each case. This would be completely impractical, because of the vast number of possible combinations of three angles.

Such lists have been made however for the relatively simple case of *right-angled triangles*, for which we need to know only the size of one angle besides the right angle. For example, all right-angled triangles with one angle of 60° are similar; their angles are 90°, 60° and 30°.

The ratios of the lengths of the sides of right-angled triangles are known as *trigonometric ratios*. *Trigonometry* is the branch of mathematics which is concerned with the relations between angles and lengths of sides of a triangle.

Naming the sides and the ratios

Corresponding sides of similar triangles are identified by their position opposite the corresponding equal angles. For example, we could name the sides of all the triangles with angles of 30°, 60° and 90°, as:

> 'the side opposite the angle of 90°', that is, the *hypotenuse*,
> 'the side opposite the angle of 60°' and
> 'the side opposite the angle of 30°'.

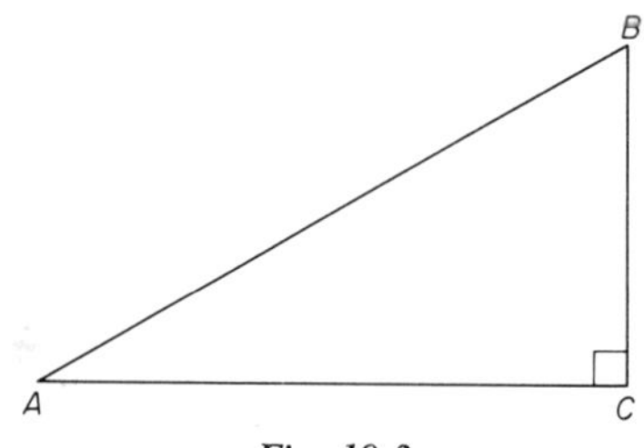

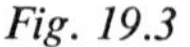

Fig. 19.3

In Fig. 19.3, AB is the hypotenuse, AC is the side opposite $\angle B$ and BC is the side opposite $\angle A$.

AC is also called the *side adjacent* to $\angle A$, and BC is the side adjacent to $\angle B$. 'Adjacent' thus means 'the side next to the angle', other than the hypotenuse.

No side can be said to be just 'opposite' or 'adjacent'; these terms only have meaning when they are related to a particular angle:

> BC is the side *opposite* $\angle A$ and is also the side *adjacent to* $\angle B$;
> AC is the side *adjacent to* $\angle A$ and is also the side *opposite* $\angle B$.

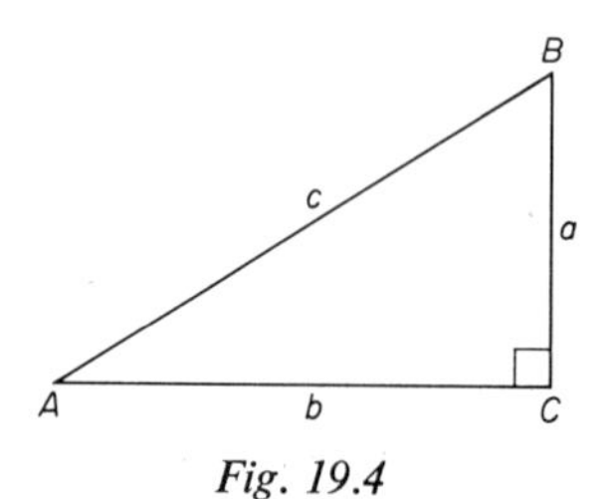

Fig. 19.4

There are basically three ratios which relate the lengths of the three sides of a $\triangle ABC$: $\dfrac{|BC|}{|AC|}$, $\dfrac{|BC|}{|AB|}$ and $\dfrac{|AC|}{|AB|}$. If we use the notation $a = |BC|$, $b = |AC|$ and $c = |AB|$, these ratios are $\dfrac{a}{b}$, $\dfrac{a}{c}$ and $\dfrac{b}{c}$.

The ratio $\dfrac{a}{b}$, or $\dfrac{\text{the length of the side opposite } \angle A}{\text{the length of the side adjacent to } \angle A}$, is named

the *tangent* of $\angle A$, abbreviated to *tan A*.

The ratio $\frac{a}{c}$, or $\frac{\text{the length of the side opposite } \angle A}{\text{the length of the hypotenuse}}$, is named

the *sine* of $\angle A$, abbreviated to *sin A*.

The ratio $\frac{b}{c}$, or $\frac{\text{the length of the side opposite } \angle B}{\text{the length of the hypotenuse}}$, is similarly

the *sine* of $\angle B$.

$\angle B$ is the complement of $\angle A$ ($\angle A + \angle B = 90°$), and this ratio is also called the *cosine* of $\angle A$, i.e. the sine of the complement of $\angle A$, or,

$\frac{b}{c} = \frac{\text{the length of the side adjacent to } \angle A}{\text{the length of the hypotenuse}}$ is

the *cosine* of $\angle A$, abbreviated to *cos A*.

In the same way, the ratio $\frac{b}{a}$, or $\frac{\text{length of the side opposite } \angle B}{\text{length of the side adjacent to } \angle B}$ (*tan B*) is called the *cotangent* of $\angle A$, i.e. the tangent of the complement of $\angle A$; or,

$\frac{b}{a} = \frac{\text{the length of the side adjacent to } \angle A}{\text{the length of the side opposite } \angle A}$, is

the *cotangent* of $\angle A$, abbreviated to *cot A*.

When you memorize these ratios, you may find it easier if you shorten the phrases 'the length of the side opposite $\angle A$' to 'side opposite A', and so on, or, for private use only, to just 'opposite', 'hypotenuse', etc. The ratios can then be summarized as:

$$\sin A = \frac{\text{side opposite } A}{\text{hypotenuse}} \quad \left(= \frac{a}{c}\right) \text{ or just } \frac{\text{opposite}}{\text{hypotenuse}},$$

$$\cos A = \frac{\text{side adjacent to } A}{\text{hypotenuse}} \quad \left(= \frac{b}{c}\right) \text{ or just } \frac{\text{adjacent}}{\text{hypotenuse}},$$

$$\tan A = \frac{\text{side opposite } A}{\text{side adjacent to } A} \quad \left(= \frac{a}{b}\right) \text{ or just } \frac{\text{opposite}}{\text{adjacent}},$$

$$\cot A = \frac{\text{side adjacent to } A}{\text{side opposite } A} \quad \left(= \frac{b}{a}\right) \text{ or just } \frac{\text{adjacent}}{\text{opposite}},$$

where the words 'opposite' and 'adjacent' are all used in relation to $\angle A$.

19.2 Exercises

1. Each of the triangles in Fig. 19.5 is right-angled. Write down the values of sin B, tan A, cos Q and cot P.

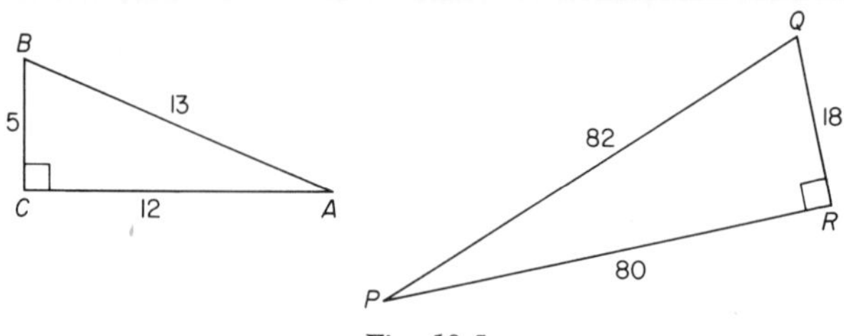

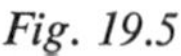

Fig. 19.5

2. Draw a triangle with sides of length 6 cm, 8 cm and 10 cm (see Pythagorean triples, Section 15.29).

   Measure the angles of this triangle, and so find approximate values of:

   (*a*) the angle whose tangent is $\frac{4}{3}$;
   (*b*) the angle whose sine is 0·8;
   (*c*) the angle whose cosine is 0·6.

19.3 The Trigonometric Ratios of Some Special Angles

The trigonometric ratios of some angles can be calculated easily.

(*a*) 45°

A right-angled triangle of which one angle is 45° is isosceles; the third angle is also 45° (Fig. 19.6).

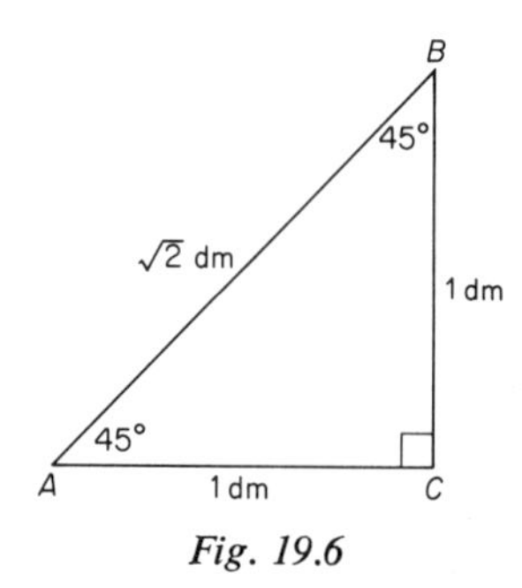

Fig. 19.6

If $|AC|$ is given to be 1 dm, then $|BC| = 1$ dm, and by Pythagoras' Theorem,

$$|AB|^2 = 1^2 + 1^2 = 2 \Rightarrow |AB| = \sqrt{2} \text{ dm}.$$

Therefore,

$$\sin 45° \left(= \sin A = \frac{|BC|}{|AB|}\right) = \frac{1}{\sqrt{2}};$$

$$\cos 45° \left(= \cos A = \frac{|AC|}{|AB|}\right) = \frac{1}{\sqrt{2}};$$

$$\tan 45° \left(= \tan A = \frac{|BC|}{|AC|}\right) = \frac{1}{1} = 1;$$

$$\cot 45° \left(= \cot A = \frac{|AC|}{|BC|}\right) = \frac{1}{1} = 1.$$

These values of the trigonometric ratios of 45° are always valid, irrespective of the lengths of the sides of the triangle. Any right-angled triangle with angles of 45° is isosceles and therefore has equal sides around the right angle.

(*b*) 30° and 60°

In $\triangle ABC$ (Fig. 19.7), $\angle A = 30°$ and $\angle B = 60°$, and $|AB| = 2$ dm.

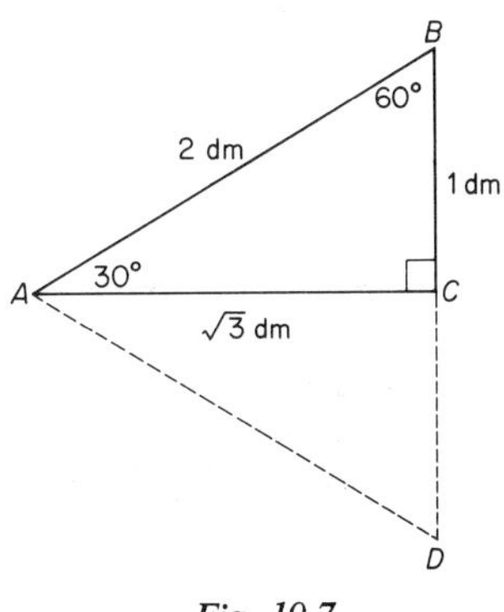

Fig. 19.7

If $\triangle ABC$ is considered as half of the equilateral triangle ABD, then clearly

$$|BC| = \tfrac{1}{2}|BD| = \tfrac{1}{2}|AB| = 1 \text{ dm.}$$

By Pythagoras' Theorem, $|AC|^2 = 2^2 - 1^2 = 3 \Rightarrow |AC| = \sqrt{3}$ dm.

Therefore, $\sin 30° \left(= \sin A = \dfrac{|BC|}{|AB|}\right) = \dfrac{1}{2} = 0{\cdot}5;$

$$\cos 30° = \frac{\sqrt{3}}{2} = 0{\cdot}866 \text{ (to 3 SF);}$$

$$\tan 30° = \frac{1}{\sqrt{3}} = 0{\cdot}577 \text{ (to 3 SF);}$$

$$\cot 30° = \frac{\sqrt{3}}{1} = \sqrt{3} = 1{\cdot}73 \text{ (to 3 SF).}$$

Also, $\sin 60° \left(= \sin B = \dfrac{|AC|}{|AB|}\right) = \dfrac{\sqrt{3}}{2} = 0{\cdot}866$ (to 3 SF);

$$\cos 60° \left(= \cos B = \frac{|BC|}{|AB|}\right) = \frac{1}{2} = 0{\cdot}5;$$

$$\tan 60° = \frac{\sqrt{3}}{1} = \sqrt{3} = 1{\cdot}73 \text{ (to 3 SF),}$$

$$\cot 60° = \frac{1}{\sqrt{3}} = 0{\cdot}577 \text{ (to 3 SF).}$$

Notice that sin 30° and cos 60° are *both* the ratio $\frac{|BC|}{|AB|}$.

Also

$$\sin 60° = \cos 30°;$$
$$\tan 30° = \cot 60°;$$
$$\tan 60° = \cot 30°.$$

From the definition of cosine and cotangent ('the sine and tangent of the complement of' respectively) it follows that, for any angle of $\alpha°$,

$$\boxed{\begin{aligned}\sin \alpha° &= \cos (90 - \alpha)°, \text{ and}\\ \tan \alpha° &= \cot (90 - \alpha)°\end{aligned}}$$

19.4 Finding Sines and Cosines by Drawing

We can find rough values of trigonometric ratios by drawing right-angled triangles and measuring the lengths of sides. For example, we can find sin 55°, cos 55° and tan 55°, by drawing a large triangle with angles 55°, 35° and 90°.

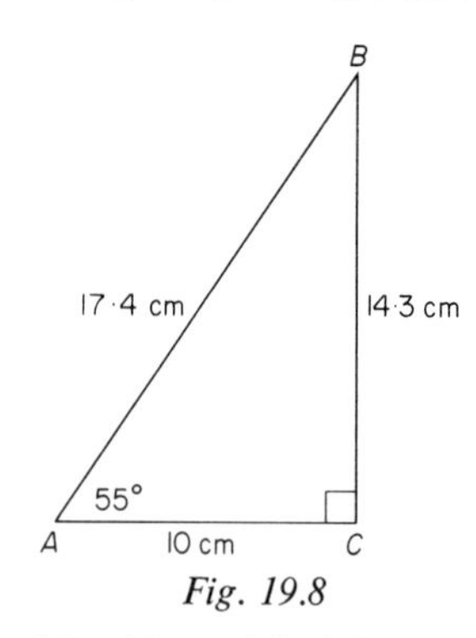

Fig. 19.8

We measure the lengths of the sides and find, for example,

$$|AB| = 17{\cdot}4 \text{ cm}, |BC| = 14{\cdot}3 \text{ cm and } |AC| = 10 \text{ cm}.$$

Then

$$\sin 55° = \frac{14{\cdot}3}{17{\cdot}4} = 0{\cdot}82 \text{ (to 2 SF)};$$

$$\cos 55° = \frac{10}{17{\cdot}4} = 0{\cdot}57 \text{ (to 2 SF)};$$

$$\tan 55° = 1{\cdot}4 \text{ (to 2 SF)}.$$

A simpler method of finding approximate values of *sines* and *cosines* of angles between 0° and 90° is as follows:

Draw a circle of radius 1 dm and centre O, or just a quadrant of this circle (i.e. a sector whose angle is 90°). Mark the radii of the sector, OX and OY (Fig. 19.9).

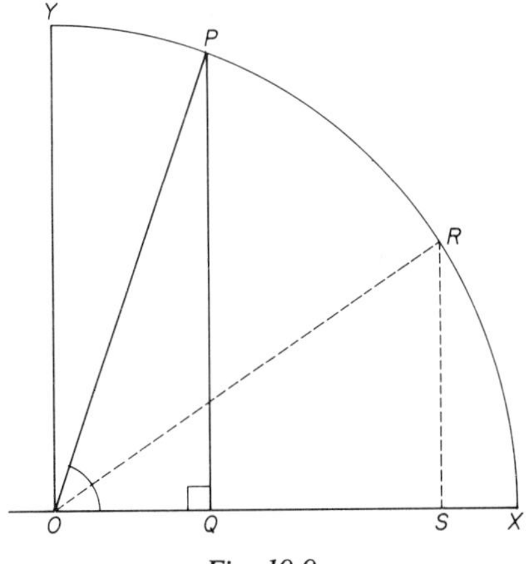

Fig. 19.9

To find, say, sin 70° and cos 70°, draw the radius OP such that $\angle XOP = 70°$, and draw the perpendicular PQ to OX. Measure $|PQ|$ and $|OQ|$; $|PQ|$ is found to be approximately 0·94 dm, and $|OQ|$ approximately 0·34 dm. The hypotenuse of $\triangle OPQ$ is the radius of the circle, which is of length 1 dm.

$$\text{Then } \sin 70° \ (= \sin \angle POQ) = \frac{|PQ|}{|OP|} = \frac{0{\cdot}94}{1} = 0{\cdot}94 \text{ (to 2 SF);}$$

$$\cos 70° \ (= \cos \angle POQ) = \frac{|OQ|}{|OP|} = \frac{0{\cdot}34}{1} = 0{\cdot}34 \text{ (to 2 SF).}$$

The length of the segment PQ represents the value of sin 70°, and the length of the segment OQ represents the value of cos 70°. The same diagram of the quadrant of the circle can be used to find approximate values of sines and cosines of other angles; for example,

sin 34° (sin $\angle ROS$) is read off as 0·56 ($|RS| = 0{\cdot}56$), and
cos 34° = 0·83 ($|OS| = 0{\cdot}83$).

If graph paper is used, the measurement of the line segments is easy.

19.5 Exercises

Use the second method in Section 19.4 to find approximate values of:

sin 25° and cos 25°, and of sin 42° and cos 42°.

19.6 Finding Tangents by Drawing

We can draw a triangle OPQ as described in Section 19.4, and measure the lengths of the sides PQ and OQ.

For example, to find tan 70°, using Fig. 19.9 again:

$$\tan 70^\circ = \frac{|PQ|}{|OQ|} = \frac{0{\cdot}94}{0{\cdot}34} = 2{\cdot}7 \text{ (approximately)},$$

or

$$\tan 70^\circ = \frac{\sin 70^\circ}{\cos 70^\circ}.$$

This relation between tan, sin and cos is always valid, whatever the size of the angle.

In any $\triangle OPQ$, right-angled at Q,

$$\sin \angle POQ = \frac{|PQ|}{|OP|} \text{ and } \cos \angle POQ = \frac{|OQ|}{|OP|}$$

$$\Rightarrow \frac{\sin \angle POQ}{\cos \angle POQ} = \frac{|PQ|}{|OP|} \Big/ \frac{|OQ|}{|OP|} = \frac{|PQ|}{|OQ|} \text{ (multiplying 'numerator' and 'denominator' by } |OP|\text{).}$$

But $\tan \angle POQ = \dfrac{|PQ|}{|OQ|}$.

Therefore,

$$\tan \angle POQ = \frac{\sin \angle POQ}{\cos \angle POQ}.$$

An alternative method of finding the approximate value of the tangent by drawing shows clearly the relation between the tangent as a trigonometric ratio and the geometrical tangent. Draw a circle of radius 1 dm, and then draw the tangent to the circle at a point X (Fig. 19.10).

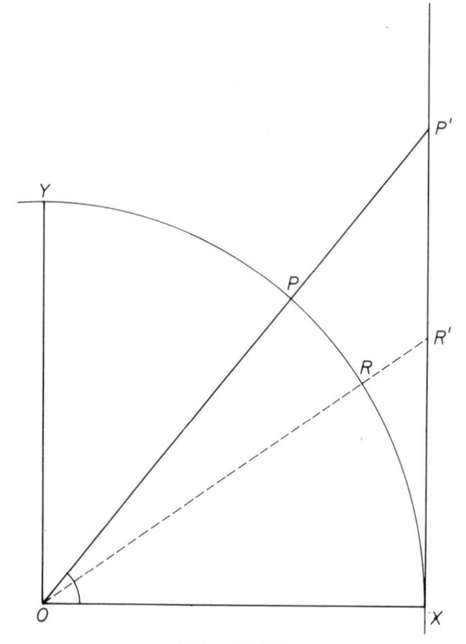

Fig. 19.10

We can find, for example, tan 50°, by drawing the radius OP, making an angle of 50° with OX, and extending it to meet the tangent at P'.

By measurement, we find $|XP'| = 1{\cdot}19$ dm;

then $\tan 50° \left(= \tan \angle POX = \dfrac{|XP'|}{|OX|}\right) = \dfrac{1{\cdot}19}{1} = 1{\cdot}19$; that is, the length of the line segment XP' represents the value of tan 50°. The same diagram can be used to find approximate values of tangents of other angles; for example, if $\angle ROX = 34°$, then $\tan 34° = |XR'| = 0{\cdot}67$.

19.7 Exercise

Find by drawing approximate values of tan 30° and tan 65°.

19.8 The Tables of Trigonometric Ratios

The trigonometric ratios for all angles between 0° and 90° have been calculated, and can be found in the *tables of natural sines, cosines and tangents* in any book of four-figure tables. (See pages 576–587.)

All trigonometric tables use an arrangement of rows and columns very like that of the arithmetic tables discussed in Unit 5.

Again, there are three major vertical divisions. The first column lists the 'whole degree' part of the angle. In the second vertical division the trigonometric ratios are given at intervals of one-tenth of a degree: in some tables these columns are headed by 0, 0·1°, 0·2°, 0·3°, and so forth, while others use the sub-division of a degree into minutes, and they are headed by the exact equivalents 0′, 6′, 12′, 18′, 24′, etc. The last set of columns gives the difference tables, which are used to calculate the trigonometric ratios at even smaller intervals. In tables giving decimal sub-divisions of a degree in point fractions, these intervals are one-hundredth of a degree; where minutes are used as sub-units of a degree, they are one-minute intervals.

Table 19.1 on pages 422 and 423 shows an extract of each of these alternative *sine tables*. To find from these tables, for example, sin 28°, look down the first column for 28 and determine the row to be used; sin 28° is found in the first column of the second division (the column headed by 0′ or 0·0), that is, the number 0·4695 which lies immediately to the right of 28 in the first column.

Notice that, for clarity's sake, the unit point is only marked in the first column of the second division. This applies to tables of all trigonometric ratios. Table 19.2 on pages 424 and 425, an extract of the *tangent table*, shows, moreover, that the *integral part* of the tangent is similarly given in the first column of the second division only; any further change in the integral part of the tangent in the middle of a row is usually indicated by a change of type face.

Table 19.1 Sines

	0′	6′	12′	18′	24′	30′	36′	42′	48′	54′	1′	2′	3′	4′	5′
0°	0·0000	0017	0035	0052	0070	0087	0105	0122	0140	0157	3	6	9	12	15
19	0·3256	3272	3289	3305	3322	3338	3355	3371	3387	3404	3	5	8	11	14
20	0·3420	3437	3453	3469	3486	3502	3518	3535	3551	3567	3	5	8	11	14
21	0·3584	3600	3616	3633	3649	3665	3681	3697	3714	3730	3	5	8	11	14
22	0·3746	3762	3778	3795	3811	3827	3843	3859	3875	3891	3	5	8	11	13
23	0·3907	3923	3939	3955	3971	3987	4003	4019	4035	4051	3	5	8	11	13
24	0·4067	4083	4099	4115	4131	4147	4163	4179	4195	4210	3	5	8	11	13
25	0·4226	4242	4258	4274	4289	4305	4321	4337	4352	4368	3	5	8	11	13
26	0·4384	4399	4415	4431	4446	4462	4478	4493	4509	4524	3	5	8	10	13
27	0·4540	4555	4571	4586	4602	4617	4633	4648	4664	4679	3	5	8	10	13
28	0·4695	4710	4726	4741	4756	4772	4787	4802	4818	4833	3	5	8	10	13
29	0·4848	4863	4879	4894	4909	4924	4939	4955	4970	4985	3	5	8	10	13
30	0·5000	5015	5030	5045	5060	5075	5090	5105	5120	5135	3	5	8	10	13
31	0·5150	5165	5180	5195	5210	5225	5240	5255	5270	5284	2	5	7	10	12
32	0·5299	5314	5329	5344	5358	5373	5388	5402	5417	5432	2	5	7	10	12
33	0·5446	5461	5476	5490	5505	5519	5534	5548	5563	5577	2	5	7	10	12

OR:

	0·0	0·1	0·2	0·3	0·4	0·5	0·6	0·7	0·8	0·9	1	2	3	4	5	6	7	8	9
0°	0·0000	0017	0035	0052	0070	0087	0105	0122	0140	0157	2	3	5	7	9	10	12	14	16
19	0·3256	3272	3289	3305	3322	3338	3355	3371	3387	3404	2	3	5	7	8	10	12	13	15
20	0·3420	3437	3453	3469	3486	3502	3518	3535	3551	3567	2	3	5	7	8	10	11	13	15
21	0·3584	3600	3616	3633	3649	3665	3681	3697	3714	3730	2	3	5	6	8	10	11	13	15
22	0·3746	3762	3778	3795	3811	3827	3843	3859	3875	3891	2	3	5	6	8	10	11	13	15
23	0·3907	3923	3939	3955	3971	3987	4003	4019	4035	4051	2	3	5	6	8	10	11	13	14
24	0·4067	4083	4099	4115	4131	4147	4163	4179	4195	4210	2	3	5	6	8	10	11	13	14
25	0·4226	4242	4258	4274	4289	4305	4321	4337	4352	4368	2	3	5	6	8	9	11	13	14
26	0·4384	4399	4415	4431	4446	4462	4478	4493	4509	4524	2	3	5	6	8	9	11	12	14
27	0·4540	4555	4571	4586	4602	4617	4633	4648	4664	4679	2	3	5	6	8	9	11	12	14
28	0·4695	4710	4726	4741	4756	4772	4787	4802	4818	4833	2	3	5	6	8	9	11	12	14
29	0·4848	4863	4879	4894	4909	4924	4939	4955	4970	4985	2	3	5	6	8	9	11	12	14
30	0·5000	5015	5030	5045	5060	5075	5090	5105	5120	5135	2	3	5	6	8	9	11	12	14
31	0·5150	5165	5180	5195	5210	5225	5240	5255	5270	5284	1	3	4	6	7	9	10	12	13
32	0·5299	5314	5329	5344	5358	5373	5388	5402	5417	5432	1	3	4	6	7	9	10	12	13
33	0·5446	5461	5476	5490	5505	5519	5534	5548	5563	5577	1	3	4	6	7	9	10	12	13

Table 19.2 Tangents

	0′	6′	12′	18′	24′	30′	36′	42′	48′	54′	1′	2′	3′	4′	5′
45°	1·0000	0035	0070	0105	0141	0176	0212	0247	0283	0319	6	12	18	24	30
46	1·0355	0392	0428	0464	0501	0538	0575	0612	0649	0686	6	12	18	25	31
47	1·0724	0761	0799	0837	0875	0913	0951	0990	1028	1067	6	13	19	25	32
48	1·1106	1145	1184	1224	1263	1303	1343	1383	1423	1463	7	13	20	26	33
49	1·1504	1544	1585	1626	1667	1708	1750	1792	1833	1875	7	14	21	28	34
50	1·1918	1960	2002	2045	2088	2131	2174	2218	2261	2305	7	14	22	29	36
51	1·2349	2393	2437	2482	2527	2572	2617	2662	2708	2753	8	15	23	30	38
52	1·2799	2846	2892	2938	2985	3032	3079	3127	3175	3222	8	16	24	31	39
53	1·3270	3319	3367	3416	3465	3514	3564	3613	3663	3713	8	16	25	33	41
54	1·3764	3814	3865	3916	3968	4019	4071	4124	4176	4229	9	17	26	34	43
55	1·4281	4335	4388	4442	4496	4550	4605	4659	4715	4770	9	18	27	36	45
56	1·4826	4882	4938	4994	5051	5108	5166	5224	5282	5340	10	19	29	38	48
57	1·5399	5458	5517	5577	5637	5697	5757	5818	5880	5941	10	20	30	40	50
58	1·6003	6066	6128	6191	6255	6319	6383	6447	6512	6577	11	21	32	43	53
59	1·6643	6709	6775	6842	6909	6977	7045	7113	7182	7251	11	23	34	45	56
60	1·7321	7391	7461	7532	7603	7675	7747	7820	7893	7966	12	24	36	48	60
61	1·8040	8115	8190	8265	8341	8418	8495	8572	8650	8728	13	26	38	51	64
62	1·8807	8887	8967	9047	9128	9210	9292	9375	9458	9542	14	27	41	55	68
63	1·9626	9711	9797	9883	9970	**0057**	**0145**	**0233**	**0323**	**0413**	15	29	44	58	73
64	2·0503	0594	0686	0778	0872	0965	1060	1155	1251	1348	16	31	47	63	78
65	2·1445	1543	1642	1742	1842	1943	2045	2148	2251	2355	17	34	51	68	85

OR:

	0·0	0·1	0·2	0·3	0·4	0·5	0·6	0·7	0·8	0·9	1	2	3	4	5	6	7	8	9
45	1·0000	0035	0070	0105	0141	0176	0212	0247	0283	0319	4	7	11	14	18	21	25	28	32
46	1·0355	0392	0428	0464	0501	0538	0575	0612	0649	0686	4	7	11	15	18	22	26	29	33
47	1·0724	0761	0799	0837	0875	0913	0951	0990	1028	1067	4	8	11	15	19	23	27	31	34
48	1·1106	1145	1184	1224	1263	1303	1343	1383	1423	1463	4	8	12	16	20	24	28	32	36
49	1·1504	1544	1585	1626	1667	1708	1750	1792	1833	1875	4	8	12	17	21	25	29	33	37
50	1·1918	1960	2002	2045	2088	2131	2174	2218	2261	2305	4	9	13	17	22	26	30	35	39
51	1·2349	2393	2437	2482	2527	2572	2617	2662	2708	2753	5	9	14	18	23	27	32	36	41
52	1·2799	2846	2892	2938	2985	3032	3079	3127	3175	3222	5	9	14	19	24	28	33	38	42
53	1·3270	3319	3367	3416	3465	3515	3564	3613	3663	3713	5	10	15	20	25	30	35	39	44
54	1·3764	3814	3865	3916	3968	4019	4071	4124	4176	4229	5	10	16	21	26	31	36	41	47
55	1·4281	4335	4388	4442	4496	4550	4605	4659	4715	4770	5	11	16	22	27	33	38	44	49
56	1·4826	4882	4938	4994	5051	5108	5166	5224	5282	5340	6	11	17	23	29	34	40	46	52
57	1·5399	5458	5517	5577	5637	5697	5757	5818	5880	5941	6	12	18	24	30	36	42	48	54
58	1·6003	6066	6128	6191	6255	6319	6383	6447	6512	6577	6	13	19	26	32	38	45	51	58
59	1·6643	6709	6775	6842	6909	6977	7045	7113	7182	7251	7	14	20	27	34	41	47	54	61
60	1·7321	7391	7461	7532	7603	7675	7747	7820	7893	7966	7	14	22	29	36	43	50	58	65
61	1·8040	8115	8190	8265	8341	8418	8495	8572	8650	8728	8	15	23	31	38	46	54	61	69
62	1·8807	8887	8967	9047	9128	9210	9292	9375	9458	9542	8	16	25	33	41	49	57	66	74
63	1·9626	9711	9797	9883	9970	**0057**	**0145**	**0233**	**0323**	**0413**	9	18	26	35	44	53	61	70	79
64	2·0503	0594	0686	0778	0872	0965	1060	1155	1251	1348	9	19	28	38	47	57	66	75	85
65	2·1445	1543	1642	1742	1842	1943	2045	2148	2251	2355	10	20	30	41	51	61	71	81	91

To find, for example, tan 62° 36′ (= tan 62·6°), look down the first column for 62 and determine the row to be used; in this row select the column headed by 36′ (or 0·6) and you find the number 9292. *Do not forget* to look in the first column to find the integral part of the ratio, 1. Then tan 62° 36′ = 1·9292.

When angles are given in degrees and minutes, the most suitable table of trigonometric ratios is the one that uses the minute as a sub-division of a degree. If angles are given in degrees and point fractions of degrees, the tables using decimal fractions of degrees are the most suitable, especially if the use of the difference columns is necessary.

To find from the tangent table the value of tan 56° 44′, look down the first column for 56 and determine the row to be used; in this row select the column headed by 42′ and find

tan 56° 42′ = 1·5224;
the difference corresponding to 2′ is found in this row to be 0·0019
tan 56° 44′ = 1·5243.

Similarly, to find tan 56·73°, look down the first column for 56 and determine the row to be used; in this row select the column headed by 0·7 and find

tan 56·7° = 1·5224;
the difference corresponding to 0·03 is found in this row to be 0·0017
tan 56·73° = 1·5241.

Note. The trigonometric ratios as given in four-figure tables are approximations. Use four figures throughout your working, but give the final result correct to three significant figures.

19.9 Exercises

1. Use four-figure tables to find the values of:

   sin 69°;
   sin 30° 36′ (or sin 30·6°);
   tan 12° 28′ (or tan 12·47°);
   tan 72° 39′ (or tan 72·63°).

2. In calculations with logarithms involving trigonometric ratios, such as for example, $\frac{253 \sin 32°}{15{\cdot}6 \sin 12°}$, you may look up sin 32° (= 0·5299) and sin 12° (= 0·2079) and then find the logarithm of 0·5299 (= $\bar{1}$·7242) and the logarithm of 0·2079 (= $\bar{1}$·3179). However, special tables are available, named *log sin*, *log cos*, etc., from which the 'logarithm of the sine of . . .' can be found immediately (pp. 588–99): we can read off log (sin 32°) = $\bar{1}$·7242, log (sin 12°) = $\bar{1}$·3179, and so on.

   Use *log sin* and *log tan* tables to find

   long sin 27° 36′ (or log sin 27·6°) and
   log tan 63° 18′ (or log tan 63·3°).

3. Add further examples of your own, until you are thoroughly familiar with the procedure of finding trigonometric ratios from tables.

19.10 The Table of Cosines

Because of the relation between *sin* and *cos*, that is cos $\alpha°$ is the sine of the complement of $\alpha°$, or cos $\alpha°$ = sin $(90 - \alpha)°$, a special table of cosines is not strictly necessary. The cosine of an angle can be found by looking up the sine of its complement; for example, cos 65° 24′ = sin (90° − 65° 24′) = sin 24° 36′ = 0·4163 (or cos 65·4° = sin (90° − 65·4°) = sin 24·6°).

Some books of tables do not have a special table of cosines; instead, the corresponding complementary angles are listed as a further column in the table of sines.

If a special *table of cosines* is used, you should take special care when using the difference columns: *the differences are to be subtracted.* You will understand the reason for this if you study the values of cosines (see Table 19.3 on p. 428): as the angle *increases* from 0° to 90°, the value of the cosine *decreases* from 1 down to 0. Indeed, a warning '*subtract*' is often printed over the difference columns of cosine tables.

To find, for example, cos 42° 20′: consider the two values cos 42° 18′ = 0·7396 and cos 42° 24′ = 0·7385; the value of cos 42° 20′ clearly lies between 0·7396 and 0·7385 and is *less* than cos 42° 18′ = 0·7396; the difference for 2′ is given in the appropriate row as 0·0004, therefore

$$\cos 42° \, 20' = 0{\cdot}7396 - 0{\cdot}0004 = 0{\cdot}7392.$$

19.11 Exercise

Use four-figure tables to find the values of

cos 27° 13′ (or 27·21°) and
cos 65° 50′ (or cos 65·83°).

19.12 Finding Angles whose Sines or Tangents are Known

We can use an inverse procedure to find the angles whose sines or tangents are known from the same tables of trigonometric ratios.

Examples

1. In $\triangle ABC$ (Fig. 19.11), right-angled at C, $c = |AB| = 10$ cm and $a = |BC| = 3{\cdot}5$ cm.

$$\text{Sin}\, A = \frac{a}{c} = \frac{3{\cdot}5}{10} = 0{\cdot}35.$$

Looking down the columns of the table of sines, we find the number nearest to 0·35 (and less than 0·35) in the row for 20° and in the column headed by 24′; thus, sin 20° 24′ = 0·3486. In the difference columns in the same row, we find the number of minutes corresponding to a difference of 14 (that is 0·0014), to

Table 19.3 Sines–Cosines

	0′	6′	12′	18′	24′	30′	36′	42′	48′	54′	1′	2′	3′	4′	5′	↓
0°	0·0000	0017	0035	0052	0070	0087	0105	0122	0140	0157	3	6	9	12	15	90°
1	0·0175	0192	0209	0227	0244	0262	0279	0297	0314	0332	3	6	9	12	15	89
2	0·0349	0366	0384	0401	0419	0436	0454	0471	0488	0506	3	6	9	12	15	88
3	0·0523	0541	0558	0576	0593	0610	0628	0645	0663	0680	3	6	9	12	15	87
4	0·0698	0715	0732	0750	0767	0785	0802	0819	0837	0854	3	6	9	12	14	86
5	0·0872	0889	0906	0924	0941	0958	0976	0993	1011	1028	3	6	9	12	14	85
6	0·1045	1063	1080	1097	1115	1132	1149	1167	1184	1201	3	6	9	12	14	84
7	0·1219	1236	1253	1271	1288	1305	1323	1340	1357	1374	3	6	9	12	14	83
8	0·1392	1409	1426	1444	1461	1478	1495	1513	1530	1547	3	6	9	12	14	82

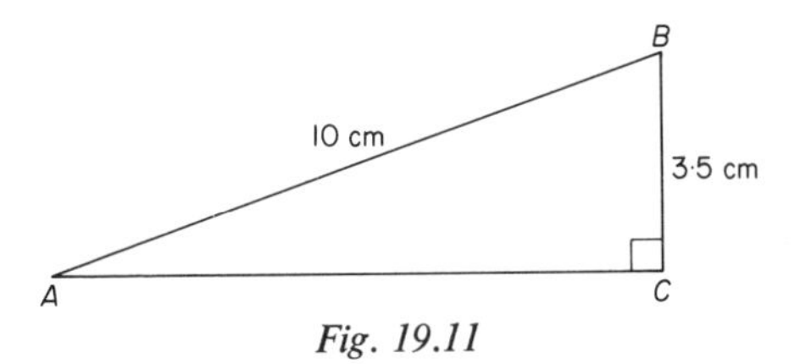

Fig. 19.11

be 5. The angle whose sine is 0·35 (that is 0·3486 + 0·0014), is therefore 20° 29′, and $\angle A = 20° 29'$.

2. In $\triangle PQR$ (Fig. 19.12), right-angled at R, $|PR| = 15$ cm and $|QR| = 8$ cm.

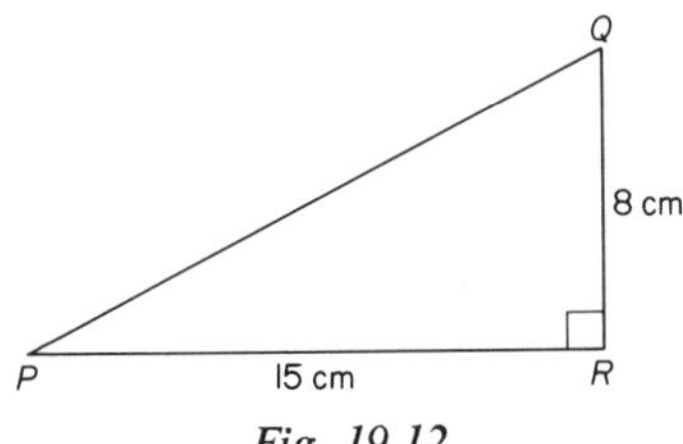

Fig. 19.12

To find $\angle P$, we use a ratio which relates the two sides of known length, the sides $|PR|$ and $|QR|$; this can be either $\tan P = \dfrac{|QR|}{|PR|}$ or $\tan Q = \dfrac{|PR|}{|QR|}$.

Tan $P = \frac{8}{15} = 0{\cdot}5333$ (to 4 SF).

The angle whose tangent is 0·5333 is found in the table of tangents to be 28° 4′.

Therefore, $\angle P = 28° 4'$.

3. $\triangle KLM$ (Fig. 19.13) is right-angled at M; $|KL| = 24$ cm and $|KM| = 11$ cm.

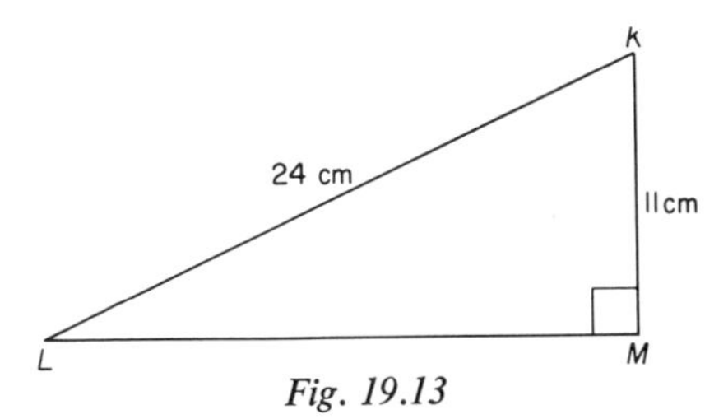

Fig. 19.13

To find $\angle K$, we use the ratio which relates the two sides of known length, $|KL|$ and $|KM|$.

The ratio $\dfrac{|KM|}{|KL|}$ is named sin L or cos K. (KM is the side opposite $\angle L$, and also the side adjacent to $\angle K$.)

Since $\angle K$ is the angle to be found, the simplest method is to use

$$\cos K = \frac{|KM|}{|KL|} = \frac{11}{24} = 0{\cdot}4583.$$

The angle whose cosine is 0·4583 is found using the table of cosines. Since differences in the cosine table are to be *subtracted*, we find the nearest number *greater* than 0·4583 which is $\cos 62^\circ 42' = 0{\cdot}4586$. A difference of 0·0003 in this row corresponds to a difference of $1'$; therefore, the angle whose cosine is $0{\cdot}4586 - 0{\cdot}0003 = 0{\cdot}4583$ is an angle of $62^\circ 42' + 1' = 62^\circ 43'$. Therefore, $\angle K = 62^\circ 43'$.

Alternatively, we may find $\angle L$ using the table of sines, and calculate $\angle K$:

$$\sin L = \frac{|KM|}{|KL|} = \frac{11}{24} = 0{\cdot}4583.$$

The angle whose sine is 0·4583 is found in the table of sines to be $27^\circ 17'$.

$$\angle L = 27^\circ 17'$$
$$\Rightarrow \angle K = 90^\circ - 27^\circ 17' = 62^\circ 43'.$$

19.13 Exercise

Use the tables of sines, cosines or tangents to calculate the sizes of the following angles in Fig. 19.14:

$\angle A$ and $\angle B$;
$\angle R$ and $\angle S$.

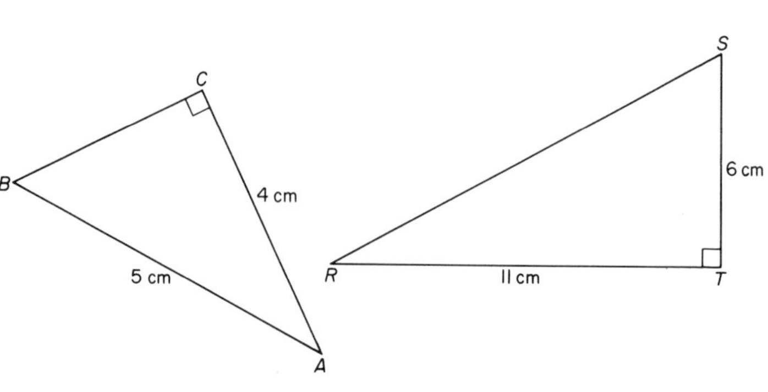

Fig. 19.14

19.14 The Graphs of Trigonometric Ratios

To plot the graph of $y = \sin x^\circ$ we write down some values from the sine table (to 2 SF).

x°	0°	10°	20°	30°	40°	50°	60°	70°	80°	90°
$\sin x^\circ$	0	0·17	0·34	0·50	0·64	0·77	0·87	0·94	0·98	1

We choose a scale of 1 cm to represent $\frac{1}{10}$ unit on the vertical axis and 1 cm to represent 10 units on the horizontal axis, plot the points and join them by a smooth curve (Fig. 19.15 is a half-scale drawing of this graph).

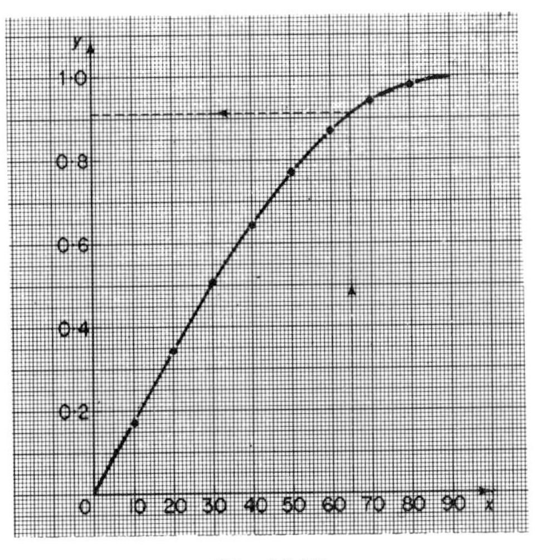

Fig. 19.15

This graph can be used to read off rough values of sin $x°$ for any value of x between 0 and 90. For example, sin 65° is found to be approximately 0·92 (marked by dotted lines in the figure). It shows clearly that the range of values of sin $x°$ is limited; the greatest value of sin $x°$ is 1 (when $x° = 90°$). This is not surprising, since

$$\sin A = \frac{\text{length of the side opposite } \angle A}{\text{length of the hypotenuse}}.$$

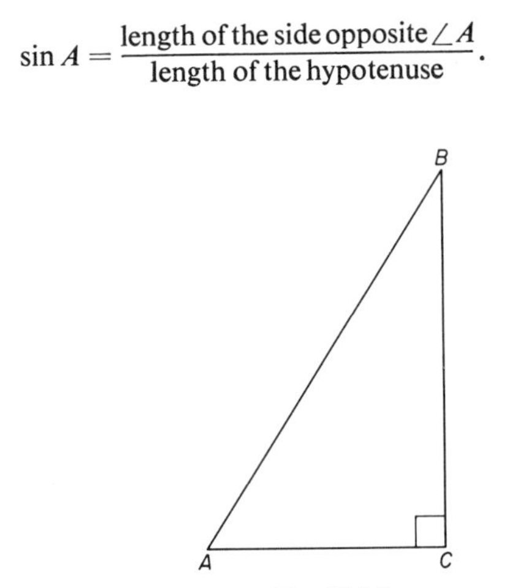

Fig. 19.16

But the hypotenuse is always the largest side, and so it is always larger than the side opposite $\angle A$; the ratio is therefore always less than 1. As $\angle A$ becomes larger, $|BC|$ also 'grows larger'; when $\angle A$ is almost 90°, $|BC|$ is almost equal to $|AB|$. When $\angle A$ reaches its limit of 90° and $|BC| = |AB|$, we cannot speak any more of $\triangle ABC$.

19.15 Exercises

1. Plot the graph of $y = \sin x°$, and on the same diagram plot the graph of $y = \cos x°$ (Fig. 19.17).

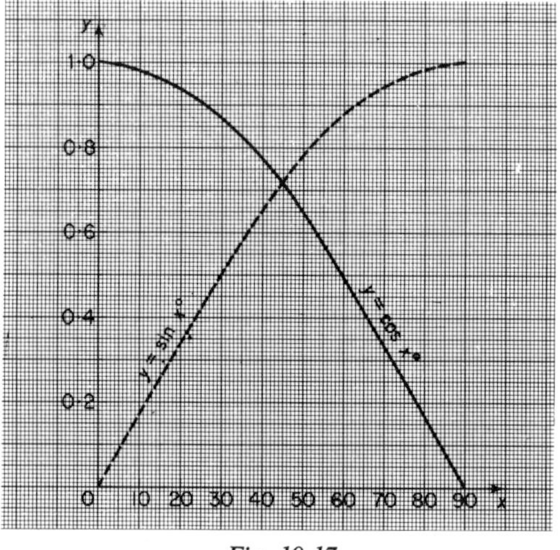

Fig. 19.17

The point of intersection of the sine and cosine graphs shows that $\sin 45° = \cos 45°$.
Fold the graph paper along the vertical line $x = 45$ and confirm the bilateral symmetry about this line. Use this symmetry to show that $\sin x° = \cos (90 - x)°$ or $\cos x° = \sin (90 - x)°$.

2. Draw the graph of $y = \tan x°$, using a scale of 1 cm to 10° on the x-axis and 5 cm to represent one unit on the vertical axis.
A quick glance at your tangent table will make it clear that for angles much greater than 75°, $\tan x°$ will be 'off your paper'.

19.16 Solving Right-angled Triangles

Most practical uses of trigonometric ratios are based on what is called *solving triangles*, that is, calculating the sizes of the angles and the lengths of the sides of a triangle from given items of information.

When we discussed congruent triangles in Section 15.11, we spoke of six 'aspects' or 'facts' of the triangle, the sizes of its three angles and the lengths of its three sides. We noticed then that any right-angled triangle is 'fixed' if three of these facts are known (unless the sizes of the three angles are all that is given).

By the same principle, the lengths of the sides of a triangle and the sizes of the three angles can be calculated from information given concerning any three of these facts (except three angles).

We are concerned in this section with right-angled triangles only, and have therefore one known fact, *the size of one angle*, 90°.

(*a*) If *the lengths of two sides are given*, we can find the length of the third side by Pythagoras' Theorem, and use tables of trigonometric ratios to find one of the angles. The third angle is the complement of this angle.

Example

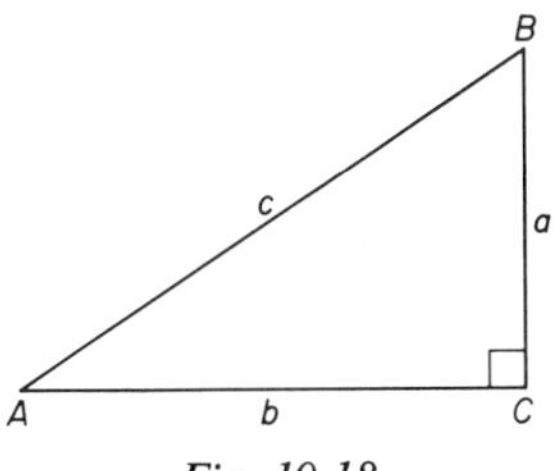

Fig. 19.18

$\triangle ABC$ is right-angled at C, $a = 5$ cm and $c = 10$ cm.

By Pythagoras' Theorem, $b^2 = 10^2 - 5^2 = 75$,

$$\text{so } b = \sqrt{75} = 8{\cdot}66 \text{ cm.}$$

Sin $A = \frac{5}{10} = 0{\cdot}5000$. The angle whose sine is $0{\cdot}5000$ is found;

$$\angle A = 30° \Rightarrow \angle B = 60°.$$

(*b*) If *the length of one side and the size of another angle are given*, the third angle is immediately known since it is the complement of the given angle. The length of each of the other two sides is found from the ratio which relates the *side wanted* and the *side known*.

Examples

1. $\triangle ABC$ is right-angled at C, $\angle A = 25°$ and $c = 12$ cm ($\angle B$ is then $90° - 25° = 65°$).

To find a, since c is known, we use a ratio involving a and c, for example sin A (cos B is the same ratio):

$$\sin A = \frac{a}{c} \Rightarrow a = c \sin A \text{ (multiplying both sides by } c\text{);}$$

substituting known values,

$$a = 12 \times \sin 25° = 12 \times 0{\cdot}4226 = 5{\cdot}07 \text{ cm (to 3 SF)}.$$

To find b we use a ratio involving b and c:

$$\sin B = \frac{b}{c} \Rightarrow b = c \sin B;$$

substituting known values,

$$b = 12 \times \sin 65° = 12 \times 0{\cdot}9063 = 10{\cdot}9 \text{ cm (to 3 SF)}.$$

2. $\triangle PQR$ (Fig. 19.19) is right-angled at R, $\angle P = 38°$ and $q = 84$ cm ($\angle P = 38° \Rightarrow \angle Q$ is 52°).

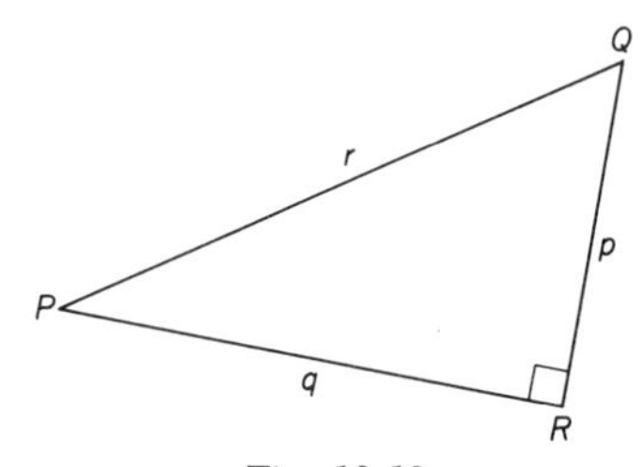

Fig. 19.19

To find p, we use the ratio which involves both p (wanted) and q (known); this ratio is $\tan P$:

$$\tan P = \frac{p}{q} \Rightarrow p = q \tan P$$

$$\Rightarrow p = 84 \times \tan 38° = 84 \times 0{\cdot}7813 = 65{\cdot}6 \text{ cm (to 3 SF)}.$$

To find r, we use the ratio which involves q (known) and r (wanted):

$$\sin Q = \frac{q}{r} \Rightarrow r = \frac{q}{\sin Q} \left(\text{multiplying both sides by } \frac{r}{\sin Q}\right);$$

substituting known values:

$$r = \frac{84}{\sin 52°} = \frac{84}{0{\cdot}7880} = 107 \text{ cm (to 3 SF)}.$$

If logarithms are needed in your calculations, use log tan and log sin tables if possible.

numbers	logs
84	1·9243
×	+
tan 38°	$\bar{1}$·8928
65·63	1·8171

= 65·6 (to 3 SF)

numbers	logs
84	1·9243
÷	−
sin 52°	$\bar{1}$·8965
106·6	2·0278

= 107 (to 3 SF)

19.17 Exercises

1. Solve the right-angled triangles in Fig. 19.20 for the unknown parts.

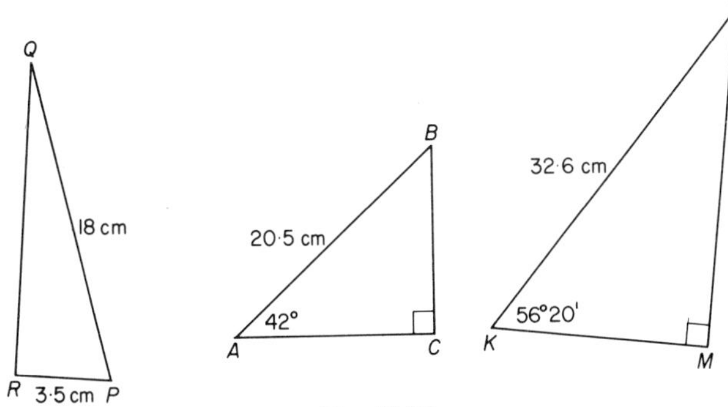

Fig. 19.20

2. $\triangle EFG$ (Fig. 19.21) is an isosceles triangle; $|EG| = |FG| = 14$ cm and $|EF| = 20$ cm. If M is the mid-point of EF, calculate $|GM|$.

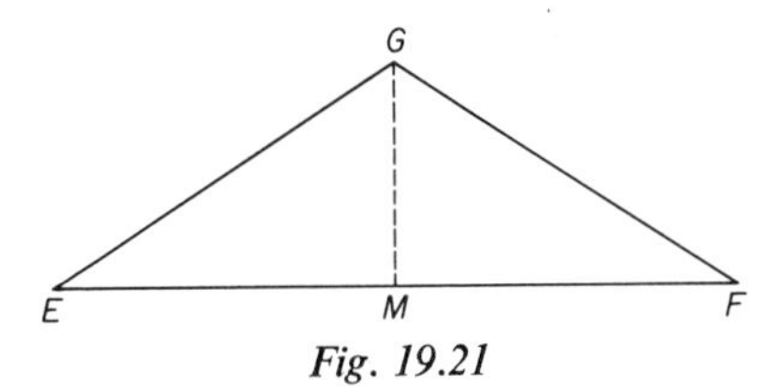

Fig. 19.21

[Hint: remember that in an isosceles triangle the median coincides with the altitude.]

3. In $\triangle ABC$ in Fig. 19.22, $|AB| = 7{\cdot}5$ cm and $\angle A = 63°$, and h is the altitude of the triangle, relative to AC. Calculate h to 3 SF.

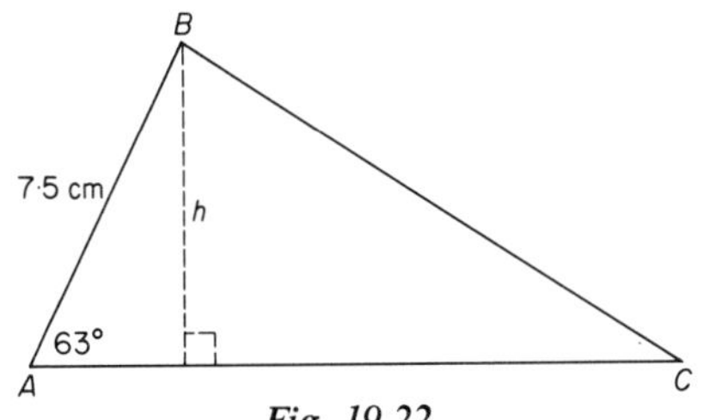

Fig. 19.22

4. $KLMN$ is a rhombus, $|KM| = 14$ cm and $|LN| = 6$ cm. Calculate $\angle KLM$.
5. $ABCDE$ is a regular pentagon inscribed in a circle centre O and radius of length 10 cm. M is the mid-point of AB. Calculate $|OM|$. [Hint: use your knowledge of angles of a regular pentagon and compare Exercise 2.]
6. The radius of the circumscribed circle of a regular octagon is 10 cm. Calculate the length of each of its sides.

19.18 Practical Applications

Practical applications of solving right-angled triangles include calculations of heights and distances (see also Section 18.13) and of angles and directions (bearings).

Examples

1. An aeroplane takes off in a straight line at an angle of 17° to the ground. What is its height after flying a distance of 1500 m in this direction?

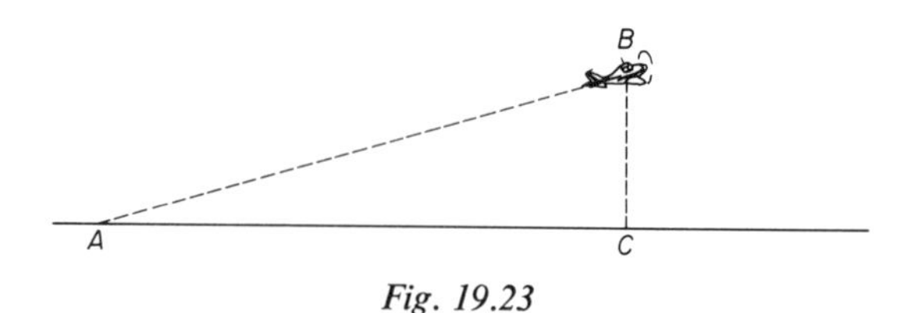

Fig. 19.23

If A is the point of take-off on the runway, B is the position of the aircraft after flying 1500 m and C is the point on the ground vertically below B, the length of the side BC of $\triangle ABC$ represents the required height. We know that $|AB| = 1500$ m.

Using the ratio of the 'side wanted' and 'side known',

$\sin A = \dfrac{|BC|}{|AB|}$ and, substituting known values:

$\sin 15° = \dfrac{|BC|}{1500\text{ m}} \Rightarrow |BC| = \sin 15° \times 1500\text{ m} = 0{\cdot}2588 \times 1500\text{ m} = 388\text{ m}$ (to 3 SF).

2. A ship leaves port P sailing on a bearing of 115°. Calculate how far 'south of P' and how far 'east of P' the ship is after travelling 70 nautical miles.

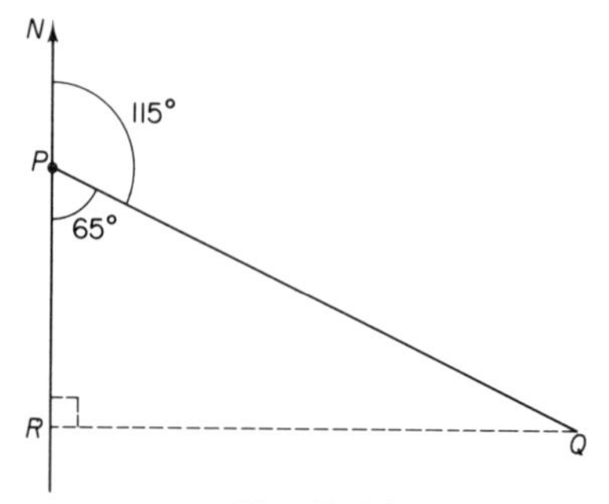

Fig. 19.24

In Fig. 19.24 the position of the ship is marked as point Q; R is the point due south of P and due west of Q.

$|PR|$ then represents the distance of the ship 'south of P' and $|QR|$ its distance 'east of P'.

Since the bearing of Q from P is 115°, $\angle QPR = 65°$.

To find $|QR|$ we use the ratio involving $|QR|$ and $|PQ|$ (known):

$$\sin \angle QPR = \frac{|QR|}{|PQ|}; \text{ substituting known values, } \sin 65° = \frac{|QR|}{70 \text{ n mile}}$$

$$\Rightarrow |QR| = \sin 65° \times 70 \text{ n mile} = 0{\cdot}9063 \times 70 \text{ n mile}$$
$$= 63{\cdot}4 \text{ n mile (to 3 SF).}$$

To find $|PR|$, we use the ratio $\frac{|PR|}{|PQ|} = \cos \angle QPR$, and substitute known values:

$$\cos 65° = \frac{|PR|}{70 \text{ n mile}} \Rightarrow |PR| = \cos 65° \times 70 \text{ n mile}$$
$$= 0{\cdot}4226 \times 70 \text{ n mile}$$
$$= 29{\cdot}6 \text{ n mile (to 3 SF).}$$

After travelling 70 nautical miles the ship is 29·6 nautical miles south of P and 63·4 nautical miles east of P.

19.19 Exercises

1. The angle of elevation of the top of a tower from a point 50 m from its foot is 24°. Calculate the height of the tower.
2. A point K is 12 km due west of point L and 25 km due south of M. Calculate the bearing of L from M.

19.20 Sines and Cosines of Obtuse Angles

If we define sines and cosines as ratios of lengths of sides of a right-angled triangle, we cannot envisage the sine or cosine of an obtuse angle, that is an angle greater than 90°: there are no obtuse angles in right-angled triangles.

Clearly, if we wish to consider sines or cosines of obtuse angles, we must find a new meaning of 'the sine of an angle' or 'the cosine of an angle', a new meaning which includes and extends our earlier definition of sines and cosines of acute angles of a right-angled triangle.

If the construction of trigonometric ratios is interpreted in graphical terms (see Section 19.4), the sines and cosines of angles between 0° and 90° are simply the y- and x-co-ordinates respectively of points in the *first quadrant*—the top right-hand quadrant, where x and y are both positive—on the circumference of the circle with unit radius (abbreviated to 'on the unit circle').
For example, the co-ordinates of point P in Fig. 19.25 are $|OQ|$ and $|QP|$. Since $|OP| = 1$ unit, and $\sin \alpha° = \frac{|PQ|}{|OP|} = \frac{|PQ|}{1} = |PQ|$ and $\cos \alpha° = \frac{|OQ|}{|OP|} = \frac{|OQ|}{1} = |OQ|$, the x-co-ordinate of P is $\cos \alpha°$ and the y-co-ordinate of P is $\sin \alpha°$.

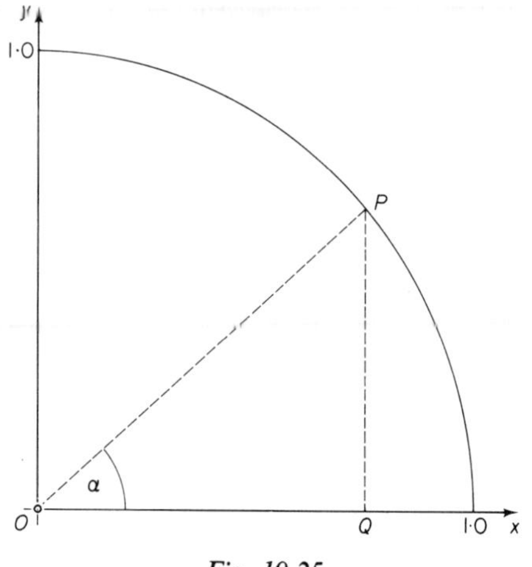

Fig. 19.25

A definition of the sine and the cosine of an acute angle is then:

If OP is a radius in the first quadrant of the unit circle centre O, and the angle between OP and the x-axis is $\alpha°$, $\cos \alpha° =$ the x-co-ordinate of P, and $\sin \alpha° =$ the y-co-ordinate of P.

This definition of sine and cosine can be extended to include angles greater than 90° by simply leaving out the restriction 'in the first quadrant' (compare Fig. 19.26).

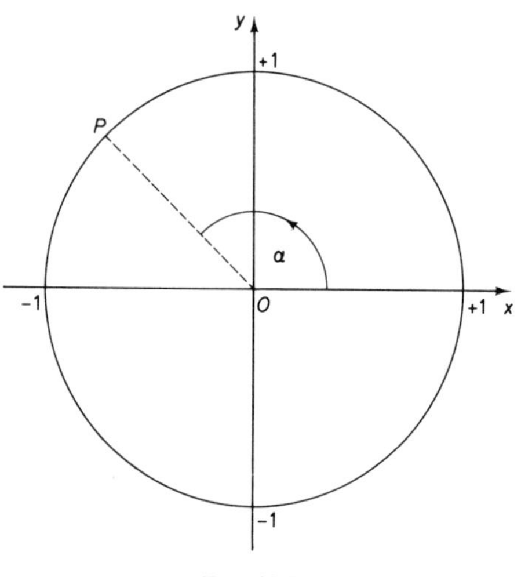

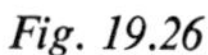

Fig. 19.26

OP is a radius of the unit circle with centre O (that is, $|OP| = 1$) and the angle formed by the positive direction of the x-axis with the radius OP measured in an anti-clockwise direction starting from OX is $\alpha°$; then

$$\sin \alpha^\circ = \frac{\text{the } y\text{-co-ordinate of } P}{|OP|} = \text{the } y\text{-co-ordinate of } P, \text{ and}$$

$$\cos \alpha^\circ = \frac{\text{the } x\text{-co-ordinate of } P}{|OP|} = \text{the } x\text{-co-ordinate of } P.$$

Since our concern is with sines and cosines of the angles of a triangle, we will only consider *sines and cosines of angles less than* 180°.

(This definition of *sine* and *cosine* does allow further extensions, including sines and cosines of angles greater than 180°, but this is beyond the scope of this book.)

If P is a point on the circumference of the unit circle in the *second quadrant* (that is the top left-hand quadrant, Fig. 19.27) its x-co-ordinate is a *negative number* (point P is to the left of the y-axis) and its y-co-ordinate is a *positive number* (point P is above the x-axis). (Look back to the discussion of co-ordinates in Unit 12, if necessary.) The co-ordinates of the point P in Fig.

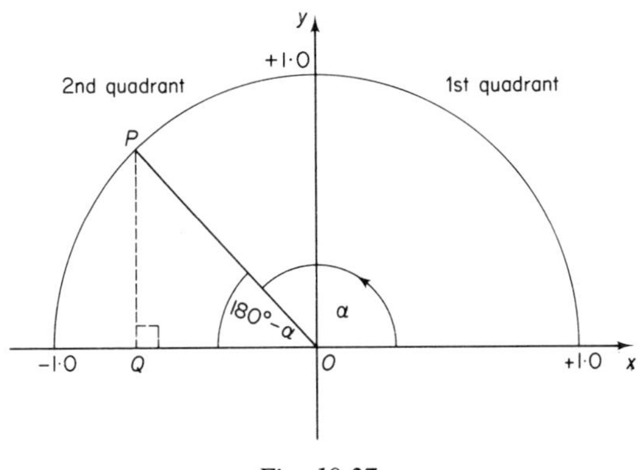

Fig. 19.27

19.27 are $-|OQ|$ and $+|QP|$, and the values of $|OQ|$ and $|QP|$ are found by considering the right-angled triangle OPQ.

$$\angle Q = 90^\circ,$$
$$|OP| = 1 \text{ unit},$$
$$\angle POQ = 180^\circ - \alpha^\circ$$
$$\Rightarrow |OQ| = \cos(180^\circ - \alpha^\circ) \text{ and } |QP| = \sin(180^\circ - \alpha^\circ).$$

If, for example, $\alpha^\circ = 150^\circ$, then $\angle POQ = 30^\circ$

$$\text{and } |OQ| = 0{\cdot}8660, \ |QP| = 0{\cdot}5000$$
$$\Rightarrow \cos 150^\circ = +0{\cdot}8660, \ \sin 150^\circ = -0{\cdot}5000.$$

In general,

$$\sin \alpha^\circ = +|QP| = +\sin(180^\circ - \alpha^\circ), \text{ and}$$
$$\cos \alpha^\circ = -|OQ| = -\cos(180^\circ - \alpha^\circ),$$

or,

> *The sine of an obtuse angle equals the sine of its supplement*, and *the cosine of an obtuse angle is the negative of the cosine of its supplement.*

To find the sine of an obtuse angle,

(*a*) calculate its supplementary angle (i.e. $180° - \alpha°$);
(*b*) look up $\sin (180° - \alpha°)$ in four-figure tables.

Examples

$$\sin 120° = \sin (180° - 120°) = \sin 60° = 0{\cdot}8660.$$
$$\sin 146° = \sin (180° - 146°) = \sin 34° = 0{\cdot}5592.$$

To find the cosine of an obtuse angle,

(*a*) calculate $180° - \alpha°$;
(*b*) look up $\cos (180° - \alpha°)$ in tables;
(*c*) write down 'the negative of' $\cos (180° - \alpha°)$.

Examples

$$\cos 110° = -\cos (180° - 110°) = -\cos 70° = -0{\cdot}3420.$$
$$\cos 167° 12' = -\cos (180° - 167° 12') = -\cos 12° 48' = -0{\cdot}9751.$$

19.21 Exercise

Write down the values of:

$\sin 125°$;
$\cos 160°$;
$\sin 137° 35'$;
$\cos 142° 16'$.

19.22 Solving Triangles which are Not Right-angled

We can solve triangles which are not right-angled by using two important relations between lengths of sides of a triangle and the sines and cosines of its angles, known as the *Sine Rule* and the *Cosine Rule.*

The Sine Rule

The Sine Rule states that *in a triangle, the ratio of the sine of each angle and the length of the side opposite that angle is constant*; in $\triangle ABC$,

$$\frac{a}{\sin A} = \frac{b}{\sin B} = \frac{c}{\sin C}$$

We will show that this constant ratio is the same as the ratio between the length of the diameter of the circumscribed circle and the unit length, or, if R is the length of the radius of the circumcircle of $\triangle ABC$, that

$$\frac{a}{\sin A} = \frac{b}{\sin B} = \frac{c}{\sin C} = \frac{2R}{1} = 2R.$$

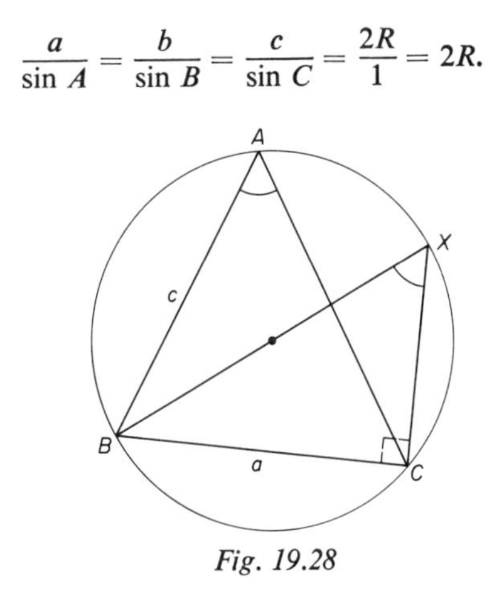

Fig. 19.28

We draw any triangle ABC and construct its circumcircle; we draw the diameter BX and join X to C (Fig. 19.28).

Then $|BX| = 2R$,
$\angle X = \angle A$ (angles at the circumference standing on the arc BC),
$\angle XCB = 90°$ (angle in a semicircle).

$$\text{In } \triangle BXC, \sin X = \frac{|BC|}{|BX|} = \frac{a}{2R}$$

$$\Rightarrow \sin A = \frac{a}{2R} \text{ (since } \angle A = \angle X).$$

Multiplying both sides by $\frac{2R}{\sin A}$,

$$\frac{2R}{\sin A} \times \sin A = \frac{2R}{\sin A} \times \frac{a}{2R}$$

$$\Rightarrow 2R = \frac{a}{\sin A}.$$

In the same way, we can show that $\frac{b}{\sin B} = 2R$ (Fig. 19.29)

$\left(\sin Y = \frac{b}{2R} \Rightarrow \sin B = \frac{b}{2R} \Rightarrow 2R = \frac{b}{\sin B}\right)$ and that $\frac{c}{\sin C} = 2R$.

Therefore, $$\frac{a}{\sin A} = \frac{b}{\sin B} = \frac{c}{\sin C} = 2R.$$

Note: the sine rule also applies to triangles of which one angle is obtuse. Show this as an exercise, using theorem 17·7 and 19·20.

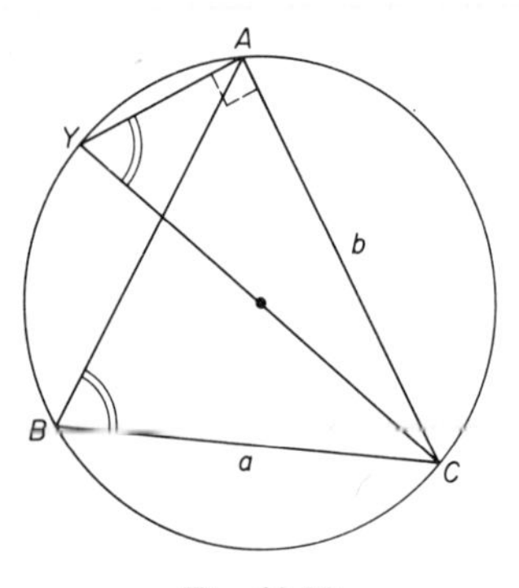

Fig. 19.29

Solving triangles by the Sine Rule

Each of the three equations $\frac{a}{\sin A} = \frac{b}{\sin B}$, $\frac{a}{\sin A} = \frac{c}{\sin C}$ and $\frac{b}{\sin B} = \frac{c}{\sin C}$

relates two angles and the lengths of the sides opposite these angles.

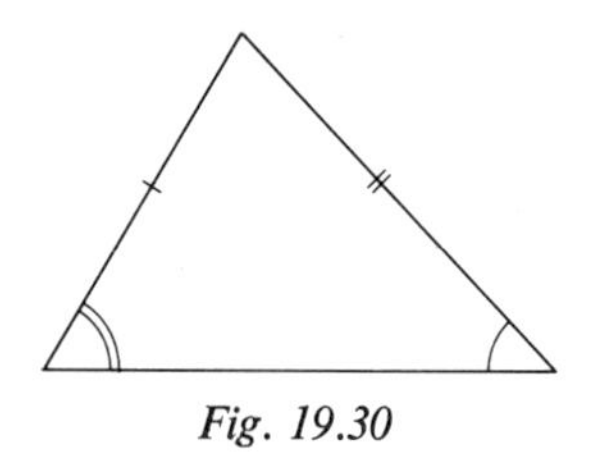

Fig. 19.30

We can therefore use the Sine Rule to solve a triangle,

(*a*) if we know the sizes of two angles and the length of one side; or
(*b*) if we know the lengths of two sides and the size of one angle opposite one of these sides.

Example

In $\triangle ABC$ (Fig. 19.31), $a = 15$ cm, $\angle A = 56°$ and $\angle B = 110°$.

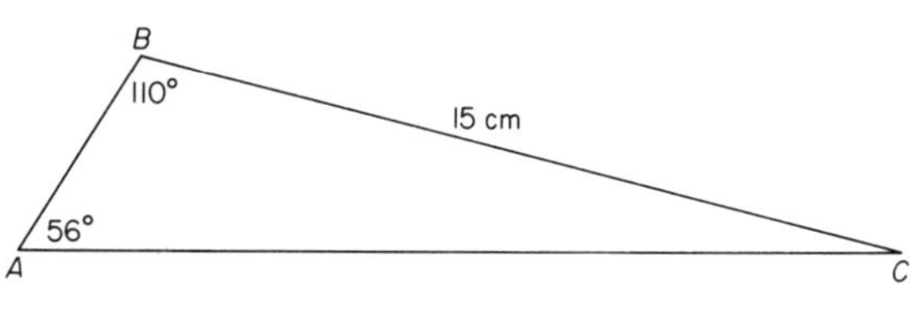

Fig. 19.31

To find b, we apply the Sine Rule using the equation which involves b and the three known 'facts', that is,

$$\frac{b}{\sin B} = \frac{a}{\sin A}.$$

Substituting known values, $\dfrac{b}{\sin 110°} = \dfrac{15 \text{ cm}}{\sin 56°}$

$$\Rightarrow b = \frac{\sin 110° \times 15 \text{ cm}}{\sin 56°} = \frac{\sin 70° \times 15 \text{ cm}}{\sin 56°} = 17{\cdot}8 \text{ cm (to 3 SF)}.$$

Note. Logarithms and log sines are used for this and similar calculations.

Similarly, we find c by applying the Sine Rule using the equation which involves c and three of the known facts. Since $\angle A$ and $\angle B$ are given, $\angle C$ can be calculated: $\angle C = 180° - (56° + 110°) = 14°$.

$$\frac{c}{\sin C} = \frac{a}{\sin A}$$

Substituting known values, $\dfrac{c}{\sin 14°} = \dfrac{15 \text{ cm}}{\sin 56°}$

$$\Rightarrow c = \frac{\sin 14° \times 15 \text{ cm}}{\sin 56°} = 4{\cdot}38 \text{ cm (to 3 SF)}.$$

Our earlier discussion on congruent triangles showed that a triangle is not necessarily 'fixed' if we know the lengths of two sides and the size of an angle opposite one of those sides (see Section 15.11).

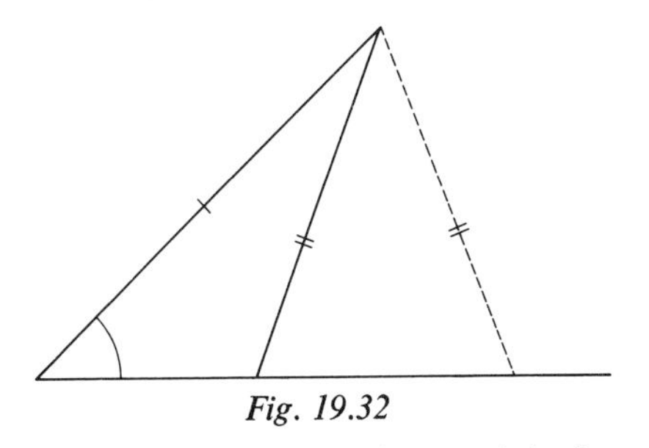

Fig. 19.32

(*a*) No such ambiguity can arise when the given angle is obtuse.

Example

In $\triangle ABC$ (Fig. 19.33), $\angle A = 130°$, $|AB| = 7$ cm and $|BC| = 18$ cm.

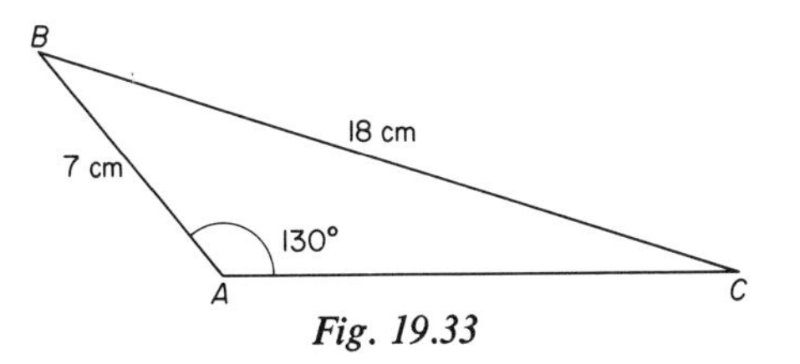

Fig. 19.33

By the Sine Rule, $\dfrac{a}{\sin A} = \dfrac{c}{\sin C}$

$$\Rightarrow \frac{18}{\sin 130^\circ} = \frac{7}{\sin C}$$

$$\Rightarrow \sin C = \frac{\sin 130^\circ \times 7}{18} = \frac{\sin 50^\circ \times 7}{18} = 0{\cdot}2979.$$

From tables, $\angle C = 17^\circ\ 20'$.
We can then calculate b as before, again using the Sine Rule.

(*b*) Fig. 19.34 shows that ambiguity *can* arise if the given angle is acute.

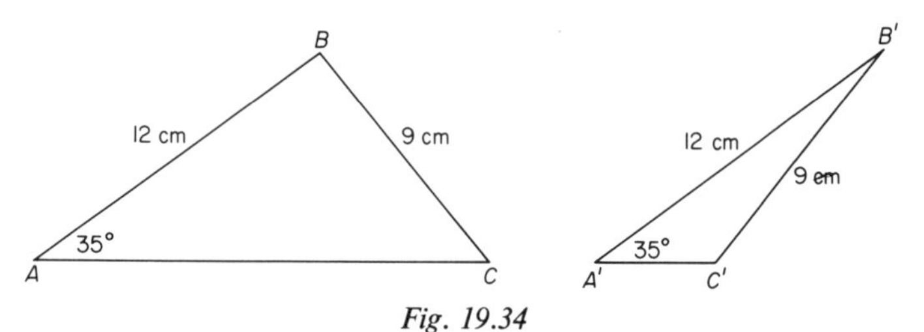

Fig. 19.34

If, in $\triangle ABC$, $\angle A = 35^\circ$, $|BC| = 9$ cm and $|AB| = 12$ cm, there are two possible constructions and therefore *two solutions* of the triangle.

By the Sine Rule, $\dfrac{|BC|}{\sin A} = \dfrac{|AB|}{\sin C}$

$$\Rightarrow \frac{9}{\sin 35^\circ} = \frac{12}{\sin C}$$

$$\Rightarrow \sin C = \frac{\sin 35^\circ \times 12}{9} = 0{\cdot}7646.$$

There are two possible values of $\angle C$, one between 0° and 90° and one between 90° and 180°.

The *acute angle* whose sine is 0·7646 is found from four-figure tables to be 49° 52′

$$\Rightarrow \angle C = 49^\circ\ 52', \text{ and } \angle B = 180^\circ - (35^\circ + 49^\circ\ 52') = 95^\circ\ 8'.$$

The *obtuse angle* whose sine is 0·7646 is $180^\circ - 49^\circ\ 52' = 130^\circ\ 8'$

$$\Rightarrow \angle C' = 130^\circ\ 8', \text{ and } \angle B' = 180^\circ - (35^\circ + 130^\circ\ 8') = 14^\circ\ 52'.$$

The Sine Rule is again used to calculate the lengths of the sides AC and $A'C'$:

$$\frac{9 \text{ cm}}{\sin 35^\circ} = \frac{|AC|}{\sin 95^\circ\ 8'} \Rightarrow |AC| = \frac{\sin 95^\circ\ 8' \times 9 \text{ cm}}{\sin 35^\circ} = \frac{\sin 84^\circ\ 52' \times 9 \text{ cm}}{\sin 35^\circ}$$

$$= 15{\cdot}6 \text{ cm (to 3 SF), and}$$

$$\frac{9\text{ cm}}{\sin 35°} = \frac{|A'C'|}{\sin 14° 52'} \Rightarrow |A'C'| = \frac{\sin 14° 52' \times 9\text{ cm}}{\sin 35°}$$
$$= 4{\cdot}03\text{ cm (to 3 SF).}$$

Notes.

(*a*) There are two possible constructions, and two solutions of the triangle, *only* if the given angle is acute *and* is opposite the shorter of the two given sides. Fig. 19.35 makes it clear that, if $a > b$, the alternative position of point B is on the other side of the side AC and $\angle CAB'$ is not the given angle, but its supplement.

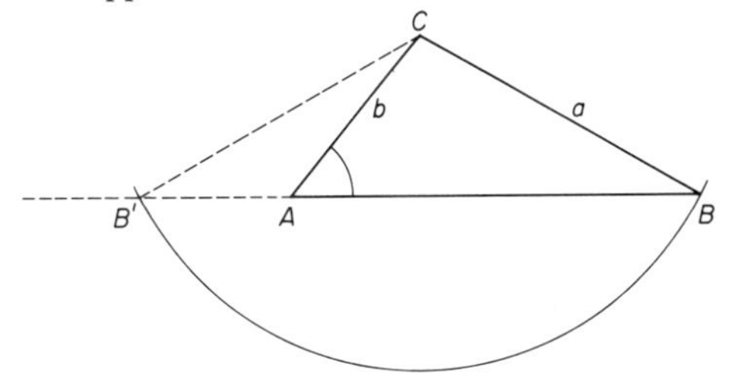

Fig. 19.35

This case is not ambiguous; there is only one possible solution.

(*b*) If $a < b$ and $\angle A$ is obtuse, there is no solution; if $\angle A$ is obtuse, it is the largest angle in the triangle and therefore a, the side opposite $\angle A$, is necessarily the largest of the three sides.

19.23 Exercises

1. Solve the following triangles:

   (*a*) $\triangle ABC$, given that $\angle A = 67°$, $\angle B = 50°$ and $a = 12$ cm;
   (*b*) $\triangle PQR$, given that $\angle P = 124°$, $p = 25$ cm and $r = 16$ cm.

2. From the data determine in each of the following cases whether one, two or no triangles can be constructed.

   (*a*) $\angle A = 67°$, $a = 8$ cm and $b = 7$ cm;
   (*b*) $\angle B = 135°$, $a = 6$ cm and $b = 5$ cm;
   (*c*) $\angle A = 37°$, $a = 7{\cdot}5$ cm and $b = 9$ cm.
   [Hint: make rough sketches.]

3. An aeroplane is observed on a bearing of 231° 42′ from a point A and at the same time on a bearing of 204° 12′ from a point B. Point B is 45 km due west of A. Use the Sine Rule to calculate how far the aeroplane is from point B.

19.24 The Cosine Rule

The Cosine Rule is an extension of Pythagoras' Theorem; it states that *in a triangle, the square of one side is equal to the sum of the squares of the other two sides minus twice the product of these two sides and the cosine of the included angle*; in any $\triangle ABC$,

$$\boxed{\begin{aligned} a^2 &= b^2 + c^2 - 2bc \cos A \\ b^2 &= a^2 + c^2 - 2ac \cos B \\ c^2 &= a^2 + b^2 - 2ab \cos C \end{aligned}}$$

We will show that $c^2 = a^2 + b^2 - 2\,ab \cos C$, and consider two alternative situations: that $\angle C$ is acute, and that $\angle C$ is obtuse.

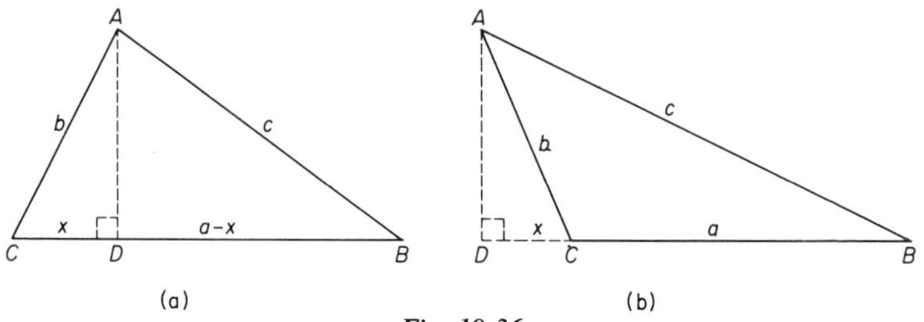

Fig. 19.36

In both cases we draw the perpendicular AD to BC.

(*a*) $\angle C < 90°$:

$|CD| = x \Rightarrow |BD| = a - x$.
In $\triangle ABD$, $c^2 = |BD|^2 + |AD|^2$ (Pythagoras' Theorem)

$$\Rightarrow c^2 = (a - x)^2 + |AD|^2 = a^2 - 2ax + x^2 + |AD|^2.$$

But $x^2 + |AD|^2 = b^2$ (Pythagoras' Theorem applied to $\triangle ACD$)

$$\text{and } \cos C = \frac{|CD|}{|CA|} = \frac{x}{b} \Rightarrow x = b \cos C.$$

Therefore,

$$\begin{aligned} c^2 &= a^2 - 2a(b \cos C) + b^2 \\ &= a^2 + b^2 - 2ab \cos C. \end{aligned}$$

(*b*) $\angle C > 90°$

$|CD| = x \Rightarrow |BD| = a + x$.
In $\triangle ABD$, $c^2 = |BD|^2 + |AD|^2$

$$= (a + x)^2 + |AD|^2 = a^2 + 2ax + x^2 + |AD|^2.$$

But $x^2 + |AD|^2 = b^2$ and $\cos C = -\cos \angle ACD = -\dfrac{|CD|}{|CA|} = -\dfrac{x}{b}$

$$\Rightarrow x = -b \cos C.$$

Therefore, $c^2 = a^2 + 2a(-\,b \cos C) + b^2 = a^2 + b^2 - 2ab \cos C$.

Solving Triangles by the Cosine Rule

The Cosine Rule is a relation between the lengths of the three sides of a triangle and one angle, and is therefore used to solve triangles if any three of these four 'facts' are known, in particular,

(*a*) if we know the lengths of three sides; or
(*b*) if we know the lengths of two sides and the size of the included angle.

Examples

1. In $\triangle ABC, a = 12{\cdot}4$ cm, $b = 15{\cdot}6$ cm and $c = 18$ cm.
To find $\angle B$, we use the equation which involves $\angle B$.

$$b^2 = a^2 + c^2 - 2ac \cos B$$
$$\Rightarrow 2ac \cos B = a^2 + c^2 - b^2$$
$$\Rightarrow \cos B = \frac{a^2 + c^2 - b^2}{2ac}$$
$$\Rightarrow \cos B = \frac{12{\cdot}4^2 + 18^2 - 15{\cdot}6^2}{2 \times 12{\cdot}4 \times 18}$$
$$= \frac{153{\cdot}8 + 324 - 243{\cdot}4}{446{\cdot}4} \text{ (using the table of squares)}$$
$$\Rightarrow \cos B = 0{\cdot}5251 \Rightarrow \angle B = 58^\circ\ 20'.$$

If $\triangle ABC$ is to be solved, that is, if the size of all the three angles of $\triangle ABC$ are to be found, the Cosine Rule is used in this way to find the size of one angle. We could use it again to find the remaining angles, but since now one of the angles is known, it is usually quicker to use the Sine Rule to complete the solution.

We find the *smaller angle*, opposite the smaller side ($a < c$), since this angle must be acute and there is no possible ambiguity.

$$\frac{a}{\sin A} = \frac{b}{\sin B} \Rightarrow \frac{12{\cdot}4}{\sin A} = \frac{15{\cdot}6}{\sin 58^\circ\ 20'}$$
$$\Rightarrow \sin A = \frac{12{\cdot}4 \times \sin 58^\circ\ 20'}{15{\cdot}6}$$
$$\Rightarrow \angle A = 42^\circ\ 34' \Rightarrow \angle C = 180^\circ - (58^\circ\ 20' + 42^\circ\ 34') = 79^\circ\ 6'.$$

2. In $\triangle ABC, \angle A = 33^\circ\ 12', b = 7{\cdot}62$ cm and $c = 9{\cdot}36$ cm.
To solve $\triangle ABC$ we must first use the Cosine Rule. The equation which involves all three known 'facts' is

$$a^2 = b^2 + c^2 - 2bc \cos A = (58{\cdot}06 + 87{\cdot}61 - 119{\cdot}3) = 26{\cdot}37$$
$$\Rightarrow a = 5{\cdot}14 \text{ cm}.$$

Now we use the Sine Rule to find $\angle B$ ($\angle B < \angle C$, since $b < c$):

$$\frac{a}{\sin A} = \frac{b}{\sin B} \Rightarrow \frac{5{\cdot}14}{\sin 33^\circ\ 12'} = \frac{7{\cdot}62}{\sin B}$$
$$\Rightarrow \sin B = \frac{7{\cdot}62 \times \sin 33^\circ\ 12'}{5{\cdot}14}$$
$$\Rightarrow \angle B = 54^\circ\ 15' \Rightarrow \angle C = 180^\circ - (33^\circ\ 12' + 54^\circ\ 15') = 92^\circ\ 33'.$$

Notes.

(*a*) When solving a triangle, the Cosine Rule *must* be used if only the lengths of the three sides of the triangle are given, or if only the lengths of two sides and the size of the included angle are given.

(*b*) Do not use the Cosine Rule twice when solving a triangle. It is always quicker to use the Sine Rule in the second operation.

(*c*) If the lengths of three sides of a triangle and the size of one angle are given, use the Sine Rule to find the *smaller* of the two remaining angles; this avoids ambiguity.

19.25 Exercises

1. Solve $\triangle ABC$ in each case, given that:

   (*a*) $b = 18{\cdot}5$ cm, $\angle A = 72^\circ\ 12'$ and $\angle C = 52^\circ\ 24'$;
   (*b*) $a = 12{\cdot}4$ cm, $b = 16{\cdot}3$ cm and $c = 9{\cdot}6$ cm.

2. Two ships leave a port at noon. One sails on a bearing of 085° and the other on a bearing of 132°. At 3 p.m. the first ship has travelled a distance of 24 nautical miles and the second ship a distance of 28 nautical miles.

   Calculate how far apart these two ships are at 3 p.m. and the bearing of the first ship from the second ship.

19.26 Relations Between Trigonometric Ratios

The alternative proof of Pythagoras' Theorem given in Section 18.15 can also be expressed in terms of trigonometric ratios.

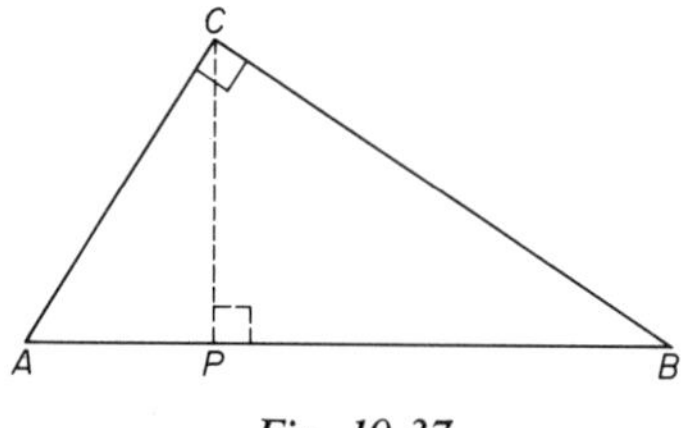

Fig. 19.37

For $\triangle APC$, $\cos A = \dfrac{|AP|}{|AC|} = \dfrac{|AP|}{b}$;

for $\triangle ABC$, $\cos A = \dfrac{b}{c}$

$$\Rightarrow \frac{|AP|}{b} = \frac{b}{c} \Rightarrow b^2 = c.\,|AP| \text{ (by cross-multiplication).}$$

Similarly, for $\triangle BPC$, $\cos B = \dfrac{|BP|}{a}$;

for $\triangle ABC$, $\cos B = \dfrac{a}{c}$

$$\Rightarrow \frac{|BP|}{a} = \frac{a}{c} \Rightarrow a^2 = c \, . \, |BP|.$$

Therefore, $a^2 + b^2 = c\,|BP| + c|AP| = c(|AP| + |BP|) = c \, . \, c = c^2$

$$\Rightarrow a^2 + b^2 = c^2.$$

Pythagoras' Theorem expresses a fundamental relation between the sine and cosine of an angle:

$$a^2 + b^2 = c^2 \Rightarrow \frac{a^2}{c^2} + \frac{b^2}{c^2} = \frac{c^2}{c^2} = 1 \text{ (dividing both sides by } c^2\text{)}$$

$$\Rightarrow \left(\frac{a}{c}\right)^2 + \left(\frac{b}{c}\right)^2 = 1$$

$$\Rightarrow (\sin A)^2 + (\cos A)^2 = 1.$$

Note. The terms $(\sin A)^2$ and $(\cos A)^2$ are usually written as $\sin^2 A$ and $\cos^2 A$, to avoid confusion with $\sin A^2$ (= the sine of A^2).

We therefore write down this fundamental relation between sine and cosine (for any angle of $\alpha°$):

$$\sin^2 \alpha° + \cos^2 \alpha° = 1.$$

This relation enables us to calculate other trigonometric ratios if the value of the sine or the cosine of an angle is known. For example, suppose that the sine of an acute angle is $\frac{3}{5}$, that is, $\sin \alpha° = \frac{3}{5}$.

Then, $\sin^2 \alpha° + \cos^2 \alpha° = 1 \Rightarrow (\frac{3}{5})^2 + \cos^2 \alpha° = 1$

$$\Rightarrow \cos^2 \alpha° = 1 - \tfrac{9}{25} = \tfrac{16}{25} \Rightarrow \cos \alpha° = \tfrac{4}{5};$$

$$\tan \alpha° = \frac{\sin \alpha°}{\cos \alpha°} = \tfrac{3}{5}/\tfrac{4}{5} = \tfrac{3}{4}.$$

19.27 Exercises

1. If $\cos \beta° = \frac{12}{13}$, find $\sin \beta°$ and $\tan \beta°$.
2. $\sin 30° = \frac{1}{2}$. Use the relations in Section 19.26 to calculate $\cos 30°$ and $\tan 30°$.

19.28 Summary

Relations between ratios

$$\sin \alpha° = \cos (90° - \alpha°);$$
$$\cos \alpha° = \sin (90° - \alpha°);$$
$$\tan \alpha° = \frac{\sin \alpha°}{\cos \alpha°};$$
$$\sin^2 \alpha° + \cos^2 \alpha° = 1.$$

In any $\triangle ABC$

$$\frac{a}{\sin A} = \frac{b}{\sin B} = \frac{c}{\sin C} = 2R \text{ (Sine Rule);}$$
$$a^2 = b^2 + c^2 - 2bc \cos A, \text{ etc. (Cosine Rule).}$$

Sines and cosines of obtuse angles

$$\sin \alpha^\circ = \sin (180^\circ - \alpha^\circ);$$
$$\cos \alpha^\circ = -\cos (180^\circ - \alpha^\circ).$$

Some important ratios

$$\sin 45^\circ = \cos 45^\circ = \frac{1}{\sqrt{2}};$$
$$\tan 45^\circ = 1;$$
$$\sin 30^\circ = \cos 60^\circ = \tfrac{1}{2};$$
$$\sin 60^\circ = \cos 30^\circ = \frac{\sqrt{3}}{2};$$
$$\tan 30^\circ = \cot 60^\circ = \frac{1}{\sqrt{3}};$$
$$\tan 60^\circ = \cot 30^\circ = \sqrt{3}.$$

Unit Twenty

Mensuration in Two Dimensions

20.1 Space and Size

One advantage of taking a motor-car on holiday, rather than touring by rail, is the car's relative freedom of motion compared with that of the train. The car is free to move along the ground in any direction in which we choose to drive it, but the train is confined to its rails and can only move forwards or backwards along the line. When we travel by air we have an even greater freedom of motion, since the aircraft can move upwards and downwards as well as forwards and sideways.

The very limited freedom of motion of the train illustrates the restrictions of what mathematicians call a *one-dimensional space*, the world of the straight line. A straight line is an infinitely long space: that is, a point moving along it meets with no restriction however far it travels. A 'closed' part of this space is the *line segment* and its size is termed *length*.

The geometrical shapes we have met so far, such as triangles, squares, circles and so on, are all closed *plane* shapes, or to put it simply, *flat* shapes, parts of a *two-dimensional space*, the plane. The size of such a plane shape, the amount of space or surface it occupies, is called the *area* of the shape. Plane geometry deals with a world of pictures; although there may be an illusion of depth there is no depth as we understand it. Everything in this world is flat.

The real world in which we live is a *three-dimensional space*. We have said already that we cannot draw or even imagine a line or a plane of 'no thickness' (see Section 14.3); all things in this world—solid, liquid or gaseous—extend in three dimensions. The amount of three-dimensional space an object occupies is called its *volume*.

We cannot evaluate the size of a shape, i.e. its length, area or volume, unless we choose specific *units of measure*. The most convenient shape for the unit of area is the square, and the most convenient shape for the unit of volume, the cube.

We discussed the internationally agreed units of measure, *standard measures*, in Unit 5, together with the problems of physically measuring shapes.

Mensuration (a term derived from the Latin word for measurement)

In this Unit we will consider the relations between the areas of two-dimensional shapes and the lengths of relevant line segments; in Unit 21 we shall be concerned with the relations between the volumes of three-dimensional shapes, and the areas of relevant 'faces' of these shapes and the lengths of line segments.

These relations, expressed as equations, provide useful formulae and methods for calculating areas and volumes. For shapes bounded by straight lines and by planes, these formulae are exact and precise, unlike the results of physical measurement or construction which must always be approximations within the limits of accuracy of the measuring instruments.

20.2 Area

If a plane shape is drawn on squared paper, its area can be found just by counting the number of unit squares within the shape, allowing also for the parts of the unit squares enclosed.

Area of the rectangle

Counting unit squares is particularly simple when the shape is a rectangle and the lengths of the sides are expressed in whole numbers. The rectangle then consists of a number of rows of unit squares; Fig. 20.1 shows a rectangle containing 3 rows of 5 unit squares,

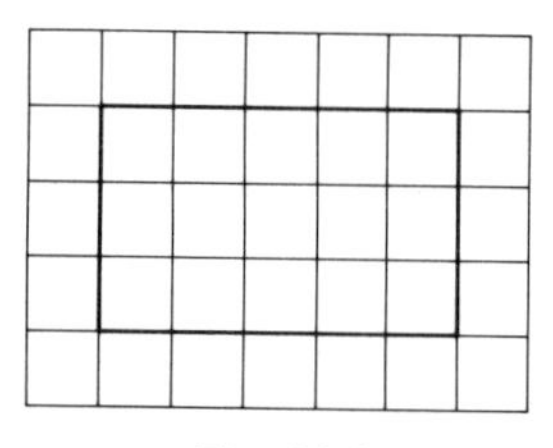

Fig. 20.1

that is, a total number of 5×3 unit squares $= 15$ unit squares. The number of units in one row is determined by the length of one side of the rectangle, and the number of rows by the length of the adjacent side. The length of the longer side is usually called the *length* of the rectangle and the length of the shorter side its *breadth*, although in calculating the area of a rectangle it does not matter how the sides are named (compare the commutative law for multiplication, mentioned in Exercise 3 of Section 3.4).

In general, *if the length of a rectangle is l unit lengths and its breadth is b unit lengths, the area of the rectangle is $l \times b$ unit squares*, or

$$\boxed{\text{area of rectangle} = l \times b}$$

It is clear that the same unit must be used to measure both sides of the rectangle, and that the unit of area is related to the unit length. For example, if the lengths of sides of the rectangle are 2 dm (= 20 cm) and 6 cm, its area is said to be either (20×6) square centimetres (written as 120 cm^2) or $(2 \times 0{\cdot}6)$ dm^2 $= 1{\cdot}2$ dm^2.

20.3 Exercises

1. Show, by drawing on squared paper and counting squares, that the area of a rectangle measuring 8 cm by $2\frac{1}{2}$ cm is $(8 \times 2\frac{1}{2})$ cm² = 20 cm².
2. In the same way, consider the area of a rectangle whose length is $6\frac{1}{2}$ cm and breadth $4\frac{1}{2}$ cm, and confirm that the formula for the area of a rectangle ($A = l \times b$) also applies when the lengths of the sides are not whole numbers of units.
3. Write down a formula for the area of a square whose sides are a unit lengths each.

20.4 Constructing a Square Equal in Area to a Given Rectangle

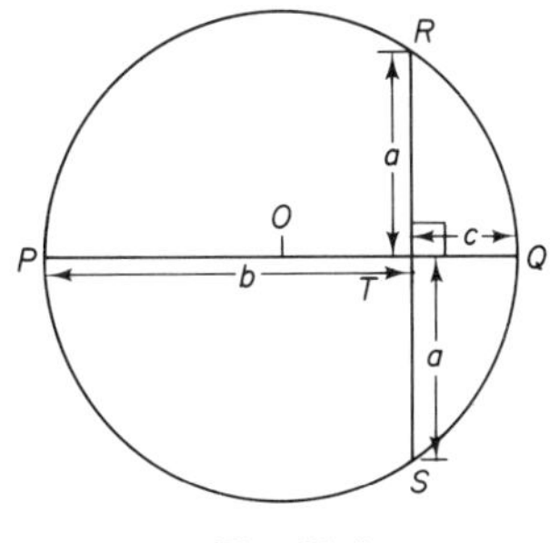

Fig. 20.2

Fig. 20.2 shows a special case of the Intersecting Chord Theorem (Theorem 18.3 in Section 18.15): PQ is a diameter of the circle and RS is a chord perpendicular to PQ. We have seen already that $|RT| = |ST|$ ($\triangle ROS$ is an isosceles triangle and OT is the altitude, median and perpendicular bisector in relation to RS).

If $|PT| = b$ unit lengths, $|TQ| = c$ unit lengths and $|RT| = |ST| = a$ unit lengths, then the Intersecting Chord Theorem states that $b \cdot c = a \cdot a = a^2$.

Therefore, a rectangle whose adjacent sides are b and c unit lengths has the same area as the square whose sides are each a unit lengths; Construction 20.1 (Fig. 20.3) is based on this relationship.

Construction 20.1: *A square equal in area to a given rectangle.*

(*a*) Mark points P, T and Q along a line so that $|PT| = b$ unit lengths and $|TQ| = c$ unit lengths.
(*b*) Find, by construction, the mid-point O of the line segment PQ.
(*c*) Draw a circle, centre O and radius $|OP|$.
(*d*) Construct a line through point T perpendicular to PQ, intersecting the circumference of the circle at R and S.
(*e*) Complete the square $RTUV$.

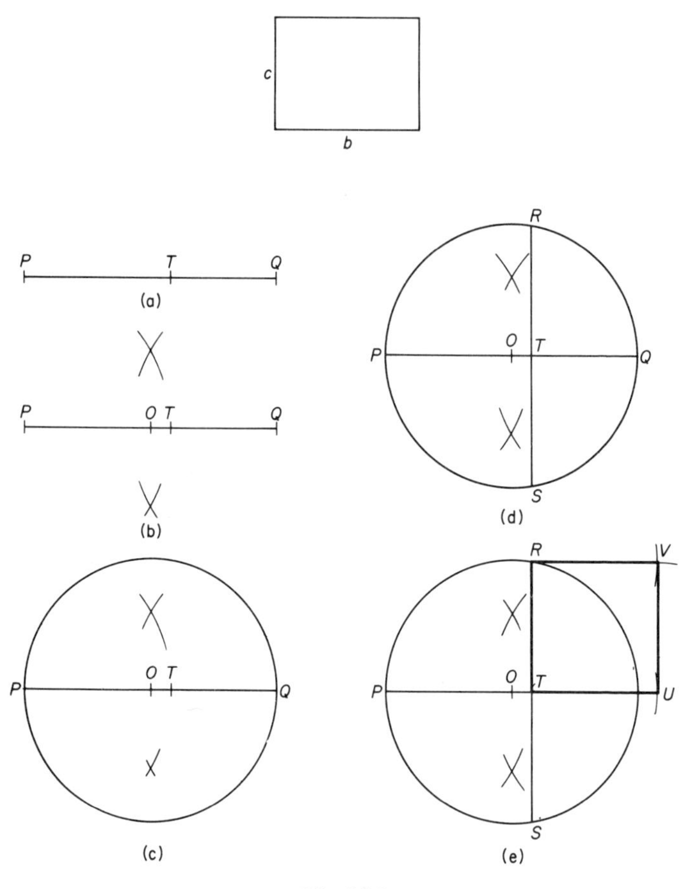

Fig. 20.3

20.5 Exercises

1. Draw any rectangle and construct a square equal in area to this rectangle.
2. Use the same construction to find a line segment measuring $\sqrt{15}$ unit lengths. [Hint: the area of a rectangle of size 5 cm by 3 cm = the area of the square whose sides are of length $\sqrt{15}$ cm each.]

20.6 The Area of a Parallelogram

In a parallelogram *PQRS*, lines *PT* and *SV* are drawn perpendicular to the line *QR*:

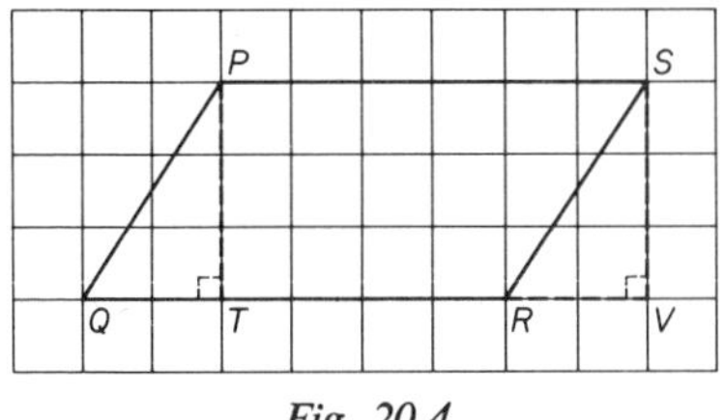

Fig. 20.4

$\triangle$s $\begin{matrix} PQT \\ SRV \end{matrix}$ are congruent (S–A–A)
since $PQ = SR$ (opposite sides of a parallelogram);
$\angle Q = \angle R$ (corresponding angles);
$\angle T = \angle V = 90°$.
We can show, by cutting out $\triangle PQT$ and placing it over $\triangle SRV$ (Exercise 2 of Section 16.18), that the area of the parallelogram $PQRS$ = area of the rectangle $PTVS$.

Fig. 20.4 also makes it clear that:
area of parallelogram $PQRS$ = area of $\triangle PQT$ + area of quadrilateral $PTRS$;
but area of $\triangle PQT$ = area of $\triangle SRV$;
therefore:
area of parallelogram $PQRS$ = area of $\triangle SRV$ + area of quadrilateral $PTRS$
= area of rectangle $PTVS$.

The 'length' of the rectangle $PTVS$ is $|TV|$ or $|PS|$ (= the length of the side of the parallelogram); its 'breadth' is $|PT|$ or $|SV|$ (= the perpendicular distance between the parallel sides PS and QR, or 'the perpendicular height' of the parallelogram, relative to the side QR as base).
In general, *if the length of a side of a parallelogram is b units of length, and the height of the parallelogram (relative to this base) is h units of length, the area of the parallelogram is b × h units of area*, or

$$\boxed{\text{area of parallelogram} = b \times h}$$

(In words, the area of a parallelogram is 'base' × 'height'.)

If the lengths of the sides of a parallelogram and the size of one of its angles are known, the height and the area of the parallelogram can be calculated *trigonometrically*.

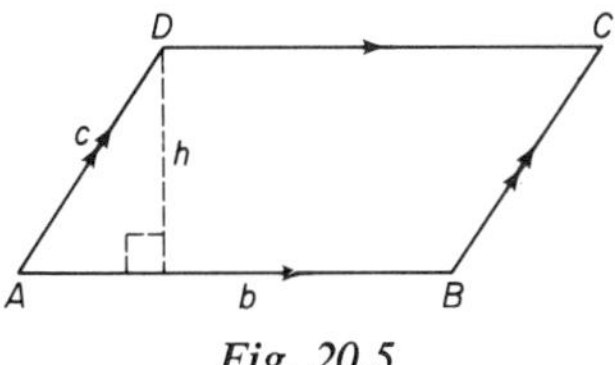

Fig. 20.5

Fig. 20.5 shows a parallelogram $ABCD$ whose sides are b and c units of length and whose 'height' is h units of length.

Then $\sin A = \frac{h}{c} \Rightarrow h = c \sin A$.

Therefore, *the area of a parallelogram is* $(b \times h)$ *units of area* $= b \times c \sin A$ *units of area*, or

$$\boxed{\text{area of parallelogram} = b \,.\, c \sin A}$$

(Since the angles of the parallelogram are $\angle A$, $\angle B$ $(= 180° - \angle A)$, $\angle C$ $(= \angle A)$ and $\angle D$ $(= 180° - \angle A)$, and $\sin A = \sin (180° - A)$, it is immaterial which angle of the parallelogram is used, but b and c must represent lengths of two *non-parallel* sides.)

20.7 Exercises

1. In parallelogram $ABCD$, $|AB| = 12{\cdot}5$ cm and the distance between the parallel sides AB and DC is 5 cm.
   Calculate the area of the parallelogram.
2. $PQRS$ is a parallelogram; $|PQ| = 8{\cdot}5$ cm, $|QR| = 6{\cdot}4$ cm and $\angle P = 52°$. Calculate the area of $PQRS$.
3. Draw a parallelogram. Construct a rectangle equal in area to this parallelogram.

20.8 Parallelograms which are Equal in Area

If the length of only one side of a parallelogram (b units of length) and the height or altitude (h units of length) relative to this side are given, the parallelogram is not uniquely determined, that is to say, not 'fixed' in shape.

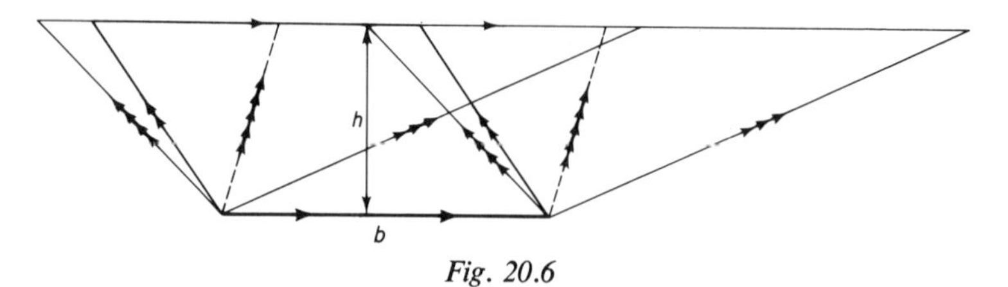

Fig. 20.6

An infinitely large number of parallelograms can be drawn with one side of b unit lengths and a 'height' of h unit lengths, and the area of each of these parallelograms is the same, that is $(b \times h)$ units of area. In general,

Theorem 20.1: *Parallelograms on the same base and between the same parallels are equal in area.*

We can demonstrate this theorem without using the formula for the area of the parallelogram.

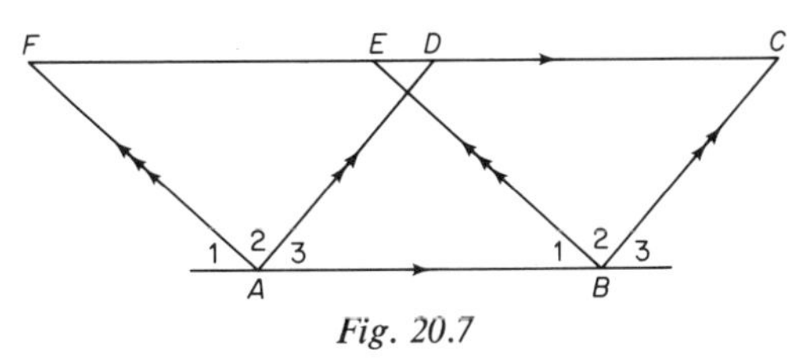

Fig. 20.7

$ABCD$ and $ABEF$ in Fig. 20.7 are parallelograms and C, D, E and F are collinear (i.e. they lie on the same straight line).

$\triangle$s $\begin{matrix} ADF \\ BCE \end{matrix}$ are congruent (S–A–S), since

$$\begin{aligned} |AF| &= |BE| \text{ (opposite sides of parallelogram } ABEF), \\ |AD| &= |BC| \text{ (opposite sides of parallelogram } ABCD), \\ \angle A_1 &= \angle B_1 \quad \text{and} \quad \angle A_3 = \angle B_3 \text{ (corresponding angles)}, \\ \angle A_2 &= 180^\circ - (\angle A_1 + \angle A_3) = 180^\circ - (\angle B_1 + \angle B_3) = \angle B_2, \end{aligned}$$

$$\Rightarrow \text{area of } \triangle ADF = \text{area of } \triangle BCE.$$

If we remove $\triangle ADF$ from the quadrilateral $ABCF$, we are left with the parallelogram $ABCD$; removing $\triangle BCE$ from the same quadrilateral, we are left with the parallelogram $ABEF$; or

$$\begin{aligned} \text{area of parallelogram } ABCD &= \text{area of } ABCF - \text{area of } \triangle ADF, \text{and} \\ \text{area of parallelogram } ABEF &= \text{area of } ABCF - \text{area of } \triangle BCE \\ &= \text{area of } ABCF - \text{area of } \triangle ADF. \end{aligned}$$

Therefore, area of parallelogram $ABCD$ = area of parallelogram $ABEF$.

20.9 The Area of a Triangle

We draw any triangle ABC, and then draw two lines through the vertices B and C, parallel to AC and AB respectively (Fig. 20.8).

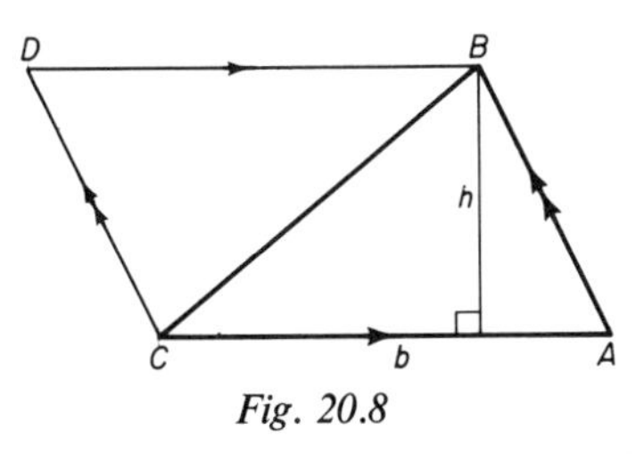

Fig. 20.8

If these two lines intersect at the point D, then $ABDC$ is a parallelogram (see Exercise 1 of Section 16.18). BC is a diagonal, and divides the parallelogram into two *congruent* triangles, $\triangle$s $\begin{matrix} ABC \\ DCB \end{matrix}$. The area of $\triangle ABC$ is therefore half the area of the parallelogram $ABDC$.

If $AC = b$ units of length and the height of the parallelogram is h units of length, the area of the parallelogram is $b \times h$ units of area. The area of $\triangle ABC$ is therefore $\frac{1}{2}(b \times h)$ units of area.

In general, *if the length of a side of a triangle is b units of length and the perpendicular height of the triangle relative to this side is h units of length, the area of the triangle is* $\frac{1}{2}(b \times h)$ *units of area*, that is, half the area of the rectangle whose sides are b and h units of length:

$$\boxed{\text{area of triangle} = \tfrac{1}{2}(b \times h)}$$

(In words, the area of a triangle is half 'base' $\times$ 'height'.)

This relation is shown clearly for several different triangles in Fig. 20.9.

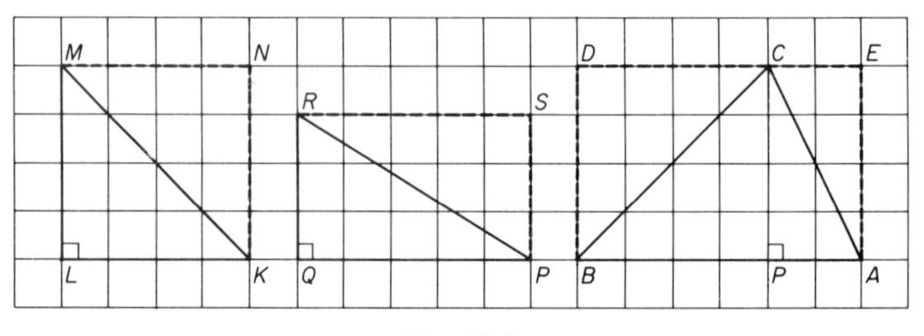

Fig. 20.9

The area of $\triangle KLM$ can be found by counting the whole and half unit squares which make it up, remembering that the hypotenuse MK divides each of the unit squares of the grid through which it passes into two congruent triangles.

The dotted lines MN and KN suggest another method:

$$\begin{aligned}\text{area of } \triangle KLM = \text{half the area of the square } KLMN &= \tfrac{1}{2} \times 16\\ &= 8 \text{ unit squares.}\end{aligned}$$

We can use this method to calculate the area of $\triangle PQR$:

area of $\triangle PQR = \frac{1}{2}$ area of rectangle $PQRS = \frac{1}{2}(5 \times 3) = 7\frac{1}{2}$ unit squares.

Any triangle ABC is divided by its altitude CP into two right-angled triangles, $\triangle BPC$ and $\triangle APC$; then

area of $\triangle BPC = \frac{1}{2}$ area of rectangle $BPCD$,

and area of $\triangle APC = \frac{1}{2}$ area of rectangle $APCE$.

Adding, area of $\triangle ABC = \frac{1}{2}$ area of rectangle $ABDE$.

20.10 Exercises

In $\triangle ABC$ in Fig. 20.10, M and N are the mid-points of the sides AC and BC respectively. The perpendiculars to AB through A, B and C intersect the line MN at the points Q, S and R.

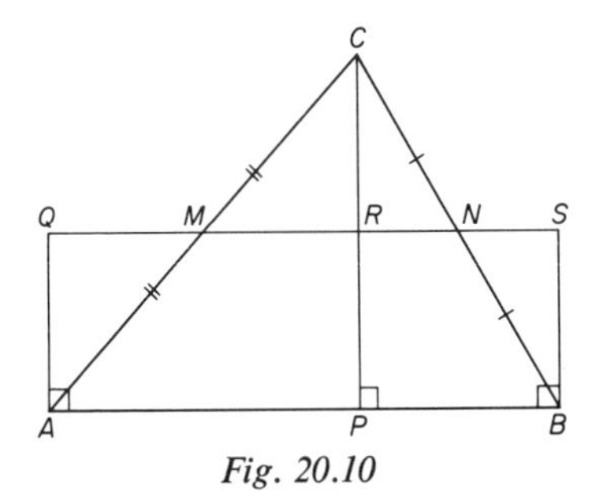

Fig. 20.10

1. Show that $\triangle$s $\frac{AMQ}{CMR}$ are congruent (A–S–A) and that therefore $|AQ| = |CR| = \frac{1}{2}|CP|$.
2. Similarly, show that $\triangle$s $\frac{BNS}{CNR}$ are congruent and that $|BS| = |CR| = \frac{1}{2}|CP|$.
3. Show that the area of $\triangle ABC$ = the area of rectangle $ABSQ$, that is, *the area of the triangle = the area of the rectangle which has the same base and half the perpendicular height of the triangle.*

   Confirm this for yourself by cutting out $\triangle ABC$, and cutting it first along the line MN, and then along the line CR, to form three pieces (Fig. 20.11).

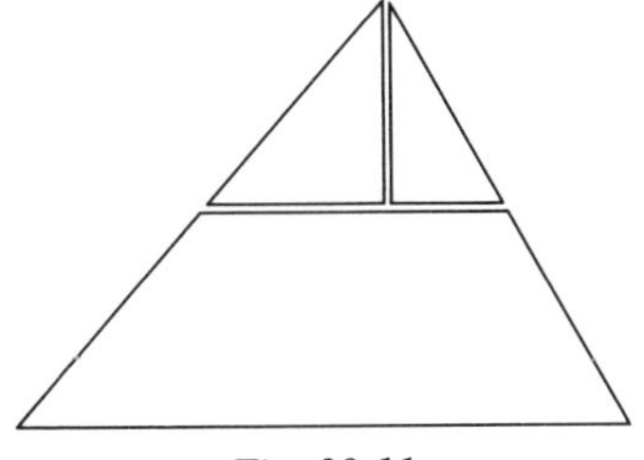

Fig. 20.11

Now put these three pieces together again so that they form a rectangle; obviously, its area is the same as that of $\triangle ABC$.

20.11 Constructing Rectangles (1): of Equal Area to a Given Triangle

Construction 20.2: *A rectangle equal in area to a given triangle ABC.*

(*a*) Find by construction the mid-point of one of the sides of the triangle—say the mid-point M of AB.
(*b*) Draw a line through the point M parallel to one of the other sides—say parallel to BC, meeting the side AC at N.
(*c*) Through the vertices B and C, draw perpendiculars to the line BC meeting the line MN at points Q and S.

The area of rectangle $BCSQ$ = area of $\triangle ABC$.

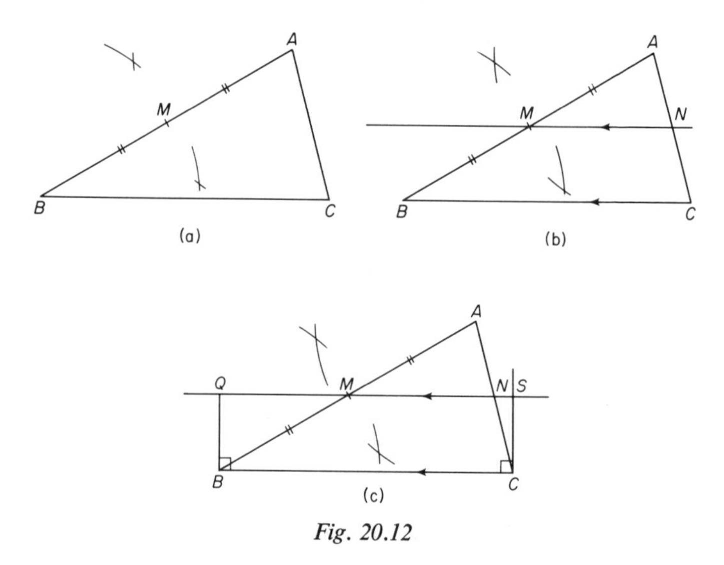

Fig. 20.12

20.12 Exercises

(*a*) Draw any triangle and construct a rectangle equal in area to this triangle.

(*b*) Show that Construction 20.2 can be used if one of the angles of $\triangle ABC$ is obtuse. Draw $\triangle ABC$, such that $\angle B > 90°$, and follow the instructions as before.

(*c*) Combine Construction 20.2 and Construction 20.1 to construct a *square* equal in area to a given triangle. [Hint: first construct a rectangle of the same area as the triangle, then construct a square equal in area to this rectangle.]

20.13 Constructing Rectangles (2): of Equal Area to a Given Quadrilateral

Construction 20.3: *A rectangle equal in area to a given quadrilateral.*

The diagonals of a quadrilateral each divide it into two triangles; for example, in Fig. 20.13 the diagonal AC divides the quadrilateral $ABCD$ into the triangles ABC and ADC.

Construction 20.2 is used to construct a rectangle equal in area to $\triangle ABC$, and another rectangle equal in area to $\triangle ADC$. These two rectangles together

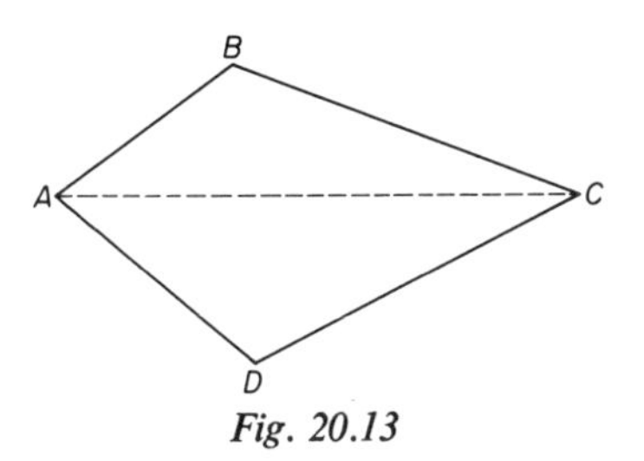

Fig. 20.13

form a third rectangle, which is equal in area to the quadrilateral *ABCD*. Fig. 20.14 shows the procedure.

(*a*) In the given quadrilateral *ABCD* draw a diagonal, for example, *AC*.
(*b*) Find the mid-points of *AB* and *AD* by construction.
(*c*) Draw lines through these two mid-points parallel to *AC*.
(*d*) Through the points *A* and *C* construct perpendiculars to *AC*.

The four lines drawn in stages (*c*) and (*d*) form a rectangle *PQRS* which is equal in area to the quadrilateral *ABCD*.

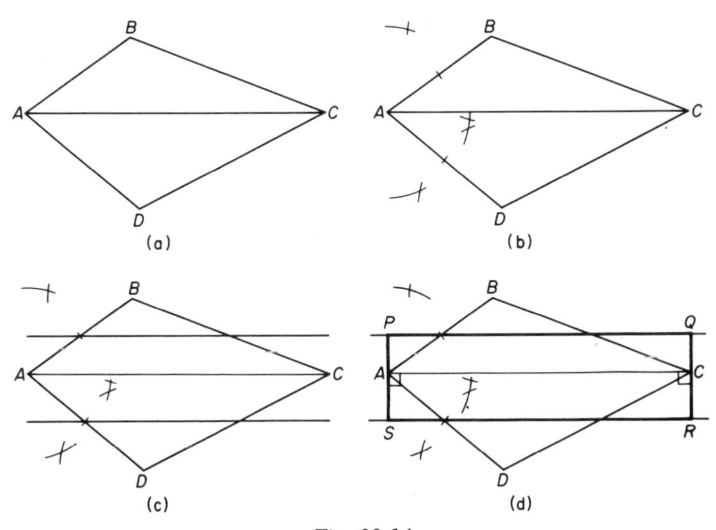

Fig. 20.14

20.14 Exercises

1. Draw a quadrilateral *ABCD*.

   (*a*) Construct a rectangle equal in area to *ABCD*.
   (*b*) Now use Construction 20.1 to construct a square equal in area to *ABCD*.

2. Draw a rhombus. Use Construction 20.3 to construct a rectangle equal in area to this rhombus.

20.15 Some Results and Applications

We can think of any triangle ABC as forming half of a parallelogram $ABCD$ of which AC is a diagonal; and we know too that the area of a parallelogram which has sides of length b and c respectively $= \frac{1}{2}b \cdot c \sin A$ (compare Section 20.6). We can therefore say two things concerning the areas of triangles:

(*a*) *If the lengths of two sides of a triangle are b and c unit lengths respectively, and A is the angle included by these sides, the area of the triangle is* $\frac{1}{2}b \cdot c \sin A$ *units of area.*

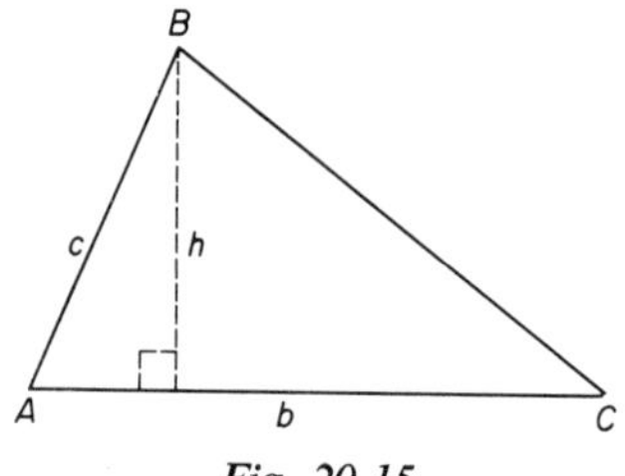

Fig. 20.15

This is really a re-statement of the formula for the area of a triangle which we discused in Section 20.9, area $= \frac{1}{2}b \times h$, in which h is expressed trigonometrically, $h = c \sin A$ (Fig. 20.15).

Then area of the triangle $= \frac{1}{2}b \times h$ units of area $= \frac{1}{2}b \cdot c \sin A$ units of area:

$$\text{area of triangle} = \tfrac{1}{2}b \cdot c \sin A$$

(*b*) *Triangles on the same base and having the same 'height' are equal in area* (compare Theorem 20.1, which states that parallelograms on the same base and between the same parallels are equal in area).

The number of possible triangles with base AB and a height of h unit lengths is unlimited; Fig. 20.16 shows just a few.

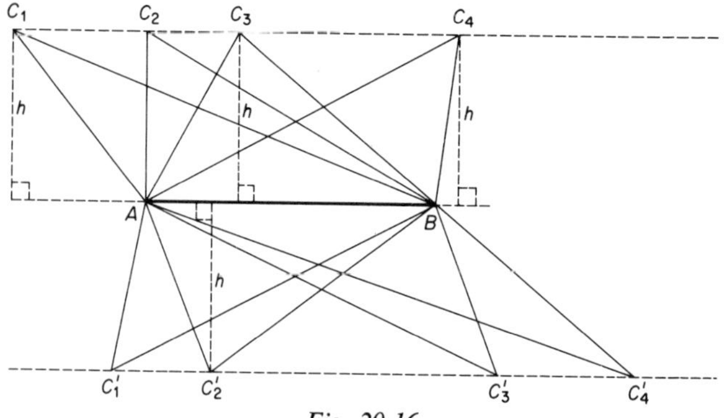

Fig. 20.16

The lines parallel to AB at a distance of h units from the line AB are the *locus* of all points C whose perpendicular distance from AB is h unit lengths. Each of these triangles ABC has the same area: $\frac{1}{2}(b \times h)$ units of area.

We can use this result to 'straighten out' corners of polygons without altering their area:

Construction 20.4: *To reduce a polygon with N sides to a polygon with* $(N - 1)$ *sides having the same area.*

As an example, we will construct a quadrilateral $APDE$ whose area equals that of a given pentagon $ABCDE$ by 'straightening out' angle C.

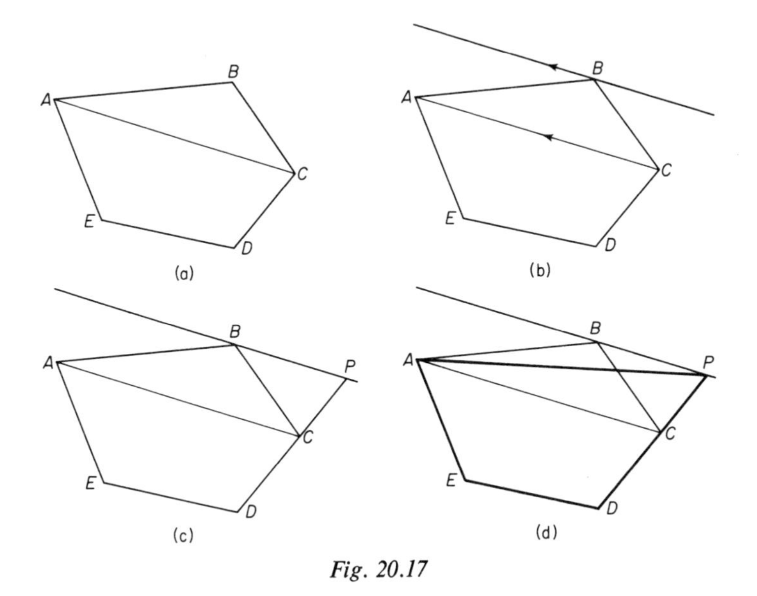

Fig. 20.17

(*a*) Draw the diagonal of the pentagon, AC.
(*b*) Draw a line through the point B parallel to AC.
(*c*) Extend the line DC to meet this line at P.
(*d*) Join points A and P.

$$\begin{aligned}
\text{Since area of } \triangle ABC &= \text{area of } \triangle APC \text{ (from (}b\text{) above)},\\
\text{area of pentagon } ABCDE &= \text{area of } \triangle ABC + \text{area of } ACDE\\
&= \text{area of } \triangle APC + \text{area of } ACDE\\
&= \text{area of quadrilateral } APDE.
\end{aligned}$$

In the same way, a hexagon can be converted to a pentagon of the same area, a heptagon to a hexagon of the same area, and so on. Indeed, we could repeat this process so as to reduce a polygon of any number of sides in the end to a triangle.

20.16 Exercises

1. Draw any hexagon; then construct a pentagon of equal area.
2. Construct a square equal in area to a given pentagon. [Hint: first reduce the number of sides from five to four (Construction 20.4); then construct a rectangle equal in area to this quadrilateral (Construction 20.3); finally convert the rectangle into a square of equal area (Construction 20.1)].
3. Two sides of a triangle are 7·5 cm and 9·2 cm long; the size of the angle included by these two sides is 47° 18′. Calculate the area of the triangle (to 3 SF). [Hint: Since two sides and the included angle are given, use the expression $\frac{1}{2}bc \sin A$, with $b = 7{\cdot}5$ cm, $c = 9{\cdot}2$ cm and $\angle A = 47° 18'$.]

Note. We can calculate the area of a triangle using any side as 'the base', provided the perpendicular height is the height *relative to this base.* Fig. 20.18 shows that:

$$\text{area of triangle } ABC = \tfrac{1}{2}a \,.\, h_1 = \tfrac{1}{2}b \,.\, h_2 = \tfrac{1}{2}c \,.\, h_3.$$

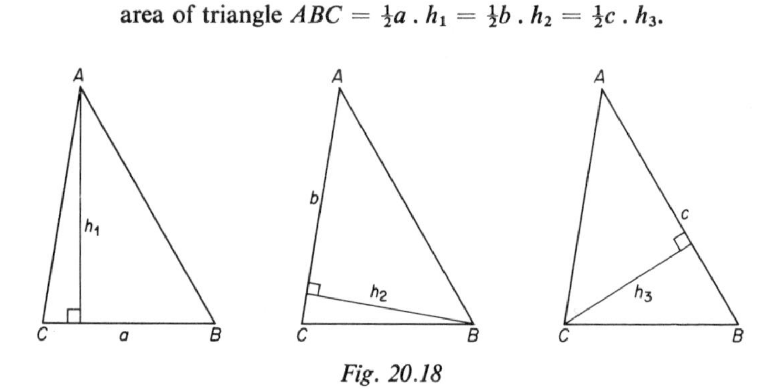

Fig. 20.18

Our choice of base and relative height depends on the information available.

Similarly, the area of the triangle $ABC = \frac{1}{2}ab \sin C = \frac{1}{2}ac \sin B = \frac{1}{2}bc \sin A$; we find the area using whichever of these relations is the most convenient.

20.17 The Area of a Trapezium

A trapezium is a special kind of quadrilateral; it is a quadrilateral which has two parallel sides. A diagonal divides a trapezium into two triangles; the area of the trapezium can be calculated as the sum of the areas of these tri-

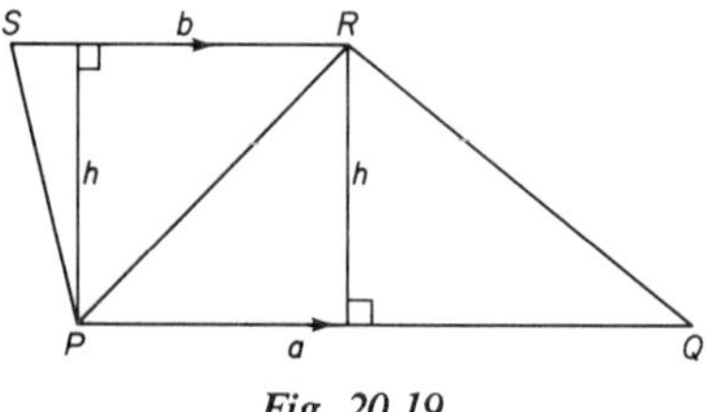

Fig. 20.19

angles if the lengths of the two parallel sides and the distance between them are known.

In the trapezium $PQRS$ in Fig. 20.19, $|PQ| = a$ units of length, $|RS| = b$ units of length and the distance between the parallel sides PQ and SR is h units of length.

$$\begin{aligned} \text{The area of } \triangle PQR &= \tfrac{1}{2}a \,.\, h \text{ units of area} \\ \text{the area of } \triangle PRS &= \tfrac{1}{2}b \,.\, h \text{ units of area} \\ \text{Adding, the area of the trapezium } PQRS &= (\tfrac{1}{2}a \,.\, h + \tfrac{1}{2}b \,.\, h) \text{ units of area} \\ &= \tfrac{1}{2}(a + b)h \text{ units of area, or,} \end{aligned}$$

$$\boxed{\text{area of trapezium} = \tfrac{1}{2}(a + b)h}$$

(In words, the area of a trapezium is 'half the sum of the parallel sides' × 'height'.)

20.18 The Area of a Rhombus

The area of a rhombus, like that of any other parallelogram, is related to the length of a side (b) and the distance between two parallel sides, by the formula

$$\text{area} = b \times h.$$

Using trigonometry,

$h = b \sin A$, and therefore area of a rhombus $= b \times b \sin A$

that is

$$\boxed{\text{area of rhombus} = b^2 \sin A}$$

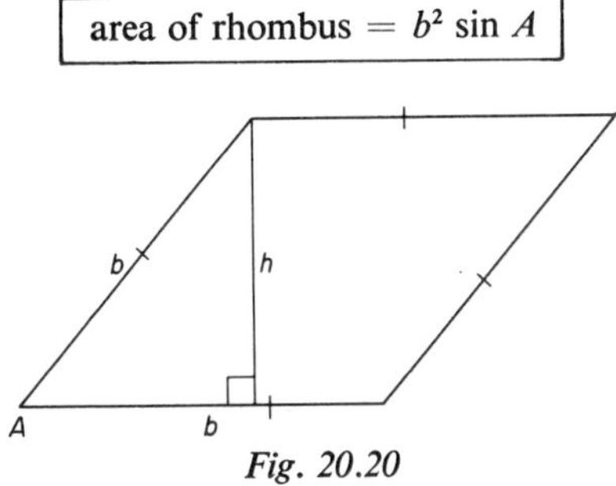

Fig. 20.20

When you constructed a rectangle equal in area to a given rhombus in Exercise 2 of Section 20.14, you may have noticed another relation: *if the lengths of the two diagonals of a rhombus are x and y units of length, the area of the rhombus is $\frac{1}{2}x \,.\, y$ units of area.*

In the rhombus $PQRS$ (Fig. 20.21) the diagonals intersect at a point O, $|PO| = |OR|$ and PR and QS intersect at right angles (Section 16.26). Therefore, PO is the altitude of $\triangle PQS$, relative to the base QS, and OR is the altitude of $\triangle RQS$, relative to the base QS.

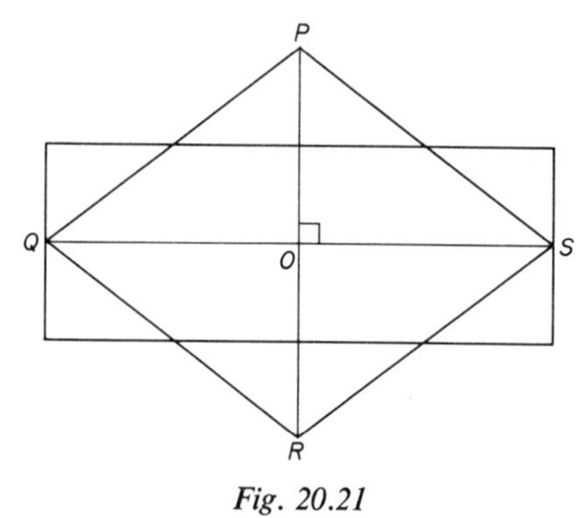

Fig. 20.21

Then area of $\triangle PQS = \frac{1}{2}|QS| \cdot |OP|$ units of area
and area of $\triangle RQS = \frac{1}{2}|QS| \cdot |OR|$ units of area
Adding, the area of the rhombus $PQRS = \frac{1}{2}|QS|(|OP| + |OR'|)$ units of area
$= \frac{1}{2}|QS| \cdot |PR|$ units of area.

If $|QS| = x$ units of length and $|PR| = y$ units of length, area of the rhombus $= \frac{1}{2}x \cdot y$ units of area.

$$\boxed{\text{area of rhombus} = \tfrac{1}{2}x \cdot y}$$

(In words, the area of the rhombus is 'half the product of the diagonals'.)

20.19 Summary

The work on areas in Sections 20.2 to 20.18 can be summarized as follows:

The area of the rectangle $= l \times b$ ('length' × 'breadth')
The area of the parallelogram $= b \times h$ ('base' × 'height')
$= b \cdot c \sin A$
The area of the triangle $= \frac{1}{2}b \times h$ (half 'base' × 'height')
$= \frac{1}{2}b \cdot c \sin A$
The area of the trapezium $= \frac{1}{2}(a + b)\, h$ ('half the sum of the parallel sides' × 'height')
The area of the rhombus $= b \times h$ (as for the parallelogram)
$= b^2 \sin A$
$= \frac{1}{2}x \cdot y$ ('half the product of the diagonals')

The area of any other polygon is found by dividing the polygon into triangles, squares, rectangles or parallelograms, and calculating the area of each part. Fig. 20.22 illustrates how some polygons could be divided for this purpose.

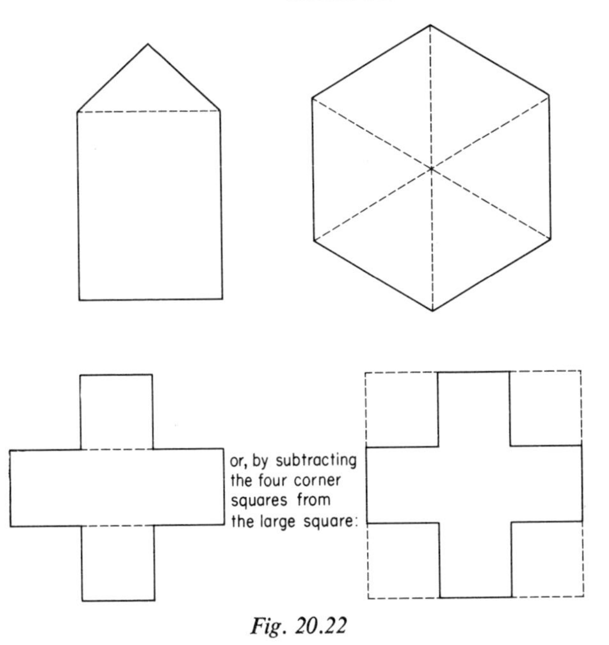

Fig. 20.22

20.20 Exercises

1. Calculate the area of each of the figures in Fig. 20.23, using the information given.

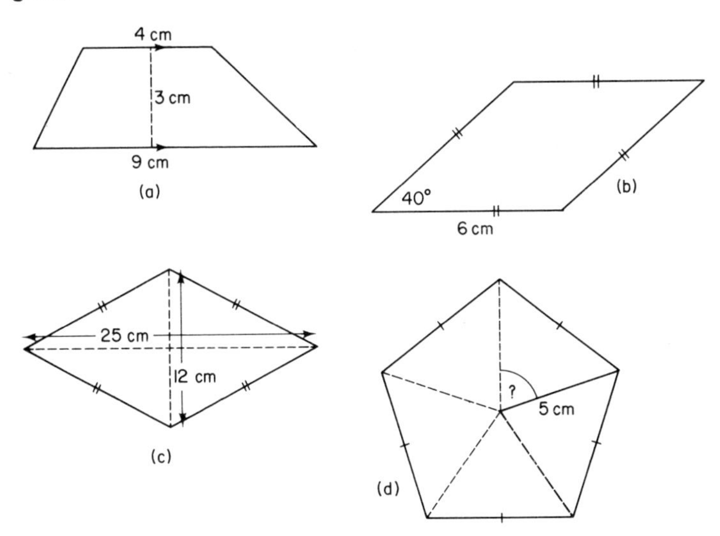

Fig. 20.23

2. Calculate the area of the shaded parts of the figures in Fig. 20.24.

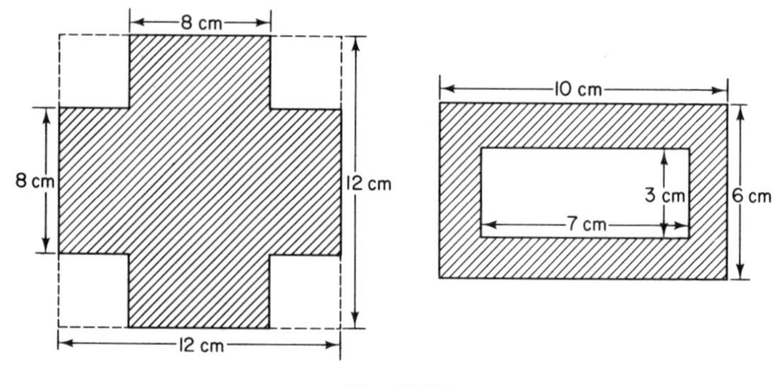

Fig. 20.24

20.21 Areas of Similar Shapes

Theorem 20.2: *If the ratio of the corresponding sides of two similar triangles is* $n{:}k$, *the ratio of their areas is* $n^2{:}k^2$. (Compare your earlier experiment in Exercise 4 of Section 18.2.)

For example, $\triangle$s $\begin{matrix} ABC \\ PQR \end{matrix}$ in Fig. 20.25 are similar and $|AB|:|PQ| = 2:3$.

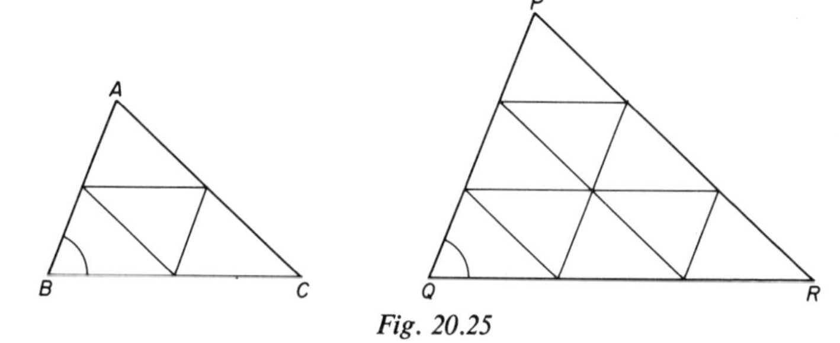

Fig. 20.25

The area of $\triangle ABC$: the area of $\triangle PQR = 2^2:3^2 = 4:9$.
We use the trigonometric formula for the area of the triangle.

$\triangle$s $\begin{matrix} ABC \\ PQR \end{matrix}$ are similar and $|AB|:|PQ| = n:k = |BC|:|QR|$

$$\Rightarrow \angle B = \angle Q;$$

$$|AB| = \frac{n}{k}|PQ|;$$

$$\text{and } |BC| = \frac{n}{k}|QR|.$$

$$\text{Area of } \triangle PQR = \tfrac{1}{2}|PQ| \,.\, |QR| \sin Q,$$
$$\text{area of } \triangle ABC = \tfrac{1}{2}|AB| \,.\, |BC| \sin B$$
$$= \tfrac{1}{2}\left(\frac{n}{k}|PQ|\right) \,.\, \left(\frac{n}{k}|QR|\right) \sin Q.$$

Then $$\frac{\text{area of } \triangle ABC}{\text{area of } \triangle PQR} = \frac{\frac{1}{2}\frac{n}{k}|PQ| \,.\, \frac{n}{k}|QR| \sin Q}{\frac{1}{2}|PQ| \,.\, |QR| \sin Q} = \frac{\frac{n}{k} \cdot \frac{n}{k}}{1} = \frac{n^2}{k^2}.$$

Since any two *similar polygons* can be divided into corresponding similar triangles, the ratio of the areas of the polygons is the ratio of the squares of their corresponding sides.

Look for example at the simple case of the square. All squares are similar; if you draw several squares on squared paper along the lines of the grid, you will find that the ratio of their areas is the ratio of the squares of the lengths of their sides.

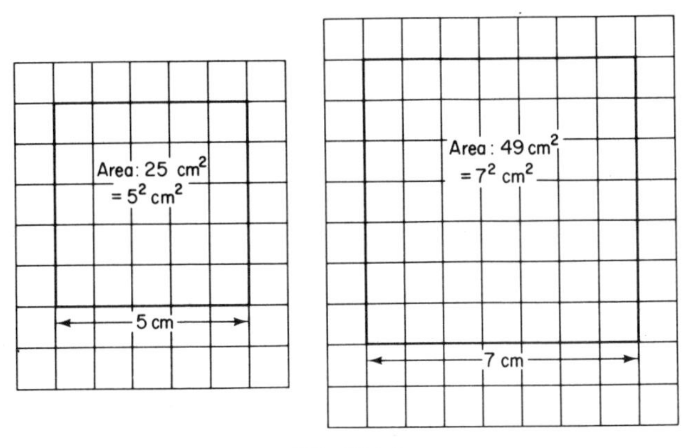

Fig. 20.26

20.22 Exercises

1. The ratio between the decimetre and the centimetre (and between any two subsequent multiple units of length in the metric system) is 10:1; confirm that 1 dm^2 = 100 cm^2 (compare Section 5.9).
2. On a scale drawing, the floor area of a room is 80 cm^2. Calculate the area of the floor if the RF (Representative Fraction, or scale; see Section 7.5) is given as $\frac{1}{50}$.

20.23 The Area of a Circle

We soon find that the methods and principles used to evaluate the areas of triangles and other polygons fail when we try to measure the areas of plane shapes bounded by *curved* lines.

For centuries, mathematicians tried in vain to 'square the circle', that is, to measure the area of a given circle by constructing a square of equal area. One method was to 'approximate' the area of the circle by drawing regular polygons both inside and outside the circle.

One very rough approximation is illustrated in Fig. 20.27: the area of the circle is greater than the area of the square inscribed in the circle ($ABCD$), but less than the area of the square circumscribing the circle ($PQRS$).

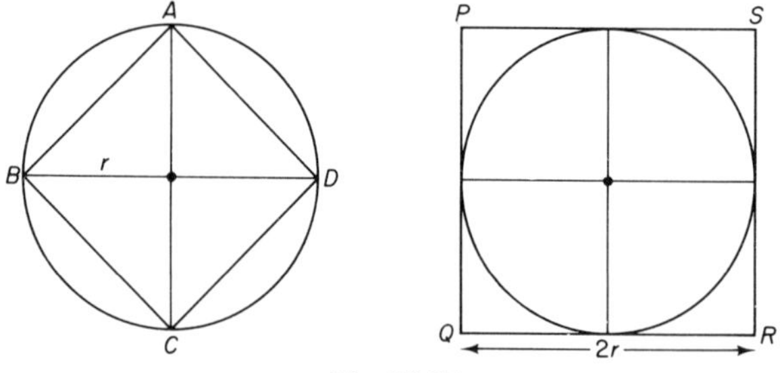

Fig. 20.27

If the length of the radius of the circle in Fig. 20.27 is r, the length of each side of the square $PQRS$ is $2r$ and the area of the square $PQRS = 2r \times 2r = 4r^2$ unit squares.

The area of the square $ABCD = \frac{1}{2}(2r \times 2r) = 2r^2$ unit squares (using the formula for the area of the rhombus, see Section 20.18).

Therefore,

$2r^2$ units of area $<$ the area of the circle of radius $r < 4r^2$ units of area.

A closer approximation can be made if we increase the number of sides of the regular polygons inscribed in, and circumscribing, the circle. For example, the area of the inscribed duodecagon (12-sided polygon) is $12(\frac{1}{2}r^2 \sin 30°) = 3r^2$ units of area, and the area of the circumscribed regular duodecagon, calculated to 3 SF, $= 3{\cdot}22r^2$ units of area. (You may try to derive these results if you wish.)

Therefore,

$3r^2$ units of area $<$ the area of the circle of radius $r < 3{\cdot}22r^2$ units of area.

A still more accurate approximation has been calculated:

$3{\cdot}141\,592r^2$ units of area $<$ area of the circle $< 3{\cdot}141\,593r^2$ units of area. These figures look familiar; in fact, the constant ratio between the area of a circle and the square of its radius is the number π that we met before.

The reason for this becomes clear if we consider a 24-sided regular polygon inscribed in a circle (Fig. 20.28).

The area of this regular polygon is nearly that of the whole area of the circle.

The area of $\triangle OAB$ (and of each of the other 23 triangles) $= \frac{1}{2}|AB| \,.\, h$.

The total area of the regular polygon $= 24 \times \frac{1}{2}|AB| \times h$

$= 24 \times |AB| \times \frac{1}{2}h$

$=$ (the perimeter of the polygon) $\times \frac{1}{2}h$.

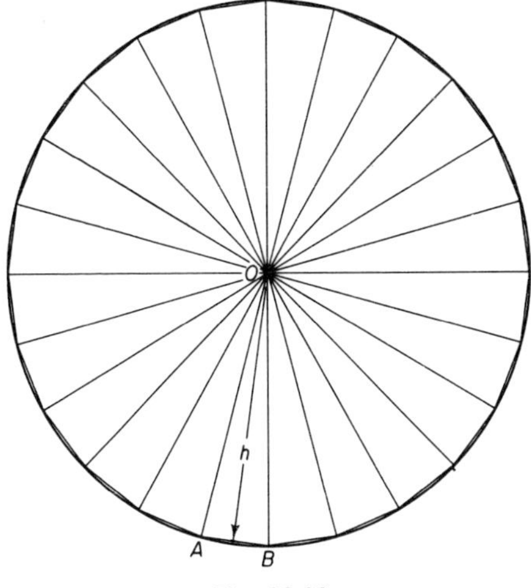

Fig. 20.28

As the number of sides of the inscribed polygon increases, the length of its perimeter approaches the length of the circumference of the circle ($2\pi r$) and the 'height' (h) of each triangle becomes nearer to the length of the radius (r). The area of the polygon therefore approaches the value $2\pi r \times \frac{1}{2}r = \pi r^2$.

area of the circle of radius $r = \pi r^2$ units of area

20.24 Exercise

On centimetre-squared paper draw a circle of radius 10 cm. Count the number of whole unit squares inside the circle, and then, by examining the divided squares, estimate the minimum number of whole unit squares which together make up the area of the circle. Use your results to find approximately the area of this circle. Divide this area by 100 (i.e. 10^2) to obtain an approximate value of π. [Hint: area of circle $= \pi r^2$; therefore $\frac{\text{area}}{r^2} = \frac{\pi r^2}{r^2} = \pi$.]

20.25 The Area of a Sector of a Circle

When a circle with radius of length r is divided into six *congruent* sectors (see Section 17.10), the angle of each sector is one-sixth of a complete turn ($\frac{360°}{6} = 60°$) and the area of each sector is one-sixth of the area of the circle, that is, $\frac{1}{6}\pi r^2$ unit squares.

Conversely, if the angle of a sector is one-sixth of a complete turn (i.e. 60°), six such sectors form a complete circle, and the area of each sector is $\frac{1}{6}\pi r^2$ unit squares.

In general terms, the area of a sector is a fraction of the area of the complete circle, and the size of the fraction is determined by the angle of the sector. For example, if the angle of the sector is 45° and the radius of the circle is r unit lengths, the area of the sector $= \frac{45^\circ}{360^\circ}\pi r^2 = \frac{1}{8}\pi r^2$.

area of the sector whose angle is α° and radius is r unit lengths $= \frac{\alpha \pi r^2}{360}$

20.26 Exercise

Calculate to 3 SF the area of each of the shapes in Fig. 20.29.

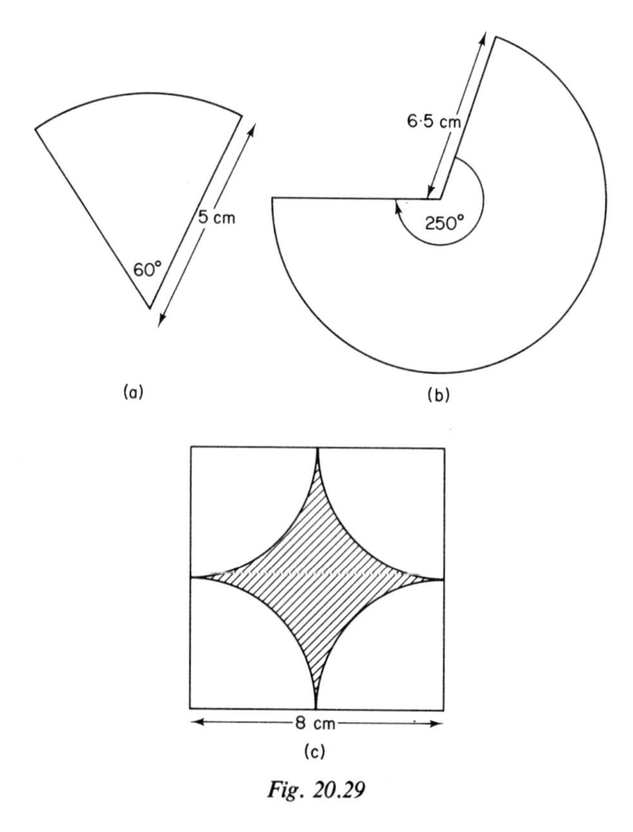

Fig. 20.29

[Hint: the area of the shaded portion of Fig. 20.29(c) is found by subtracting the areas of the four sectors (in this case quarter-circles or quadrants) from the area of the square.]

Unit Twenty-one

Mensuration in Three Dimensions

21.1 Three-dimensional Geometry

A detailed study of three-dimensional shapes, known as three-dimensional geometry or *solid geometry*, is outside the scope of this book. We are only concerned with the mensuration of the most familiar and simple three-dimensional shapes, and with the few basic 'facts' of three-dimensional geometry which we need for this purpose.

Drawings of these shapes on paper, are of course, two-dimensional, although careful use of the laws of perspective may produce an illusion of a three-dimensional shape. Sometimes the real three-dimensional situation is better understood if we use simple three-dimensional models, with cardboard or paper to represent planes and 'faces' of solids, and stiff wire, knitting needles, drinking straws or sticks to represent lines.

Angles and distances in three dimensions

One everyday description of a direction is contained in the familiar word *vertical*: if the Earth's surface is thought of as perfectly flat, this is the direction in which most plants and trees grow, the direction of the plumb line in relation to the ground.

If a rod is perfectly vertical, it is perfectly *perpendicular to the plane of the Earth's surface*, and makes a right angle with every straight line in that plane which passes through the point where the rod meets the ground (Fig. 21.1).

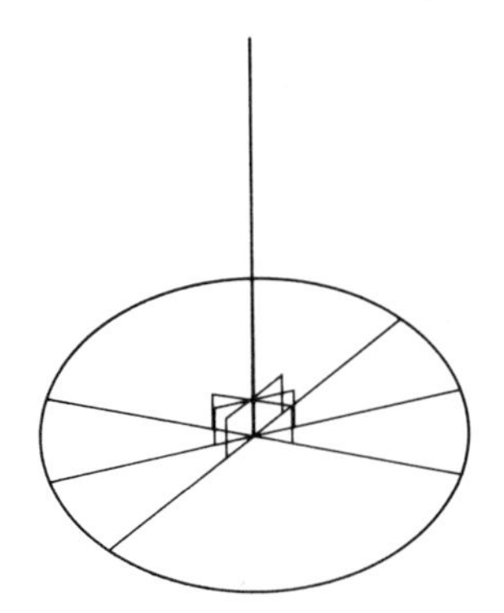

Fig. 21.1

Test this for yourself: place a set-square upright on a table against a vertical rod, and rotate it about the rod (Fig. 21.2).

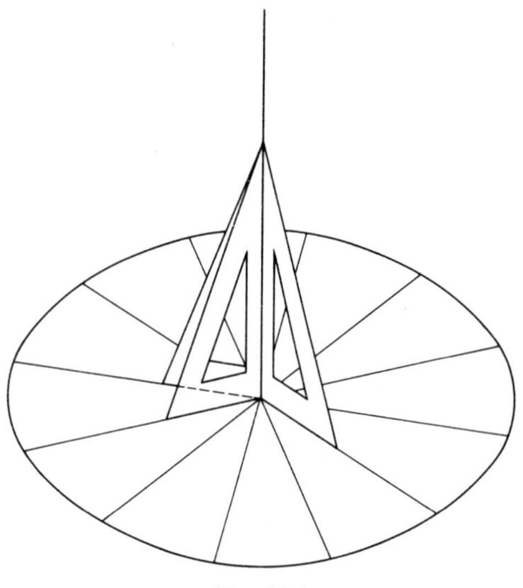

Fig. 21.2

Whichever way it stands, the set-square confirms that a right angle is formed by the rod and a line in the plane of the table.

A line perpendicular to a plane forms a right angle with every line in the plane through the point where the line meets the plane.

The *perpendicular distance of a point P from a plane* is measured from point P along the line perpendicular to the plane. If this line meets the plane at a point S, the *foot of the perpendicular*, the distance of P from the plane is $|PS|$ (Fig. 21.3).

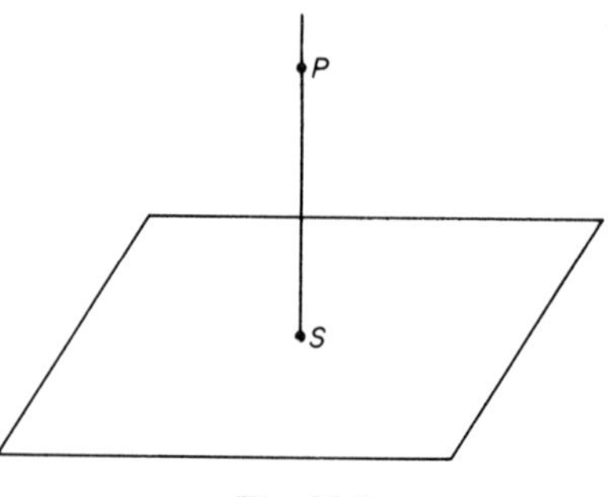

Fig. 21.3

This point S, the foot of the perpendicular from a point to a plane, is sometimes called the *projection of the point on the plane*. You might think of this projection as the shadow of the point P on the plane, thrown by a lamp placed vertically above P (Fig. 21.4).

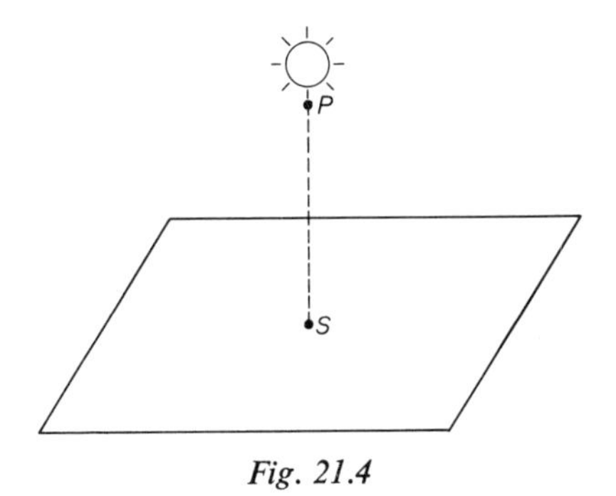

Fig. 21.4

The *projection on a plane of a line l* (not perpendicular to the plane) is another line l'; again, we can think of l' as the shadow of l, thrown on to the plane by a lamp placed vertically above l. The lines l and l' intersect at the point I where the line l meets the plane (Fig. 21.5).

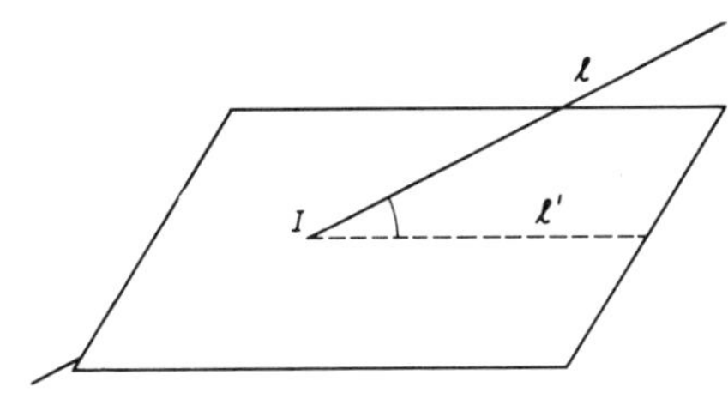

Fig. 21.5

When we speak of the *angle between a line l and a plane*, we mean the angle between the line l and its projection on the plane, that is, the angle between l and l' in Fig. 21.5. It is the smallest angle which the line l makes with any of the lines in the plane which pass through the point I.

If every point on the line l is joined to its own projection on the plane, the lines together form another plane, a 'wall' of straight lines all perpendicular to the first plane and all perpendicular to the line l'; l' is then the *line of intersection* of the two planes. The whole 'wall' is then perpendicular to the first plane, like the walls of buildings in relation to the Earth's surface.

We could check whether a wall is perpendicular to a floor by using a large set-square.

If the vertices of the set-square are marked A, B and C, as in Fig. 21.6, and the wall intersects the floor in the line l, the set-square must be placed so that both the line AB on the wall and the line BC on the floor are perpendicular to l. The angle of the set-square ($\angle ABC$) is the angle between the wall and the floor—in this case, 90°.

The wall and floor, however, are a special case, and can only be dealt with

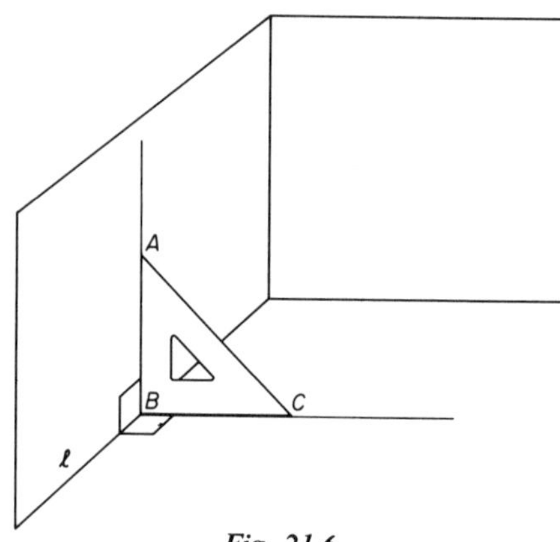

Fig. 21.6

like this because we happen to have a set-square with an appropriate angle. To find the angle *between any two planes* p_1 and p_2, intersecting in the line l, first choose a point P somewhere along l. Now draw a line PQ in the plane p_1 and a line PR in the plane p_2, both perpendicular to l; the angle between the planes is $\angle QPR$ (Fig. 21.7).

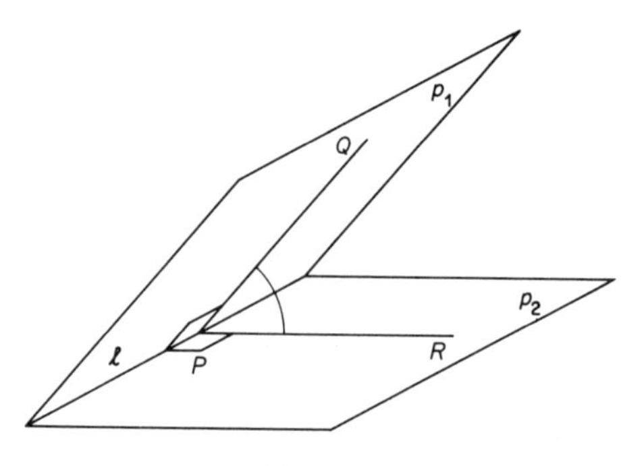

Fig. 21.7

Two planes p_1 and p_2 are called *parallel planes* if the perpendicular distances of *all* the points in the plane p_1 from the plane p_2 are the same.

21.2 The Cuboid

A *polyhedron* is a three-dimensional shape bounded only by (plane) polygons, the *faces* of the polyhedron. Cubes, cuboids and pyramids are all special kinds of polyhedron. The lines of intersection of the faces of the polyhedron are called its *edges*, and the points of intersection of three or more edges are the *vertices* of the polyhedron.

A *cuboid* (Fig. 21.8) is a polyhedron bounded by six rectangles. Any rectangular box—a shoe-box or a cigarette packet, for example—is a cuboid.

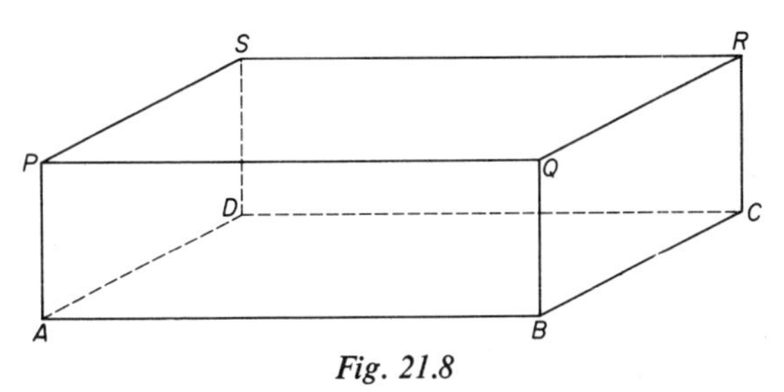

Fig. 21.8

Look at such a box, and confirm the following properties of the cuboid:

(*a*) A cuboid has six faces, twelve edges and eight vertices.
(*b*) The opposite faces of a cuboid are parallel.
(*c*) The opposite faces of a cuboid are congruent rectangles; there are, therefore, three pairs of congruent rectangles.
(*d*) At every vertex, each edge intersects the other two edges at right angles.
(*e*) Each edge is perpendicular to two opposite faces—for example, *PQ* in Fig. 21.8 is perpendicular to *PSDA* and *QRCB*—and its length is the distance between these two faces.
(*f*) Every pair of opposite faces is joined by four parallel edges of equal length; for example, the 'top' and the 'bottom' of the box in Fig. 21.8 are joined by the edges *AP*, *BQ*, *CR* and *DS*, and $|AP| = |BQ| = |CR| = |DS|$; therefore, if the lengths of any three edges intersecting at one vertex are given (the *dimensions* of the cuboid), we know the lengths of all the other edges as well.
(*g*) The angle between any two intersecting faces is a right angle.

When a paper model of a cuboid or a rectangular cardboard box is cut along the edges so that it can be laid flat (this can be done in various ways), the resulting plane shape is called a *net* of the cuboid (Fig. 21.9).

You can make a paper model of a cuboid, by drawing a net like this on stiff paper, cutting it out, and joining the faces with sellotape.

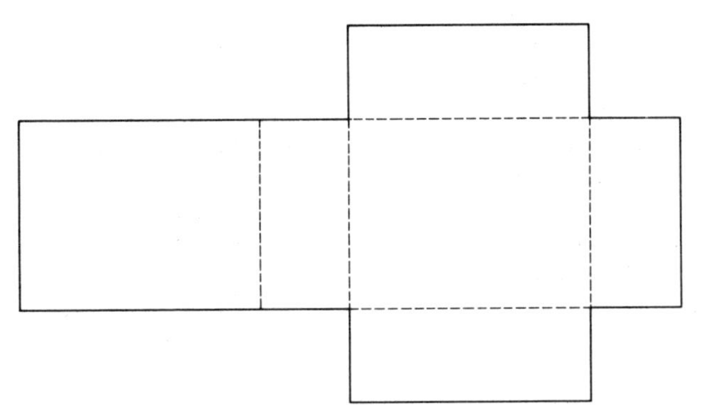

Fig. 21.9

Surface area of a cuboid

The total area of all the faces of a polyhedron is called its *surface area.* It is easy to calculate the surface area of a cuboid because of the congruence of the opposite faces.

21.3 Exercises

1. The dimensions of a cuboid $ABCDPQRS$ are given as $|AB| = 8$ cm, $|AD| = 6$ cm and $|AP| = 3$ cm.

   (*a*) Make a sketch, and mark the lengths of all the edges.
   (*b*) Draw a net of this cuboid.
   (*c*) Calculate the surface area of the cuboid.

2. The dimensions of another cuboid $A'B'C'D'P'Q'R'S'$ are given as

   $$|A'B'| = 16 \text{ cm}, |A'D'| = 12 \text{ cm and } |A'P'| = 6 \text{ cm}.$$

   Comparing this cuboid to the cuboid in Exercise 1, we find that all the corresponding faces are similar. Confirm that the ratio of the surface areas of the two cuboids is the ratio of the squares of the lengths of corresponding edges.
3. Leonhard Euler (1736) discovered a relation between the number of faces (F), the number of edges (E) and the number of vertices (V) of *any* polyhedron, the formula

   $$V - E + F = 2 \text{ (Euler's formula).}$$

   Check this formula for the cuboid, and for other polyhedra, by counting the faces, edges and vertices in each case.

21.4 The Volume of a Cuboid

A *cube* is a special cuboid: all its six faces are squares, and all its edges are of the same length.

A *unit cube* is a cube whose edges are of unit length; for example, the cubic centimetre (1 cm^3) is the cube whose edges are 1 cm each. The unit cube is the unit used for measuring volume.

Evaluating the volume of a three-dimensional shape by counting unit cubes is not a very practical proposition. We can, however, think of a cuboid with dimensions of, for example, 8 cm, 6 cm and 3 cm (Fig. 21.10), as consisting of 3 layers, each of 48 unit cubes (i.e. 48 cm^3), and each layer made up of 8 rows of 6 cubes each (see also Section 5.9).

We might think of this cuboid as a rectangular box, and the rectangle $ABCD$ as the floor of the box, the *base* of the cuboid. The area of this base determines the number of unit cubes that can be placed on it in one layer. The area of the rectangle $ABCD = (6 \times 8)$ unit squares, so the number of unit cubes which will cover it is 48.

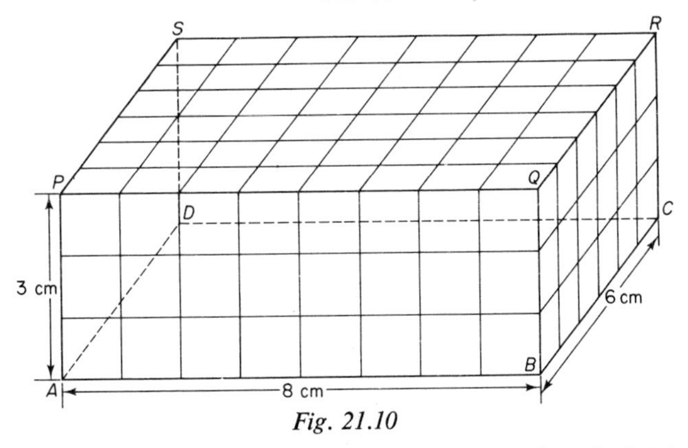

Fig. 21.10

The number of layers is determined by the third dimension, the *height* of the cuboid relative to the base $ABCD$, in this example the length of AP, i.e. 3 cm. The volume of this cuboid is therefore $3 \times 48\ \text{cm}^3 = 144\ \text{cm}^3$.
In general, if the dimensions of a cuboid are l, b and h units of length, the volume of the cuboid $= (l \times b \times h)$ units of volume.

volume of cuboid $= l \times b \times h$ units of volume

(In words, the volume of a cuboid is the product of its three dimensions, or, the volume of a cuboid = 'length' × 'breadth' × 'height'.)

But $l \times b$ is the number of units of area in the base; so we can also say that the volume of the cuboid is expressed by the formula

volume of cuboid = area of the base × height

21.5 Exercises

1. Measure the three dimensions of a rectangular room and calculate its volume.
2. Consider two cubes whose edges are respectively 4 cm and 5 cm. They are *similar in shape*. Calculate their volumes and confirm that the ratio of their volumes is the ratio of the cubes of the lengths of corresponding edges.

21.6 Lines and Angles in a Cuboid

We can use all the principles and methods of plane geometry to study plane shapes, lines and angles on the surfaces of three-dimensional shapes. For example, we can find the length of the diagonal DB of the face $ABCD$ of the cuboid in Fig. 21.10 by applying Pythagoras' Theorem to the right-angled triangle ABD:

$$|DB| = \sqrt{|AD|^2 + |AB|^2} = \sqrt{6^2 + 8^2}\ \text{cm} = 10\ \text{cm}.$$

When we are concerned with lines or angles which do not lie in the surface planes of a solid, we must consider the plane *within* the solid in which these lines lie.

For example, we will find the length of the diagonal DQ of the cuboid in Fig. 21.11.

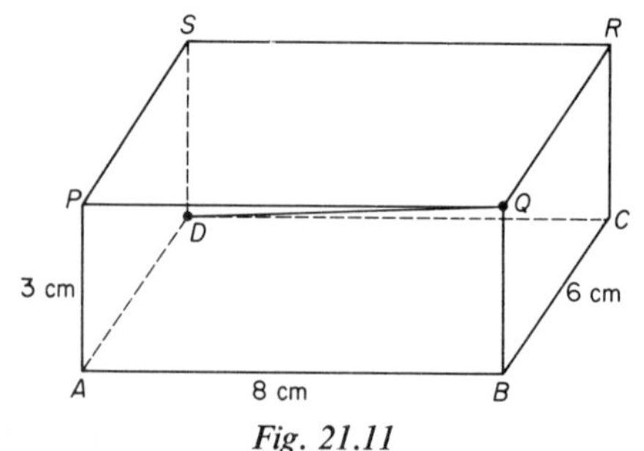

Fig. 21.11

Think of the cuboid as an empty box; then a rectangular piece of card, $DBQS$, placed across the inside of the box as in Fig. 21.12, can represent a plane containing the line segment DQ.

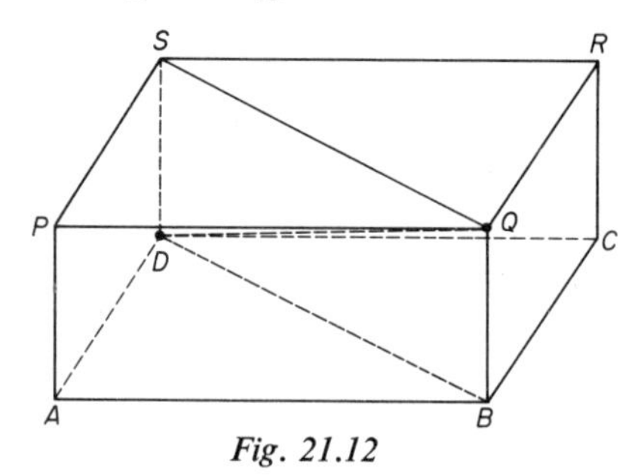

Fig. 21.12

The 'breadth' of the piece of card is $|BQ| = 3$ cm (given), and the 'length', $|DB|$, is the length of the hypotenuse of $\triangle DAB$, already calculated as 10 cm. Now we can calculate $|DQ|$ using Pythagoras' Theorem again, this time in $\triangle DQB$ where $\angle DBQ = 90°$.

$$|DQ| = \sqrt{|DB|^2 + |BQ|^2} = \sqrt{10^2 + 3^2}\text{ cm} = 10{\cdot}5\text{ cm (to 3 SF)}.$$

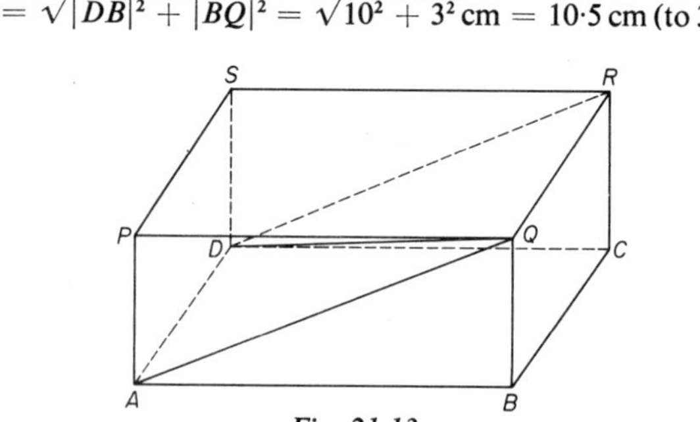

Fig. 21.13

Notice that the same line segment DQ is also the diagonal of the rectangle $ADRQ$ in the 'slanting' plane (Fig. 21.13). Its length could therefore be calculated equally well by applying Pythagoras' Theorem to $\triangle ABQ$ where $\angle ABQ = 90°$ and then to $\triangle DAQ$ in which $\angle DAQ = 90°$.

21.7 Exercises

Fig. 21.14 represents a cuboid whose dimensions are 20 cm, 8 cm and 10 cm. The points K, L, M and N are the mid-points of AD, AB, PQ and PS respectively.

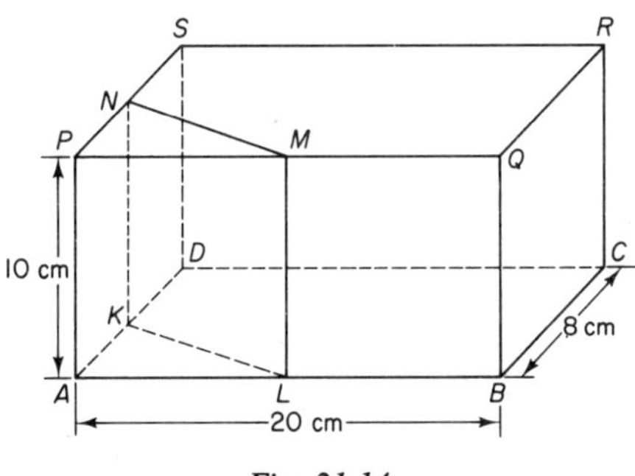

Fig. 21.14

1. Find $|AL|$ and $|AK|$ and hence calculate $|KL|$. [Hint: in $\triangle KLA$, $\angle A = 90°$. Therefore use Pythagoras' Theorem to find the length of the hypotenuse KL.]
2. Calculate the angle between the face $ABQP$ and the plane $KLMN$. [Hint: the line PQ in the face $ABQP$ and the line MN in the plane $KLMN$ are both perpendicular to ML, the line of intersection of the planes; so $\angle NMP$ (or $\angle KLA$) is the angle between the planes. Therefore, use a trigonometric ratio in $\triangle NMP$ to find $\angle NMP$.]
3. Point O is the point of intersection of the diagonals AC and BD. Calculate $|RO|$. [Hint: first find $|AC|$, by Pythagoras' Theorem, and hence $|OC|$. Then draw $\triangle RCO$ separately, remembering $\angle RCO = 90°$, and again use Pythagoras' Theorem to find $|RO|$.]

21.8 The Prism

If a polyhedron has two faces which are congruent polygons in two parallel planes, and all the remaining faces are rectangles, the polyhedron is called a *right prism*. These two parallel congruent faces are called the *bases* or *end faces* of the prism. The cuboid is a special kind of prism, in which the two end faces are rectangles. Fig. 21.15 shows two prisms, one whose base is a triangle, a triangular prism, and a hexagonal prism whose base is a hexagon.

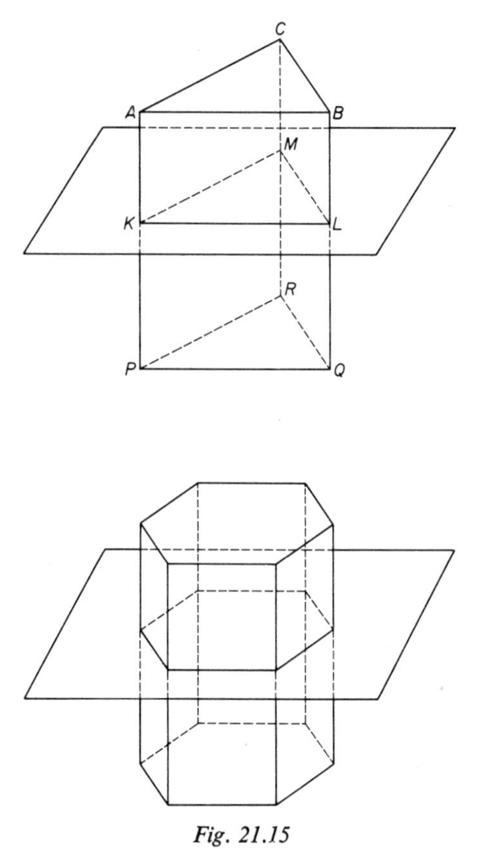

Fig. 21.15

The prisms in Fig. 21.15 are each intersected by a plane parallel to the two end faces. Clearly, the intersection of this plane with the prism is a polygon congruent with the end faces; for example, $\triangle KLM$ is congruent with both $\triangle ABC$ and $\triangle PQR$. We say that prisms have a *uniform cross-section*. The edges AP, BQ and CR in the triangular prism in Fig. 21.15 are called the *lateral edges* of the prism. They are each one of a pair of opposite sides of a rectangle; they are therefore parallel, of equal length and perpendicular to both end faces. The lateral edges of *any* right prism are perpendicular to the bases; their length is the distance between the two bases, or the *height* of the prism.

The volume of a prism

All prisms, like cuboids, can be considered to be made up of layers of unit height, each layer consisting of the same number of units of volume (Fig. 21.16).

The number of units of area in the base—that is, the area of the base—determines the number of units of volume in each layer. If, for example, the area of the base of a prism is 24 cm^2 and the height of the prism is 10 cm, there will be 24 cm^3 in each layer and 10 layers; the volume of this prism is therefore 10×24 cm^3 $=$ 240 cm^3.

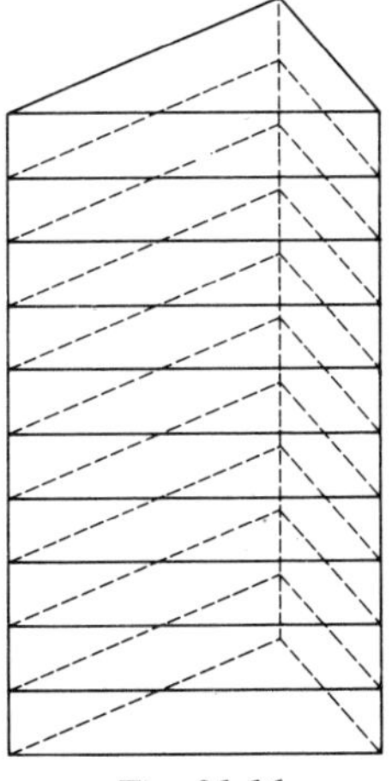

Fig. 21.16

In general, if the area of the base of a prism is B units of area and its height is h units of length, the volume of the prism $= B \times h$ units of volume:

$$\text{volume of prism} = B \times h$$

(In words, the volume of a prism is 'base area' $\times$ 'height'.)

21.9 Exercises

1. The base of a right prism is a right-angled triangle whose sides are 3 cm, 4 cm and 5 cm; the height of the prism is 10 cm.

   (*a*) Calculate the volume of the prism.
   (*b*) Make a freehand sketch of the prism and construct a *net* (see Section 21.2) of the prism.
   (*c*) Calculate the surface area of the prism.

2. Each of the figures in Fig. 21.17 (page 484) shows a right prism. Use the information given in the diagram to calculate the area of the base of each prism, and then the volume of each prism. [Hint: the base of the hexagonal prism is a regular hexagon; to find its area, see Section 20.19.]

21.10 The Cylinder

The right circular cylinder (Fig. 21.18), usually simply called a cylinder, is a three-dimensional shape which resembles the prism in many ways.

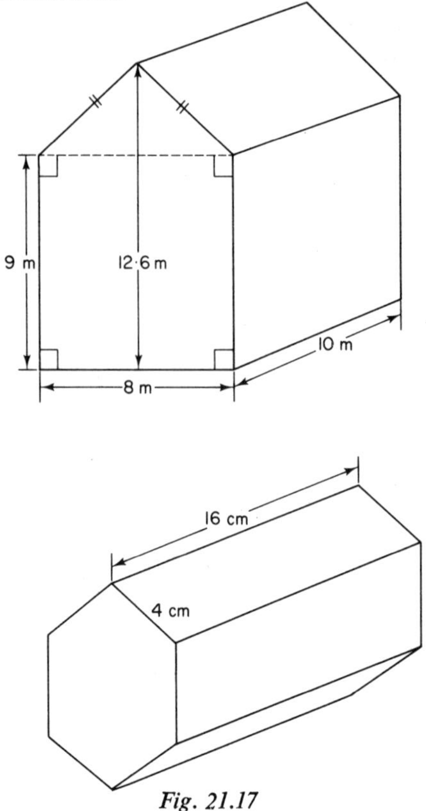

Fig. 21.17

Two faces, the bases of the cylinder, are congruent and parallel; any intersecting plane parallel to the base cuts a plane shape congruent with the base.

The cylinder differs from the prism in only two ways: the bases are not polygons but circles, and the remaining boundary of the cylinder is a closed curved surface, not a set of plane (flat) faces.

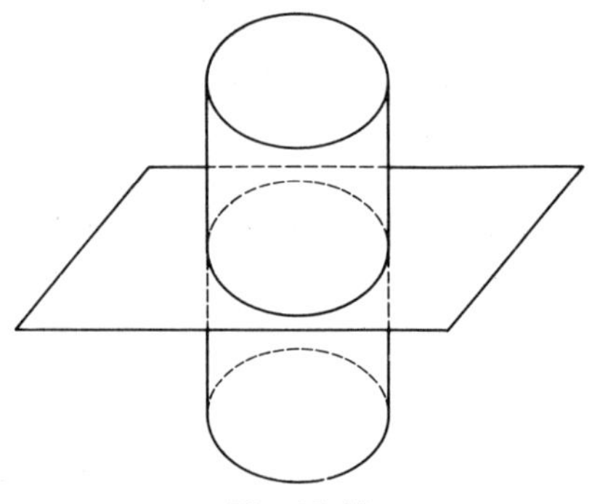

Fig. 21.18

The volume of a cylinder

To find the volume of a cylinder in terms of unit cubes, we again divide the cylinder into layers, each of which is 1 unit length thick. The height of the cylinder determines the number of layers. The number of unit cubes in each layer is determined by the number of unit squares in the circular base, that is, by its area. For example, we will suppose that the base of a cylinder is a circle of radius 5 cm, and that its height is 10 cm; the area of the base of the cylinder is $\pi \times 5^2 = 25\pi$ cm^2; the number of unit cubes in each layer is therefore 25π, i.e. the volume of each layer $= 25\pi$ cm^3. Since there are 10 such layers, the volume of the cylinder $= 10 \times 25\pi = 785$ cm^3 (to 3 SF).

In general, the volume of a cylinder is related to the area, B, of its base and its height, h, in the same way that of the right prism:

$$\text{volume of cylinder} = B \times h$$

If the length of the radius of the base of a cylinder is r unit lengths, the volume of the cylinder is also given by the formula:

$$\begin{aligned}\text{volume of cylinder} &= \pi \times r^2 \times h \\ &= \pi r^2 h\end{aligned}$$

21.11 Exercises

1. If the curved surface of a cylinder is 'cut' along a line perpendicular to the end faces of the cylinder, and is 'flattened out', it forms a rectangle whose length equals the length of the circumference of the end face and whose breadth is the height of the cylinder. Fig. 21.19 shows a net of the cylinder.

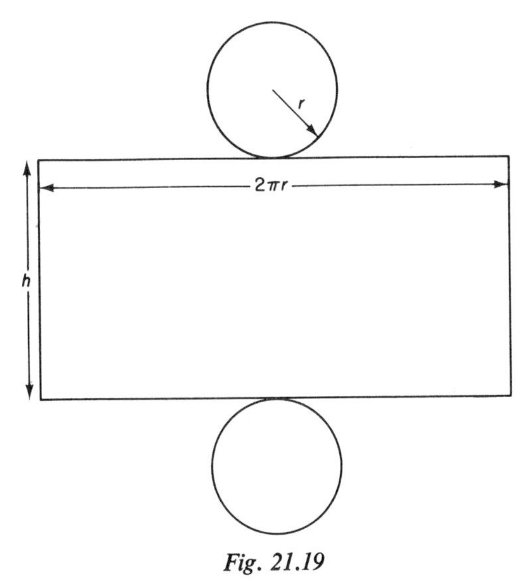

Fig. 21.19

(*a*) On squared paper, draw the net of a cylinder whose height is 8 cm and whose base is a circle of radius 6 cm.

(*b*) Confirm the following formula for the total surface area of a cylinder whose base is a circle of r unit lengths and height is h unit lengths:

$$\text{surface area of a cylinder} = 2\pi r\,(r + h) \text{ units of area}$$

[Hint: first calculate the area of the curved surface of the cylinder, and then the area of the two plane end faces.]

2. A 12 m long straight tube of internal diameter 3 cm is filled with water. Calculate in litres the amount of water in this tube (1 litre = 1000 cm³).

21.12 The Pyramid

A polyhedron of which one face (the base) is a regular polygon and the other faces are all congruent isosceles triangles is called a *regular pyramid.* Fig. 21.20 shows a *square pyramid* and a *triangular pyramid.*

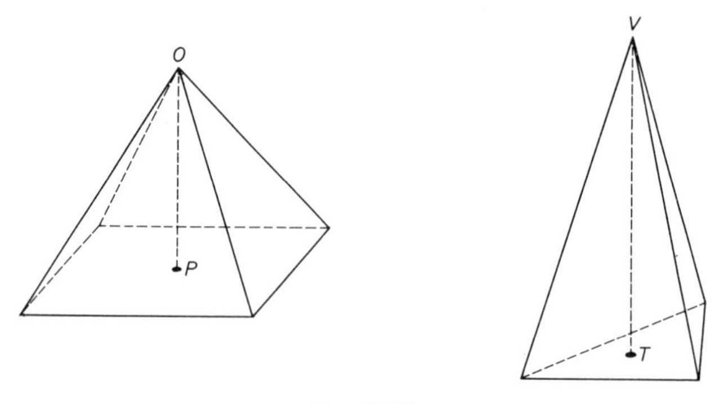

Fig. 21.20

The lateral edges of a regular pyramid are all of the same length and meet at one point, the *vertex* of the pyramid. The *height* of the pyramid is the length of the perpendicular from the vertex to the base of the pyramid; $|OP|$ and $|VT|$ in Fig. 21.20 are the heights of the pyramids shown.

The volume of a pyramid

The volume of any pyramid is related to the area, B, of its base and its height, h, by the formula:

$$\text{volume of pyramid} = \tfrac{1}{3}B \times h$$

(In words, the volume of a pyramid is 'one-third of base area' × 'height'.)

This formula can be derived using mathematical methods beyond the scope of this book; it can be demonstrated for a particular square pyramid by dividing a cube into six congruent square pyramids, as in Fig. 21.21.

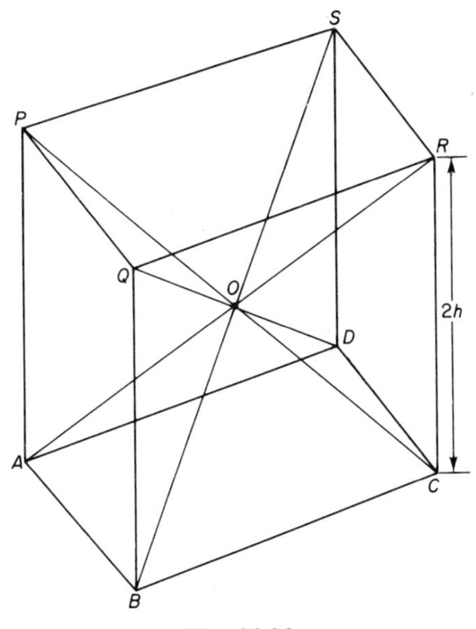

Fig. 21.21

A three-dimensional model can be made out of stiff wire, or from drinking straws joined by pipe cleaners.

The four diagonals, AR, BS, CP and DQ, all meet at the centre of the cube (point O). Each of the six faces of the cube can be considered in turn as the base of a pyramid whose vertex is the point O. Together, these six pyramids form a cube; the volume of each pyramid equals one-sixth of the volume of the cube.

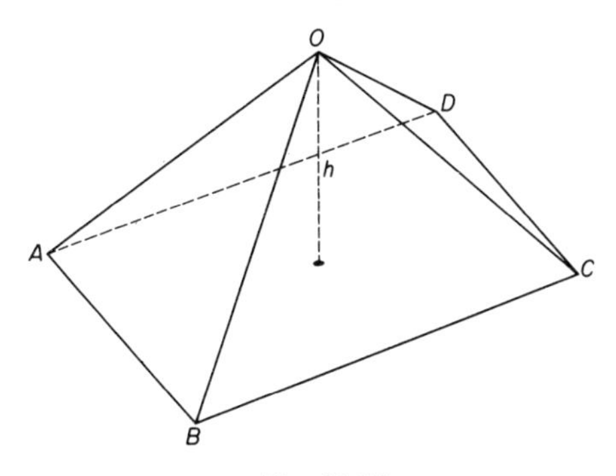

Fig. 21.22

If the height of the pyramid $OABCD$ in Fig. 21.22 is h units of length, the height of the original cube is $2h$ units of length. The volume of the cube is therefore (the area of the base $ABCD$) $\times$ $2h$, and the volume of a single pyramid $= \frac{1}{6} \times$ (area of base) $\times$ $2h = \frac{1}{3}$ (base area) $\times$ h.

21.13 Exercises

Fig. 21.23 shows a pyramid with vertex O, whose base is a square $ABCD$ of side 4 cm and whose lateral edges are 5 cm each, together with a net of the same pyramid.

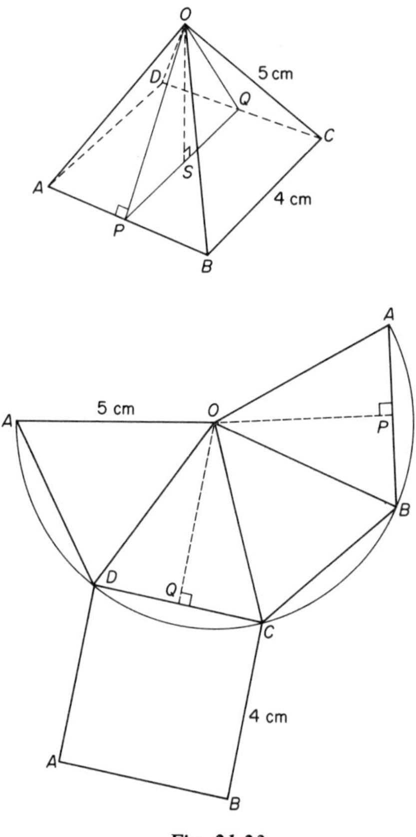

Fig. 21.23

Construct this net accurately, using ruler and compasses.

1. In $\triangle OAB$, construct the altitude OP, perpendicular to AB. [Hint: since $\triangle OAB$ is an isosceles triangle, the foot of the perpendicular, point P, is also the mid-point of AB.] Similarly in $\triangle OCD$, construct the altitude OQ.
   From this construction, you can now find, by measurement, the lengths of the sides of $\triangle OPQ$. (Note: $|PQ| = |AD| = |BC| = 4$ cm.) Now construct $\triangle OPQ$, using Construction 15.1, in Section 15.8. (If the net of the pyramid is cut out and folded, the constructed triangle OPQ can be erected within the paper model, perpendicular to the base.) Measure $\angle OPQ$ with a protractor.

2. Apply Pythagoras' Theorem to $\triangle OAP$ to calculate $|OP|$ $(= |OQ|)$ and apply the Cosine Rule to $\triangle OPQ$ to find $\angle OPQ$. [Hint: $|OA| = 5$ cm and $|AP| = \frac{1}{2}|AB| = 2$ cm; $\angle APO = 90°$. Therefore $|OP|^2 = |OA|^2 - |AP|^2$.] Check that this calculation confirms your measurement of $\angle OPQ$ in Exercise 1.
   Since both lines OP and OQ are perpendicular to AB, the angle OPQ is also the angle between the base of the pyramid and the face OAB.

3. In $\triangle OPQ$ draw the altitude OS (S is the mid-point of PQ). Calculate $|OS|$, which is the height of $\triangle OPQ$ and the *height of the pyramid.* [Hint: use Pythagoras' Theorem in $\triangle OPS$, where $\angle OSP = 90°$; that is, $|OS|^2 = |OP|^2 - |PS|^2$.]
   We can also find $\angle OPQ$ $(= \angle OPS)$ yet another way, since $\cos \angle OPS = \dfrac{|PS|}{|OP|}$, or $\sin \angle OPS = \dfrac{|OS|}{|OP|}$. Find $\angle OPQ$ in both these ways to check your results in Exercises 1 and 2 above.

21.14 The Cone

A *right circular cone* is a three-dimensional shape bounded by a (plane) circle, the *base* of the cone, and a curved surface which tapers to a point, called the *vertex* of the cone (Fig. 21.24).

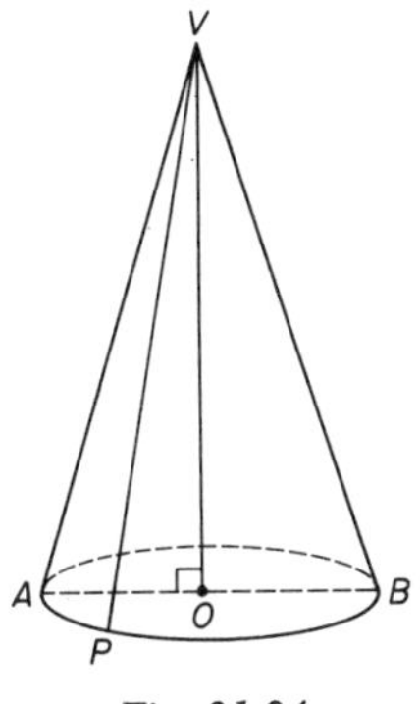

Fig. 21.24

The line through the centre O of the base and the vertex V of the cone is perpendicular to the base; the length OV is the *height* of the cone. If the end-points A and B of any diameter of the base (the circle centre O) are joined to the vertex by straight lines, these lines AV and BV are lines in the curved surface of the cone, and $\triangle ABV$ is an isosceles triangle. We can think of the right circular cone as being generated by the triangle ABV rotating about the line OV (or equally well by the right-angled triangle AOV rotating about the line OV). The cone has a three-dimensional *rotational symmetry*; compare the similar case of the cylinder, generated by a rectangle rotating about one of its sides.

Any straight line segment joining the vertex V to a point P on the circumference of the base lies in the curved surface of the cone. All these straight line segments are of equal length; their length is called the *lateral height* or *slant height* of the cone.

We saw earlier that the cylinder resembles the prism to some extent; similarly, the cone is rather like the pyramid in many ways. In Section 23.11, we will use calculus methods to prove that the formula for the *volume of the cone* is the same as the formula for the volume of the pyramid:

$$\boxed{\text{volume of cone} = \tfrac{1}{3}\text{ base area} \times \text{height}}$$

If the area, B, of the base is expressed in terms of the radius, r, of the circle, i.e. $B = \pi r^2$, and the height of the cone is h:

$$\boxed{\text{volume of cone} = \tfrac{1}{3}\pi r^2 h}$$

The surface area of a cone

Imagine the curved surface of a cone, 'cut' along a straight line joining the vertex to a point on the circumference of the base, and flattened out (Fig. 21.25).

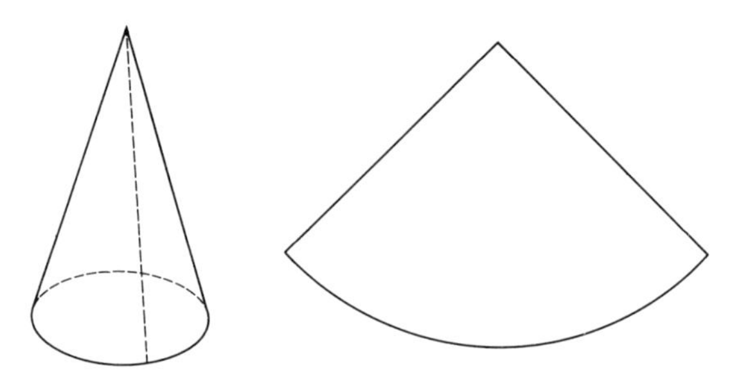

Fig. 21.25

Since all the points on the circumference of the base are at the same distance from the vertex, the flattened surface of the cone has the shape of a sector of a circle. The length of the radius of this sector is equal to the slant height of the cone, and the length of the arc of the sector is equal to the circumference of the circular base of the cone.

Fig. 21.26 shows the complete net of a cone.

To find the surface area of a cone, whose base is a circle of radius r, and whose slant height is l:

$$\begin{aligned}\text{the length of the arc of the sector} &= \text{the circumference of the circular base}\\ &= 2\pi r.\end{aligned}$$

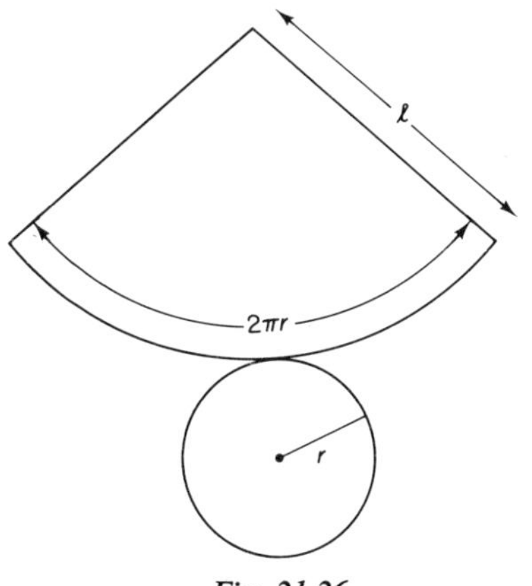

Fig. 21.26

Therefore, the area of the sector $= \frac{1}{2}$(length of its arc $\times$ slant height) (compare Section 20.23) $= \frac{1}{2}(2\pi r \times l) = \pi rl$, and the area of the circular base $= \pi r^2$.

The total surface area of the cone is, therefore, area of base + area of curved surface $= \pi r^2 + \pi rl = \pi r(r + l)$.

total surface area of cone $= \pi r(r + l)$

21.15 Exercises

The slant height of a cone is 9 cm, and the radius of its circular base is 5 cm.

1. Construct a net of this cone, cut it out and use it to make a paper model of the cone.
2. Calculate the height of the cone. [Hint: think of the cone as being generated by rotating an isosceles triangle about its altitude, the sides of the triangle being a diameter of the base of the cone, and two line segments each 9 cm long.]
3. Calculate the volume and the total surface area of the cone.

21.16 The Sphere

The *sphere* is a three-dimensional shape bounded by a closed curved surface, every point on this surface being at the same distance from one fixed point called the *centre* of the sphere.

Some of the most useful properties of the sphere are as follows:

(*a*) The intersection of a sphere and any plane cutting it is a circle (Fig. 21.27).

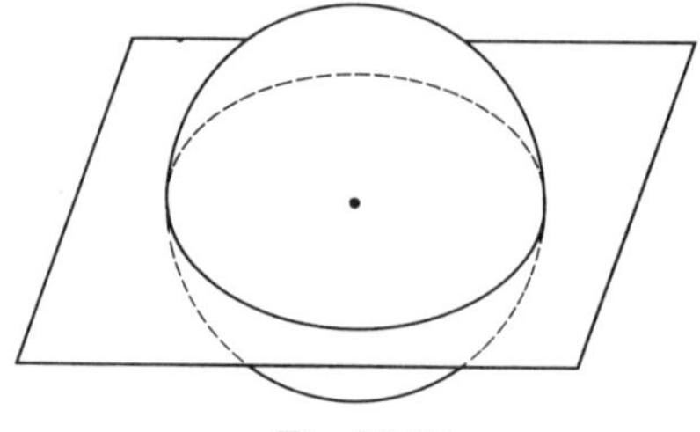

Fig. 21.27

(*b*) The intersection of a sphere and a plane through the centre of the sphere is called a *great circle. All great circles of a sphere are congruent.*

(*c*) A line through the centre of the sphere, perpendicular to a great circle E, is called the *axis of the sphere* relative to the great circle E; the points where the axis cuts the surface of the sphere are called the *poles* of the great circle E (Fig. 21.28).

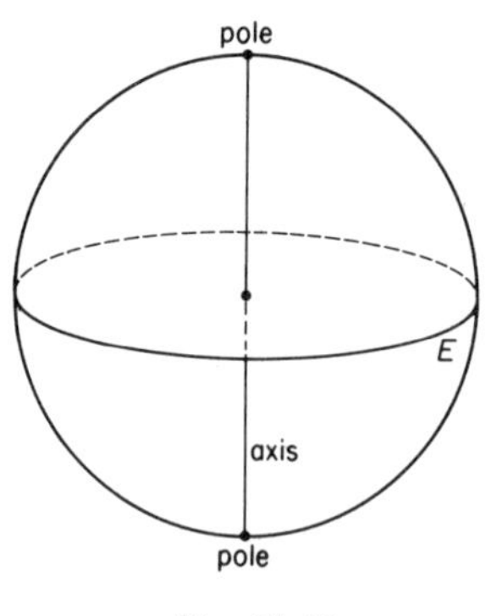

Fig. 21.28

(*d*) Every great circle divides the sphere into two congruent halves called *hemispheres.*

The surface area of a sphere

Guldin (1635) was able to show that the surface area of a sphere equals the area of the curved surface of the cylinder 'circumscribing' the sphere (Fig. 21.29).

If the radius of the sphere is of length r, then the height of the cylinder is $2r$, and the area of the curved surface of the cylinder $= 2\pi r \times 2r = 4\pi r^2$ units of area.

The *surface area of a sphere of radius r* is therefore $4\pi r^2$ units of area.

$$\boxed{\text{surface area of sphere} = 4\pi r^2}$$

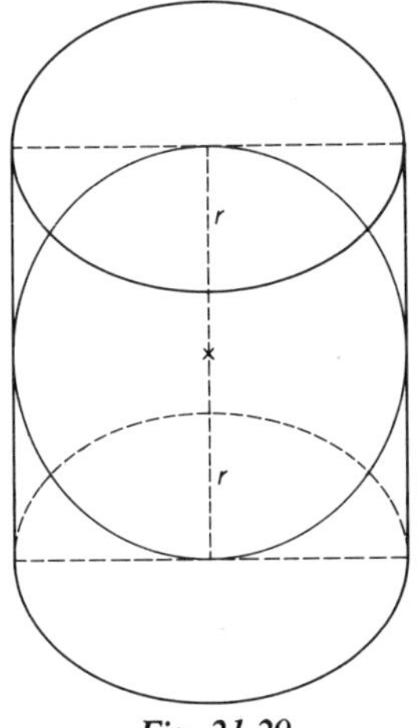

Fig. 21.29

The volume of a sphere

In Section 23.11 we shall show by calculus methods that the *volume of a sphere* of radius r is $\frac{4}{3}\pi r^3$ units of volume.

volume of sphere $= \frac{4}{3}\pi r^3$

You can confirm this by an experiment with a sphere and its circumscribing cylinder.

Choose a ball—say a ping-pong ball—and then find (or make out of stiff paper) a cylinder in which this ball just fits; the height of the cylinder should be the same as the diameter of the ball. Fill the space between the ball and the cylinder with fine sand, without packing tightly; then pour out the sand into a graduated vessel, and measure its volume. Repeat the experiment using the empty cylinder only; the *difference in the two volumes* represents the volume of the ball, and you will find that it is two-thirds of the volume of the cylinder.
Volume of the cylinder $= \pi r^2 h$ (see Section 21.10) $= \pi r^2 \times 2r = 2\pi r^3$;
volume of sphere $= \frac{2}{3} \times 2\pi r^3 = \frac{4}{3}\pi r^3$.
A similar experiment will show that the volume of a cone is one-third of the volume of its circumscribing cylinder.

21.17 Exercise

Calculate the volume and the surface area of the three-dimensional shape shown in Fig. 21.30.

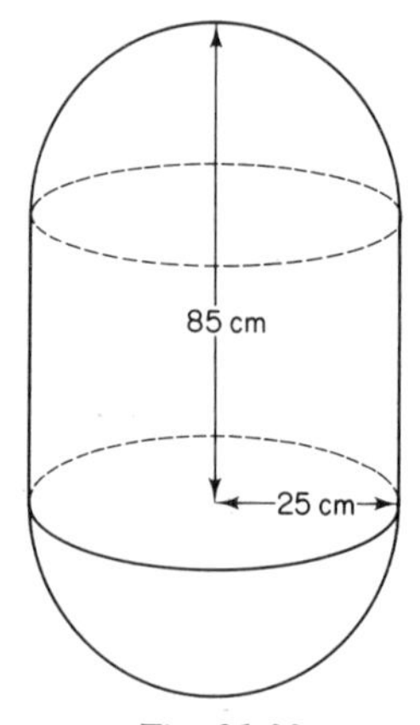

Fig. 21.30

21.18 Geometry as Earth Measurement

The rectangular grid and the co-ordinate system used in maps (see Unit 12) partly depend on the assumption that the Earth's surface is flat. We can usually neglect the inaccuracies which arise from this false assumption, especially when we are concerned with large-scale maps of limited areas.

When we consider the Earth as a sphere, however, the rectangular grid convention cannot be used to define the position of a point on its surface. Instead, in common with every other nation, we use a convention of angular measurement relative to two great circles: the *equator* and the great circle through the poles passing through Greenwich, called the Greenwich *meridian*.

Latitude

Even if this system is conventional and arbitrary, the axis chosen is a natural one: the axis of rotation of the Earth, passing through the *north pole* and the *south pole*.

The equator is the great circle relative to this axis, dividing the Earth into the Northern and Southern Hemispheres.

The line between the centre of the Earth and a point on the surface of the Earth forms an angle with the equator, called the *latitude* of the point (Fig. 21.31). It is the angle of elevation (or depression) from the centre of the Earth (imagine yourself standing at the centre of the Earth, on the plane of the equator, 'looking up', for example, to London).

This angle of elevation, the latitude, of London is $51\frac{1}{2}$°N (north); the latitude of any point on the equator is 0°, and the latitude of the north pole is 90°N. If you were able to stand at the centre of the Earth, you could see that London is not the only place whose latitude is $51\frac{1}{2}$°; you could turn round, looking up at an angle of $51\frac{1}{2}$°, and 'see' a whole range of places, all on the circumference of a *small circle* parallel to the equator, a *circle of latitude* (Fig. 21.32).

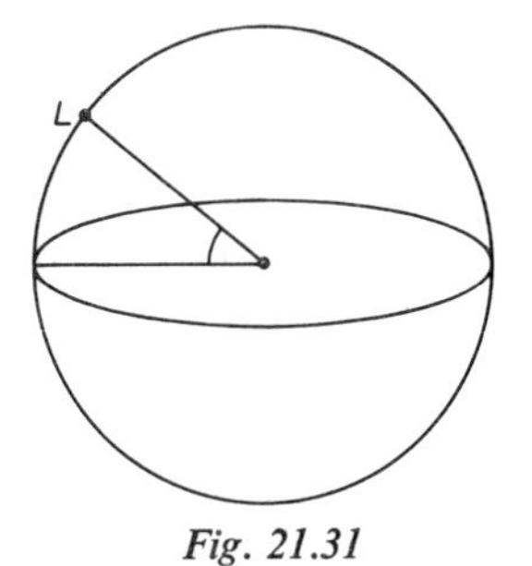

Fig. 21.31

For places on the equator, the circle of latitude is a great circle of the Earth; its radius, i.e. the radius of the Earth, is approximately 6400 km. Further north, or south, the circles of latitude become smaller and smaller; at the north (or south) pole, the circle of latitude dwindles to a point.

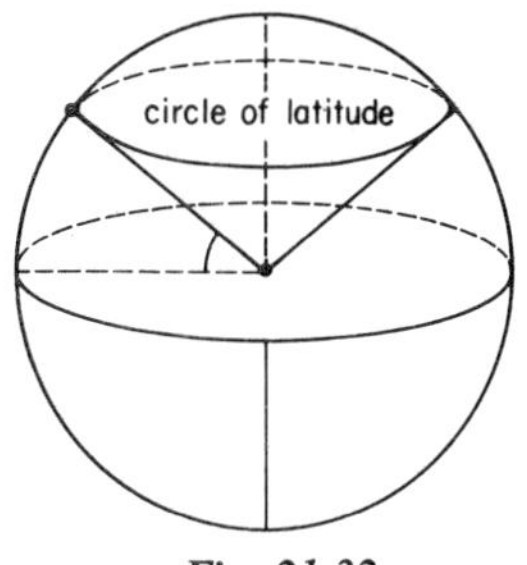

Fig. 21.32

The *radius* of a circle of latitude can be calculated trigonometrically. For example, we can find the radius of the circle of latitude passing through London ($51\frac{1}{2}$°N)

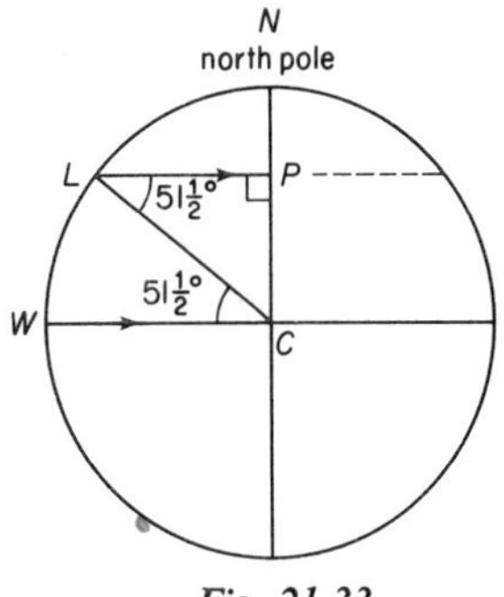

Fig. 21.33

by considering the triangle formed by the axis of the Earth (the line *CN* in Fig. 21.33), the radius of the Earth which passes through London—the line *LC*—and the line segment *LP* perpendicular to *CN*, the radius of the circle of latitude.

Since

$$|CL| = 6400\text{ km and}$$

$$\angle PLC\ (= \angle LCW) = 51\tfrac{1}{2}^\circ\ \text{(alternate angles)},$$

$$\cos \angle PLC = \frac{|LP|}{|CL|}$$

$$\Rightarrow \cos 51\tfrac{1}{2}^\circ = \frac{|LP|}{6400\text{ km}}$$

$$\Rightarrow |LP| = \cos 51\tfrac{1}{2}^\circ \times 6400\text{ km} = 3980\text{ km (to 3 SF)}.$$

21.19 Exercises

1. Study a globe, a three-dimensional model of the Earth, and notice some of the circles of latitude that are marked. Choose some places between these circles of latitude and estimate their latitudes. Compare your results with the information given in geography books, atlases or encyclopedias.
2. Calculate the radius of the circle of latitude which passes through Milan (45°N).

21.20 Longitude

Section 21.18 makes it clear that a given latitude alone is not sufficient to fix the position of a point on the Earth's surface: there are infinitely many points which all have the same latitude.

A *different* system of circles, all perpendicular to the equator and to the circles of latitude, provide a 'second co-ordinate' for each point on the Earth's surface, the *meridians* or *circles of longitude.*

The circles of longitude are all great circles which pass through both the north and the south poles, and which contain the Earth's axis of rotation (Fig. 21.34).

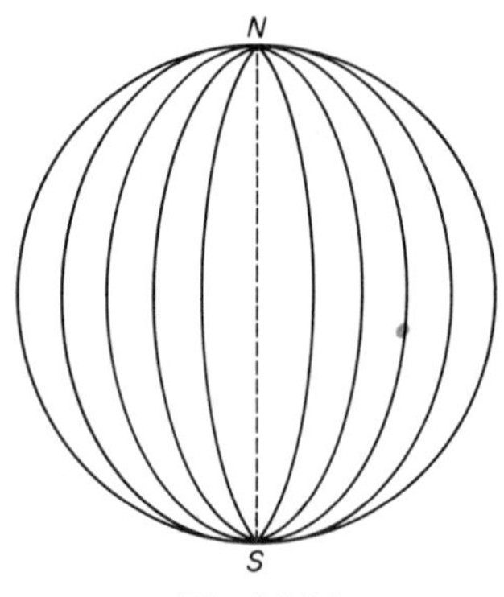

Fig. 21.34

Every point on the Earth's surface lies on such a great circle, one of which has been chosen—quite arbitrarily—as a plane of reference like the co-ordinate axes of a plane co-ordinate system. This circle of longitude is the one which passes through Greenwich, i.e. the Greenwich meridian.

Every other circle of longitude is then named by the angle it makes with the Greenwich meridian—the angle between the two planes.

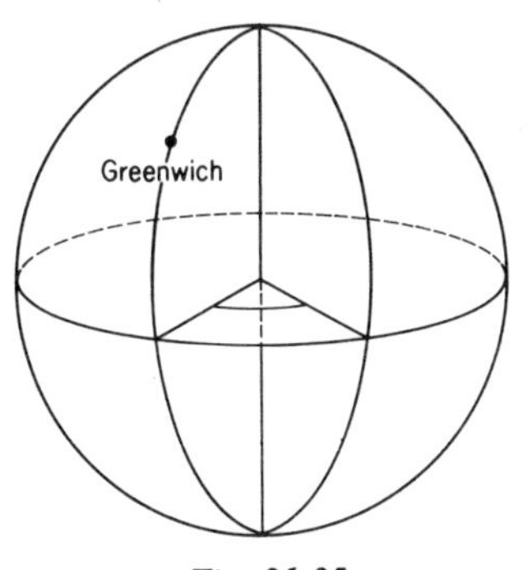

Fig. 21.35

A three-dimensional model will show this more clearly than the sketch in Fig. 21.35. Again, imagine yourself standing at the centre of the Earth on the plane of the equator: you can 'see' the equator as a giant protractor on which the Greenwich meridian is the 'initial position', the zero point (Fig. 21.36).

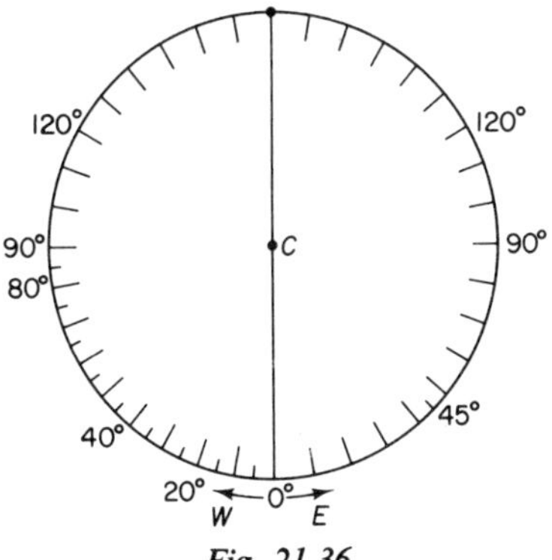

Fig. 21.36

The angle between a circle of longitude and the Greenwich meridian is measured in degrees 'west or east of Greenwich'.

For example, Moscow lies on a circle of longitude which makes an angle of 37° with the Greenwich meridian, and which lies to the east of Greenwich; we say that the longitude of Moscow is 37° East (37°E). Rio de Janeiro lies on a circle of longitude which makes an angle of 43° with the Greenwich meridian, to the west of Greenwich; we say that the longitude of Rio de Janeiro is 43° West (43°W).

The position of a point on the Earth's surface is precisely fixed if both its latitude and longitude are known. Just as the position of a point in a plane can be described as the point of intersection of two lines parallel to the x-axis and y-axis (see Section 12.2), the position of a point on the Earth's surface can be regarded as the intersection of two circles, the circle of latitude and the circle of longitude. For example, the position of Moscow is stated exactly and unambiguously as (56°N, 37°E).

Distances along circles of longitude and latitude

The distance between two points on the same circle of longitude is the length of the arc of this circle between these two points. We use the formula for the length of the arc of a circle (determined by the angle of $\alpha°$ at the centre):

$$\text{length of arc} = \frac{\alpha}{360} \times 2\pi r \text{ (see Section 17.10).}$$

Example

The positions of two places P and Q are given as (25°N, 42°E) and (72°N, 42°E) respectively. This means that the two places are on the same circle of longitude (42°E); the angle 'at the centre' is $72° - 25° = 47°$. Taking the radius of the Earth as 6400 km, the length of the arc PQ on the circle of longitude 42°E =

$$\frac{47}{360} \times 2\pi \times 6400 \text{ km} = 5250 \text{ km (to 3 SF).}$$

Note. If the distance between two places on different hemispheres is required, remember to *add* the latitudes. For example, if the positions of points A and B are given as (40°N, 10°E) and (20°S, 10°E), respectively, the 'angle at the centre' is $40° + 20° = 60°$.

To find the *shortest* ('great circle') distance between two places on the same circle of latitude is more complicated. But we can find the distance *along* the circle of latitude as follows: first find the radius of the circle of latitude by the method in Section 21.18, and then calculate the length of the arc as we have just described.

Example

Find the distance between two places A (50°N, 12°W) and B (50°N, 32°W):

Both places lie on the same circle of latitude (50°N); the length of the radius of this 'small circle' is

$$\cos 50° \times 6400 \text{ km} = 4115 \text{ km.}$$

The length of the arc AB

$$= \frac{20}{360} \times 2\pi \times 4115 \text{ km} = 1440 \text{ km (to 3 SF).}$$

If you are using a slide rule or tables, you will find it simpler to combine these calculations: the distance between the places A and B in this example

$$= \frac{20}{360} \times 2\pi \times \cos 50° \times 6400 \text{ km} = 1440 \text{ km (to 3 SF)}.$$

The nautical mile

One possible unit of measure for distances on the surface of the Earth is the *nautical mile*: it is the length of an arc of a great circle which subtends an angle of 1′ (one minute) at the centre.

Therefore, 1 nautical mile (abbreviated to 1 n mile) $= \dfrac{1}{360 \times 60} \times 2\pi R$, where R is the length of the radius of the Earth,

$$\Rightarrow 1 \text{ n mile} \approx \frac{1}{360 \times 60} \times 2\pi \times 6400 \text{ km} = 1{\cdot}862 \text{ km}.$$

$$\boxed{1 \text{ n mile} \approx 1862 \text{ m}}$$

The *knot* is a measure of speed along the Earth's surface:

$$\boxed{1 \text{ knot} = 1 \text{ n mile per hour}}$$

These units are used mainly by navigators in their calculations.

21.21 Exercises

1. Make yourself familiar with the system of locating points on the Earth's surface; study maps or, if possible, a globe, by finding the position of a place, given its latitude and longitude, and vice versa.
2. Find the distance between the following pairs of points on the Earth's surface:

   A (12°N, 27°W) and B (48°N, 27°W);
   K (41°S, 46°E) and L (19°N, 46°E);
   P (36°N, 25°W) and Q (36°N, 15°E along the circle of latitude).

Summary of Theorems

Theorem		*page*
14.1	The bisectors of the angles formed by two intersecting lines are perpendicular to one another.	261
15.1	Each exterior angle of a triangle is equal in size to the sum of the interior opposite angles.	272
15.2	If two sides of a triangle have the same length, the triangle is isosceles and the base angles are equal.	292
15.3	If two angles of a triangle are equal, the triangle is isosceles and therefore two sides are of equal length.	293
15.4	If, in a triangle, the altitude from a vertex coincides with the median from that vertex, the triangle is isosceles.	293
15.5	If the bisector of the angle C of a triangle ABC coincides with the median through the vertex C, the triangle ABC is isosceles.	294
15.6	If the three sides of a triangle are of equal length, the triangle is equilateral and therefore the angles are equal.	297
15.7	If the three angles of a triangle are equal, the triangle is equilateral and therefore the sides are of equal length.	298
15.8	Pythagoras' Theorem:	
	In a right-angled triangle, the square on the hypotenuse equals the sum of the squares on the other two sides.	301
	Alternative proof of Pythagoras' Theorem using similarity.	408
	Alternative proof of Pythagoras' Theorem using trigonometric ratios.	448
16.1	The sum of the angles of any quadrilateral is 360°.	308
16.2	If the four sides of a quadrilateral are two pairs of adjacent sides of equal length, the quadrilateral is a kite.	313
16.3	If a quadrilateral is symmetrical about one of its diagonals it is a kite, and therefore two pairs of adjacent sides are of equal lengths.	314
16.4	If the opposite sides of a quadrilateral are of equal length, the quadrilateral is a parallelogram.	322
16.5	If the opposite sides of a quadrilateral are parallel, the quadrilateral is a parallelogram.	322
16.6	If the diagonals of a quadrilateral bisect each other, the quadrilateral is a parallelogram.	323
16.7	If one pair of opposite sides of a quadrilateral are parallel and of equal length, the quadrilateral is a parallelogram.	324
16.8	If the opposite angles of a quadrilateral are equal, the quadrilateral is a parallelogram.	324
16.9	The line drawn through the mid-point of one of the sides of a triangle, parallel to another side, is half the length of that side and bisects the third side. (See also the more general Intercept Theorem 18.1, p. 386.)	327

Theorem *page*

20.2 If the ratio of the corresponding sides of two similar triangles is $n:k$, the ratio of their areas is $n^2:k^2$. 468

The Sine Rule 440

In a triangle, the ratio of the sine of each angle and the length of the side opposite that angle is constant: in $\triangle ABC$,

$$\frac{a}{\sin A} = \frac{b}{\sin B} = \frac{c}{\sin C}.$$

The Cosine Rule 445

(extension of Pythagoras' Theorem):
In $\triangle ABC$,

$$a^2 = b^2 + c^2 - 2bc \cos A;$$
$$b^2 = a^2 + c^2 - 2ac \cos B;$$
$$c^2 = a^2 + b^2 - 2ab \cos \mathrm{C}.$$

Summary of Constructions

Part Five

The Calculus

Unit Twenty-two

The Differential Calculus

22.1 The Measurement of Change

Some of the greatest Greek mathematicians described the world of mathematics as a strange heaven where numbers and geometrical shapes lead a divine existence, and can be described by terms such as 'perfect', 'timeless', 'motionless' and 'unchangeable'.

They were acutely aware of the great contrast between this static world of geometry and number, and the real world where everything moves and changes with time.

Change, so typical of our own world, proved to be one of the greatest mysteries they encountered, a mathematical problem far harder to grapple with than the relative simplicity of the idealized, motionless world of geometry. For centuries, mathematicians searched for the tools to deal with this most characteristic aspect of their own experience. The study of simple motion led in the end to the mathematical solution of the problem of change, and to an entirely new branch of mathematics called the *calculus*.

Our world is constantly changing and varying in many ways, and so the calculus has proved itself the most powerful mathematical tool in all the physical sciences, from the examination of the motion of stars and planets in the universe to the study of the physics of the atomic nucleus, and even in the social sciences as well.

In this Unit and the next, we will introduce some of the basic principles and methods of the calculus, and illustrate them by considering the various aspects of *simple motion*: displacement, velocity and acceleration. Before embarking on this Unit, be sure that you are thoroughly familiar with Units 12 and 13 on *graphs*, especially Sections 12.16 to 12.19 on *gradients*.

The branches of the calculus

There are two main branches of the calculus: the differential calculus and the integral calculus.

The *differential calculus* is a method of calculating the rate at which a given quantity is changing. In a graph which relates varying quantities, change is represented by the *slope* or *gradient* of the graph: the differential calculus enables us to calculate this gradient, even if the graph is not a straight line but a curve.

The *integral calculus* may be loosely described as a method of summing up quantities that are changing; for example, we can use it to calculate the distance covered by a car whose speed is changing continually. It also provides methods of measuring the areas and volumes of shapes bounded by curves or curved surfaces, i.e. lines or surfaces that continually change their direction. In terms of graphs, the integral calculus is a method of calculating the *area under the graph*, that is the area between the graph and the x-axis over a given range of values of x.

22.2 Gradients

The gradient of a straight-line graph over a certain interval was described in Section 12.16 as the ratio of the lengths of two sides of a right-angled triangle. For example, the gradient of the graph $y = 2x + 3$ in the interval $x = 1$ to $x = 2$ is the ratio of the lengths of the two sides PR and QR in Fig. 22.1, that is the ratio $\frac{|PR|}{|QR|} = \frac{2}{1}$: in trigonometrical terms, tan Q.

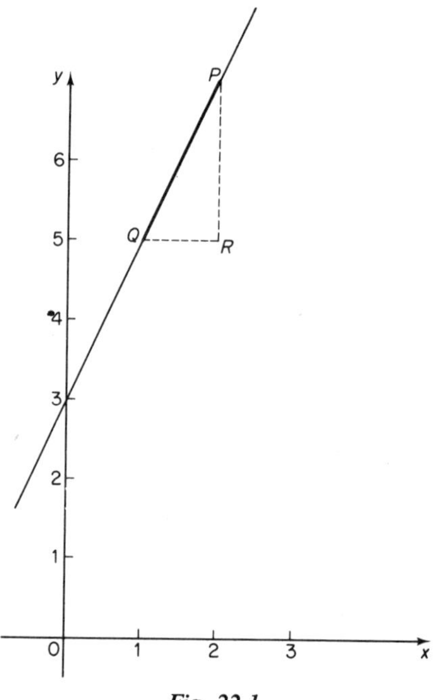

Fig. 22.1

In general terms, this ratio was described as

$$\frac{\text{the increase in the value of } y}{\text{the increase in the value of } x}$$

within the interval.

We found as well from the table of the graph—and from the nature of the straight line itself—that the gradient of a straight-line graph is always the same, whatever interval is considered. Moreover, the gradient can be read off immediately if the equation of the graph is written in the form $y = mx + c$; it is the value m (see Section 12.16). In a straight-line graph, the section of the graph to be considered is the hypotenuse of a right-angled triangle (in Fig. 22.1, $\triangle PQR$). If the graph is not a straight line but a curve, we cannot find such a right-angled triangle, or indeed speak of a gradient in the same sense that we have so far.

We may, however, join two points on the curve by a straight line, say PQ in Fig. 22.2.

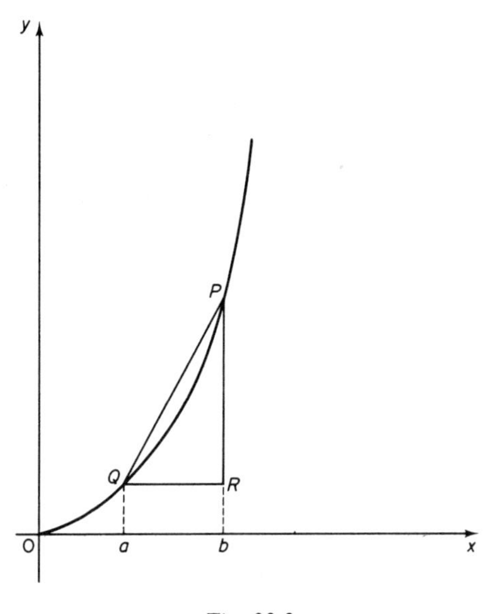

Fig. 22.2

The gradient of this straight line can be expressed in the usual way, the ratio $\frac{|PR|}{|QR|}$, and this is the *average gradient* of the graph over the interval from $x = a$ to $x = b$.

We could think of the graph in Fig. 22.3 as the outline of a mountain. The average gradient over the interval from F to C is then the gradient of the dotted line FT, the ratio $\frac{|TC|}{|FC|}$. Imagine a mountaineer climbing this mountain starting from point F; the average gradient of his climb is the ratio between the total increase in height ($|TC|$) and the horizontal distance travelled ($|FC|$).

But clearly the average gradient of his ascent is not necessarily the same as the average gradient over a small part of the climb. Compare, for example, the average gradient over the interval from A to B with the average gradient over the interval from F to C.

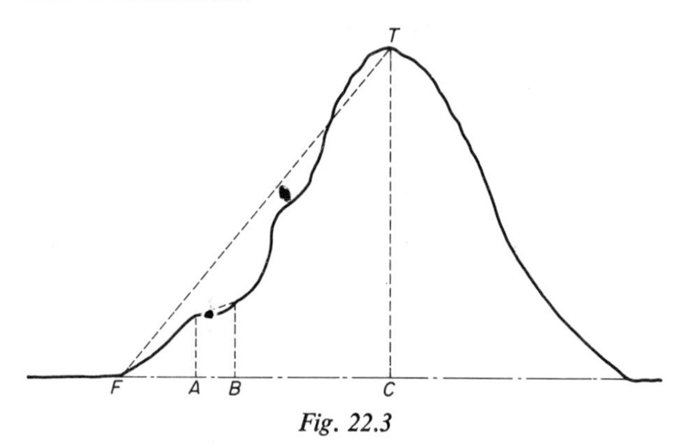

Fig. 22.3

22.3 Exercises

1. Sketch the graph of $y = x^2$ for values of x between 0 and $^+5$ using a scale of 2 cm to represent one unit on the x-axis and $\frac{1}{2}$ cm to represent one unit on the y-axis. (Set up a table first, as we did in Section 13.1.)
2. Join the points (0, 0) and (1, 1) by a straight line and write down the average gradient over the interval $x = 0$ to $x = 1$. Repeat this for the intervals

$$x = 1 \text{ to } x = 2,$$
$$x = 2 \text{ to } x = 3,$$
$$x = 3 \text{ to } x = 4 \quad \text{and}$$
$$x = 4 \text{ to } x = 5,$$

and compare the values of the 'gradient of the graph' over these intervals.
3. Using coloured pencils for clarity, draw straight lines to join the points:

(0, 0) and (2, 4),
(0, 0) and (3, 9),
(0, 0) and (4, 16),
(0, 0) and (5, 25),

and confirm that the values of the gradient also vary when a larger interval is considered.

22.4 The Gradient of a Curve at a Point

If a car travels a total distance of 60 km in one hour, we say that its *average speed* is 60 km per hour. Every driver knows that this does not necessarily mean that throughout the journey the speedometer needle was pointing at 60 km, nor that the car travelled 1 km during each minute of that hour; its speed may well have been much higher during some periods and it may have slowed down or even stopped altogether at other times. The average speed calculated over one hour is no guarantee that at any particular instant during that hour—say at 11 a.m.—the car was moving at that speed.

A better indication of the actual speed of the car at that instant is the average speed calculated over a much shorter interval, for example, by considering the 3 km travelled between 10.59 a.m. and 11.01 a.m. Indeed, the shorter the interval of time, the more accurately the average speed indicates the actual speed at a given moment during that interval.

If we consider the notion of speed remembering that

$$\text{speed} = \frac{\text{distance travelled}}{\textit{period} \text{ of time}},$$

we may even conclude that it is impossible to speak of speed at a given point in time; we must always consider a *period* of time. Some philosophers drew the conclusion that speed—and similarly any other form of change—is impossible, which starkly contradicts the experiences of real life.

The invention of the calculus overcame this difficulty. Accepting that an interval of time *is* necessary for the description of speed, we then recognize that we may make this interval infinitesimally small, that is to say, as small as we please. As the intervals are made smaller and smaller, the average speed gets nearer and nearer, or *approaches*, a value which mathematicians call the *limit*, or the *limiting value*. In the mathematical model, that is the graph, of this distance–time relation, the limit at a given point is something very real; it is the gradient of the tangent to the curve at that point.

Fig. 22.4 shows part of the graph of $y = x^2$.

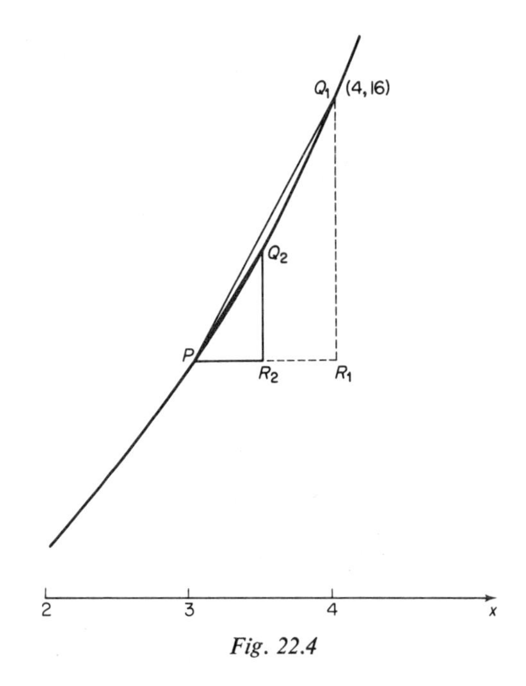

Fig. 22.4

Suppose that we need to find the gradient of this graph near the point where $x = 3$; we first consider the interval between $x = 3$ and $x = 4$; the average gradient over this interval is read off as $\frac{7}{1}$, the gradient of the line PQ_1 (see exercises in Section 22.3). A better approximation of the gradient at the point $x = 3$ is the average gradient over the interval $x = 3$ to $x = 3\frac{1}{2}$, the gradient of the line PQ_2 in Fig. 22.4.

$$\text{The gradient of } PQ_2 = \frac{|Q_2R_2|}{|PR_2|};$$

$$|PR_2| = 3\tfrac{1}{2} - 3 = \tfrac{1}{2},$$

$$|Q_2R_2| = (3\tfrac{1}{2})^2 - 3^2$$

$$= 12\tfrac{1}{4} - 9 = 3\tfrac{1}{4}.$$

$$\text{Therefore, the gradient of } PQ_2 = \frac{3\frac{1}{4}}{\frac{1}{2}} = 6\tfrac{1}{2}.$$

Now we consider an even smaller interval, between $x = 3$ and $x = 3\frac{1}{3}$:

$$\text{the gradient of } PQ_3 = \frac{|Q_3R_3|}{|PR_3|}$$

$$= \frac{(3\frac{1}{3})^2 - 3^2}{3\frac{1}{3} - 3}$$

$$= \frac{2\frac{1}{9}}{\frac{1}{3}} = 6\tfrac{1}{3}.$$

In the same way we can calculate the average gradient over the interval between $x = 3$ and $x = 3\frac{1}{4}$, the gradient of PQ_4, and will find that the gradient of $PQ_4 = 6\frac{1}{4}$; similarly, the gradient of $PQ_5 = 6\frac{1}{5}$, the gradient of $PQ_6 = 6\frac{1}{6}$, and so on. (Check this, as an exercise.)

The reason for this pattern becomes clear if instead of a fraction $\frac{1}{2}$, or $\frac{1}{3}$, we use a small value k, and analyse the calculation of the average gradient over the interval between $x = 3$ and $x = 3 + k$ (Fig. 22.5).

As before,

$$\text{the gradient of } PQ = \frac{|QR|}{|PR|};$$

$$|PR| = (3 + k) - 3 = k,$$

$$|QR| = (3 + k)^2 - 3^2$$

$$= (3^2 + 6k + k^2) - 3^2$$

$$= 6k + k^2.$$

Therefore,

$$\text{the gradient of } PQ = \frac{6k + k^2}{k}$$

$$= \frac{6k}{k} + \frac{k^2}{k}$$

$$= 6 + k.$$

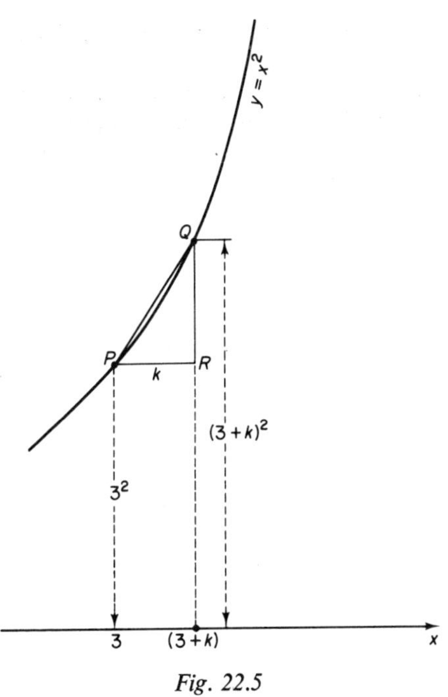

Fig. 22.5

If we consider the interval between $x = 3$ and $x = 3\frac{1}{7}$, the average gradient of the curve over this interval is $6 + \frac{1}{7} = 6\frac{1}{7}$.

The smaller the length, k, of the interval, the nearer the average gradient of the line PQ is to the value 6. We say that 6 is the *limit* or *limiting value* of the gradient of this curve at the point $x = 3$. It is the precise value (not an approximation) of the gradient of the *tangent* to the curve at that point.

22.5 Exercises

1. Sketch the graph of $y = x^2$.
   Calculate the average gradient of this curve over the intervals between:

$$x = 4 \text{ and } x = 4\tfrac{1}{2};$$
$$x = 4 \text{ and } x = 4\tfrac{1}{3};$$
$$x = 4 \text{ and } x = 4 + k.$$

   State the limiting value of the gradient of $y = x^2$ at the point $x = 4$.
2. In the same way, find the limiting value of the gradient of $y = x^2$ at the point $x = 5$.

22.6 Differentiation

Another pattern emerges from the last exercise: the limiting value of the gradient of $y = x^2$ at a given point x is twice the value of x. Thus at the point

$x = 3$, the limiting value of the gradient is 6; at the point $x = 4$, the limiting value of the gradient is 8, and so forth.

Again the reason for this pattern becomes clear if we think of the situation in general terms, instead of specifying any particular value of x.

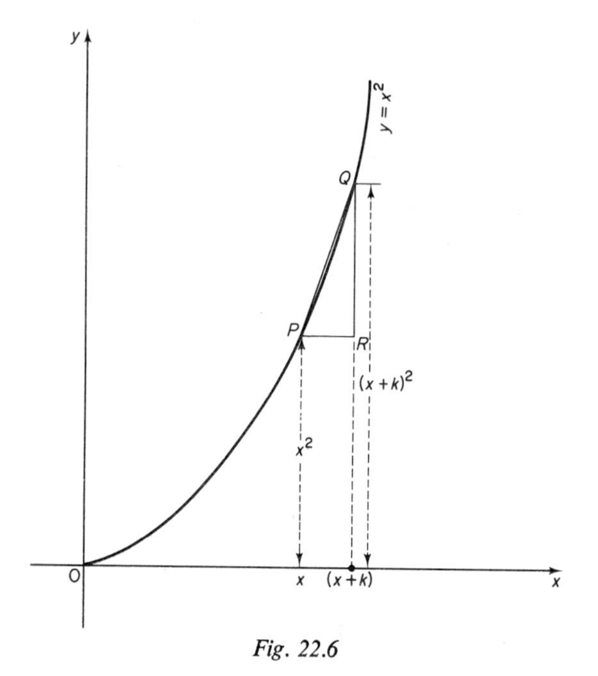

Fig. 22.6

The average gradient of the curve $y = x^2$ over the interval between x and $(x + k)$ is the gradient of the straight line marked PQ in Fig. 22.6.

$$\text{The gradient of } PQ = \frac{|QR|}{|PR|} = \frac{(x + k)^2 - x^2}{(x + k) - x}$$

$$= \frac{(x^2 + 2xk + k^2) - x^2}{k}$$

$$= \frac{2xk + k^2}{k}$$

$$= 2x + k.$$

As k becomes smaller and smaller (that is, as k approaches or *tends to* zero) the average gradient approaches the limiting value $2x$.

This process of calculating the limiting value of the gradient of a curve is called *differentiation* or *derivation*. This limiting value, $2x$, for the function $y = x^2$ is called 'the *derivative* of the function y with respect to x', written as $\frac{dy}{dx}$.

This notation, $\frac{dy}{dx}$, should be considered as a single abbreviation in which the two letters 'd' always remain the same, and the two remaining letters refer to the two variables of the function. For example, 'differentiating' $u = v^2$,

we write
$$\frac{du}{dv} = 2v;$$

or
$$s = t^2 \Rightarrow \frac{ds}{dt} = 2t, \text{ and so on.}$$

22.7 Exercises

1. Knowing that the gradient of the graph of $y = x^2$ is given by the formula $\frac{dy}{dx} = 2x$ for any value of x, find the gradient of $y = x^2$ at the points:

$$x = 6;$$
$$x = 2\tfrac{1}{2};$$
$$x = -2.$$

2. Sketch the graph of $y = x^2$ over the range $x = -4$ to $x = +4$ and confirm that the gradient of $y = x^2$ is negative when x is negative (compare Section 12.18).

22.8 The Derivative of $y = x^n$

We can find the limiting value of the gradient at any point of the graph $y = x^3$, using a similar procedure ('from first principles').

The average gradient of the curve over the interval between x and $x + k$ is the gradient of the line marked as PQ in Fig. 22.7.

$$\text{The gradient of } PQ = \frac{|QR|}{|PR|} = \frac{(x+k)^3 - x^3}{(x+k) - x}$$
$$= \frac{x^3 + 3x^2k + 3xk^2 + k^3 - x^3}{k}$$
$$= \frac{3x^2k}{k} + \frac{3xk^2}{k} + \frac{k^3}{k}$$
$$= 3x^2 + 3xk + k^2.$$

As the value of k becomes smaller and smaller, that is, as k tends to zero, the value of $3x^2 + 3xk + k^2$ approaches the value $3x^2$ ($3xk$ and k^2 become smaller and smaller as k becomes smaller and smaller, and therefore tend to zero as k does).

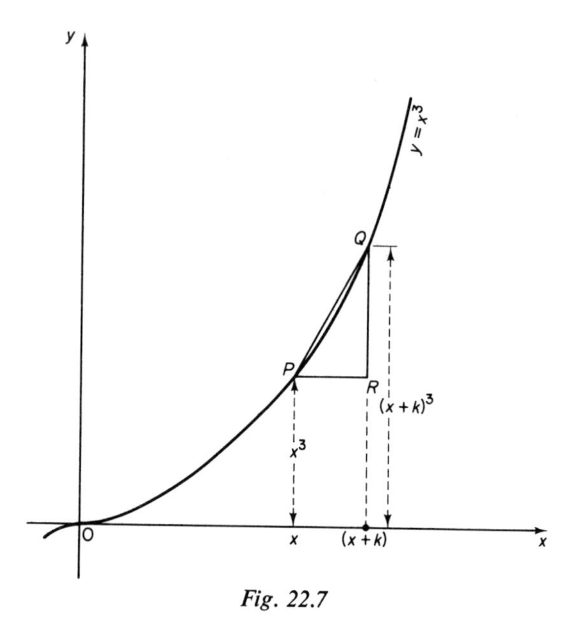

Fig. 22.7

The limiting value of the gradient of the graph of $y = x^3$ is therefore $3x^2$, or

$$y = x^3 \Rightarrow \frac{dy}{dx} = 3x^2.$$

We can use the same method to show that

$$\text{if } y = x^4 \text{ then } \frac{dy}{dx} = 4x^3;$$

$$\text{if } y = x^5 \text{ then } \frac{dy}{dx} = 5x^4;$$

$$\text{if } y = x^6 \text{ then } \frac{dy}{dx} = 6x^5, \text{ and so forth.}$$

In general,

$$\boxed{y = x^n \Rightarrow \frac{dy}{dx} = nx^{n-1}}$$

22.9 Exercises

1. Use the formula to write down the derivative of each of the following functions:

$$y = x^{10};$$
$$s = t^3;$$
$$v = t^4.$$

2. When we apply the general formula for differentiating a function $y = x^n$ to the function $y = x^1$ (a straight line), we find $\frac{dy}{dx}$ to be $1 \times x^{1-1} = 1 \times x^0 = 1$ (see Section 9.16). Compare this with our earlier result, that the gradient of the graph of $y = mx + c$ is m (see Section 12.16).
3. Draw the graph of $y = 3$ (a *constant function*) and state its gradient. Remembering that the zero power of any number (except 0) equals 1—that is, $x^0 = 1$—we may write this function as $y = 3x^0$.

   Test whether the general formula for differentiating powers of x also applies when the power is zero. Confirm that the derivative of a constant function is zero.
4. The general formula

$$y = x^n \Rightarrow \frac{dy}{dx} = nx^{n-1}$$

also applies when n is a negative number or a fraction. For example,

$$\text{if } y = \frac{1}{x^2} = x^{-2}, \text{ then } \frac{dy}{dx} = -2 \times x^{-2-1}$$

$$= -2x^{-3}$$

$$= \frac{-2}{x^3}.$$

Similarly, $\quad$ if $y = x^{1\frac{1}{2}}$, then $\frac{dy}{dx} = 1\frac{1}{2} \times x^{1\frac{1}{2}-1} = 1\frac{1}{2}x^{\frac{1}{2}}$

$$= 1\tfrac{1}{2}\sqrt{x} \text{ (see Section 9.19).}$$

Write down the derivative of each of the following functions:

$$y = \frac{1}{x} (= x^{-1});$$

$$y = \frac{1}{x^3};$$

$$y = \sqrt{x} (= x^{\frac{1}{2}});$$

$$y = \frac{1}{\sqrt{x}} (= x^{-\frac{1}{2}});$$

$$y = x^{\frac{1}{3}}.$$

22.10 The Derivative of $y = ax^n$

When we differentiate a function such as $y = 5x^2$ 'from first principles', we find that the constant coefficient, the number 5, does not alter during differentiation, that is $\frac{dy}{dx} = 5 \times 2x$.

Using the same notation as before, the gradient of

$$PQ = \frac{|QR|}{|PR|} = \frac{5(x+k)^2 - 5x^2}{k}$$

$$= 5\left(\frac{(x+k)^2 - x^2}{k}\right)$$

$$= 5(2x + k).$$

As the value of k approaches 0 (that is, as we consider smaller and smaller intervals), the value of $5(2x + k)$ approaches $5 \times 2x$. The gradient of the tangent to the curve, or the derivative, of $y = 5x^2$ is therefore $5 \times 2x = 10x$, or $\frac{dy}{dx} = 5 \times 2x = 10x$.

We chose the coefficient 5 in this example quite arbitrarily; the same rule applies to any number, and to any power of x:

$$y = ax^n \Rightarrow \frac{dy}{dx} = a \times nx^{n-1}$$

For example, if $y = 3x^5$ then $\frac{dy}{dx} = 3 \times 5x^4 = 15x^4$;

or, if $y = {}^-2x^4$ then $\frac{dy}{dx} = ({}^-2) \times 4x^3 = {}^-8x^3$.

22.11 Derivatives of Sums

We could consider the function $y = 3x^5$ as a *sum function*, for example as $y = x^5 + x^5 + x^5$, and its derivative as the sum of the derivatives of each term, that is,

$$\frac{dy}{dx} = 3 \times 5x^4$$

$$= 5x^4 + 5x^4 + 5x^4.$$

This is a particular instance of a more general rule which applies to any function consisting of sums or differences of powers of x (for example, $y = 3x^4 + 2x^3 + 5x - 2$):

The derivative of a sum is the sum of the derivatives of the terms

For example, we can differentiate the function $y = 5x^3 + 3x^2 - 2x$ by differentiating each term in succession: $\frac{dy}{dx} = 5 \times 3x^2 + 3 \times 2x - 2(1)$

$$= 15x^2 + 6x - 2.$$

Since the derivative of a constant function is zero (compare Exercise 3 in Section 22.9), the constant term in any function 'vanishes' when we differentiate.

For example, if $y = 3x^2 - 2x \boxed{+ 4,}$

then $\frac{dy}{dx} = 3 \times 2x - 2 \times 1 \boxed{+ 0}$

$$= 6x - 2.$$

This confirms the results in Section 12.16: that the gradients of the linear graphs of $y = 3x$, $y = 3x + 1$, $y = 3x - 2$, etc., are all the same (3 in each case) and that the gradient of the graph of $y = mx + c$ is m.

22.12 Summary

We can summarize the results of Sections 22.2 to 22.11:

(*a*) 'The gradient of a curve at a point' means 'the gradient of the tangent to the curve at that point'.

(*b*) The gradient of a curve is different at different points; the derivative is a formula for the calculation of the gradient at each point on the curve.

(*c*) If $y = x^n$, then $\frac{dy}{dx} = nx^{n-1}$;

this general rule applies for all values of n, including negative numbers and fractions.

(*d*) In differentiating $y = ax^n$, the coefficient a is unchanged, and $\frac{dy}{dx} = a \times nx^{n-1}$.

(*e*) We differentiate a sum function by differentiating each of its terms in succession.

(*f*) A constant term vanishes in differentiation.

22.13 Exercises

1. Differentiate the following functions:

(*a*) $y = 5x^3 + 2x^2 - 3x + 2$;
(*b*) $y = 2t^2 + 3t - 4$;
(*c*) $y = x^4 - 2\sqrt{x}$ [Hint: $\sqrt{x} = x^{\frac{1}{2}}$];
(*d*) $y = x^2 + \frac{3}{x} + 6$.

2. Find the gradient of the graph of $y = x^2 - x + 2$, at the point $x = 2$, and also at the point $x = {}^-1$. [Hint: first find the derivative of this function; then use it as a formula to calculate the gradient at the required points by substituting the appropriate values for x.] Compare your findings with a sketch of this graph (see Exercise 1 of Section 13.11).

22.14 The Distance–Time Graph

The first stages of the launching of a moon rocket are usually recorded in a table relating the distance the rocket has travelled to the time after lift-off, the 'zero hour', and the values listed in such a table can be plotted graphically. This kind of graph, which relates the distance from a starting point and the time elapsed since the starting time, is called a *distance–time graph*; usually, the time t is measured along the horizontal axis, and the distance s along the vertical axis, as in Fig. 22.8.

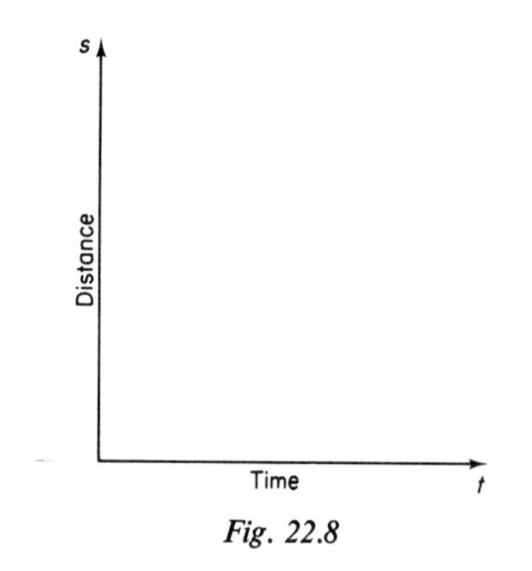

Fig. 22.8

The graph in Fig. 22.9 represents the movements of a cyclist who leaves a point P at 12 noon precisely and travels along a straight road. The units on the horizontal t-axis are minutes (after 12 noon) and each unit on the vertical s-axis corresponds to a distance of 100 m (measured from P).

Reading this graph, we see that during the first minute the cyclist is continually moving further away from P, and after 1 minute (at 12.01) he is at a point 250 m from P.

After another minute, at 12.02, he is still 250 m from P. Indeed, throughout the time between 12.01 and 12.02 he remains 250 m from P: the horizontal direction of the graph—the zero gradient of this section—indicates that there is no movement during that time, that is, no speed. After this minute's rest, the cyclist sets off again; in fact the graph shows that during the third minute he covers 350 m, a greater distance than during the first minute: he is moving faster, and we say that his speed is therefore greater.

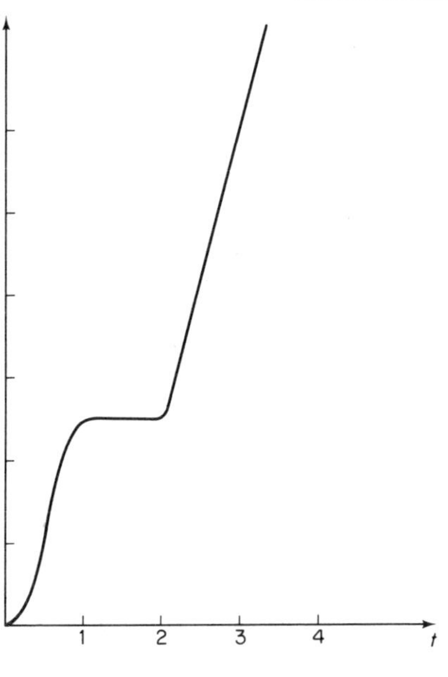

Fig. 22.9

In a distance–time graph, the total distance travelled is given by the second co-ordinate, the s-co-ordinate; the velocity, or speed, is represented by the gradient of the graph, $\frac{ds}{dt}$, that is, the rate at which the total distance travelled changes with the time.

Note. We will use the word *velocity* in the ordinary sense of *speed.*

If the relation between 'distance from point P' and 'time after the start' is expressed by a function, for example, $s = t^2 + 2t$,

(*a*) the *distance travelled* at a particular time is found by substituting that time for t; thus, for example, if s is measured in metres and t in seconds, the distance from P after 3 seconds is $(3^2 + 2 \times 3)$ m $= 15$ m;

(*b*) the *velocity* is given by the derivative of the function, $\frac{ds}{dt} = 2t + 2$ (the gradient of the tangent at the point t); the velocity at a particular time is found by substituting that time for t in the expression $2t + 2$.

For example, after 3 seconds the velocity is $(2 \times 3 + 2)$ m/s $= 8$ m/s. In general terms,

$$\text{velocity} = \frac{ds}{dt}$$

22.15 Exercises

1. The distance (s) of a moving object from a point P after t seconds is given by $s = t^3 + 3t^2 + 2t$. Find a general formula relating the velocity (v) of the object and the time (t).
2. The distance, s metres, of a moving object from a point P, t seconds after the starting time is given by $s = t^2 + t$. Plot the distance–time graph, using a scale of 1 cm to represent one unit (1 second) on the t-axis and 1 cm to represent one unit (1 m) on the s-axis.
   By differentiation and substitution, calculate the velocity of the object after 4 seconds.
   Sketch the tangent to the curve at the point $t = 4$.

22.16 The Velocity–Time Graph

In the last exercise we saw that the velocity of an object can be found, or *derived*, from the distance–time relation. From the relation $s = t^3 + 3t^2 + 2t$ we established, by differentiation, the relation $v = 3t^2 + 6t + 2$.

We can also read off the velocity of the object at any time from the distance–time graph, by considering the gradient of the graph at that time.

The *velocity–time* graph gives the relation between velocity and time more immediately: velocity is plotted against time, i.e. the second co-ordinate, v of any point (t, v) on the graph is the velocity at the time represented by the first co-ordinate (t) (Fig. 22.10).

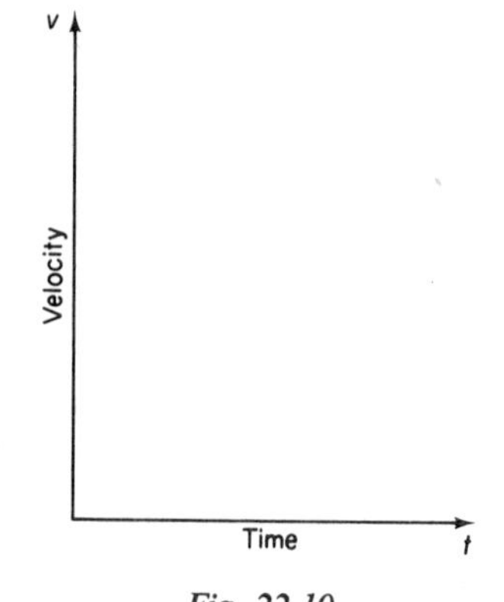

Fig. 22.10

In Fig. 22.11, each unit of the v-axis represents 1 m/s, and each unit on the t-axis represents 1 second.

If the relation between velocity and time is expressed by an equation, for example, $v = 4t^2 + t + 3$ (v is the velocity in m/s, and t is the time in seconds),

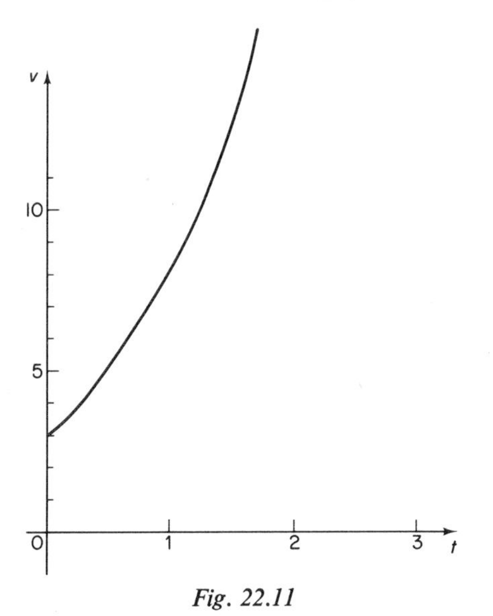

Fig. 22.11

(*a*) the velocity at any instant is found by substitution; for example, the velocity after 2 seconds $= (4 \times 2^2 + 2 + 3) = 21$ m/s.

(*b*) by differentiating $v = 4t^2 + t + 3$ we derive a general formula for *acceleration*, that is, the rate at which the velocity is changing, the gradient of the tangent at t:

$$\frac{dv}{dt} = 8t + 1.$$

The acceleration at a given time is found by substituting this time for t in the expression $8t + 1$; for example, after 5 seconds the acceleration $=$ $(8 \times 2 + 1)$ m/s/s, always written as 17 m/s^2, and read as '17 metres per second per second'.

Note that mathematicians use the term 'acceleration' to describe any kind of change in velocity, not just increasing velocity as is usual in ordinary conversation. Acceleration in the mathematical sense can have either a positive value (speeding up) or a negative value (slowing down).

22.17 Exercises

1. The velocity (v metres per second) of a rocket at t seconds after the start is given by the formula $v = 4t^3 + 2t^2$. Calculate its acceleration after 2 seconds and after 10 seconds.
2. The distance (s metres) that a moon rocket has travelled is given by the formula

$$s = 3t^4 + 2t^3,$$

where t is the time (measured in seconds) after launching.

(*a*) By differentiating, find a general formula relating its velocity (v) and time.
(*b*) Having established this formula, find, by differentiation, a formula relating the rocket's acceleration and time.
(*c*) Calculate the acceleration of the rocket after 4 seconds.

22.18 Second Derivatives

The formula for acceleration in the last exercise was derived from the equation relating distance and time by differentiating *twice.*

Such a 'derivative of a derivative' is called a *second derivative*. A special notation for the second derivative of any function s with respect to t is $\frac{d^2s}{dt^2}$.

Note. $\frac{d^2s}{dt^2}$ is *not* the square of $\frac{ds}{dt}$, i.e. $\left(\frac{ds}{dt}\right)^2$; it is a single notation for the result of differentiating a function s with respect to t twice in succession.

In general, if s is a function expressing distance in terms of time t, then

$$\text{velocity} = \frac{ds}{dt}$$

$$\text{acceleration} = \frac{d^2s}{dt^2}$$

22.19 Maxima and Minima

The derivation of formulae for velocity and acceleration from a formula relating distance and time is only one of the many applications of differentiation. Another important application is a geometrical one: the derivative of a function tells us something about the general shape of the graph of that function, in particular whether a graph has a *maximum* or a *minimum* (see Section 13.14), and if so, where it occurs.

We will only consider maxima and minima of *quadratic functions,* that is, functions such as $y = x^2 + 3x + 1$ or in more general terms, a function of the form $y = ax^2 + bx + c$ where a, b and c are constants and a is not zero ($a \neq 0$).

If we consider the general appearance of the graph of $y = ax^2 + bx + c$ after the graph had been plotted, we find that whenever a—the coefficient of x^2—is not zero, the graph is a curve.

If $a > 0$, the graph is U-shaped and has a lowest point, a *minimum*;
if $a < 0$, the graph has an inverted U-shape, and has a highest point, a *maximum.*

Fig. 22.12 shows two examples.

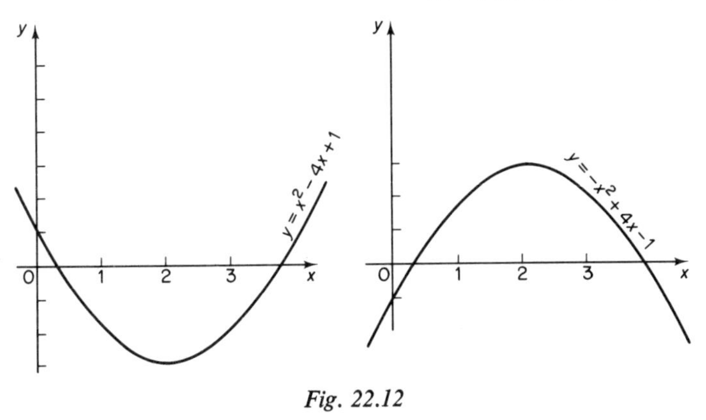

Fig. 22.12

Now we will consider the graph of $y = x^2 - 4x + 1$ more closely, studying its gradient (Fig. 22.13).

Moving along the graph to the right, the general direction is at first downward—a *negative gradient*; at the point whose x-co-ordinate is 2 the graph flattens out, and the gradient is 0; further to the right of $x = 2$, the general direction of the curve is upward, and the gradient is *positive*.

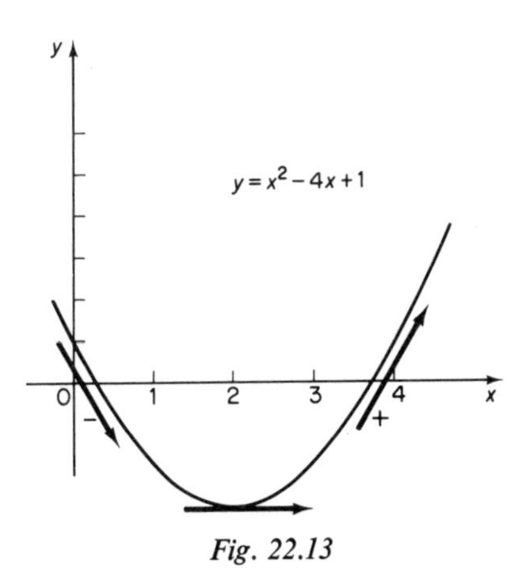

Fig. 22.13

Since the derivative of $y = x^2 - 4x + 1$ is a formula giving the gradient of the graph *at any point*, we should be able to learn something about the general appearance of the curve itself, that is from the derivative, $\frac{dy}{dx} = 2x - 4$.

This derivative is another function, and its graph can also be drawn. Fig. 22.14 is *not* the graph of $y = x^2 - 4x + 1$, but the graph of its gradient-function, $\frac{dy}{dx} = 2x - 4$, and it tells us that:

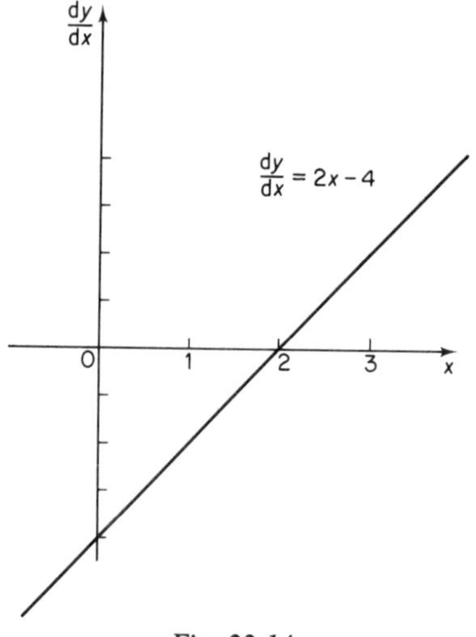

Fig. 22.14

at the point $x = 2$, the value of the gradient, $\frac{dy}{dx}$, $= 0$;

at the point $x = 3$, the value of the gradient, $\frac{dy}{dx}$, $= 2$;

at the point $x = 1$, the value of the gradient, $\frac{dy}{dx}$, $= -2$.

We have already seen that the curve reaches its lowest point, its ***minimum***, when the gradient is zero, that is, when

$$2x - 4 = 0.$$

By solving this equation, we can calculate the value of x at the point where the graph reaches its minimum value without drawing graphs at all.

$$\begin{aligned} 2x - 4 &= 0 \\ \Rightarrow 2x &= 4 \\ \Rightarrow x &= 2. \end{aligned}$$

The minimum value of y is found by substituting 2 for x in the original function, $y = x^2 - 4x + 1$:

$$y = 2^2 - 4 \times 2 + 1 = -3.$$

Fig. 22.15 shows the graph of $y = -x^2 + 2x + 3$ and, immediately below, the graph of its gradient-function, $\frac{dy}{dx} = -2x + 2$.

Moving along the curve from left to right, we notice that at first the direction of the curve is upwards; the value of the gradient is positive. At the point $x = 1$ the gradient of the curve is zero, and to the right of this point the direction of the curve is downwards; the gradient of the curve is negative.

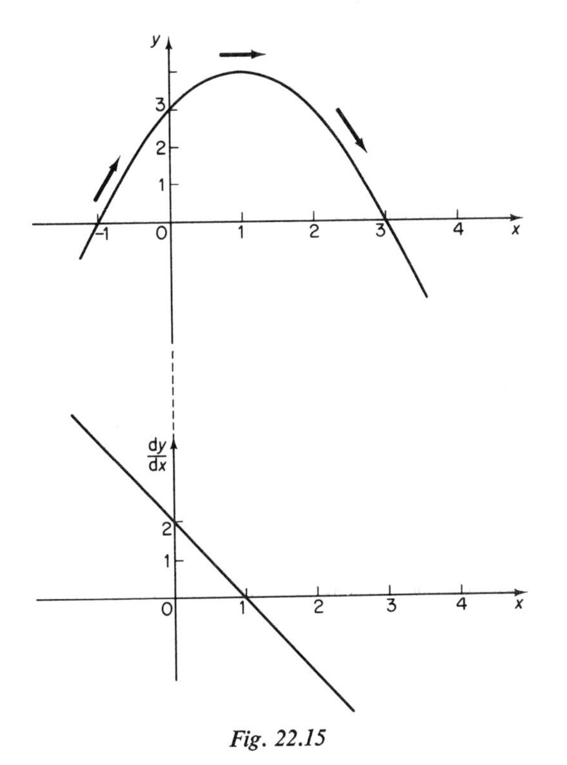

Fig. 22.15

The curve reaches its highest point, its *maximum*, at the point $x = 1$ where its gradient is zero, that is where $\frac{dy}{dx} = -2x + 2 = 0$. This value of x is confirmed by solving the equation:

$$\begin{aligned} -2x + 2 &= 0 \\ \Rightarrow 2x &= 2 \\ \Rightarrow x &= 1. \end{aligned}$$

To find the maximum or minimum of any function which has the form

$$y = ax^2 + bx + c:$$

(*a*) differentiate the function (calculate $\frac{dy}{dx}$);
(*b*) equate the derivative to zero;
(*c*) solve this equation to find x (the first co-ordinate of the maximum or minimum point);
(*d*) substitute this value of x in $y = ax^2 + bx + c$ to calculate y, the second co-ordinate of the maximum or minimum point.

Example

$$y = x^2 - 3x + 4.$$

Find the minimum value of y by first finding the value of x for which $y = x^2 - 3x + 4$ is a minimum.

(*a*) $\dfrac{dy}{dx} = 2x - 3;$

(*b*) $\dfrac{dy}{dx} = 0$ if $2x - 3 = 0$

(*c*) $\Rightarrow 2x = 3$

$\Rightarrow \quad x = 1\tfrac{1}{2};$

(*d*) the corresponding value of $y = (1\tfrac{1}{2})^2 - 3(1\tfrac{1}{2}) + 4 = 1\tfrac{3}{4}.$

The minimum value of y is $1\tfrac{3}{4}$; it has this value when $x = 1\tfrac{1}{2}$.

We can *distinguish between a maximum and a minimum* by considering the value of the gradient on either side of the point where the gradient equals zero.

In the case of a maximum (Fig. 22.16), the value of the gradient passes through three stages: positive–zero–negative as we move from left to right.

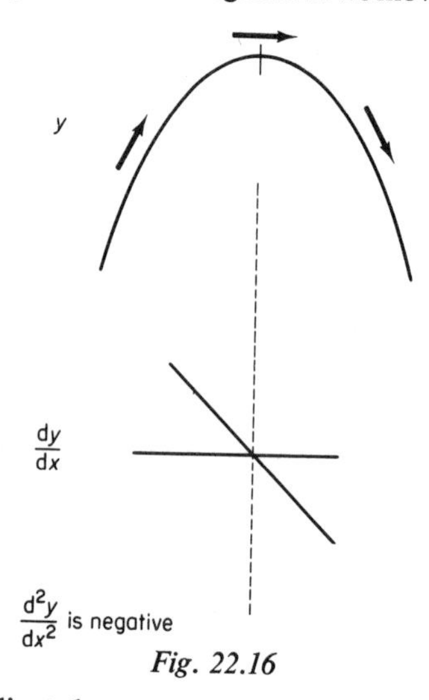

Fig. 22.16

The value of the gradient decreases as we proceed to the right; that is, the gradient of the 'graph of the gradient' is *negative*, or $\dfrac{d^2y}{dx^2} < 0$.

Clearly, this happens when the coefficient of the quadratic term in the function is negative.

$$\text{For example, } y = -3x^2 + 2x - 4$$

$$\Rightarrow \frac{dy}{dx} = -6x + 2$$

$$\Rightarrow \frac{d^2y}{dx^2} = -6.$$

In the case of a minimum (Fig. 22.17), the value of the gradient passes through three stages: negative–zero–positive as we proceed to the right.

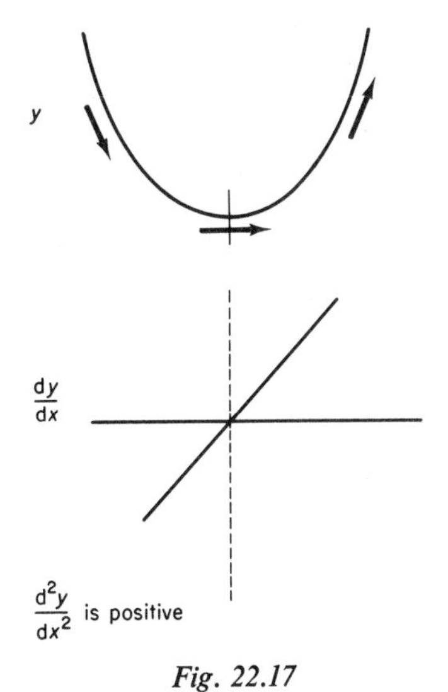

Fig. 22.17

The gradient of the graph of the gradient is *positive*, or $\frac{d^2y}{dx^2} > 0$.

$$\text{For example, } y = 3x^2 + 5x - 2$$
$$\Rightarrow \frac{dy}{dx} = +6x + 5$$
$$\Rightarrow \frac{d^2y}{dx^2} = +6.$$

We can establish whether the graph of $y = ax^2 + bx + c$ has a minimum or a maximum as follows:

> if a is a negative number ($a < 0$), the graph has a maximum;
> if $a > 0$, the graph has a minimum.

22.20 Exercises

1. Without drawing the graph, establish whether the following functions have a maximum or a minimum:

$$y = 3x^2 + 2x - 4;$$
$$y = -2x^2 + 2;$$
$$y + 3x^2 - 2x - 4 = 0.$$

[Hint: first write this equation in the 'right' form.]

2. Find out whether the graph of $y = x^2 + 2x + 1$ has a maximum or a minimum.
   If so, calculate this maximum or minimum value of y, and the value of x where this maximum or minimum is reached.
3. Find the maximum area of a rectangle whose perimeter is 24 cm. [Hint: let the length of one side of the rectangle be x cm; its breadth is then $(12 - x)$ cm. The area of the rectangle, A, is then $A = x(12 - x)$ $\Rightarrow A = 12x - x^2 = -x^2 + 12x$. Then find $\frac{dA}{dx}$ and calculate the maximum value of A.]

Unit Twenty-three

The Integral Calculus

23.1 Integration as the Inverse Process of Differentiation

In Unit 13, we established the shape of the graph of a function such as $y = x^2$ by plotting points and sketching a smooth curve through the plotted points. Then in Section 22.19 on maxima and minima, we found that the general shape of a graph can also be deduced from its gradient-function, in this case $\frac{dy}{dx} = 2x$, also written $y' = 2x$.

Using the original function $y = x^2$, we can find the precise value of y for any value of x, and so the exact position of each of the points; for example,

$$x = 3 \Rightarrow y = 3^2 = 9.$$

Using the gradient-function $y' = 2x$, we can establish in what direction the graph is proceeding, the 'trend' of the graph, at any stage. For example, at the point where $x = 3$, the gradient of the graph, y', $= 2 \times 3 = {}^{+}6$, that is, the graph is moving 'upwards'. The gradient-function $y' = 2x$, moreover, tells us that, as we move to the right of this point, the graph becomes steadily steeper (y' gets larger as x gets larger). The gradient-function, y', is an excellent indicator of the general shape of the graph of y; but it tells us nothing about its exact position.

We have already noticed that we need to know more than the gradient of a straight-line graph to be able to draw it; compare, for example, the linear graphs given in Exercise 2 of Section 12.16, whose gradients are all 3 (whose gradient-functions are all $y' = 3$). The straight-line graph could be drawn once we knew the gradient and the co-ordinates of any one point on the graph, for example the intercept on the y-axis.

In the same way, $y' = 2x$ is the gradient-function of $y = x^2$, $y = x^2 + 3$, $y = x^2 - 2$, and so on (Fig. 23.1).

When we differentiate any of these functions, the added constants vanish; however, we can 'reconstruct' part of the function y from the gradient-function $y' = 2x$ by *reversing the process of differentiation*. We can conclude that $y = x^2 +$ some constant; usually we write $y = x^2 + c$.

This process of reversing differentiation is known as *integration*;

$y = x^2 + c$ is the *general integral of* $y' = 2x$
and the constant c is the *constant of integration*.

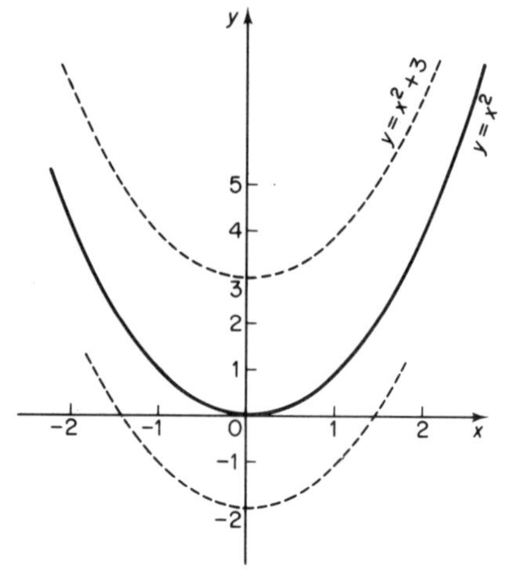

Fig. 23.1

23.2 The Integral of x^n

When we said that $2x$ is the result of differentiating x^2, we relied to some extent on memory. We can, however, establish rules for integrating $2x$ by tracing the steps we took when we differentiated x^2.

To differentiate $y = x^2$ with respect to x (or simply to differentiate x^2 with respect to x):

(*a*) multiply by the 'old' power of x, that is, 2: $(2x^2)$;
(*b*) subtract 1 from the power of x: $(2x^{2-1} = 2x^1 = 2x)$.

Reversing this process, we integrate $2x^1$ by

(*a*) adding 1 to the power of x: $(2x^{1+1} = 2x^2)$;
(*b*) dividing by the 'new' power, that is, by 2: $\left(\frac{2x^2}{2} = x^2\right)$.

In general, to find the integral of x^n with respect to x:

(*a*) add 1 to the power of x: (x^{n+1});
(*b*) divide by the 'new' power: $\left(\frac{x^{n+1}}{n+1}\right)$.

Note. This procedure is not valid for $n = -1$.

After the constant of integration has been added, the integral of x^n is then $\frac{x^{n+1}}{n+1} + c$. We do not know the value of c, unless we are given some more information; this general integration is therefore called *indefinite* integration.

Examples

The integral of x^5 is $\frac{x^{5+1}}{5+1} + c = \frac{x^6}{6} + c$;

the integral of $3x^7$ is $\frac{3x^{7+1}}{7+1} + c = \frac{3x^8}{8} + c$;

the integral of 3 ($= 3x^0$) is $\frac{3x^{0+1}}{1} + c = 3x + c$;

the integral of $x^4 + 3x^2 + 2x + 1$ is

$$\frac{x^5}{5} + \frac{3x^3}{3} + \frac{2x^2}{2} + x + c = \frac{x^5}{5} + x^3 + x^2 + x + c.$$

Checking. Integration is easily checked by differentiating the result. Using the examples above,
the derivative of $\frac{x^6}{6} + c$ is $\frac{6x^{6-1}}{6} + 0 = x^5$;
the derivative of $3x^1 + c$ is $1 \times 3x^{1-1} = 3x^0 = 3$.

23.3 Exercises

Integrate the following functions with respect to x:

x^4;
$3x^2 + x + 4$;
$\sqrt{x}\ (= x^{\frac{1}{2}})$;
$\frac{1}{x^2}\ (= x^{-2})$.

Add more examples of your own and check your answers by differentiation.

23.4 The Integration Sign

A standard notation for 'the integral of . . . with respect to x' is $\int \ldots \mathrm{d}x$. The combination of the symbol $\int$, an elongated S, and $\mathrm{d}x$, is to be taken as a single notation. For example:

$\int x^2 \mathrm{d}x$ = the integral of x^2 with respect to x;

$\int (x^2 + 3x - 2)\, \mathrm{d}x$ = the integral of $x^2 + 3x - 2$ with respect to x.

23.5 Exercises

Find the following indefinite integrals:

$$\int x^3 \mathrm{d}x;$$

$$\int (x^4 + 2x + 1)\mathrm{d}x;$$

$$\int (x - 1)^2 \mathrm{d}x \; [= \int (x^2 - 2x + 1)\mathrm{d}x];$$

$$\int \frac{\mathrm{d}x}{x^4} \text{ [another way of writing } \int \frac{1}{x^4}\mathrm{d}x \text{ is } \int x^{-4}\mathrm{d}x];$$

$$\int t^2 \mathrm{d}t.$$

Do not forget the constant of integration.

23.6 The Constant of Integration

We have described integration as a process of reconstructing a function (y) from its gradient-function (y'). By integrating the gradient-function y' we determine the whole of the function y, except that the added constant, c, remains unknown.

An example will show that this constant of integration is nothing but the *intercept* on the y-axis (as we found in the equation of the straight line). For example, the graph of $y = x^2 + x - 2$ (Fig. 23.2) meets the y-axis at the point where $x = 0$; substituting 0 for x, we find $y = -2$ at this point.

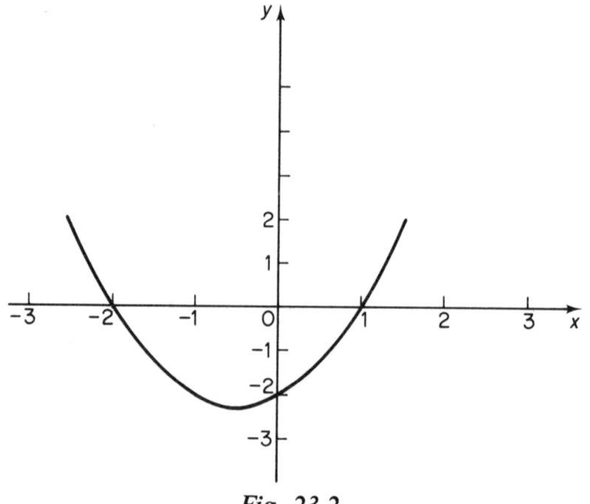

Fig. 23.2

Therefore, if the intercept of a graph of a function on the y-axis is known *and* its gradient-function (y') is known, we can completely reconstruct the function y.

For example, suppose that a curve passes through the point (0, 1) and its gradient-function is known to be $y' = 2x - 1$. Integrating $2x - 1$, we find $y = \int (2x - 1)\mathrm{d}x = x^2 - x + c$. Substituting the values $x = 0$ and $y = 1$ in $y = x^2 - x + c$,

$$1 = 0 - 0 + c$$
$$\Rightarrow c = 1.$$

Therefore, the equation of this curve is $y = x^2 - x + 1$.

By the same method of substitution, we can determine the constant of integration if *any* point on the curve is known.

Example

Find the equation of the graph which passes through the point (2, 5), whose gradient-function is $y' = 6x + 2$.
By integration,

$$y = \int (6x + 2)\mathrm{d}x$$
$$= 3x^2 + 2x + c.$$

Since $y = 5$ when $x = 2$,

$$5 = 3(2)^2 + 2 \times 2 + c$$
$$\Rightarrow 5 = 12 + 4 + c$$
$$\Rightarrow c = -11.$$

Therefore, the required equation is $y = 3x^2 + 2x - 11$.

23.7 Exercises

1. Find the equation of the curve through the point (0, 2), whose gradient-function is $y' = 2x - 3$.
2. Find the equation of the curve through the point (3, 1), whose gradient-function is $y' = x + 2$.
3. An object travels in a straight line and passes a point O at a speed of 3 m/s. Its acceleration t seconds after passing point O is given by $(3t + 1)$ m/s². Find a general formula for its speed v in terms of the time t.

23.8 The Area Under the Graph

Many applications of the integral calculus concern the measurement of shapes; they depend on an aspect of integration which at first sight seems to have little connexion with the reversal of the differentiation process.

So far, we have considered the integration of a function such as $y' = 2x$ as a process of determining *another* function, y, of which $y' = 2x$ is the gradient-function; geometrically, the function $y' = 2x$ is represented by the gradient of the graph of $y = x^2 + c$, the integral of $y' = 2x$. But the integral of $y' = 2x$ has an interpretation in the graph of $y' = 2x$ itself: the *integral of* $y' = 2x$ *is represented by the area under the graph* of $y' = 2x$, that is the area between the graph and the x-axis. Since it is only meaningful to speak of the area of a *closed* shape, we must consider the area under the graph within a definite interval of values of x, for example, the area bounded by the graph, the x-axis and the lines $x = 1$ and $x = 3$ in Fig. 23.3.

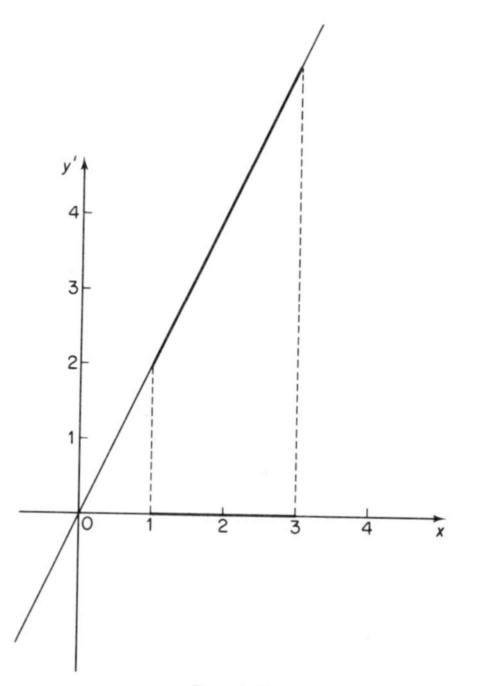

Fig. 23.3

An integral considered as the area under the graph between definite limits is therefore called a *definite integral.* In the notation of a definite integral, the limits of the interval under consideration are written in the integration sign, the larger number at the top and the smaller number at the bottom, for example, $\int_1^3 2x\mathrm{d}x$.

The area bounded by the graph of $y' = 2x$, the x-axis and the lines $x = 1$ and $x = 3$ is completely determined by the graph of $y' = 2x$; it does not depend on a constant of integration. The definite integral, therefore, has *no constant of integration.*

We will demonstrate that the area under the graph of a function does represent the integral of that function, using as an example $y = 2x$ (Fig. 23.4).

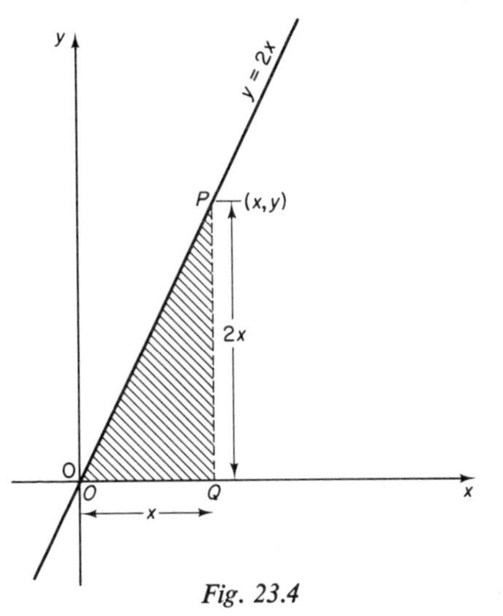

Fig. 23.4

Let P be any point (x, y) on the graph of $y = 2x$. We will consider in turn the area under the graph between the limits 0 and x, and the integral of $y = 2x$ in this interval.

The area under the graph is the area of the right-angled triangle OPQ, in which the lengths of the perpendicular sides OQ and PQ are x and $2x$ respectively.

The area under the graph in this interval = area of $\triangle OPQ$

$$= \tfrac{1}{2}x \times 2x = x^2.$$

By the usual method of integration, we calculate the integral of $y = 2x$, leaving out the constant of integration, and find $\int 2x\mathrm{d}x = x^2$.

By integrating the function $y = 2x$ with respect to x we have obtained a formula for the area under the graph between the limits 0 and x. The area under the graph *within specific limits* is found by substituting the appropriate values for x in this formula. For example, between the limits 0 and a, the area under the graph of $y = 2x$

$$= \int_0^a 2x\mathrm{d}x = a^2.$$

$$\text{If } a = 3, \text{ area} = \int_0^3 2x\mathrm{d}x = 3^2 = 9;$$

$$\text{if } a = 2, \text{ area} = \int_0^2 2x\mathrm{d}x = 2^2 = 4.$$

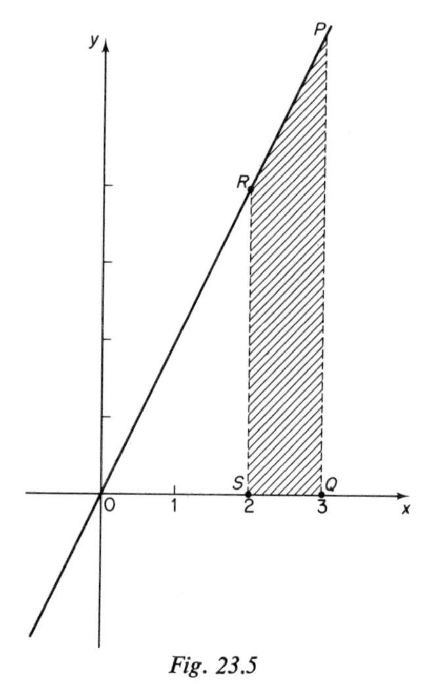

Fig. 23.5

The area under the graph between the limits $x = 2$ and $x = 3$ (the area of the trapezium $PQSR$ in Fig. 23.5) $=$ (area of $\triangle OPQ$) $-$ (area of $\triangle ORS$),

or
$$\int_2^3 2x\,dx = \int_0^3 2x\,dx - \int_0^2 2x\,dx$$
$$= 3^2 - 2^2$$
$$= 9 - 4$$
$$= 5.$$

It can be shown that for any function y (whether its graph is a straight line or a curve), the area under the graph between the limits $x = a$ and $x = b$ represents the value of the definite integral:

$$\int_a^b y\,dx = \int_0^b y\,dx - \int_0^a y\,dx, \text{ or}$$

$$\boxed{\text{area} = \int_a^b y\,dx}$$

Integration is therefore a powerful tool for the calculation of areas of shapes bounded by curves.

23.9 Evaluating Definite Integrals (Calculating Areas)

To evaluate a definite integral such as $\int_2^5 (x^2)\mathrm{d}x$, for example, use the following procedure and notation:

(*a*) Use the usual method to integrate the function:

write $$\int_2^5 x^2\mathrm{d}x = \left[\frac{x^3}{3}\right]_2^5;$$

(*b*) substitute the upper limit (5), and the lower limit (2), for x, and find the difference:

$$\left(\frac{5^3}{3}\right) - \left(\frac{2^3}{3}\right) = \frac{125}{3} - \frac{8}{3}$$
$$= \frac{117}{3} = 39.$$

Make yourself familiar with this notation and method of working, and use them in all calculations of this kind.

Example

$$\int_1^2 (x^3 + 2x + 1)\mathrm{d}x$$
$$= \left[\frac{x^4}{4} + x^2 + x\right]_1^2$$
$$= \left(\frac{2^4}{4} + 2^2 + 2\right) - \left(\tfrac{1}{4} + 1 + 1\right)$$
$$= 10 - 2\tfrac{1}{4}$$
$$= 7\tfrac{3}{4}.$$

2. Calculate the area bounded by the x-axis and the curve $y = -x^2 + 4x - 3$ (Fig. 23.6).
   First we solve the quadratic equation $-x^2 + 4x - 3 = 0$ to find the two points where the curve cuts the x-axis.
   Factorizing LHS, $-(x - 1)(x - 3) = 0$
   $\Rightarrow x = 1$ or $x = 3$.
   The area of the shaded portion is found by evaluating the definite integral:

$$\int_1^3 (-x^2 + 4x - 3)\mathrm{d}x$$
$$= \left[-\frac{x^3}{3} + \frac{4x^2}{2} - 3x\right]_1^3$$
$$= \left(-\frac{27}{3} + \frac{36}{2} - 9\right) - \left(-\frac{1}{3} + 2 - 3\right)$$
$$= 0 - (-1\tfrac{1}{3})$$
$$= +1\tfrac{1}{3} \text{ units of area.}$$

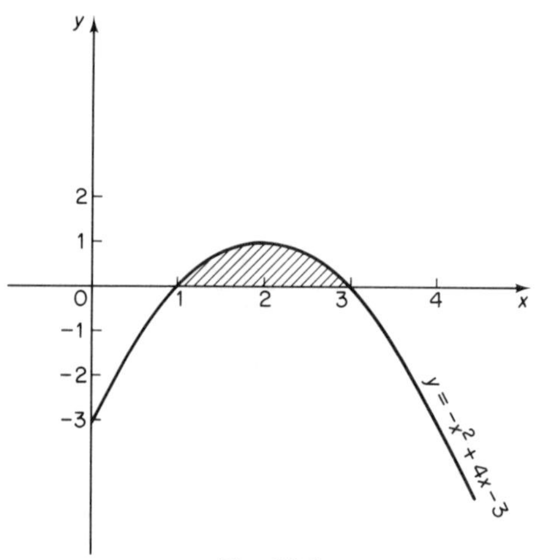

Fig. 23.6

An area must of course be measured in *units*, and we usually write 'square units' or 'units of area' after a result obtained in this way. But if the units of length are given as, say, cm, then the area should be given in cm^2.

23.10 Exercises

1. Evaluate the following definite integrals:

$$\int_2^8 x^2 \mathrm{d}x;$$

$$\int_2^3 (x^3 - 2)\mathrm{d}x;$$

$$\int_1^4 (x^2 + 4x + 2)\mathrm{d}x.$$

2. Calculate the areas cut off the following graphs by the x-axis:

   (*a*) $y = -(x - 1)(x - 4)$;
   (*b*) $y = -x^2 + 5x - 6$.

3. The velocity v of a rocket t seconds after its launching is given by the formula $v = (t^3 + t)$ m/s. Calculate the distance it travels during the first 10 seconds after take-off. $\left[\text{Hint: since } v = \frac{\mathrm{d}s}{\mathrm{d}t}, s = \int v\mathrm{d}t.\right]$

23.11 Volume of Rotation About the x-axis

If the triangle formed by the graph of $y = 2x$, the x-axis and the line $x = 3$ is made to rotate about the x-axis, it generates a cone (see Section 21.14).

The volume of this cone, generated by 'the area under the graph' of y, is given by the formula:

$$V = \pi\int_0^3 y^2 dx$$

$$\Rightarrow V = \pi\int_0^3 (2x)^2 dx$$

$$= \pi\int_0^3 4x^2 dx$$

$$= \pi\left[\frac{4x^3}{3}\right]_0^3$$

$$= \pi\left(\frac{4 \times 27}{3}\right) - (0)$$

$$= 36\pi.$$

If the units of length are cm, the volume of the cone is 36π cm^3. Again, if no units are stated, work in general units, in this case 'cubic units' or 'units of volume'.

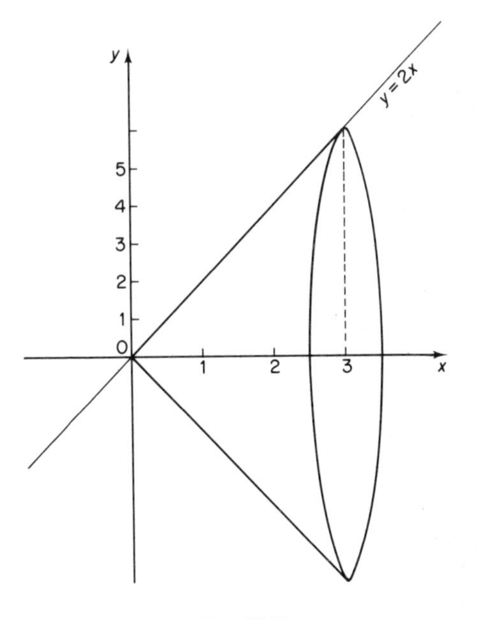

Fig. 23.7

It can be shown that the volume of any solid—that is, three-dimensional shape—generated by rotating about the x-axis a section of the area under a graph of a function y between the limits $x = a$ and $x = b$ is given by the formula:

$$\text{volume} = \pi\int_a^b y^2 dx$$

Examples

1. The area bounded by the curve $y = x^2$, the x-axis and the lines $x = 1$ and $x = 2$ is rotated about the x-axis. The volume generated is

$$\begin{aligned}\pi\int_1^2 y^2 dx &= \pi\int_1^2 (x^2)^2 dx \\ &= \pi\int_1^2 x^4 dx \\ &= \pi\left[\frac{x^5}{5}\right]_1^2 \\ &= \pi\left(\frac{2^5}{5}\right) - \pi\left(\frac{1^5}{5}\right) \\ &= \left(\frac{32}{5} - \frac{1}{5}\right)\pi \\ &= 6\tfrac{1}{5}\pi \text{ units of volume.}\end{aligned}$$

2. The equation of the circumference of the circle whose centre is the origin and whose radius is 5 is $y^2 + x^2 = 5^2$ or $y^2 = 5^2 - x^2$ (Fig. 23.8).

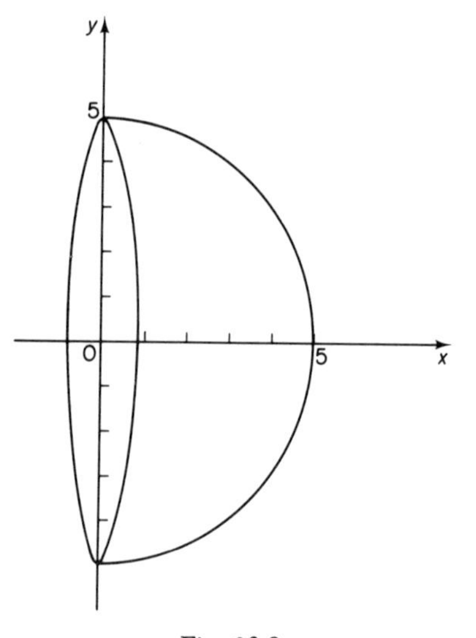

Fig. 23.8

Rotating about the x-axis the sector (quadrant) above the x-axis, between the limits $x = 0$ and $x = 5$, generates a hemisphere; its volume is

$$\begin{aligned}\pi\int_0^5 y^2 dx &= \pi\int_0^5 (5^2 - x^2)dx \\ &= \pi\left[5^2x - \frac{x^3}{3}\right]_0^5 \\ &= \pi\left(5^3 - \frac{5^3}{3}\right) \\ &= \tfrac{2}{3}\pi 5^3 \text{ units of volume.}\end{aligned}$$

The volume of the whole of the sphere of radius 5 is

$$2 \times \tfrac{2}{3}\pi \times 5^3 = \tfrac{4}{3}\pi \,.\, 5^3.$$

(Compare this with the earlier result in Section 21.16.)

23.12 Exercises

1. Calculate the volume of a solid generated by rotating about the x-axis the section under the curve of $y = x^2 - x$ between the limits $x = 2$ and $x = 3$.
2. Calculate the volume of a solid generated by rotating about the x-axis the area bounded by the curve of $y = x^2$, the x-axis and the line $x = 1$. Similarly, calculate the volume of a solid generated by rotating about the x-axis the area bounded by the curve $y = x^3$, the x-axis and the line $x = 1$.

   By subtraction, find the volume of the solid generated by rotating about the x-axis the area bounded by the curves of $y = x^2$ and $y = x^3$.

Answers to Exercises

(where appropriate)

Unit 1

1.7 *1.* $8:12 = 16:24 = 2:3$; $12:36 = 36:108 = 1:3$; $3:4 = 9:12 = 36:48$, etc.

2. 1/3, 2/3 and 2/5.

Unit 2

2.6 *1.* $^{+}1$; $^{+}8$; $^{-}6$.

2.11 *1.* $^{+}4 - {}^{-}2 = {}^{+}6$, $^{-}3 - {}^{-}2 = {}^{-}1$.

2. $-{}^{-}2$: to the right.

2.13 *1.* $^{+}6$; 0; $^{-}4$; $^{+}2$.

2.17 *1.* $^{-}5$; $^{-}3$; $^{+}2$; $^{-}4$; $^{+}7$.

Unit 3

3.13 *1.* 1, 2, 3, 5, 7, 11, 13, 17, 19, 23, 29, 31, 37, 41, 43, 47, 53, 59, 61, 67.

3.17 *1.* $24 = 2^3 \times 3$; $210 = 2 \times 3 \times 5 \times 7$; $396 = 2^2 \times 3^2 \times 11$.

2. Common factors $= 2 \times 3 \Rightarrow$ HCF $= 6$.

3. $10 = 2 \times 5$; $12 = 2^2 \times 3$; $18 = 2 \times 3^2$.
LCM $= 2^2 \times 3^2 \times 5 = 180$.

Unit 4

4.2 $\frac{3}{4}$; $\frac{5}{8}$; $^{+}\frac{3}{4}$; $^{-}\frac{1}{4}$; $^{-}\frac{5}{8}$.

4.5 *1.* $\frac{2}{21}$; $\frac{2}{5}$; $\frac{1}{3}$; 2; $4\frac{3}{5}$.

4.7 $^{+}\frac{1}{4}$; $^{-}\frac{5}{6}$.

4.9 $\frac{1}{4}$; $3\frac{1}{3}$; $1\frac{1}{14}$.

4.11 $\frac{1}{4} = \frac{25}{100}$; $\frac{1}{8} = \frac{125}{1000}$; $\frac{1}{20} = \frac{5}{100}$; $\frac{3}{5} = \frac{6}{10}$.

4.14 *1.* $\frac{12}{100} = 0{\cdot}12$; $\frac{125}{1000} = 0{\cdot}125$; $3\frac{75}{100} = 3{\cdot}75$.

2. $\frac{2}{5} = \frac{4}{10} = 0{\cdot}4$; $\frac{3}{8} = \frac{375}{1000} = 0{\cdot}375$; $\frac{3}{4} = \frac{75}{100} = 0{\cdot}75$;
$\frac{5}{16} = \frac{3125}{10\,000} = 0{\cdot}3125$.

4.17 *1.* 3·574; 102·366.

4.19 *1.* $\dfrac{745 \times 45{\cdot}5}{325 \times 56}$.

2. 3·72; 3·7; 0·32; 15; 5625.

4.22 $8{\cdot}235 \times 10^{11}$; $7{\cdot}245 \times 10^2$; $2{\cdot}937 \times 10$; $6{\cdot}84 \times 10^{-3}$; $1{\cdot}36 \times 10^{-1}$.

Unit 5

5.10 *2.* 1·67 m = 16·7 dm = 167 cm;
500 g = 0·5 kg;
345 cm = 34·5 dm;
0·45 dm = 4·5 cm.

3. 3 m^3; 4·3 t.

5.13 *1.* 20·5 cm to 21·5 cm; $\pm 0{\cdot}5$ cm;
34·5 mm to 35·5 mm; $\pm 0{\cdot}5$ mm;
123·5 m to 124·5 m; $\pm 0{\cdot}5$ m.
2. 113 000; 30 000; 10 000.
3. 37 400; 890 000.

5.15 3·16; 296; 0·0147; 5·70.

5.18 4·32; 0·06; 3·20.

5.21 *1.* 5265; 4115; 1403.
2. 2876; 1069; 4606.

5.23 *1.* 144; 1225; 11·56; 2435; 0·2857.
2. $(3{\cdot}456 \times 10)^2 = 1194$; $(1{\cdot}134 \times 10^3)^2 = 1\,286\,000$;
$(7{\cdot}843 \times 10^2)^2 = 615\,200$; $(5{\cdot}642 \times 10^3)^2 = 31\,830\,000$;
$(8{\cdot}423 \times 10^{-3})^2 = 70{\cdot}95 \times 10^{-6} = 0{\cdot}000\,070\,95$.

5.25 19·40; 9·616; 0·050 12.

Unit 6

6.2 *1.* 11, 18, 25, 32, 39, 46, 53, 60.
2. $^-2$, 6, 14, 22, 30, 38, 46, 54.
3. $1\frac{1}{3}$, 2, $2\frac{2}{3}$, $3\frac{1}{3}$, 4, $4\frac{2}{3}$, $5\frac{1}{3}$, 6.
4. 12, 8, 4, 0, $^-4$, $^-8$, $^-12$, $^-16$.
5. $^-5$, $^-7$, $^-9$, $^-11$, $^-13$, $^-15$, $^-17$, $^-19$.

6.4 *1.* 36.
2. 97.
3. 5, 8, 11, 14, 17, 20.
4. $2 + 4n$; $15 - 3n$.
5. $^-7$; 7.

6.6 *1.* 156.
2. 325.
3. $\frac{n}{2}(n + 1)$.
4. 4; 70; 750.

6.8 *1.* 3, 12, 48, 192, 768, 3072.
2. $^-5$, $^-10$, $^-20$, $^-40$, $^-80$, $^-160$.
3. 5, $^-10$, $^+20$, $^-40$, $^+80$, $^-160$.
4. $\frac{1}{2}$, 2, 8, 32, 128, 512.
5. 48, 24, 12, 6, 3, $1\frac{1}{2}$.

6.10 *1.* 12 288; $^-320$; $^+320$; 2048; $\frac{3}{4}$.
2. 3; 9; 9; 27.
3. 3.
4. 16.

6.12 2; $\sqrt{225} = 15$.

6.16 *1.* 12; 4·5; 4; 2·5; 10; 3; 3; 3·4.

6.19 *2.* 8·99; 10·3; 22·4; 12·8; 5·21; 5·8.

6.23 59·5; 155 000; 14 000 000; 2·64; 0·0464; 0·627; 0·118.

6.25 *1.* 7·82.
2. 73 200.

Unit 7

7.4 *1.* 2.
2. 2.

7.6 *1.* 149 km.

2. 5:8; 4:1; 20:73.

7.9 *1.* 1:8:12.

2. £40; 2:7; 2:5.

3. £1250; £250 each.

7.11 *1.* 60%; $6\frac{1}{4}$%; 8%; 5%; $7\frac{1}{2}$%.

£500 is $62\frac{1}{2}$% of £800; £2 is 400% of 50p; 100 days is $27\frac{29}{73}$% of one year.

2. $12\frac{1}{2}$%.

7.13 *1.* 6 h.

2. $12\frac{1}{2}$ m/s.

3. £262 500.

7.15 *1.* £37·50.

2. £15·625.

7.18 *1.* £562·432.

2. £18.

7.20 *1.* 3%; 4%.

2. $\frac{112}{100} = \frac{28}{25}$; or 1·12; $\frac{104\frac{1}{2}}{100}$ or 1·045; $\frac{47}{50}$ or 0·94; $\frac{37}{40}$ or 0·925.

7.22 *2.* £608·3265 (to 4 decimal places); £108·3265.

7.24 *1.* £200; £160; £128.

2. Initial value of plant = £P ⇒ value after n years = £$P \times \left(1 - \frac{R}{100}\right)^n$.

7.26 £80.

Unit 8

8.8 *1.* A' = {students in the college not studying mathematics}

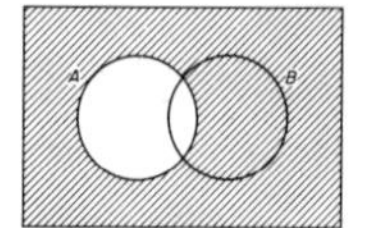

B' = {students in the college not studying English}

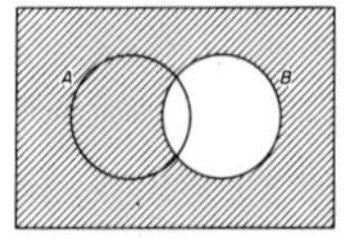

$A \cup B$ = {students in the college studying maths or English or both}

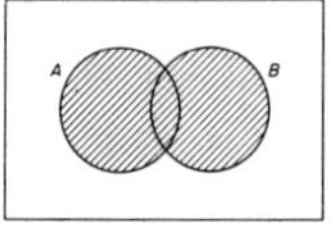

$A \cap B$ = {students in the college studying both maths and English}

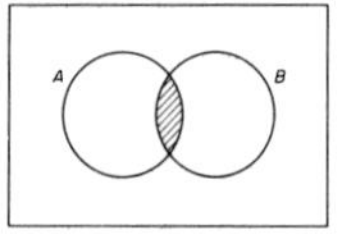

$A' \cap B =$ {students in the college studying English but not maths}

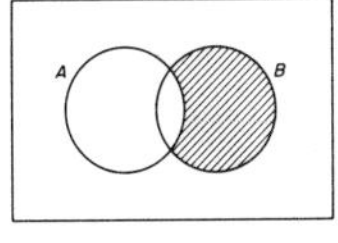

$A \cap B' =$ {students in the college studying maths but not English}

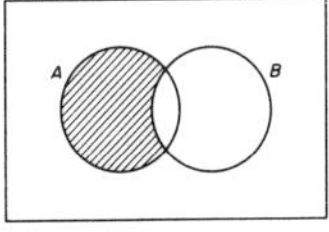

$A' \cap B' =$ {students in the college who are not studying English or maths}

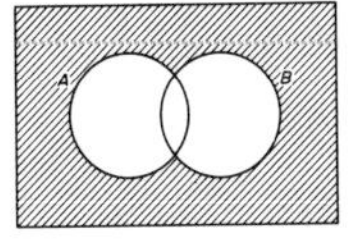

$(A \cap B)' =$ {students in the college who are not studying both English and maths}

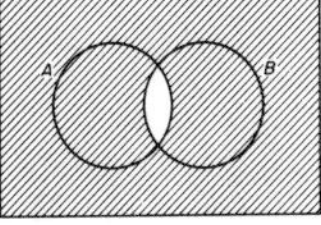

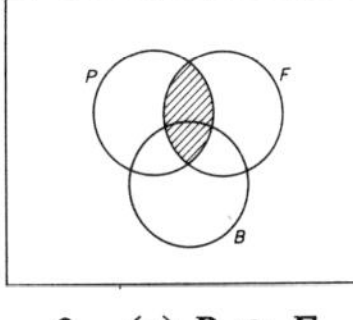

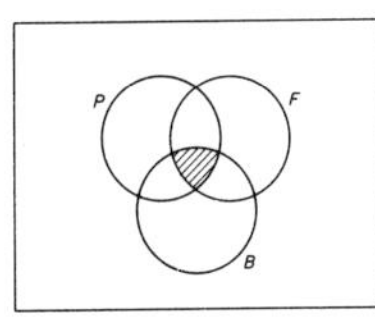

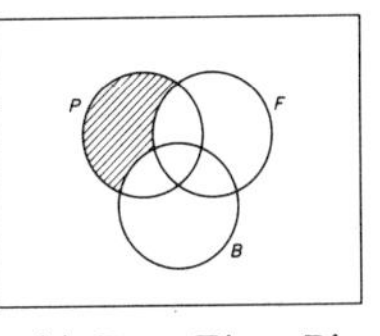

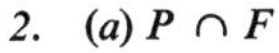

2. (*a*) $P \cap F$ (*b*) $P \cap F \cap B$ (*c*) $P \cap F' \cap B'$

8.10 *1.* 22; 9.
2. 2600; 8000; 200; 4400.

Unit 9

9.5 *2.* $4a - 8b + 4c$; $-6x + 6y$; $-k + l - m$; $-2xa + 6xb + 4xc$.
3. $9 \times (100 - 5) = 900 - 45 = 855$;
$4 \times 996 = 4 \times (1000 - 4) = 4000 - 16 = 3984$.

9.7 $ac - ad + bc - bd$; $3\,km + 6\,kn + lm + 2ln$.

9.9 $x^2 + 5x + 6$; $x^2 + x - 20$; $x^2 - 9x + 20$; $6x^2 + 16x + 8$; $a^2 + 6a + 9$; $x^2 - 4x + 4$.

9.11 *1.* $x^2 + 2xy + y^2$; $4x^2 + 12x + 9$; $9k^2 + 6k + 1$; $4k^2 + 12kl + 9l^2$; $x^2 - 4xy + 4y^2$; $4x^2 - 4xy + y^2$; $16x^2 - 24xy + 9y^2$.
2. 169; 10 404; $110\frac{1}{4}$; 1 006 009.

9.13 $a^2 - 9$; $4x^2 - 25$; $9m^2 - 4n^2$; 9999; $9999\frac{3}{4}$.

9.15 *1.* $x^5 + 2x^4 + 3x^3$; $a^3b^2 + 3a^2b^3 + ab^4$.
2. a^3; a^3; $x^3y + 3xy^2$.

9.17 $\frac{1}{16}$; $\frac{1}{27}$; $\frac{1}{16}$; 1.

9.20 *1.* a^2; x^{-6}; y^2x.
2. 4; 18; 5; 9; $\frac{1}{4}$.

9.23 *1.* $10^{0\cdot5356}$; $10^{0\cdot7260}$; $10^{0\cdot9366}$.
2. 0·5356; 0·7260; 0·9366.

9.25 2·5476; 4·7198; 1·0927.

9.28 *1.* $\bar{2}$·6264; $\bar{1}$·8288; $\bar{4}$·6402.
2. 1·9751; $\bar{3}$·2179.
3. 0·7463; $\bar{2}$·9013; $\bar{3}$·1006; 0·7043.

9.30 211·1; 4·813; 0·2329; 0·002 641.

9.32 *1.* $\bar{1}$·8409; $\bar{1}$·6652; $\bar{1}$·4293.
2. 121·4; 1992; 3·347; 7·299; 0·05176; 199 300; 0·000 382 5; 2·915; 0·7743; 0·3194.

9.34 *1.* 31·76.
2. 0·61.

Unit 10

10.3 $3(a + 2b)$; $2x(x + 3y)$; $3x(x^2 + 4x + 2)$.

10.5 $(a + 2b)(c + d)$; $(k + l)(m - n)$; $(a + 2)(c + 4)$; $(2x + 3)(y - 2)$.

10.7 $(x + 4)^2$; $(x + 10)(x - 10)$; $(a - 5)^2$; $(x + 3y)^2$; $5(2x + 3y)(2x - 3y)$.

10.9 $(x + 3)(x + 6)$; $(a + 7)(a - 2)$; $(m + 4)(m - 5)$; $(x - 3)(x - 5)$.

10.11 $(2y + 1)(y + 2)$; $(3x - 1)(x - 2)$; $(3y + 4)(2y - 1)$.

10.13 $\frac{3a + 4}{a - 2}$; $\frac{1}{x + 1}$.

10.16 $\frac{3b + 2a}{ab}$; $\frac{10 - 7y}{2y^2}$; $\frac{2 + 5a - 5b}{a^2 - b^2}$; $\frac{2x^2 + 5x - 4}{(x - 3)(x + 2)(x + 5)}$; $\frac{mk}{12ln}$.

10.18 $x + 2$; $y - 9$; $2b - 5$, remainder 12; $3x - 4$; $x - 4$, remainder $-10x - 27$.

10.20 *1.* 26; -102.
2. $x - 2$ is a factor of $x^3 - 42x + 76$;
$x + 3$ is *not* a factor of $2x^3 - 3x^2 + 54$.
3. $k = -92$.

Unit 11

11.2 *1.* (*a*) $N + N + 1 + N + 2 + N + 3 = 26 \Rightarrow 4N + 6 = 26 \Rightarrow N = 5$.
(*b*) $x + x + 15 + x + 30 = 180 \Rightarrow 3x + 45 = 180 \Rightarrow x = 45$.
45°, 60°, 75°.
2. $t = 2$; $s = 4$; $x = {}^-9$; $z = 7\frac{1}{2}$; $y = 17$.

11.5 *1.* (*a*) $a = 3$ or ${}^-3$.
(*b*) $b = 6$ or ${}^-6$.
2. 1·2 m.

11.7 $-9, 4$; $-1, -3$; $-3, 2$; $-\frac{1}{3}, -3$; $+\frac{2}{5}, {}^-3$; ${}^+3, -3$.

11.9 8 km/h.

11.11 *1.* $4;\ \frac{49}{4};\ 4;\ 1;\ a^2;\ \frac{a^2}{4}.$

2. $1, -3;\ 9, -1;\ \frac{1}{3}, -\frac{5}{3};\ 0, -4.$

11.13 $\frac{1}{2}, -\frac{2}{3};\ 1, -\frac{1}{5}.$

11.15 *1.* 1·449, −3·449;
3, $\frac{1}{2}$;
4·075, −2·575;
1;
1·422, −0·422.

2. (*a*) $x^2 + 2x - 10 = 0$; $x = 2{\cdot}317, -4{\cdot}317$; sides are 2·317 cm and 4·317 cm.
(*b*) 0·13 s and 7·87 s.

11.17 Factorize, $x = 2$; take square root of both sides, $y = 7$ or $y = {}^-7$; formula, $y = -0{\cdot}088, -1{\cdot}626$ (to 3 decimal places); factorize, $m = -3, -7$.

11.19 *1.* No roots; one root; two roots.

2. (*a*) $k < 4$; (*b*) $k = 4$; (*c*) $k > 4$.

11.22 *1.* $t = -4, v = 26.$

2. $a = 6, b = 3.$

11.24 *1.* (*a*) $p = -5, q = 3.$
(*b*) $x = 4, y = 3.$

2. $5s + 4f = 96,\ 8\frac{1}{2}s + 2f = 96 \Rightarrow s = 8, f = 14.$

11.26 *1.* $\frac{8}{3}, \frac{8}{3}$ and −8, 8.

2. 3, 1 and −2, 6.

3. $-\frac{1}{2}, 2\frac{1}{2}$ and −1, 2.

11.28 *1.* $c = \frac{5}{9}(f - 32).$

2. $R = \frac{E}{C}.$

3. $r = \frac{c}{2\pi}.$

4. $u = \frac{2s}{t} - v.$

Unit 12

12.9 *1.* Mrs Johnson's age = m years; son's age = s years, $m = s + 25$;

s	0	1	2	3
m	25	26	27	28

2. $y = 5x$;

x	0	1	2	3	4	5	6
y	0	5	10	15	20	25	30

$y = 5x + 2$;

x	0	1	2	3	4	5	6
y	2	7	12	17	22	27	32

$y = 5x - 3$;

x	0	1	2	3	4	5	6
y	−3	2	7	12	17	22	27

12.11 *1.* (0, 3), (1, 4), (5, 8) and so on. The points all lie on one straight line.

2. Yes.

3. Yes.

12.15 *1.* (*a*) 32°F; 75·2°F; 89·6°F; 96·8°F; 105·8°F.
(*b*) 4·4°C; 18·3°C; 26·7°C; 38·9°C.

2. 18·8 cm; 7·5 cm; 6·4 cm; 2·2 cm.

12.17 *3.* $5;\ 3;\ \frac{5}{4}.$

12.19 $-5; \frac{3}{2}; -4.$

12.22

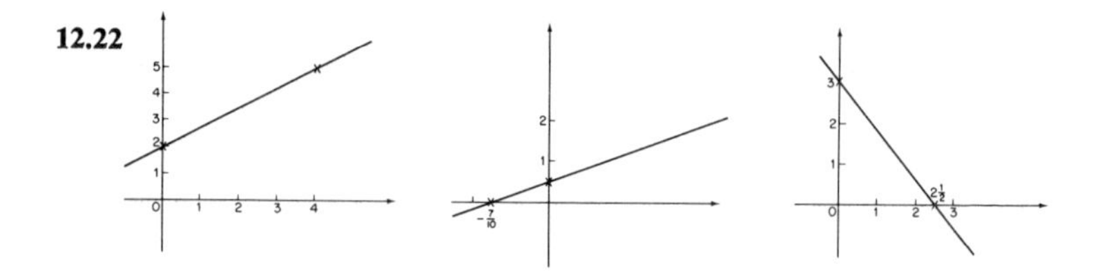

12.24 *1.* $x = 1{\cdot}3;\ y = 3{\cdot}7.$

2. Shorter side $= x$ cm; longer side $= y$ cm $\Rightarrow 2x + 2y = 48$ and $y = x + 10$; $x = 7$; $y = 17$.

12.26 *1.* (*a*)

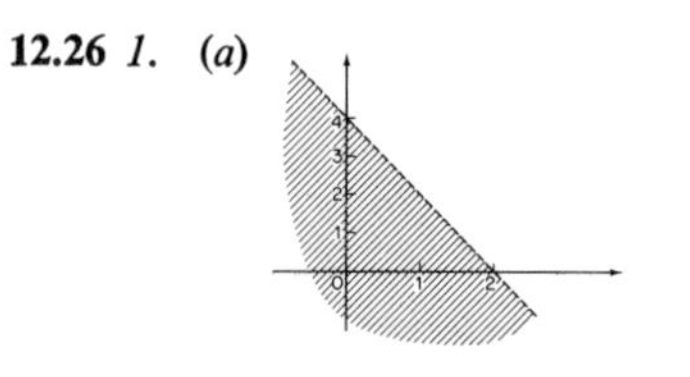

(*b*)

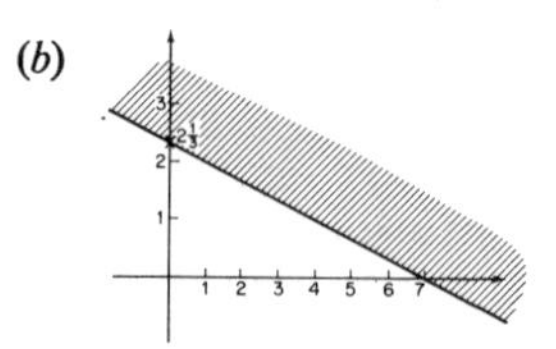

2.

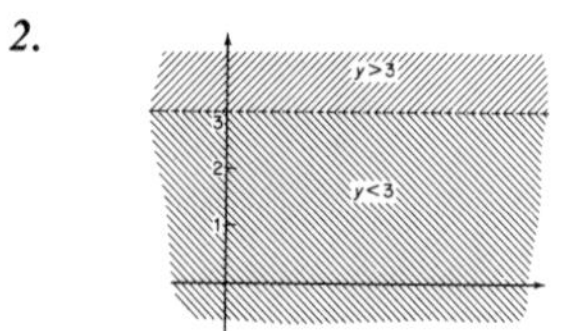

3. (*a*)

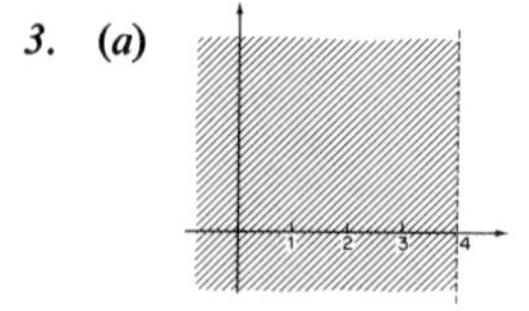

(*b*)

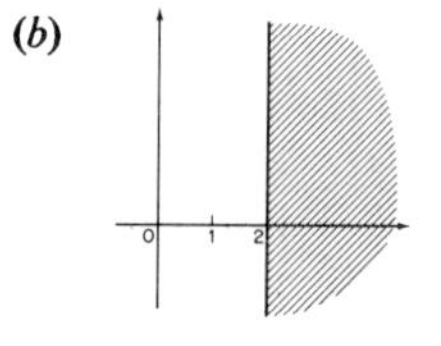

12.28

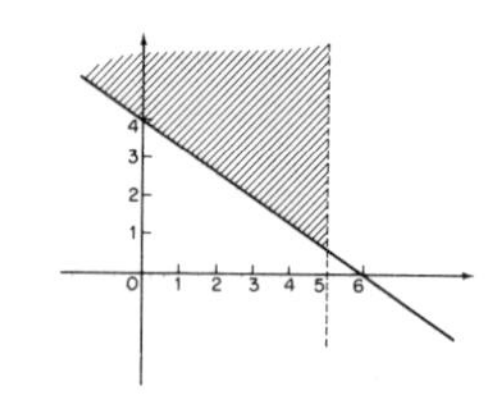

12.30

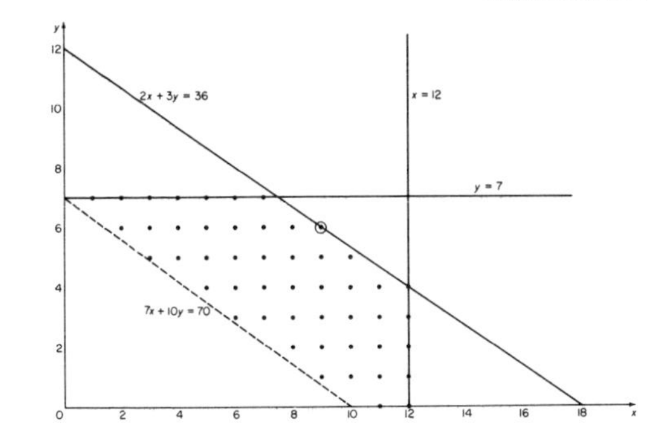

Dot marked ⊙ is furthest from $7x + 10y = 70$ in region common to all inequalities, i.e. $x = 9$, $y = 6 \Rightarrow$ maximum net profit $= (9 \times 70 + 6 \times 100 - 700)$p $= (1230 - 700)$p $=$ £5·30 per hour.

Unit 13

13.3 $^{+}1$, $^{+}\frac{1}{16}$, 0, $^{+}\frac{1}{16}$, $^{+}1$, $^{+}16$

13.4 *1.*

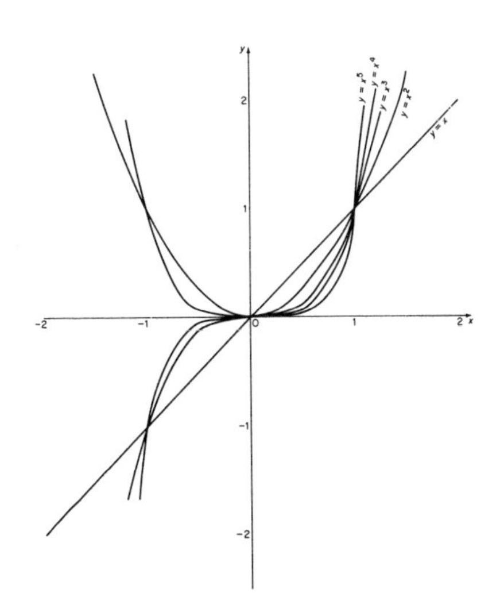

2.

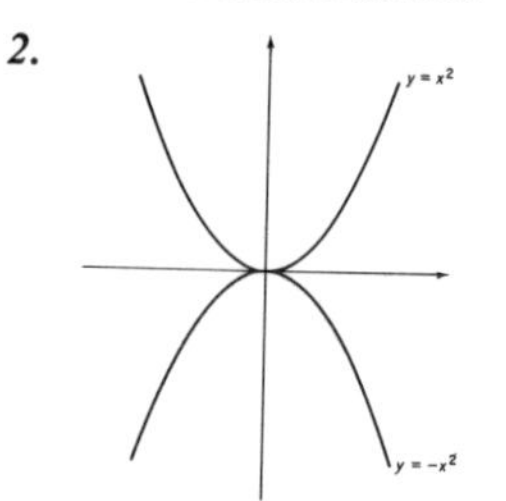

3.

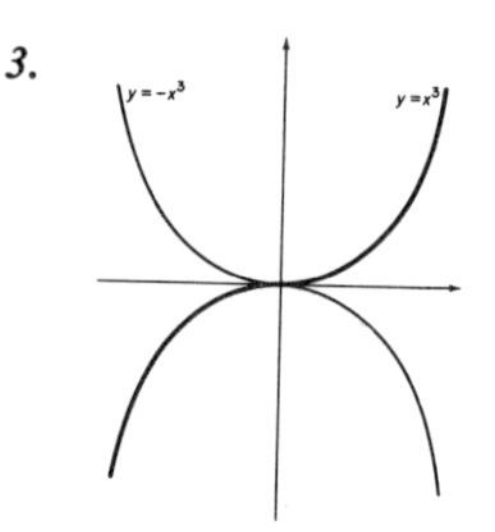

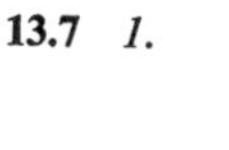

13.7 *1.*

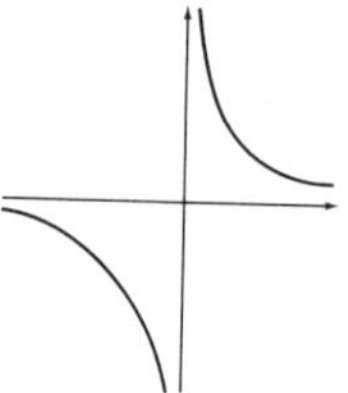

2.

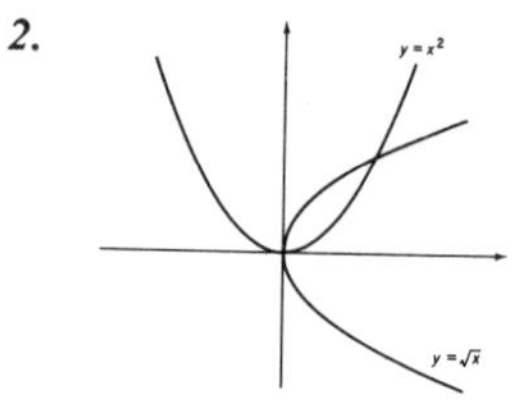

13.9 $x = 0{\cdot}6,\ y = 1{\cdot}4;\ x = -1{\cdot}6,\ y = 3{\cdot}6.$

13.11 *1.*

x	-3	-2	-1	0	1	2	3
x^2	9	4	1	0	1	4	9
$-x$	3	2	1	0	-1	-2	-3
$+2$	2	2	2	2	2	2	2
y	14	8	4	2	2	4	8

$x = 3;\ -2.$

2. (*a*) $3{\cdot}24,\ -1{\cdot}24.$
(*b*) $3{\cdot}45,\ -1{\cdot}45.$
(*c*) Yes; $x = 3,\ -1.$

13.13 Draw the line $y = 2x + 1$ on the graph, and read off points of intersection. $x = 5{\cdot}16,\ -1{\cdot}16.$

13.15 (*a*) 5.
(*b*) 1.
(*c*) $-1{\cdot}24 < x < 3{\cdot}24.$

13.17 *1.* Draw the graph of $y = x(x + 5)$, i.e. $y = x^2 + 5x$. Draw on this graph the line $y = 84$ and read off the intersection points. The sides are of length 7 cm and $(7 + 5)$ cm, i.e. 7 cm, 12 cm.
Draw, on the same graph, the line $y = 24$ and again read off the intersections with the curve. The sides are of length 3 cm, 8 cm.

2. (*a*) 6 m after $3\frac{1}{2}$ s.
(*b*) $2{\cdot}1$ s and $4{\cdot}9$ s.
(*c*) 7 s.

Unit 14

14.14 *1.* 135°, 45°, 135°.

14.16 *3.* (*b*) 90°.

14.20 *2.* Only one line can be drawn through *P* parallel to the given line.

Unit 15

15.5 *1.* (*a*) 60°.
(*b*) 70°.
(*c*) 60°.
(*d*) 45°.
2. No.

15.7 *1.* $\angle A = 50°$; $\angle B = 70°$; $\angle S = 50°$; $\angle K = 120°$.

15.13 *3.* (*a*) Triangles are congruent (*S–A–S*).
(*b*) $\angle ASP + \angle APS = 90° \Rightarrow \angle ASP + \angle DSR = 90° = \angle PSR = 90°$; etc.

15.23 *2.* Two angles of 60° $\Rightarrow$ third angle = 60°.

15.25 *2.* Three triangles are congruent (*A–S–A*).
30°; 30°; 120°.
3. Triangles are congruent (*A–S–A*).

15.30 *1.* $\sqrt{13}$ cm; $\sqrt{39}$ cm; 16 cm; $3\sqrt{13}$ cm.
2. $26^2 = 676 = 100 + 576 = 10^2 + 24^2 \Rightarrow$ triangle is right-angled.
3. 7·42 cm (to 3 SF).

Unit 16

16.3 540°; 720°; 900°; 1080°.

16.20 Triangles are congruent (*S–A–S*).

16.22 Triangles are congruent (*S–A–S*).

16.29 *3.* 10 cm.

16.31 *3.* 7·07 cm (to 3 SF).

16.36 $128\frac{4}{7}°$; 140°; 144°; 150°.

16.38 $128\frac{4}{7}°$, $51\frac{3}{7}°$;
135°, 45°;
140°, 40°;
144°, 36°;
150°, 30°.

16.40 *1.* 120°; 90°; 72°; 60°; $51\frac{3}{7}°$; 45°; 40°; 36°; 30°.
2. (*a*) 360°.
(*b*) 360°.
(*c*) 360°.
3. $\frac{360°}{n} \times n = 360°$.

16.46 *1.* Triangles are congruent (*S–A–S*).

Unit 17

17.7 *4.* Triangles are congruent (*S–S–S*).

17.9 *1.* 31·4 cm.
3. 1·88 km; 531 (to 3 SF).
4. 9·6 cm.

17.11 *1.* 40 000 km.
2. 6370 km.

17.13 *1.* $\angle BPC = \angle BQC = \angle BRC$, etc.

17.15 *2.*

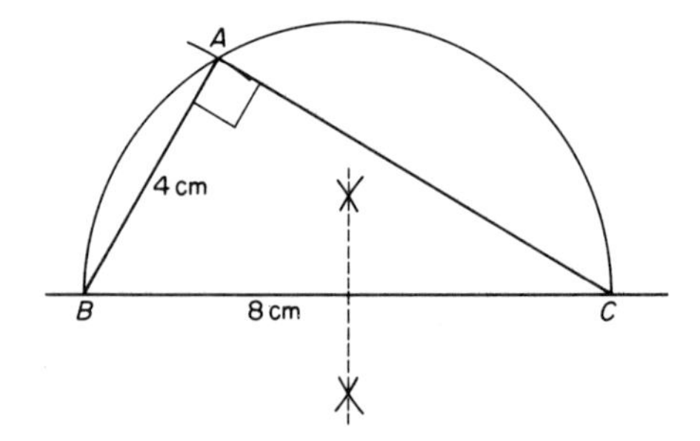

4. 100°; 100°; 160°; 80°.

17.17 *1.* 120°; 80°; 30°; 100°; 120°; 60°.

17.19 (*a*) concyclic.
(*b*) not concyclic.
(*c*) not necessarily concyclic; we are given insufficient information.

17.24 90°; 80°; 30°; 70°; 70°; 95°; $72\frac{1}{2}$°; 65°; $42\frac{1}{2}$°.

Unit 18

18.2 *3.* 2:1; 3:2; 4:3.
4. 9:1; 4:1; 9:4; 16:9.

18.8 *1.* (*a*) 5·92 cm; 5·08 cm.
(*b*) 77 cm; 66 cm.

18.10 *1.* (*a*) not similar.
(*b*) similar.
2. $\angle A = 55°$; $\angle APC = 90°$; $\angle ACP = 35°$; $\angle BCP = 55°$.

18.12 *1.* (*a*) $\triangle$s $\begin{matrix} ABC \\ DBE \end{matrix}$ similar.
(*b*) $\triangle$s $\begin{matrix} PQR \\ STR \end{matrix}$ similar.
2. $2\frac{1}{4}$ cm; $4\frac{1}{2}$ cm; 6 cm; 4 cm; 3 cm; $7\frac{1}{2}$ cm.

18.14 *2.* 249 m.

18.16 3; 1; 4; 2.

Unit 19

19.2 *1.* $\frac{12}{13}$; $\frac{5}{12}$; $\frac{9}{41}$; $\frac{40}{9}$ ($= 4\frac{4}{9}$).
2. (*a*) 53°.
(*b*) 53°.
(*c*) 53°.

19.5 0·42; 0·91; 0·67; 0·74.

19.7 0·58; 2·14.

19.9 *1.* 0·9336; 0·5090; 0·2211; 3·2007.
2. $\bar{1}$·6659; 0·2985.

19.11 0·8893; 0·4094.

19.13 $\angle A = 36° \, 52'$, $\angle B = 53° \, 8'$; $\angle R = 28° \, 37'$, $\angle S = 61° \, 23'$.

19.17 *1.* (*a*) $\angle P = 78° 47'$; $\angle Q = 11° 13'$; $|RQ| = 17{\cdot}7$ cm.
(*b*) $\angle B = 48°$; $|AC| = 15{\cdot}2$ cm; $|BC| = 13{\cdot}7$ cm.
(*c*) $\angle L = 33° 40'$; $|KM| = 18{\cdot}1$ cm; $|LM| = 27{\cdot}1$ cm.
2. 9·80 cm.
3. 6·68 cm.
4. 133° 36′.
5. 8·09 cm.
6. 7·65 cm.
19.19 *1.* 22·26 m.
2. 154° 22′.
19.21 0·8192; −0·9397; 0·6745; −0·7908.
19.23 *1.* (*a*) $b = 9{\cdot}99$ cm; $c = 11{\cdot}6$ cm; $\angle C = 63°$.
(*b*) $q = 12{\cdot}2$ cm; $\angle Q = 23° 57'$; $\angle R = 32° 3'$.
2. (*a*) one triangle.
(*b*) no triangle.
(*c*) two triangles.
3. 60·4 km.
19.25 *1.* (*a*) $\angle B = 55° 24'$; $a = 21{\cdot}4$ cm; $c = 17{\cdot}8$ cm.
(*b*) $\angle A = 49° 18'$; $\angle B = 94° 46'$; $\angle C = 35° 56'$.
2. 21·1 n mile; 008° 28′.
19.27 *1.* $\frac{5}{13}$; $\frac{5}{12}$.
2. $\frac{\sqrt{3}}{2}$; $\frac{1}{\sqrt{3}}$.

Unit 20

20.3 *3.* Area $= a^2$.
20.7 *1.* 62·5 cm².
2. 42·9 cm².
20.16 *3.* 25·4 cm².
20.20 *1.* (*a*) 19·5 cm².
(*b*) 23·1 cm².
(*c*) 150 cm².
(*d*) 59·4 cm².
2. 128 cm²; 39 cm².
20.22 *2.* 20 m².
20.26 (*a*) 13·1 cm².
(*b*) 92·1 cm².
(*c*) 13·8 cm².

Unit 21

21.3 *1.* (*c*) 180 cm².
21.5 *2.* 64 cm³; 125 cm³. $64{:}125 = 4^3{:}5^3$.
21.7 *1.* 10 cm; 4 cm; 10·8 cm.
2. 21° 48′.
3. 14·7 cm.
21.9 *1.* (*a*) 60 cm³.
(*c*) 132 cm².
2. 86·4 m²; 864 m³; 41·6 cm²; 665 cm³.
21.11 *2.* 8·48*l*.
21.13 *2.* $|OP| = 4{\cdot}58$ cm; $\angle OPQ = 64° 7'$.

21.15 *2.* 7·48 cm.

3. 196 cm^3; 220 cm^2.

21.17 183 000 cm^3 (or 0·183 m^3); 17 300 cm^2 (or 1·73 m^2).

21.19 *2.* 4530 km.

21.21 *2.* 4019 km; 6700 km; 3610 km.

Unit 22

22.3 *2.* 1; 3; 5; 7; 9.

22.5 *1.* $8\frac{1}{2}$; $8\frac{1}{3}$; $8 + k$; 8.

2. $10\frac{1}{2}$; $10\frac{1}{3}$; $10 + k$; 10.

22.7 *1.* 12; 5; −4.

22.9 *1.* $10x^9$; $3t^2$; $4t^3$.

3. Zero.

4. $-\frac{1}{x^2}\ (= x^{-2})$; $-\frac{3}{x^4}\ (= -\ 3x^{-4})$; $\frac{1}{2x^{\frac{1}{2}}}\ (= \frac{1}{2}x^{-\frac{1}{2}})$;

$-\frac{1}{2x^{\frac{3}{2}}}\ (= -\ \frac{1}{2}x^{-\frac{3}{2}})$; $\frac{1}{3x^{\frac{2}{3}}}\ (= \frac{1}{3}x^{-\frac{2}{3}})$.

22.13 *1.* (*a*) $\frac{dy}{dx} = 15x^2 + 4x - 3$.

(*b*) $\frac{dy}{dt} = 4t + 3$.

(*c*) $\frac{dy}{dx} = 4x^3 - \frac{1}{\sqrt{x}}$.

(*d*) $\frac{dy}{dx} = 2x - \frac{3}{x^2}$.

2. 3; −3.

22.15 *1.* $v = \frac{ds}{dt} = 3t^2 + 6t + 2$.

2. $v = \frac{ds}{dt} = 2t + 1 \Rightarrow$ velocity after 4 seconds = 9 m/s.

22.17 *1.* 56 m/s^2; 1240 m/s^2.

2. (*a*) $v = 12t^3 + 6t^2$.

(*b*) acceleration $= \frac{dv}{dt} = 36t^2 + 12t$.

(*c*) 624 m/s^2.

22.20 *1.* A minimum; a maximum; a maximum.

2. A minimum; at the minimum $x = -1$ and $y = 0$.

3. If area of rectangle = y cm^2, $y = -x^2 + 12x$; maximum area = 36 cm^2, when $x = 6$ cm (i.e. a square).

Unit 23

23.3 $\frac{x^5}{5} + c$; $x^3 + \frac{x^2}{2} + 4x + c$; $\frac{2x^{1\frac{1}{2}}}{3} + c\ (= \frac{x^{1\frac{1}{2}}}{1\frac{1}{2}} + c)$;

$-x^{-1} + c\left(= -\frac{1}{x} + c\right)$.

23.5 $\frac{x^4}{4} + c$; $\frac{x^5}{5} + x^2 + x + c$; $\frac{x^3}{3} - x^2 + x + c$;

$-\frac{1}{3x^3} + c\left(= \frac{x^{-3}}{-3} + c\right)$; $\frac{t^3}{3} + c$.

23.7 *1.* $y = x^2 - 3x + 2$.

2. $y = \frac{x^2}{2} + 2x - 9\frac{1}{2}$.

3. $v = \frac{3}{2}t^2 + t + 3$.

23.10 *1.* 168; $14\frac{1}{4}$; 57.

2. (*a*) $4\frac{1}{2}$.

(*b*) $\frac{1}{6}$.

3. 2550 m (= 2·55 km).

23.12 *1.* $16\frac{1}{30}\pi$ units of volume.

2. $\frac{\pi}{5}$ units of volume; $\frac{\pi}{7}$ units of volume; $\frac{2\pi}{35}$ units of volume.

Four-figure Tables

Squares

	0	1	2	3	4	5	6	7	8	9	1	2	3	4	5	6	7	8	9
10	1000	1020	1040	1061	1082	1103	1124	1145	1166	1188	2	4	6	8	10	13	15	17	19
11	1210	1232	1254	1277	1300	1323	1346	1369	1392	1416	2	5	7	9	11	14	16	18	21
12	1440	1464	1488	1513	1538	1563	1588	1613	1638	1664	2	5	7	10	12	15	17	20	22
13	1690	1716	1742	1769	1796	1823	1850	1877	1904	1932	3	5	8	11	13	16	19	22	24
14	1960	1988	2016	2045	2074	2103	2132	2161	2190	2220	3	6	9	12	14	17	20	23	26
15	2250	2280	2310	2341	2372	2403	2434	2465	2496	2528	3	6	9	12	15	19	22	25	28
16	2560	2592	2624	2657	2690	2723	2756	2789	2822	2856	3	7	10	13	16	20	23	26	30
17	2890	2924	2958	2993	3028	3063	3098	3133	3168	3204	3	7	10	14	17	21	24	28	31
18	3240	3276	3312	3349	3386	3423	3460	3497	3534	3572	4	7	11	15	18	22	26	30	33
19	3610	3648	3686	3725	3764	3803	3842	3881	3920	3960	4	8	12	16	19	23	27	31	35
20	4000	4040	4080	4121	4162	4203	4244	4285	4326	4368	4	8	12	16	20	25	29	33	37
21	4410	4452	4494	4537	4580	4623	4666	4709	4752	4796	4	9	13	17	21	26	30	34	39
22	4840	4884	4928	4973	5018	5063	5108	5153	5198	5244	4	9	13	18	22	27	31	36	40
23	5290	5336	5382	5429	5476	5523	5570	5617	5664	5712	5	9	14	19	23	28	33	38	42
24	5760	5808	5856	5905	5954	6003	6052	6101	6150	6200	5	10	15	20	24	29	34	39	44
25	6250	6300	6350	6401	6452	6503	6554	6605	6656	6708	5	10	15	20	25	31	36	41	46
26	6760	6812	6864	6917	6970	7023	7076	7129	7182	7236	5	11	16	21	26	32	37	42	48
27	7290	7344	7398	7453	7508	7563	7618	7673	7728	7784	5	11	16	22	28	33	38	44	49
28	7840	7896	7952	8009	8066	8123	8180	8237	8294	8352	6	11	17	23	28	34	40	46	51
29	8410	8468	8526	8585	8644	8703	8762	8821	8880	8940	6	12	18	24	30	35	41	47	53
30	9000	9060	9120	9181	9242	9303	9364	9425	9486	9548	6	12	18	24	31	37	43	49	55
31	9610	9672	9734	9797	9860	9923	9986	1005	1011	1018	1	1	2	3	3	4	5	5	6

	0	1	2	3	4	5	6	7	8	9	1	2	3	4	5	6	7	8	9
32	1024	1030	1037	1043	1050	1056	1063	1069	1076	1082	1	1	2	3	3	4	5	5	6
33	1089	1096	1102	1109	1116	1122	1129	1136	1142	1149	1	1	2	3	3	4	5	5	6
34	1156	1163	1170	1176	1183	1190	1197	1204	1211	1218	1	1	2	3	3	4	5	6	6
35	1225	1232	1239	1246	1253	1260	1267	1274	1282	1289	1	1	2	3	4	4	5	6	6
36	1296	1303	1310	1318	1325	1332	1340	1347	1354	1362	1	1	2	3	4	4	5	6	7
37	1369	1376	1384	1391	1399	1406	1414	1421	1429	1436	1	2	2	3	4	5	5	6	7
38	1444	1452	1459	1467	1475	1482	1490	1498	1505	1513	1	2	2	3	4	5	5	6	7
39	1521	1529	1537	1544	1552	1560	1568	1576	1584	1592	1	2	2	3	4	5	6	6	7
40	1600	1608	1616	1624	1632	1640	1648	1656	1665	1673	1	2	2	3	4	5	6	6	7
41	1681	1689	1697	1706	1714	1722	1731	1739	1747	1756	1	2	2	3	4	5	6	7	7
42	1764	1772	1781	1789	1798	1806	1815	1823	1832	1840	1	2	3	3	4	5	6	7	8
43	1849	1858	1866	1875	1884	1892	1901	1910	1918	1927	1	2	3	3	4	5	6	7	8
44	1936	1945	1954	1962	1971	1980	1989	1998	2007	2016	1	2	3	4	5	5	6	7	8
45	2025	2034	2043	2052	2061	2070	2079	2088	2098	2107	1	2	3	4	5	5	6	7	8
46	2116	2125	2134	2144	2153	2162	2172	2181	2190	2200	1	2	3	4	5	6	7	7	8
47	2209	2218	2228	2237	2247	2256	2266	2275	2285	2294	1	2	3	4	5	6	7	8	9
48	2304	2314	2323	2333	2343	2352	2362	2372	2381	2391	1	2	3	4	5	6	7	8	9
49	2401	2411	2421	2430	2440	2450	2460	2470	2480	2490	1	2	3	4	5	6	7	8	9
50	2500	2510	2520	2530	2540	2550	2560	2570	2581	2591	1	2	3	4	5	6	7	8	9
51	2601	2611	2621	2632	2642	2652	2663	2673	2683	2694	1	2	3	4	5	6	7	8	9
52	2704	2714	2725	2735	2746	2756	2767	2777	2788	2798	1	2	3	4	5	6	7	8	9
53	2809	2820	2830	2841	2852	2862	2873	2884	2894	2905	1	2	3	4	5	6	7	9	10
54	2916	2927	2938	2948	2959	2970	2981	2992	3003	3014	1	2	3	4	6	7	8	9	10

Squares—*continued*

	0	1	2	3	4	5	6	7	8	9	1	2	3	4	5	6	7	8	9
55	3025	3036	3047	3058	3069	3080	3091	3102	3114	3125	1	2	3	4	6	7	8	9	10
56	3136	3147	3158	3170	3181	3192	3204	3215	3226	3238	1	2	3	5	6	7	8	9	10
57	3249	3260	3272	3283	3295	3306	3318	3329	3341	3352	1	2	3	5	6	7	8	9	10
58	3364	3376	3387	3399	3411	3422	3434	3446	3457	3469	1	2	4	5	6	7	8	9	11
59	3481	3493	3505	3516	3528	3540	3552	3564	3576	3588	1	2	4	5	6	7	8	10	11
60	3600	3612	3624	3636	3648	3660	3672	3684	3697	3709	1	2	4	5	6	7	8	10	11
61	3721	3733	3745	3758	3770	3782	3795	3807	3819	3832	1	2	4	5	6	7	9	10	11
62	3844	3856	3869	3881	3894	3906	3919	3931	3944	3956	1	3	4	5	6	8	9	10	11
63	3969	3982	3994	4007	4020	4032	4045	4058	4070	4083	1	3	4	5	6	8	9	10	11
64	4096	4109	4122	4134	4147	4160	4173	4186	4199	4212	1	3	4	5	6	8	9	10	12
65	4225	4238	4251	4264	4277	4290	4303	4316	4330	4343	1	3	4	5	7	8	9	10	12
66	4356	4369	4382	4396	4409	4422	4436	4449	4462	4476	1	3	4	5	7	8	9	11	12
67	4489	4502	4516	4529	4543	4556	4570	4583	4597	4610	1	3	4	5	7	8	9	11	12
68	4624	4638	4651	4665	4679	4692	4706	4720	4733	4747	1	3	4	5	7	8	10	11	12
69	4761	4775	4789	4802	4816	4830	4844	4858	4872	4886	1	3	4	6	7	8	10	11	13
70	4900	4914	4928	4942	4956	4970	4984	4998	5013	5027	1	3	4	6	7	8	10	11	13
71	5041	5055	5069	5084	5098	5112	5127	5141	5155	5170	1	3	4	6	7	9	10	11	13
72	5184	5198	5213	5227	5242	5256	5271	5285	5300	5314	1	3	4	6	7	9	10	11	13
73	5239	5344	5358	5373	5388	5402	5417	5432	5446	5461	1	3	4	6	7	9	10	12	13
74	5476	5491	5506	5520	5535	5550	5565	5580	5595	5610	1	3	4	6	7	9	10	12	13
75	5625	5640	5655	5670	5685	5700	5715	5730	5746	5761	2	3	5	6	8	9	11	12	14
76	5776	5791	5806	5822	5837	5852	5868	5883	5898	5914	2	3	5	6	8	9	11	12	14

	0	1	2	3	4	5	6	7	8	9	2	2	3	4	5	6	7	8	9
77	5929	5944	5960	5975	5991	6006	6022	6037	6053	6068	2	3	5	6	8	9	11	12	14
78	6084	6100	6115	6131	6147	6162	6178	6194	6209	6225	2	3	5	6	8	9	11	13	14
79	6241	6257	6273	6288	6304	6320	6336	6352	6368	6384	2	3	5	6	8	10	11	13	14
80	6400	6416	6432	6448	6464	6480	6496	6512	6529	6545	2	3	5	6	8	10	11	13	14
81	6561	6577	6593	6610	6626	6642	6659	6675	6691	6708	2	3	5	7	8	10	11	13	15
82	6724	6740	6757	6773	6790	6806	6823	6839	6856	6872	2	3	5	7	8	10	12	13	15
83	6889	6906	6922	6939	6956	6972	6989	7006	7022	7039	2	3	5	7	8	10	12	13	15
84	7056	7073	7090	7106	7123	7140	7157	7174	7191	7208	2	3	5	7	8	10	12	14	15
85	7225	7242	7259	7276	7293	7310	7327	7344	7362	7379	2	3	5	7	9	10	12	14	15
86	7396	7413	7430	7448	7465	7482	7500	7517	7534	7552	2	3	5	7	9	10	12	14	16
87	7569	7586	7604	7621	7639	7656	7674	7691	7709	7726	2	4	5	7	9	11	12	14	16
88	7744	7762	7779	7797	7815	7832	7850	7868	7885	7903	2	4	5	7	9	11	12	14	16
89	7921	7939	7957	7974	7992	8010	8028	8046	8064	8082	2	4	5	7	9	11	13	14	16
90	8100	8118	8136	8154	8172	8190	8208	8226	8245	8263	2	4	5	7	9	11	13	14	16
91	8281	8299	8317	8336	8354	8372	8391	8409	8427	8446	2	4	5	7	9	11	13	15	16
92	8464	8482	8501	8519	8538	8556	8575	8593	8612	8630	2	4	6	7	9	11	13	15	17
93	8649	8668	8686	8705	8724	8742	8761	8780	8798	8817	2	4	6	7	9	11	13	15	17
94	8836	8855	8874	8892	8911	8930	8949	8968	8987	9006	2	4	6	8	9	11	13	15	17
95	9025	9044	9063	9082	9101	9120	9139	9158	9178	9197	2	4	6	8	10	11	13	15	17
96	9216	9235	9254	9274	9293	9312	9332	9351	9370	9390	2	4	6	8	10	12	14	15	17
97	9409	9428	9448	9467	9487	9506	9526	9545	9565	9584	2	4	6	8	10	12	14	16	18
98	9604	9624	9643	9663	9683	9702	9722	9742	9761	9781	2	4	6	8	10	12	14	16	18
99	9801	9821	9841	9860	9880	9900	9920	9940	9960	9980	2	4	6	8	10	12	14	16	18

Square Roots

	0	1	2	3	4	5	6	7	8	9	1	2	3	4	5	6	7	8	9
10	1000 3162	1005 3178	1010 3194	1015 3209	1020 3225	1025 3240	1030 3256	1034 3271	1039 3286	1044 3302	0 2	1 3	1 5	2 6	2 8	3 9	3 11	4 12	4 14
11	1049 3317	1054 3332	1058 3347	1063 3362	1068 3376	1072 3391	1077 3406	1082 3421	1086 3435	1091 3450	0 1	1 3	1 4	2 6	2 7	3 9	3 10	4 12	4 13
12	1095 3464	1100 3479	1105 3493	1109 3507	1114 3521	1118 3536	1122 3550	1127 3564	1131 3578	1136 3592	0 1	1 3	1 4	2 6	2 7	3 8	3 10	4 11	4 13
13	1140 3606	1145 3619	1149 3633	1153 3647	1158 3661	1162 3674	1166 3688	1170 3701	1175 3715	1179 3728	0 1	1 3	1 4	2 5	2 7	3 8	3 10	3 11	4 12
14	1183 3742	1187 3755	1192 3768	1196 3782	1200 3795	1204 3808	1208 3821	1212 3834	1217 3847	1221 3860	0 1	1 3	1 4	2 5	2 7	3 8	3 9	3 11	4 12
15	1225 3873	1229 3886	1233 3899	1237 3912	1241 3924	1245 3937	1249 3950	1253 3962	1257 3975	1261 3987	0 1	1 3	1 4	2 5	2 6	3 8	3 9	3 10	4 11
16	1265 4000	1269 4012	1273 4025	1277 4037	1281 4050	1285 4062	1288 4074	1292 4087	1296 4099	1300 4111	0 1	1 2	1 4	2 5	2 6	3 7	3 9	3 10	4 11
17	1304 4123	1308 4135	1311 4147	1315 4159	1319 4171	1323 4183	1327 4195	1330 4207	1334 4219	1338 4231	0 1	1 2	1 4	2 5	2 6	2 7	3 8	3 10	3 11
18	1342 4243	1345 4254	1349 4266	1353 4278	1356 4290	1360 4301	1364 4313	1367 4324	1371 4336	1375 4347	0 1	1 2	1 3	1 5	2 6	2 7	3 8	3 9	3 10
19	1378 4359	1382 4370	1386 4382	1389 4393	1393 4405	1396 4416	1400 4427	1404 4438	1407 4450	1411 4461	0 1	1 2	1 3	1 5	2 6	2 7	3 8	3 9	3 10
20	1414 4472	1418 4483	1421 4494	1425 4506	1428 4517	1432 4528	1435 4539	1439 4550	1442 4561	1446 4572	0 1	1 2	1 3	1 4	2 5	2 7	2 8	3 9	3 10

	0	1	2	3	4	5	6	7	8	9	1	2	3	4	5	6	7	8	9
21	1449	1453	1456	1459	1463	1466	1470	1473	1476	1480	0	1	1	1	2	2	2	3	3
	4583	4593	4604	4615	4626	4637	4648	4658	4669	4680	1	2	3	4	5	6	8	9	10
22	1483	1487	1490	1493	1497	1500	1503	1507	1510	1513	0	1	1	1	2	2	2	3	3
	4690	4701	4712	4722	4733	4743	4754	4764	4775	4785	1	2	3	4	5	6	7	8	9
23	1517	1520	1523	1526	1530	1533	1536	1539	1543	1546	0	1	1	1	2	2	2	3	3
	4796	4806	4817	4827	4837	4848	4858	4868	4879	4889	1	2	3	4	5	6	7	8	9
24	1549	1552	1556	1559	1562	1565	1568	1572	1575	1578	0	1	1	1	2	2	2	3	3
	4899	4909	4919	4930	4940	4950	4960	4970	4980	4990	1	2	3	4	5	6	7	8	9
25	1581	1584	1587	1591	1594	1597	1600	1603	1606	1609	0	1	1	1	2	2	2	3	3
	5000	5010	5020	5030	5040	5050	5060	5070	5079	5089	1	2	3	4	5	6	7	8	9
26	1612	1616	1619	1622	1625	1628	1631	1634	1637	1640	0	1	1	1	2	2	2	2	3
	5099	5109	5119	5128	5138	5148	5158	5167	5177	5187	1	2	3	4	5	6	7	8	9
27	1643	1646	1649	1652	1655	1658	1661	1664	1667	1670	0	1	1	1	2	2	2	2	3
	5196	5206	5215	5225	5235	5244	5254	5263	5273	5282	1	2	3	4	5	6	7	8	9
28	1673	1676	1679	1682	1685	1688	1691	1694	1697	1700	0	1	1	1	1	2	2	2	3
	5292	5301	5310	5320	5329	5339	5348	5357	5367	5376	1	2	3	4	5	6	7	7	8
29	1703	1706	1709	1712	1715	1718	1720	1723	1726	1729	0	1	1	1	1	2	2	2	3
	5385	5394	5404	5413	5422	5431	5441	5450	5459	5468	1	2	3	4	5	5	6	7	8
30	1732	1735	1738	1741	1744	1746	1749	1752	1755	1758	0	1	1	1	1	2	2	2	3
	5477	5486	5495	5505	5514	5523	5532	5541	5550	5559	1	2	3	4	4	5	6	7	8
31	1761	1764	1766	1769	1772	1775	1778	1780	1783	1786	0	1	1	1	1	2	2	2	3
	5568	5577	5586	5595	5604	5612	5621	5630	5639	5648	1	2	3	3	4	5	6	7	8
32	1789	1792	1794	1797	1800	1803	1806	1808	1811	1814	0	1	1	1	1	2	2	2	2
	5657	5666	5675	5683	5692	5701	5710	5718	5727	5736	1	2	3	3	4	5	6	7	8

Square Roots—*continued*

	0	1	2	3	4	5	6	7	8	9	1	2	3	4	5	6	7	8	9
33	1817 5745	1819 5753	1822 5762	1825 5771	1828 5779	1830 5788	1833 5797	1836 5805	1838 5814	1841 5822	0 1	1 2	1 3	1 3	1 4	2 5	2 6	2 7	2 8
34	1844 5831	1847 5840	1849 5848	1852 5857	1855 5865	1857 5874	1860 5882	1863 5891	1865 5899	1868 5908	0 1	1 2	1 3	1 3	1 4	2 5	2 6	2 7	2 8
35	1871 5916	1873 5925	1876 5933	1879 5941	1881 5950	1884 5958	1887 5967	1889 5975	1892 5983	1895 5992	0 1	1 2	1 2	1 3	1 4	2 5	2 6	2 7	2 8
36	1897 6000	1900 6008	1903 6017	1905 6025	1908 6033	1910 6042	1913 6050	1916 6058	1918 6066	1921 6075	0 1	1 2	1 2	1 3	1 4	2 5	2 6	2 7	2 7
37	1924 6083	1926 6091	1929 6099	1931 6107	1934 6116	1936 6124	1939 6132	1942 6140	1944 6148	1947 6156	0 1	1 2	1 2	1 3	1 4	2 5	2 6	2 7	2 7
38	1949 6164	1952 6173	1954 6181	1957 6189	1960 6197	1962 6205	1965 6213	1967 6221	1970 6229	1972 6237	0 1	1 2	1 2	1 3	1 4	2 5	2 6	2 6	2 7
39	1975 6245	1977 6253	1980 6261	1982 6269	1985 6277	1987 6285	1990 6293	1992 6301	1995 6309	1997 6317	0 1	1 2	1 2	1 3	1 4	2 5	2 6	2 6	2 7
40	2000 6325	2002 6332	2005 6340	2007 6348	2010 6356	2012 6364	2015 6372	2017 6380	2020 6387	2022 6395	0 1	0 2	1 2	1 3	1 4	1 5	2 6	2 6	2 7
41	2025 6403	2027 6411	2030 6419	2032 6427	2035 6434	2037 6442	2040 6450	2042 6458	2045 6465	2047 6473	0 1	0 2	1 2	1 3	1 4	1 5	2 5	2 6	2 7
42	2049 6481	2052 6488	2054 6496	2057 6504	2059 6512	2062 6519	2064 6527	2066 6535	2069 6542	2071 6550	0 1	0 2	1 2	1 3	1 4	1 5	2 5	2 6	2 7
43	2074 6557	2076 6565	2078 6573	2081 6580	2083 6588	2086 6595	2088 6603	2090 6611	2093 6618	2095 6626	0 1	0 2	1 2	1 3	1 4	1 5	2 5	2 6	2 7

	0	1	2	3	4	5	6	7	8	9	1	2	3	4	5	6	7	8	9
44	2098 6633	2100 6641	2102 6648	2105 6656	2107 6663	2110 6671	2112 6678	2114 6686	2117 6693	2119 6701	0 1	0 2	1 2	1 3	1 4	1 4	2 5	2 6	2 7
45	2121 6708	2124 6716	2126 6723	2128 6731	2131 6738	2133 6745	2135 6753	2138 6760	2140 6768	2142 6775	0 1	0 1	1 2	1 3	1 4	1 4	2 5	2 6	2 7
46	2145 6782	2147 6790	2149 6797	2152 6804	2154 6812	2156 6819	2159 6826	2161 6834	2163 6841	2166 6848	0 1	0 1	1 2	1 3	1 4	1 4	2 5	2 6	2 7
47	2168 6856	2170 6863	2173 6870	2175 6877	2177 6885	2179 6892	2182 6899	2184 6907	2186 6914	2189 6921	0 1	0 1	1 2	1 3	1 4	1 4	2 5	2 6	2 7
48	2191 6928	2193 6935	2195 6943	2198 6950	2200 6957	2202 6964	2205 6971	2207 6979	2209 6986	2211 6993	0 1	0 1	1 2	1 3	1 4	1 4	2 5	2 6	2 6
49	2214 7000	2216 7007	2218 7014	2220 7021	2223 7029	2225 7036	2227 7043	2229 7050	2232 7057	2234 7064	0 1	0 1	1 2	1 3	1 4	1 4	2 5	2 6	2 6
50	2236 7071	2238 7078	2241 7085	2243 7092	2245 7099	2247 7106	2249 7113	2252 7120	2254 7127	2256 7134	0 1	0 1	1 2	1 3	1 4	1 4	2 5	2 6	2 6
51	2258 7141	2261 7148	2263 7155	2265 7162	2267 7169	2269 7176	2272 7183	2274 7190	2276 7197	2278 7204	0 1	0 1	1 2	1 3	1 4	1 4	2 5	2 6	2 6
52	2280 7211	2283 7218	2285 7225	2287 7232	2289 7239	2291 7246	2293 7253	2296 7259	2298 7266	2300 7273	0 1	0 1	1 2	1 3	1 3	1 4	2 5	2 6	2 6
53	2302 7280	2304 7287	2307 7294	2309 7301	2311 7308	2313 7314	2315 7321	2317 7328	2319 7335	2322 7342	0 1	0 1	1 2	1 3	1 3	1 4	2 5	2 5	2 6
54	2324 7348	2326 7355	2328 7362	2330 7369	2332 7376	2335 7382	2337 7389	2339 7396	2341 7403	2343 7409	0 1	0 1	1 2	1 3	1 3	1 4	1 5	2 5	2 6

Square Roots—*continued*

	0	1	2	3	4	5	6	7	8	9	1	2	3	4	5	6	7	8	9
55	2345 7416	2347 7423	2349 7430	2352 7436	2354 7443	2356 7450	2358 7457	2360 7463	2362 7470	2364 7477	0 1	0 1	1 2	1 3	1 3	1 4	1 5	2 5	2 6
56	2366 7483	2369 7490	2371 7497	2373 7503	2375 7510	2377 7517	2379 7523	2381 7530	2383 7537	2385 7543	0 1	0 1	1 2	1 3	1 3	1 4	1 5	2 5	2 6
57	2387 7550	2390 7556	2392 7563	2394 7570	2396 7576	2398 7583	2400 7589	2402 7596	2404 7603	2406 7609	0 1	0 1	1 2	1 3	1 3	1 4	1 5	2 5	2 6
58	2408 7616	2410 7622	2412 7629	2415 7635	2417 7642	2419 7649	2421 7655	2423 7662	2425 7668	2427 7675	0 1	0 1	1 2	1 3	1 3	1 4	1 5	2 5	2 6
59	2429 7681	2431 7688	2433 7694	2435 7701	2437 7707	2439 7714	2441 7720	2443 7727	2445 7733	2447 7740	0 1	0 1	1 2	1 3	1 3	1 4	1 5	2 5	2 6
60	2449 7746	2452 7752	2454 7759	2456 7765	2458 7772	2460 7778	2462 7785	2464 7791	2466 7797	2468 7804	0 1	0 1	1 2	1 3	1 3	1 4	1 4	2 5	2 6
61	2470 7810	2472 7817	2474 7823	2476 7829	2478 7836	2480 7842	2482 7849	2484 7855	2486 7861	2488 7868	0 1	0 1	1 2	1 3	1 3	1 4	1 4	2 5	2 6
62	2490 7874	2492 7880	2494 7887	2496 7893	2498 7899	2500 7906	2502 7912	2504 7918	2506 7925	2508 7931	0 1	0 1	1 2	1 3	1 3	1 4	1 4	2 5	2 6
63	2510 7937	2512 7944	2514 7950	2516 7956	2518 7962	2520 7969	2522 7975	2524 7981	2526 7987	2528 7994	0 1	0 1	1 2	1 3	1 3	1 4	1 4	2 5	2 6
64	2530 8000	2532 8006	2534 8012	2536 8019	2538 8025	2540 8031	2542 8037	2544 8044	2546 8050	2548 8056	0 1	0 1	1 2	1 2	1 3	1 4	1 4	2 5	2 6
65	2550 8062	2551 8068	2553 8075	2555 8081	2557 8087	2559 8093	2561 8099	2563 8106	2565 8112	2567 8118	0 1	0 1	1 2	1 2	1 3	1 4	1 4	2 5	2 5

	0	1	2	3	4	5	6	7	8	9	1	2	3	4	5	6	7	8	9
66	2569 8124	2571 8130	2573 8136	2575 8142	2577 8149	2579 8155	2581 8161	2583 8167	2585 8173	2587 8179	0 1	0 1	1 2	1 2	1 3	1 4	1 4	2 5	2 5
67	2588 8185	2590 8191	2592 8198	2594 8204	2596 8210	2598 8216	2600 8222	2602 8228	2604 8234	2606 8240	0 1	0 1	1 2	1 2	1 3	1 4	1 4	2 5	2 5
68	2608 8246	2610 8252	2612 8258	2613 8264	2615 8270	2617 8276	2619 8283	2621 8289	2623 8295	2625 8301	0 1	0 1	1 2	1 2	1 3	1 4	1 4	2 5	2 5
69	2627 8307	2629 8313	2631 8319	2632 8325	2634 8331	2636 8337	2638 8343	2640 8349	2642 8355	2644 8361	0 1	0 1	1 2	1 2	1 3	1 4	1 4	2 5	2 5
70	2646 8367	2648 8373	2650 8379	2651 8385	2653 8390	2655 8396	2657 8402	2659 8408	2661 8414	2663 8420	0 1	0 1	1 2	1 2	1 3	1 4	1 4	2 5	2 5
71	2665 8426	2666 8432	2668 8438	2670 8444	2672 8450	2674 8456	2676 8462	2678 8468	2680 8473	2681 8479	0 1	0 1	1 2	1 2	1 3	1 3	1 4	1 5	2 5
72	2683 8485	2685 8491	2687 8497	2689 8503	2691 8509	2693 8515	2694 8521	2696 8526	2698 8532	2700 8538	0 1	0 1	1 2	1 2	1 3	1 3	1 4	1 5	2 5
73	2702 8544	2704 8550	2706 8556	2707 8562	2709 8567	2711 8573	2713 8579	2715 8585	2717 8591	2718 8597	0 1	0 1	1 2	1 2	1 3	1 3	1 4	1 5	2 5
74	2720 8602	2722 8608	2724 8614	2726 8620	2728 8626	2729 8631	2731 8637	2733 8643	2735 8649	2737 8654	0 1	0 1	1 2	1 2	1 3	1 3	1 4	1 5	2 5
75	2739 8660	2740 8666	2742 8672	2744 8678	2746 8683	2748 8689	2750 8695	2751 8701	2753 8706	2755 8712	0 1	0 1	1 2	1 2	1 3	1 3	1 4	1 5	2 5
76	2757 8718	2759 8724	2760 8729	2762 8735	2764 8741	2766 8746	2768 8752	2769 8758	2771 8764	2773 8769	0 1	0 1	1 2	1 2	1 3	1 3	1 4	1 5	2 5
77	2775 8775	2777 8781	2778 8786	2780 8792	2782 8798	2784 8803	2786 8809	2787 8815	2789 8820	2791 8826	0 1	0 1	1 2	1 2	1 3	1 3	1 4	1 4	2 5

Square Roots—*continued*

	0	1	2	3	4	5	6	7	8	9	1	2	3	4	5	6	7	8	9
78	2793 8832	2795 8837	2796 8843	2798 8849	2800 8854	2802 8860	2804 8866	2805 8871	2807 8877	2809 8883	0 1	0 1	1 2	1 2	1 3	1 3	1 4	1 4	2 5
79	2811 8888	2812 8894	2814 8899	2816 8905	2818 8911	2820 8916	2821 8922	2823 8927	2825 8933	2827 8939	0 1	0 1	1 2	1 2	1 3	1 3	1 4	1 4	2 5
80	2828 8944	2830 8950	2832 8955	2834 8961	2835 8967	2837 8972	2839 8978	2841 8983	2843 8989	2844 8994	0 1	0 1	1 2	1 2	1 3	1 3	1 4	1 4	2 5
81	2846 9000	2848 9006	2850 9011	2851 9017	2853 9022	2855 9028	2857 9033	2858 9039	2860 9044	2862 9050	0 1	0 1	1 2	1 2	1 3	1 3	1 4	1 4	2 5
82	2864 9055	2865 9061	2867 9066	2869 9072	2871 9077	2872 9083	2874 9088	2876 9094	2877 9099	2879 9105	0 1	0 1	1 2	1 2	1 3	1 3	1 4	1 4	2 5
83	2881 9110	2883 9116	2884 9121	2886 9127	2888 9132	2890 9138	2891 9143	2893 9149	2895 9154	2897 9160	0 1	0 1	1 2	1 2	1 3	1 3	1 4	1 4	2 5
84	2898 9165	2900 9171	2902 9176	2903 9182	2905 9187	2907 9192	2909 9198	2910 9203	2912 9209	2914 9214	0 1	0 1	1 2	1 2	1 3	1 3	1 4	1 4	2 5
85	2915 9220	2917 9225	2919 9230	2921 9236	2922 9241	2924 9247	2926 9252	2927 9257	2929 9263	2931 9268	0 1	0 1	1 2	1 2	1 3	1 3	1 4	1 4	2 5
86	2933 9274	2934 9279	2936 9284	2938 9290	2939 9295	2941 9301	2943 9306	2944 9311	2946 9317	2948 9322	0 1	0 1	1 2	1 2	1 3	1 3	1 4	1 4	2 5
87	2950 9327	2951 9333	2953 9338	2955 9343	2956 9349	2958 9354	2960 9359	2961 9365	2963 9370	2965 9375	0 1	0 1	1 2	1 2	1 3	1 3	1 4	1 4	2 5
88	2966 9381	2968 9386	2970 9391	2972 9397	2973 9402	2975 9407	2977 9413	2978 9418	2980 9423	2982 9429	0 1	0 1	1 2	1 2	1 3	1 3	1 4	1 4	2 5

	0	1	2	3	4	5	6	7	8	9	1	2	3	4	5	6	7	8	9
89	2983	2985	2987	2988	2990	2992	2993	2995	2997	2998	0	0	1	1	1	1	1	1	2
	9434	9439	9445	9450	9455	9460	9466	9471	9476	9482	1	1	2	2	3	3	4	4	5
90	3000	3002	3003	3005	3007	3008	3010	3012	3013	3015	0	0	0	1	1	1	1	1	1
	9487	9492	9497	9503	9508	9513	9518	9524	9529	9534	1	1	2	2	3	3	4	4	5
91	3017	3018	3020	3022	3023	3025	3027	3028	3030	3032	0	0	0	1	1	1	1	1	1
	9539	9545	9550	9555	9560	9566	9571	9576	9581	9586	1	1	2	2	3	3	4	4	5
92	3033	3035	3036	3038	3040	3041	3043	3045	3046	3048	0	0	0	1	1	1	1	1	1
	9592	9597	9602	9607	9612	9618	9623	9628	9633	9638	1	1	2	2	3	3	4	4	5
93	3050	3051	3053	3055	3056	3058	3059	3061	3063	3064	0	0	0	1	1	1	1	1	1
	9644	9649	9654	9659	9664	9670	9675	9680	9685	9690	1	1	2	2	3	3	4	4	5
94	3066	3068	3069	3071	3072	3074	3076	3077	3079	3081	0	0	0	1	1	1	1	1	1
	9695	9701	9706	9711	9716	9721	9726	9731	9737	9742	1	1	2	2	3	3	4	4	5
95	3082	3084	3085	3087	3089	3090	3092	3094	3095	3097	0	0	0	1	1	1	1	1	1
	9747	9752	9757	9762	9767	9772	9778	9783	9788	9793	1	1	2	2	3	3	4	4	5
96	3098	3100	3102	3103	3105	3106	3108	3110	3111	3113	0	0	0	1	1	1	1	1	1
	9798	9803	9808	9813	9818	9823	9829	9834	9839	9844	1	1	2	2	3	3	4	4	5
97	3114	3116	3118	3119	3121	3122	3124	3126	3127	3129	0	0	0	1	1	1	1	1	1
	9849	9854	9859	9864	9869	9874	9879	9884	9889	9894	1	1	2	2	3	3	4	4	5
98	3130	3132	3134	3135	3137	3138	3140	3142	3143	3145	0	0	0	1	1	1	1	1	1
	9899	9905	9910	9915	9920	9925	9930	9935	9940	9945	0	1	1	2	2	3	3	4	4
99	3146	3148	3150	3151	3153	3154	3156	3158	3159	3161	0	0	0	1	1	1	1	1	1
	9950	9955	9960	9965	9970	9975	9980	9985	9990	9995	0	1	1	2	2	3	3	4	4

Logarithms

	0	1	2	3	4	5	6	7	8	9	1	2	3	4	5	6	7	8	9
10	0000	0043	0086	0128	0170	0212	0253	0294	0334	0374	4	8	12	17	21	25	29	33	37
11	0414	0453	0492	0531	0569	0607	0645	0682	0719	0755	4	8	11	15	19	23	26	30	34
12	0792	0828	0864	0899	0934	0969	1004	1038	1072	1106	3	7	10	14	17	21	24	28	31
13	1139	1173	1206	1239	1271	1303	1335	1367	1399	1430	3	6	10	13	16	19	23	26	29
14	1461	1492	1523	1553	1584	1614	1644	1673	1703	1732	3	6	9	12	15	18	21	24	27
15	1761	1790	1818	1847	1875	1903	1931	1959	1987	2014	3	6	8	11	14	17	20	22	25
16	2041	2068	2095	2122	2148	2175	2201	2227	2253	2279	3	5	8	11	13	16	18	21	24
17	2304	2330	2355	2380	2405	2430	2455	2480	2504	2529	2	5	7	10	12	15	17	20	22
18	2553	2577	2601	2625	2648	2672	2695	2718	2742	2765	2	5	7	9	12	14	16	19	21
19	2788	2810	2833	2856	2878	2900	2923	2945	2967	2989	2	4	7	9	11	13	16	18	20
20	3010	3032	3054	3075	3096	3118	3139	3160	3181	3201	2	4	6	8	11	13	15	17	19
21	3222	3243	3263	3284	3304	3324	3345	3365	3385	3404	2	4	6	8	10	12	14	16	18
22	3424	3444	3464	3483	3502	3522	3541	3560	3579	3598	2	4	6	8	10	12	14	15	17
23	3617	3636	3655	3674	3692	3711	3729	3747	3766	3784	2	4	6	7	9	11	13	15	17
24	3802	3820	3838	3856	3874	3892	3909	3927	3945	3962	2	4	5	7	9	11	12	14	16
25	3979	3997	4014	4031	4048	4065	4082	4099	4116	4133	2	3	5	7	9	10	12	14	15
26	4150	4166	4183	4200	4216	4232	4249	4265	4281	4298	2	3	5	7	8	10	11	13	15
27	4314	4330	4346	4362	4378	4393	4409	4425	4440	4456	2	3	5	6	8	9	11	13	14
28	4472	4487	4502	4518	4533	4548	4564	4579	4594	4609	2	3	5	6	8	9	11	12	14
29	4624	4639	4654	4669	4683	4698	4713	4728	4742	4757	1	3	4	6	7	9	10	12	13
30	4771	4786	4800	4814	4829	4843	4857	4871	4886	4900	1	3	4	6	7	9	10	11	13
31	4914	4928	4942	4955	4969	4983	4997	5011	5024	5038	1	3	4	6	7	8	10	11	12
32	5051	5065	5079	5092	5105	5119	5132	5145	5159	5172	1	3	4	5	7	8	9	11	12

	0	1	2	3	4	5	6	7	8	9	1	2	3	4	5	6	7	8	9
33	5185	5198	5211	5224	5237	5250	5263	5276	5289	5302	1	3	4	5	6	8	9	10	12
34	5315	5328	5340	5353	5366	5378	5391	5403	5416	5428	1	3	4	5	6	8	9	10	11
35	5441	5453	5465	5478	5490	5502	5514	5527	5539	5551	1	2	4	5	6	7	9	10	11
36	5563	5575	5587	5599	5611	5623	5635	5647	5658	5670	1	2	4	5	6	7	8	10	11
37	5682	5694	5705	5717	5729	5740	5752	5763	5775	5786	1	2	3	5	6	7	8	9	10
38	5798	5809	5821	5832	5843	5855	5866	5877	5888	5899	1	2	3	5	6	7	8	9	10
39	5911	5922	5933	5944	5955	5966	5977	5988	5999	6010	1	2	3	4	5	7	8	9	10
40	6021	6031	6042	6053	6064	6075	6085	6096	6107	6117	1	2	3	4	5	6	8	9	10
41	6128	6138	6149	6160	6170	6180	6191	6201	6212	6222	1	2	3	4	5	6	7	8	9
42	6232	6243	6253	6263	6274	6284	6294	6304	6314	6325	1	2	3	4	5	6	7	8	9
43	6335	6345	6355	6365	6375	6385	6395	6405	6415	6425	1	2	3	4	5	6	7	8	9
44	6435	6444	6454	6464	6474	6484	6493	6503	6513	6522	1	2	3	4	5	6	7	8	9
45	6532	6542	6551	6561	6571	6580	6590	6599	6609	6618	1	2	3	4	5	6	7	8	9
46	6628	6637	6646	6656	6665	6675	6684	6693	6702	6712	1	2	3	4	5	6	7	7	8
47	6721	6730	6739	6749	6758	6767	6776	6785	6794	6803	1	2	3	4	5	5	6	7	8
48	6812	6821	6830	6839	6848	6857	6866	6875	6884	6893	1	2	3	4	4	5	6	7	8
49	6902	6911	6920	6928	6937	6946	6955	6964	6972	6981	1	2	3	4	4	5	6	7	8
50	6990	6998	7007	7016	7024	7033	7042	7050	7059	7067	1	2	3	3	4	5	6	7	8
51	7076	7084	7093	7101	7110	7118	7126	7135	7143	7152	1	2	3	3	4	5	6	7	8
52	7160	7168	7177	7185	7193	7202	7210	7218	7226	7235	1	2	2	3	4	5	6	7	7
53	7243	7251	7259	7267	7275	7284	7292	7300	7308	7316	1	2	2	3	4	5	6	6	7
54	7324	7332	7340	7348	7356	7364	7372	7380	7388	7396	1	2	2	3	4	5	6	6	7
55	7404	7412	7419	7427	7435	7443	7451	7459	7466	7474	1	2	2	3	4	5	5	6	7

Logarithms (*continued*)

	0	1	2	3	4	5	6	7	8	9	1	2	3	4	5	6	7	8	9
56	7482	7490	7497	7505	7513	7520	7528	7536	7543	7551	1	2	2	3	4	5	5	6	7
57	7559	7566	7574	7582	7589	7597	7604	7612	7619	7627	1	2	2	3	4	5	5	6	7
58	7634	7642	7649	7657	7664	7672	7679	7686	7694	7701	1	1	2	3	4	4	5	6	7
59	7709	7716	7723	7731	7738	7745	7752	7760	7767	7774	1	1	2	3	4	4	5	6	7
60	7782	7789	7796	7803	7810	7818	7825	7832	7839	7846	1	1	2	3	4	4	5	6	6
61	7853	7860	7868	7875	7882	7889	7896	7903	7910	7917	1	1	2	3	4	4	5	6	6
62	7924	7931	7938	7945	7952	7959	7966	7973	7980	7987	1	1	2	3	3	4	5	6	6
63	7993	8000	8007	8014	8021	8028	8035	8041	8048	8055	1	1	2	3	3	4	5	5	6
64	8062	8069	8075	8082	8089	8096	8102	8109	8116	8122	1	1	2	3	3	4	5	5	6
65	8129	8136	8142	8149	8156	8162	8169	8176	8182	8189	1	1	2	3	3	4	5	5	6
66	8195	8202	8209	8215	8222	8228	8235	8241	8248	8254	1	1	2	3	3	4	5	5	6
67	8261	8267	8274	8280	8287	8293	8299	8306	8312	8319	1	1	2	3	3	4	5	5	6
68	8325	8331	8338	8344	8351	8357	8363	8370	8376	8382	1	1	2	3	3	4	4	5	6
69	8388	8395	8401	8407	8414	8420	8426	8432	8439	8445	1	1	2	2	3	4	4	5	6
70	8451	8457	8463	8470	8476	8482	8488	8494	8500	8506	1	1	2	2	3	4	4	5	6
71	8513	8519	8525	8531	8537	8543	8549	8555	8561	8567	1	1	2	2	3	4	4	5	5
72	8573	8579	8585	8591	8597	8603	8609	8615	8621	8627	1	1	2	2	3	4	4	5	5
73	8633	8639	8645	8651	8657	8663	8669	8675	8681	8686	1	1	2	2	3	4	4	5	5
74	8692	8698	8704	8710	8716	8722	8727	8733	8739	8745	1	1	2	2	3	4	4	5	5
75	8751	8756	8762	8768	8774	8779	8785	8791	8797	8802	1	1	2	2	3	3	4	5	5
76	8808	8814	8820	8825	8831	8837	8842	8848	8854	8859	1	1	2	2	3	3	4	5	5
77	8865	8871	8876	8882	8887	8893	8899	8904	8910	8915	1	1	2	2	3	3	4	4	5

	0	1	2	3	4	5	6	7	8	9	1	2	3	4	5	6	7	8	9
78	8921	8927	8932	8938	8943	8949	8954	8960	8965	8971	1	1	2	2	3	3	4	4	5
79	8976	8982	8987	8993	8998	9004	9009	9015	9020	9025	1	1	2	2	3	3	4	4	5
80	9031	9036	9042	9047	9053	9058	9063	9069	9074	9079	1	1	2	2	3	3	4	4	5
81	9085	9090	9096	9101	9106	9112	9117	9122	9128	9133	1	1	2	2	3	3	4	4	5
82	9138	9143	9149	9154	9159	9165	9170	9175	9180	9186	1	1	2	2	3	3	4	4	5
83	9191	9196	9201	9206	9212	9217	9222	9227	9232	9238	1	1	2	2	3	3	4	4	5
84	9243	9248	9253	9258	9263	9269	9274	9279	9284	9289	1	1	2	2	3	3	4	4	5
85	9294	9299	9304	9309	9315	9320	9325	9330	9335	9325	1	1	2	2	3	3	4	4	5
86	9345	9350	9355	9360	9365	9370	9375	9380	9385	9390	1	1	2	2	3	3	4	4	5
87	9395	9400	9405	9410	9415	9420	9425	9430	9435	9440	0	1	1	2	2	3	3	4	4
88	9445	9450	9455	9460	9465	9469	9474	9479	9484	9489	0	1	1	2	2	3	3	4	4
89	9494	9499	9504	9509	9513	9518	9523	9528	9533	9538	0	1	1	2	2	3	3	4	4
90	9542	9547	9552	9557	9562	9566	9571	9576	9581	9586	0	1	1	2	2	3	3	4	4
91	9590	9595	9600	9605	9609	9614	9619	9624	9628	9633	0	1	1	2	2	3	3	4	4
92	9638	9643	9647	9652	9657	9661	9666	9671	9675	9680	0	1	1	2	2	3	3	4	4
93	9685	9689	9694	9699	9703	9708	9713	9717	9722	9727	0	1	1	2	2	3	3	4	4
94	9731	9736	9741	9745	9750	9754	9759	9763	9768	9773	0	1	1	2	2	3	3	4	4
95	9777	9782	9786	9791	9795	9800	9805	9809	9814	9818	0	1	1	2	2	3	3	4	4
96	9823	9827	9832	9836	9841	9845	9850	9854	9859	9863	0	1	1	2	2	3	3	4	4
97	9868	9872	9877	9881	9886	9890	9894	9899	9903	9908	0	1	1	2	2	3	3	4	4
98	9912	9917	9921	9926	9930	9934	9939	9943	9948	9952	0	1	1	2	2	3	3	4	4
99	9956	9961	9965	9969	9974	9978	9983	9987	9991	9996	0	1	1	2	2	3	3	3	4

Sines

	0′	6′	12′	18′	24′	30′	36′	42′	48′	54′	1′	2′	3′	4′	5′
0°	0·0000	0017	0035	0052	0070	0087	0105	0122	0140	0157	3	6	9	12	15
1	0·0175	0192	0209	0227	0244	0262	0279	0297	0314	0332	3	6	9	12	15
2	0·0349	0366	0384	0401	0419	0436	0454	0471	0488	0506	3	6	9	12	15
3	0·0523	0541	0558	0576	0593	0610	0628	0645	0663	0680	3	6	9	12	15
4	0·0698	0715	0732	0750	0767	0785	0802	0819	0837	0854	3	6	9	12	14
5	0·0872	0889	0906	0924	0941	0958	0976	0993	1011	1028	3	6	9	12	14
6	0·1045	1063	1080	1097	1115	1132	1149	1167	1184	1201	3	6	9	12	14
7	0·1219	1236	1253	1271	1288	1305	1323	1340	1357	1374	3	6	9	12	14
8	0·1392	1409	1426	1444	1461	1478	1495	1513	1530	1547	3	6	9	12	14
9	0·1564	1582	1599	1616	1633	1650	1668	1685	1702	1719	3	6	9	11	14
10	0·1736	1754	1771	1788	1805	1822	1840	1857	1874	1891	3	6	9	11	14
11	0·1908	1925	1942	1959	1977	1994	2011	2028	2045	2062	3	6	9	11	14
12	0·2079	2096	2113	2130	2147	2164	2181	2198	2215	2233	3	6	9	11	14
13	0·2250	2267	2284	2300	2317	2334	2351	2368	2385	2402	3	6	8	11	14
14	0·2419	2436	2453	2470	2487	2504	2521	2538	2554	2571	3	6	8	11	14
15	0·2588	2605	2622	2639	2656	2672	2689	2706	2723	2740	3	6	8	11	14
16	0·2756	2773	2790	2807	2823	2840	2857	2874	2890	2907	3	6	8	11	14
17	0·2924	2940	2957	2974	2990	3007	3024	3040	3057	3074	3	6	8	11	14
18	0·3090	3107	3123	3140	3156	3173	3190	3206	3223	3239	3	6	8	11	14
19	0·3256	3272	3289	3305	3322	3338	3355	3371	3387	3404	3	5	8	11	14
20	0·3420	3437	3453	3469	3486	3502	3518	3535	3551	3567	3	5	8	11	14
21	0·3584	3600	3616	3633	3649	3665	3681	3697	3714	3730	3	5	8	11	14

	0′	6′	12′	18′	24′	30′	36′	42′	48′	54′	1′	2′	3′	4′	5′
22°	0·3746	3762	3778	3795	3811	3827	3843	3859	3875	3891	3	5	8	11	13
23	0·3907	3923	3939	3955	3971	3987	4003	4019	4035	4051	3	5	8	11	13
24	0·4067	4083	4099	4115	4131	4147	4163	4179	4195	4210	3	5	8	11	13
25	0·4226	4242	4258	4274	4289	4305	4321	4337	4352	4368	3	5	8	11	13
26	0·4384	4399	4415	4431	4446	4462	4478	4493	4509	4524	3	5	8	10	13
27	0·4540	4555	4571	4586	4602	4617	4633	4648	4664	4679	3	5	8	10	13
28	0·4695	4710	4726	4741	4756	4772	4787	4802	4818	4833	3	5	8	10	13
29	0·4848	4863	4879	4894	4909	4924	4939	4955	4970	4985	3	5	8	10	13
30	0·5000	5015	5030	5045	5060	5075	5090	5105	5120	5135	3	5	8	10	13
31	0·5150	5165	5180	5195	5210	5225	5240	5255	5270	5284	2	5	7	10	12
32	0·5299	5314	5329	5344	5358	5373	5388	5402	5417	5432	2	5	7	10	12
33	0·5446	5461	5476	5490	5505	5519	5534	5548	5563	5577	2	5	7	10	12
34	0·5592	5606	5621	5635	5650	5664	5678	5693	5707	5721	2	5	7	10	12
35	0·5736	5750	5764	5779	5793	5807	5821	5835	5850	5864	2	5	7	9	12
36	0·5878	5892	5906	5920	5934	5948	5962	5976	5990	6004	2	5	7	9	12
37	0·6018	6032	6046	6060	6074	6088	6101	6115	6129	6143	2	5	7	9	12
38	0·6157	6170	6184	6198	6211	6225	6239	6252	6266	6280	2	5	7	9	11
39	0·6293	6307	6320	6334	6347	6361	6374	6388	6401	6414	2	4	7	9	11
40	0·6428	6441	6455	6468	6481	6494	6508	6521	6534	6547	2	4	7	9	11
41	0·6561	6574	6587	6600	6613	6626	6639	6652	6665	6678	2	4	7	9	11
42	0·6691	6704	6717	6730	6743	6756	6769	6782	6794	6807	2	4	6	9	11
43	0·6820	6833	6845	6858	6871	6884	6896	6909	6921	6934	2	4	6	8	11
44	0·6947	6959	6972	6984	6997	7009	7022	7034	7046	7059	2	4	6	8	10

Sines (*continued*)

	0′	6′	12′	18′	24′	30′	36′	42′	48′	54′	1′	2′	3′	4′	5′
45°	0·7071	7083	7096	7108	7120	7133	7145	7157	7169	7181	2	4	6	8	10
46	0·7193	7206	7218	7230	7242	7254	7266	7278	7290	7302	2	4	6	8	10
47	0·7314	7325	7337	7349	7361	7373	7385	7396	7408	7420	2	4	6	8	10
48	0·7431	7443	7455	7466	7478	7490	7501	7513	7524	7536	2	4	6	8	10
49	0·7547	7559	7570	7581	7593	7604	7615	7627	7638	7649	2	4	6	8	9
50	0·7660	7672	7683	7694	7705	7716	7727	7738	7749	7760	2	4	6	7	9
51	0·7771	7782	7793	7804	7815	7826	7837	7848	7859	7869	2	4	5	7	9
52	0·7880	7891	7902	7912	7923	7934	7944	7955	7965	7976	2	4	5	7	9
53	0·7986	7997	8007	8018	8028	8039	8049	8059	8070	8080	2	3	5	7	9
54	0·8090	8100	8111	8121	8131	8141	8151	8161	8171	8181	2	3	5	7	8
55	0·8192	8202	8211	8221	8231	8241	8251	8261	8271	8281	2	3	5	7	8
56	0·8290	8300	8310	8320	8329	8339	8348	8358	8368	8377	2	3	5	6	8
57	0·8387	8396	8406	8415	8425	8434	8443	8453	8462	8471	2	3	5	6	8
58	0·8480	8490	8499	8508	8517	8526	8536	8545	8554	8563	2	3	5	6	8
59	0·8572	8581	8590	8599	8607	8616	8625	8634	8643	8652	1	3	4	6	7
60	0·8660	8669	8678	8686	8695	8704	8712	8721	8729	8738	1	3	4	6	7
61	0·8746	8755	8763	8771	8780	8788	8796	8805	8813	8821	1	3	4	6	7
62	0·8829	8838	8846	8854	8862	8870	8878	8886	8894	8902	1	3	4	5	7
63	0·8910	8918	8926	8934	8942	8949	8957	8965	8973	8980	1	3	4	5	6
64	0·8988	8996	9003	9011	9018	9026	9033	9041	9048	9056	1	3	4	5	6
65	0·9063	9070	9078	9085	9092	9100	9107	9114	9121	9128	1	2	4	5	6
66	0·9135	9143	9150	9157	9164	9171	9178	9184	9191	9198	1	2	3	5	6

	0′	6′	12′	18′	24′	30′	36′	42′	48′	54′	1′	2′	3′	4′	5′
67°	0·9205	9212	9219	9225	9232	9239	9245	9252	9259	9265	1	2	3	4	6
68	0·9272	9278	9285	9291	9298	9304	9311	9317	9323	9330	1	2	3	4	5
69	0·9336	9342	9348	9354	9361	9367	9373	9379	9385	9391	1	2	3	4	5
70	0·9397	9403	9409	9415	9421	9426	9432	9438	9444	9449	1	2	3	4	5
71	0·9455	9461	9466	9472	9478	9483	9489	9494	9500	9505	1	2	3	4	5
72	0·9511	9516	9521	9527	9532	9537	9542	9548	9553	9558	1	2	3	4	4
73	0·9563	9568	9573	9578	9583	9588	9593	9598	9603	9608	1	2	2	3	4
74	0·9613	9617	9622	9627	9632	9636	9641	9646	9650	9655	1	2	2	3	4
75	0·9659	9664	9668	9673	9677	9681	9686	9690	9694	9699	1	1	2	3	4
76	0·9703	9707	9711	9715	9720	9724	9728	9732	9736	9740	1	1	2	3	3
77	0·9744	9748	9751	9755	9759	9763	9767	9770	9774	9778	1	1	2	3	3
78	0·9781	9785	9789	9792	9796	9799	9803	9806	9810	9813	1	1	2	2	3
79	0·9816	9820	9823	9826	9829	9833	9836	9839	9842	9845	1	1	2	2	3
80	0·9848	9851	9854	9857	9860	9863	9866	9869	9871	9874	0	1	1	2	2
81	0·9877	9880	9882	9885	9888	9890	9893	9895	9898	9900	0	1	1	2	2
82	0·9903	9905	9907	9910	9912	9914	9917	9919	9921	9923	0	1	1	2	2
83	0·9925	9928	9930	9932	9934	9936	9938	9940	9942	9943	0	1	1	1	2
84	0·9945	9947	9949	9951	9952	9954	9956	9957	9959	9960	0	1	1	1	1
85	0·9962	9963	9965	9966	9968	9969	9971	9972	9973	9974	0	0	1	1	1
86	0·9976	9977	9978	9979	9980	9981	9982	9983	9984	9985	0	0	1	1	1
87	0·9986	9987	9988	9989	9990	9990	9991	9992	9993	9993	0	0	0	0	1
88	0·9994	9995	9995	9996	9996	9997	9997	9997	9998	9998					
89	0·9998	9999	9999	9999	9999	1·000	1·000	1·000	1·000	1·000					

Cosines

	0′	6′	12′	18′	24′	30′	36′	42′	48′	54′	Subtract 1′	2′	3′	4′	5′
0°	1·0000	1·000	1·000	1·000	1·000	1·000	**9999**	**9999**	**9999**	**9999**					
1	0·9998	9998	9998	9997	9997	9997	9996	9996	9995	9995					
2	0·9994	9993	9993	9992	9991	9990	9990	9989	9988	9987	0	0	0	0	1
3	0·9986	9985	9984	9983	9982	9981	9980	9979	9978	9977	0	0	1	1	1
4	0·9976	9974	9973	9972	9971	9969	9968	9966	9965	9963	0	0	1	1	1
5	0·9962	9960	9959	9957	9956	9954	9952	9951	9949	9947	0	1	1	1	1
6	0·9945	9943	9942	9940	9938	9936	9934	9932	9930	9928	0	1	1	1	2
7	0·9925	9923	9921	9919	9917	9914	9912	9910	9907	9905	0	1	1	2	2
8	0·9903	9900	9898	9895	9893	9890	9888	9885	9882	9880	0	1	1	2	2
9	0·9877	9874	9871	9869	9866	9863	9860	9857	9854	9851	0	1	1	2	2
10	0·9848	9845	9842	9839	9836	9833	9829	9826	9823	9820	1	1	2	2	3
11	0·9816	9813	9810	9806	9803	9799	9796	9792	9789	9785	1	1	2	2	3
12	0·9781	9778	9774	9770	9767	9763	9759	9755	9751	9748	1	1	2	3	3
13	0·9744	9740	9736	9732	9728	9724	9720	9715	9711	9707	1	1	2	3	3
14	0·9703	9699	9694	9690	9686	9681	9677	9673	9668	9664	1	1	2	3	4
15	0·9659	9655	9650	9646	9641	9636	9632	9627	9622	9617	1	2	2	3	4
16	0·9613	9608	9603	9598	9593	9588	9583	9578	9573	9568	1	2	2	3	4
17	0·9563	9558	9553	9548	9542	9537	9532	9527	9521	9516	1	2	3	4	4
18	0·9511	9505	9500	9494	9489	9483	9478	9472	9466	9461	1	2	3	4	5
19	0·9455	9449	9444	9438	9432	9426	9421	9415	9409	9403	1	2	3	4	5
20	0·9397	9391	9385	9379	9373	9367	9361	9354	9348	9342	1	2	3	4	5
21	0·9336	9330	9323	9317	9311	9304	9298	9291	9285	9278	1	2	3	4	5

	0′	6′	12′	18′	24′	30′	36′	42′	48′	54′	1′	2′	3′	4′	5′
22	0·9272	9265	9259	9252	9245	9239	9232	9225	9219	9212	1	2	3	4	6
23	0·9205	9198	9191	9184	9178	9171	9164	9157	9150	9143	1	2	3	5	6
24	0·9135	9128	9121	9114	9107	9100	9092	9085	9078	9070	1	2	4	5	6
25	0·9063	9056	9048	9041	9033	9026	9018	9011	9003	8996	1	3	4	5	6
26	0·8988	8980	8973	8965	8957	8949	8942	8934	8926	8918	1	3	4	5	6
27	0·8910	8902	8894	8886	8878	8870	8862	8854	8846	8838	1	3	4	5	7
28	0·8829	8821	8813	8805	8796	8788	8780	8771	8763	8755	1	3	4	6	7
29	0·8746	8738	8729	8721	8712	8704	8695	8686	8678	8669	1	3	4	6	7
30	0·8660	8652	8643	8634	8625	8616	8607	8599	8590	8581	1	3	4	6	7
31	0·8572	8563	8554	8545	8536	8526	8517	8508	8499	8490	2	3	5	6	8
32	0·8480	8471	8462	8453	8443	8434	8425	8415	8406	8396	2	3	5	6	8
33	0·8387	8377	8368	8358	8348	8339	8329	8320	8310	8300	2	3	5	6	8
34	0·8290	8281	8271	8261	8251	8241	8231	8221	8211	8202	2	3	5	7	8
35	0·8192	8181	8171	8161	8151	8141	8131	8121	8111	8100	2	3	5	7	8
36	0·8090	8080	8070	8059	8049	8039	8028	8018	8007	7997	2	3	5	7	9
37	0·7986	7976	7965	7955	7944	7934	7923	7912	7902	7891	2	4	5	7	9
38	0·7880	7869	7859	7848	7837	7826	7815	7804	7793	7782	2	4	5	7	9
39	0·7771	7760	7749	7738	7727	7716	7705	7694	7683	7672	2	4	6	7	9
40	0·7660	7649	7638	7627	7615	7604	7593	7581	7570	7559	2	4	6	8	9
41	0·7547	7536	7524	7513	7501	7490	7478	7466	7455	7443	2	4	6	8	10
42	0·7431	7420	7408	7396	7385	7373	7361	7349	7337	7325	2	4	6	8	10
43	0·7314	7302	7290	7278	7266	7254	7242	7230	7218	7206	2	4	6	8	10
44	0·7193	7181	7169	7157	7145	7133	7120	7108	7096	7083	2	4	6	8	10

Cosines (*continued*)

	0′	6′	12′	18′	24′	30′	36′	42′	48′	54′	Subtract 1′	2′	3′	4′	5′
45°	0·7071	7059	7046	7034	7022	7009	6997	6984	6972	6959	2	4	6	8	10
46	0·6947	6934	6921	6909	6896	6884	6871	6858	6845	6833	2	4	6	8	11
47	0·6820	6807	6794	6782	6769	6756	6743	6730	6717	6704	2	4	6	9	11
48	0·6691	6678	6665	6652	6639	6626	6613	6600	6587	6574	2	4	7	9	11
49	0·6561	6547	6534	6521	6508	6494	6481	6468	6455	6441	2	4	7	9	11
50	0·6428	6414	6401	6388	6374	6361	6347	6334	6320	6307	2	4	7	9	11
51	0·6293	6280	6266	6252	6239	6225	6211	6198	6184	6170	2	5	7	9	11
52	0·6157	6143	6129	6115	6101	6088	6074	6060	6046	6032	2	5	7	9	12
53	0·6018	6004	5990	5976	5962	5948	5934	5920	5906	5892	2	5	7	9	12
54	0·5878	5864	5850	5835	5821	5807	5793	5779	5764	5750	2	5	7	9	12
55	0·5736	5721	5707	5693	5678	5664	5650	5635	5621	5606	2	5	7	10	12
56	0·5592	5577	5563	5548	5534	5519	5505	5490	5476	5461	2	5	7	10	12
57	0·5446	5432	5417	5402	5388	5373	5358	5344	5329	5314	2	5	7	10	12
58	0·5299	5284	5270	5255	5240	5225	5210	5195	5180	5165	2	5	7	10	12
59	0·5150	5135	5120	5105	5090	5075	5060	5045	5030	5015	3	5	8	10	13
60	0·5000	4985	4970	4955	4939	4924	4909	4894	4879	4863	3	5	8	10	13
61	0·4848	4833	4818	4802	4787	4772	4756	4741	4726	4710	3	5	8	10	13
62	0·4695	4679	4664	4648	4633	4617	4602	4586	4571	4555	3	5	8	10	13
63	0·4540	4524	4509	4493	4478	4462	4446	4431	4415	4399	3	5	8	10	13
64	0·4384	4368	4352	4337	4321	4305	4289	4274	4258	4242	3	5	8	11	13
65	0·4226	4210	4195	4179	4163	4147	4131	4115	4099	4083	3	5	8	11	13
66	0·4067	4051	4035	4019	4003	3987	3971	3955	3939	3923	3	5	8	11	13

	0′	6′	12′	18′	24′	30′	36′	42′	48′	54′	1′	2′	3′	4′	5′
67	0·3907	3891	3875	3859	3843	3827	3811	3795	3778	3762	3	5	8	11	13
68	0·3746	3730	3714	3697	3681	3665	3649	3633	3616	3600	3	5	8	11	14
69	0·3584	3567	3551	3535	3518	3502	3486	3469	3453	3437	3	5	8	11	14
70	0·3420	3404	3387	3371	3355	3338	3322	3305	3289	3272	3	5	8	11	14
71	0·3256	3239	3223	3206	3190	3173	3156	3140	3123	3107	3	6	8	11	14
72	0·3090	3074	3057	3040	3024	3007	2990	2974	2957	2940	3	6	8	11	14
73	0·2924	2907	2890	2874	2857	2840	2823	2807	2790	2773	3	6	8	11	14
74	0·2756	2740	2723	2706	2689	2672	2656	2639	2622	2605	3	6	8	11	14
75	0·2588	2571	2554	2538	2521	2504	2487	2470	2453	2436	3	6	8	11	14
76	0·2419	2402	2385	2368	2351	2334	2317	2300	2284	2267	3	6	8	11	14
77	0·2250	2233	2215	2198	2181	2164	2147	2130	2113	2096	3	6	9	11	14
78	0·2079	2062	2045	2028	2011	1994	1977	1959	1942	1925	3	6	9	11	14
79	0·1908	1891	1874	1857	1840	1822	1805	1788	1771	1754	3	6	9	11	14
80	0·1736	1719	1702	1685	1668	1650	1633	1616	1599	1582	3	6	9	11	14
81	0·1564	1547	1530	1513	1495	1478	1461	1444	1426	1409	3	6	9	12	14
82	0·1392	1374	1357	1340	1323	1305	1288	1271	1253	1236	3	6	9	12	14
83	0·1219	1201	1184	1167	1149	1132	1115	1097	1080	1063	3	6	9	12	14
84	0·1045	1028	1011	0993	0976	0958	0941	0924	0906	0889	3	6	9	12	14
85	0·0872	0854	0837	0819	0802	0785	0767	0750	0732	0715	3	6	9	12	14
86	0·0698	0680	0663	0645	0628	0610	0593	0576	0558	0541	3	6	9	12	15
87	0·0523	0506	0488	0471	0454	0436	0419	0401	0384	0366	3	6	9	12	15
88	0·0349	0332	0314	0297	0279	0262	0244	0227	0209	0192	3	6	9	12	15
89	0·0175	0157	0140	0122	0105	0087	0070	0052	0035	0017	3	6	9	12	15

Tangents

	0′	6′	12′	18′	24′	30′	36′	42′	48′	54′	1′	2′	3′	4′	5′
0°	0·0000	0017	0035	0052	0070	0087	0105	0122	0140	0157	3	6	9	12	15
1	0·0175	0192	0209	0227	0244	0262	0279	0297	0314	0332	3	6	9	12	15
2	0·0349	0367	0384	0402	0419	0437	0454	0472	0489	0507	3	6	9	12	15
3	0·0524	0542	0559	0577	0594	0612	0629	0647	0664	0682	3	6	9	12	15
4	0·0699	0717	0734	0752	0769	0787	0805	0822	0840	0857	3	6	9	12	15
5	0·0875	0892	0910	0928	0945	0963	0981	0998	1016	1033	3	6	9	12	15
6	0·1051	1069	1086	1104	1122	1139	1157	1175	1192	1210	3	6	9	12	15
7	0·1228	1246	1263	1281	1299	1317	1334	1352	1370	1388	3	6	9	12	15
8	0·1405	1423	1441	1459	1477	1495	1512	1530	1548	1566	3	6	9	12	15
9	0·1584	1602	1620	1638	1655	1673	1691	1709	1727	1745	3	6	9	12	15
10	0·1763	1781	1799	1817	1835	1853	1871	1890	1908	1926	3	6	9	12	15
11	0·1944	1962	1980	1998	2016	2035	2053	2071	2089	2107	3	6	9	12	15
12	0·2126	2144	2162	2180	2199	2217	2235	2254	2272	2290	3	6	9	12	15
13	0·2309	2327	2345	2364	2382	2401	2419	2438	2456	2475	3	6	9	12	15
14	0·2493	2512	2530	2549	2568	2586	2605	2623	2642	2661	3	6	9	12	16
15	0·2679	2698	2717	2736	2754	2773	2792	2811	2830	2849	3	6	9	13	16
16	0·2867	2886	2905	2924	2943	2962	2981	3000	3019	3038	3	6	9	13	16
17	0·3057	3076	3096	3115	3134	3153	3172	3191	3211	3230	3	6	10	13	16
18	0·3249	3269	3288	3307	3327	3346	3365	3385	3404	3424	3	6	10	13	16
19	0·3443	3463	3482	3502	3522	3541	3561	3581	3600	3620	3	7	10	13	16
20	0·3640	3659	3679	3699	3719	3739	3759	3779	3799	3819	3	7	10	13	17
21	0·3839	3859	3879	3899	3919	3939	3959	3979	4000	4020	3	7	10	13	17

	0′	6′	12′	18′	24′	30′	36′	42′	48′	54′	1′	2′	3′	4′	5′
22	0·4040	4061	4081	4101	4122	4142	4163	4183	4204	4224	3	7	10	14	17
23	0·4245	4265	4286	4307	4327	4348	4369	4390	4411	4431	3	7	10	14	17
24	0·4452	4473	4494	4515	4536	4557	4578	4599	4621	4642	4	7	11	14	18
25	0·4663	4684	4706	4727	4748	4770	4791	4813	4834	4856	4	7	11	14	18
26	0·4877	4899	4921	4942	4964	4986	5008	5029	5051	5073	4	7	11	15	18
27	0·5095	5117	5139	5161	5184	5206	5228	5250	5272	5295	4	7	11	15	18
28	0·5317	5340	5362	5384	5407	5430	5452	5475	5498	5520	4	8	11	15	19
29	0·5543	5566	5589	5612	5635	5658	5681	5704	5727	5750	4	8	12	15	19
30	0·5774	5797	5820	5844	5867	5890	5914	5938	5961	5985	4	8	12	16	20
31	0·6009	6032	6056	6080	6104	6128	6152	6176	6200	6224	4	8	12	16	20
32	0·6249	6273	6297	6322	6346	6371	6395	6420	6445	6469	4	8	12	16	20
33	0·6494	6519	6544	6569	6594	6619	6644	6669	6694	6720	4	8	13	17	21
34	0·6745	6771	6796	6822	6847	6873	6899	6924	6950	6976	4	9	13	17	21
35	0·7002	7028	7054	7080	7107	7133	7159	7186	7212	7239	4	9	13	18	22
36	0·7265	7292	7319	7346	7373	7400	7427	7454	7481	7508	5	9	14	18	23
37	0·7536	7563	7590	7618	7646	7673	7701	7729	7757	7785	5	9	14	18	23
38	0·7813	7841	7869	7898	7926	7954	7983	8012	8040	8069	5	9	14	19	24
39	0·8098	8127	8156	8185	8214	8243	8273	8302	8332	8361	5	10	15	20	24
40	0·8391	8421	8451	8481	8511	8541	8571	8601	8632	8662	5	10	15	20	25
41	0·8693	8724	8754	8785	8816	8847	8878	8910	8941	8972	5	10	16	21	26
42	0·9004	9036	9067	9099	9131	9163	9195	9228	9260	9293	5	11	16	21	27
43	0·9325	9358	9391	9424	9457	9490	9523	9556	9590	9623	6	11	17	22	28
44	0·9657	9691	9725	9759	9793	9827	9861	9869	9930	9965	6	11	17	23	29

Tangents (*continued*)

	0′	6′	12′	18′	24′	30′	36′	42′	48′	54′	1′	2′	3′	4′	5′
45°	1·0000	0035	0070	0105	0141	0176	0212	0247	0283	0319	6	12	18	24	30
46	1·0355	0392	0428	0464	0501	0538	0575	0612	0649	0686	6	12	18	25	31
47	1·0724	0761	0799	0837	0875	0913	0951	0990	1028	1067	6	13	19	25	32
48	1·1106	1145	1184	1224	1263	1303	1343	1383	1423	1463	7	13	20	26	33
49	1·1504	1544	1585	1626	1667	1708	1750	1792	1833	1875	7	14	21	28	34
50	1·1918	1960	2002	2045	2088	2131	2174	2218	2261	2305	7	14	22	29	36
51	1·2349	2393	2437	2482	2527	2572	2617	2662	2708	2753	8	15	23	30	38
52	1·2799	2846	2892	2938	2985	3032	3079	3127	3175	3222	8	16	24	31	39
53	1·3270	3319	3367	3416	3465	3514	3564	3613	3663	3713	8	16	25	33	41
54	1·3764	3814	3865	3916	3968	4019	4071	4124	4176	4229	9	17	26	34	43
55	1·4281	4335	4388	4442	4496	4550	4605	4659	4715	4770	9	18	27	36	45
56	1·4826	4882	4938	4994	5051	5108	5166	5224	5282	5340	10	19	29	38	48
57	1·5399	5458	5517	5577	5637	5697	5757	5818	5880	5941	10	20	30	40	50
58	1·6003	6066	6128	6191	6255	6319	6383	6447	6512	6577	11	21	32	43	53
59	1·6643	6709	6775	6842	6909	6977	7045	7113	7182	7251	11	23	34	45	56
60	1·7321	7391	7461	7532	7603	7675	7747	7820	7893	7966	12	24	36	48	60
61	1·8040	8115	8190	8265	8341	8418	8495	8572	8650	8728	13	26	38	51	64
62	1·8807	8887	8967	9047	9128	9210	9292	9375	9458	9542	14	27	41	55	68
63	1·9626	9711	9797	9883	9970	**0057**	**0145**	**0233**	**0323**	**0413**	15	29	44	58	73
64	2·0503	0594	0686	0778	0872	0965	1060	1155	1251	1348	16	31	47	63	78
65	2·1445	1543	1642	1742	1842	1943	2045	2148	2251	2355	17	34	51	68	85
66	2·2460	2566	2673	2781	2889	2998	3109	3220	3332	3445	18	37	55	73	91

	0′	6′	12′	18′	24′	30′	36′	42′	48′	54′	1′	2′	3′	4′	5′
67	2·3559	3673	3789	3906	4023	4142	4262	4383	4504	4627	20	40	60	79	99
68	2·4751	4876	5002	5129	5257	5386	5517	5649	5782	5916	22	43	65	87	108
69	2·6051	6187	6325	6464	6605	6746	6889	7034	7179	7326	24	47	71	95	119
70	2·7475	7625	7776	7929	8083	8239	8397	8556	8716	8878	26	52	78	104	130
71	2·9042	9208	9375	9544	9714	9887	**0061**	**0237**	**0415**	**0595**	29	58	87	116	144
72	3·0777	0961	1146	1334	1524	1716	1910	2106	2305	2506	32	64	97	129	161
73	3·2709	2914	3122	3332	3544	3759	3977	4197	4420	4646	36	72	108	144	180
74	3·4874	5105	5339	5576	5816	6059	6305	6554	6806	7062	41	81	122	163	203
75	3·7321	7583	7848	8118	8391	8667	8947	9232	9520	9812	46	93	139	186	232
76	4·0108	0408	0713	1022	1335	1653	1976	2303	2635	2972	53	107	160	214	267
77	4·3315	3662	4015	4373	4737	5107	5483	5864	6252	6646	62	124	186	248	310
78	4·7046	7453	7867	8288	8716	9152	9594	**0045**	**0504**	**0970**	73	146	220	293	366
79	5·1446	1929	2422	2924	3435	3955	4486	5026	5578	6140	87	175	263	350	438
80	5·671	5·730	5·789	5·850	5·912	5·976	6·041	6·107	6·174	6·243					
81	6·314	6·386	6·460	6·535	6·612	6·691	6·772	6·855	6·940	7·026					
82	7·115	7·207	7·300	7·396	7·495	7·596	7·700	7·806	7·916	8·028					
83	8·144	8·264	8·386	8·513	8·643	8·777	8·915	9·058	9·205	9·357					
84	9·51	9·68	9·84	10·02	10·20	10·39	10·58	10·78	10·99	11·20					
85	11·43	11·66	11·91	12·16	12·43	12·71	13·00	13·30	13·62	13·95					
86	14·30	14·67	15·06	15·46	15·89	16·35	16·83	17·34	17·89	18·46					
87	19·08	19·74	20·45	21·20	22·02	22·90	23·86	24·90	26·03	27·27					
88	28·64	30·14	31·82	33·69	35·80	38·19	40·92	44·07	47·74	52·08					
89	57·29	63·66	71·62	81·85	95·49	114·6	143·2	191·0	286·5	573·0					

The **bold** typeface indicates that the integral part changes

Logarithms of Sines

	0′	6′	12′	18′	24′	30′	36′	42′	48′	54′	1′	2′	3′	4′	5′
0°	$-\infty$	$\bar{3}$·242	$\bar{3}$·543	$\bar{3}$·719	$\bar{3}$·844	$\bar{3}$·941	$\bar{2}$·020	$\bar{2}$·087	$\bar{2}$·145	$\bar{2}$·196					
1	$\bar{2}$·2419	2832	3210	3558	3880	4179	4459	4723	4971	5206					
2	$\bar{2}$·5428	5640	5842	6035	6220	6397	6567	6731	6889	7041					
3	$\bar{2}$·7188	7330	7468	7602	7731	7857	7979	8098	8213	8326	21	41	62	83	103
4	$\bar{2}$·8436	8543	8647	8749	8849	8946	9042	9135	9226	9315	16	32	48	64	81
5	$\bar{2}$·9403	9489	9573	9655	9736	9816	9894	9970	**0046**	**0120**	13	26	39	53	66
6	$\bar{1}$·0192	0264	0334	0403	0472	0539	0605	0670	0734	0797	11	22	33	44	55
7	$\bar{1}$·0859	0920	0981	1040	1099	1157	1214	1271	1326	1381	10	19	29	38	48
8	$\bar{1}$·1436	1489	1542	1594	1646	1697	1747	1797	1847	1895	8	17	25	34	42
9	$\bar{1}$·1943	1991	2038	2085	2131	2176	2221	2266	2310	2353	8	15	23	30	38
10	$\bar{1}$·2397	2439	2482	2524	2565	2606	2647	2687	2727	2767	7	14	20	27	34
11	$\bar{1}$·2806	2845	2883	2921	2959	2997	3034	3070	3107	3143	6	12	19	25	31
12	$\bar{1}$·3179	3214	3250	3284	3319	3353	3387	3421	3455	3488	6	11	17	23	28
13	$\bar{1}$·3521	3554	3586	3618	3650	3682	3713	3745	3775	3806	5	11	16	21	26
14	$\bar{1}$·3837	3867	3897	3927	3957	3986	4015	4044	4073	4102	5	10	15	20	24
15	$\bar{1}$·4130	4158	4186	4214	4242	4269	4296	4323	4350	4377	5	9	14	18	23
16	$\bar{1}$·4403	4430	4456	4482	4508	4533	4559	4584	4609	4634	4	9	13	17	21
17	$\bar{1}$·4659	4684	4709	4733	4757	4781	4805	4829	4853	4876	4	8	12	16	20
18	$\bar{1}$·4900	4923	4946	4969	4992	5015	5037	5060	5082	5104	4	8	11	15	19
19	$\bar{1}$·5126	5148	5170	5192	5213	5235	5256	5278	5299	5320	4	7	11	14	18
20	$\bar{1}$·5341	5361	5382	5402	5423	5443	5463	5484	5504	5523	3	7	10	14	17
21	$\bar{1}$·5543	5563	5583	5602	5621	5641	5660	5679	5698	5717	3	6	10	13	16

	0′	6′	12′	18′	24′	30′	36′	42′	48′	54′	1′	2′	3′	4′	5′
22	$\bar{1}$·5736	5754	5773	5792	5810	5828	5847	5865	5883	5901	3	6	9	12	15
23	$\bar{1}$·5919	5937	5954	5972	5990	6007	6024	6042	6059	6076	3	6	9	12	15
24	$\bar{1}$·6093	6110	6127	6144	6161	6177	6194	6210	6227	6243	3	6	8	11	14
25	$\bar{1}$·6259	6276	6292	6308	6324	6340	6356	6371	6387	6403	3	5	8	11	13
26	$\bar{1}$·6418	6434	6449	6465	6480	6495	6510	6526	6541	6556	3	5	8	10	13
27	$\bar{1}$·6570	6585	6600	6615	6629	6644	6659	6673	6687	6702	2	5	7	10	12
28	$\bar{1}$·6716	6730	6744	6759	6773	6787	6801	6814	6828	6842	2	5	7	9	12
29	$\bar{1}$·6856	6869	6883	6896	6910	6923	6937	6950	6963	6977	2	4	7	9	11
30	$\bar{1}$·6990	7003	7016	7029	7042	7055	7068	7080	7093	7106	2	4	6	9	11
31	$\bar{1}$·7118	7131	7144	7156	7168	7181	7193	7205	7218	7230	2	4	6	8	10
32	$\bar{1}$·7242	7254	7266	7278	7290	7302	7314	7326	7338	7349	2	4	6	8	10
33	$\bar{1}$·7361	7373	7384	7396	7407	7419	7430	7442	7453	7464	2	4	6	8	10
34	$\bar{1}$·7476	7487	7498	7509	7520	7531	7542	7553	7564	7575	2	4	6	7	9
35	$\bar{1}$·7586	7597	7607	7618	7629	7640	7650	7661	7671	7682	2	4	5	7	9
36	$\bar{1}$·7692	7703	7713	7723	7734	7744	7754	7764	7774	7785	2	3	5	7	9
37	$\bar{1}$·7795	7805	7815	7825	7835	7844	7854	7864	7874	7884	2	3	5	7	8
38	$\bar{1}$·7893	7903	7913	7922	7932	7941	7951	7960	7970	7979	2	3	5	6	8
39	$\bar{1}$·7989	7998	8007	8017	8026	8035	8044	8053	8063	8072	2	3	5	6	8
40	$\bar{1}$·8081	8090	8099	8108	8117	8125	8134	8143	8152	8161	1	3	4	6	7
41	$\bar{1}$·8169	8178	8187	8195	8204	8213	8221	8230	8238	8247	1	3	4	6	7
42	$\bar{1}$·8255	8264	8272	8280	8289	8297	8305	8313	8322	8330	1	3	4	6	7
43	$\bar{1}$·8338	8346	8354	8362	8370	8378	8386	8394	8402	8410	1	3	4	5	7
44	$\bar{1}$·8418	8426	8433	8441	8449	8457	8464	8472	8480	8487	1	3	4	5	6

The **bold** typeface indicates that the integral part changes

Logarithms of Sines (*continued*)

	0′	6′	12′	18′	24′	30′	36′	42′	48′	54′	1′	2′	3′	4′	5′
45°	$\bar{1}$·8495	8502	8510	8517	8525	8532	8540	8547	8555	8562	1	2	4	5	6
46	$\bar{1}$·8569	8577	8584	8591	8598	8606	8613	8620	8627	8634	1	2	4	5	6
47	$\bar{1}$·8641	8648	8655	8662	8669	8676	8683	8690	8697	8704	1	2	3	5	6
48	$\bar{1}$·8711	8718	8724	8731	8738	8745	8751	8758	8765	8771	1	2	3	4	6
49	$\bar{1}$·8778	8784	8791	8797	8804	8810	8817	8823	8830	8836	1	2	3	4	5
50	$\bar{1}$·8843	8849	8855	8862	8868	8874	8880	8887	8893	8899	1	2	3	4	5
51	$\bar{1}$·8905	8911	8917	8923	8929	8935	8941	8947	8953	8959	1	2	3	4	5
52	$\bar{1}$·8965	8971	8977	8983	8989	8995	9000	9006	9012	9018	1	2	3	4	5
53	$\bar{1}$·9023	9029	9035	9041	9046	9052	9057	9063	9069	9074	1	2	3	4	5
54	$\bar{1}$·9080	9085	9091	9096	9101	9107	9112	9118	9123	9128	1	2	3	4	5
55	$\bar{1}$·9134	9139	9144	9149	9155	9160	9165	9170	9175	9181	1	2	3	3	4
56	$\bar{1}$·9186	9191	9196	9201	9206	9211	9216	9221	9226	9231	1	2	3	3	4
57	$\bar{1}$·9236	9241	9246	9251	9255	9260	9265	9270	9275	9279	1	2	2	3	4
58	$\bar{1}$·9284	9289	9294	9298	9303	9308	9312	9317	9322	9326	1	2	2	3	4
59	$\bar{1}$·9331	9335	9340	9344	9349	9353	9358	9362	9367	9371	1	1	2	3	4
60	$\bar{1}$·9375	9380	9384	9388	9393	9397	9401	9406	9410	9414	1	1	2	3	4
61	$\bar{1}$·9418	9422	9427	9431	9435	9439	9443	9447	9451	9455	1	1	2	3	3
62	$\bar{1}$·9459	9463	9467	9471	9475	9479	9483	9487	9491	9495	1	1	2	3	3
63	$\bar{1}$·9499	9503	9506	9510	9514	9518	9522	9525	9529	9533	1	1	2	3	3
64	$\bar{1}$·9537	9540	9544	9548	9551	9555	9558	9562	9566	9569	1	1	2	2	3
65	$\bar{1}$·9573	9576	9580	9583	9587	9590	9594	9597	9601	9604	1	1	2	2	3
66	$\bar{1}$·9607	9611	9614	9617	9621	9624	9627	9631	9634	9637	1	1	2	2	3

	0′	6′	12′	18′	24′	30′	36′	42′	48′	54′	1′	2′	3′	4′	5′
67	$\bar{1}$·9640	9643	9647	9650	9653	9656	9659	9662	9666	9669	1	1	2	2	3
68	$\bar{1}$·9672	9675	9678	9681	9684	9687	9690	9693	9696	9699	0	1	1	2	2
69	$\bar{1}$·9702	9704	9707	9710	9713	9716	9719	9722	9724	9727	0	1	1	2	2
70	$\bar{1}$·9730	9733	9735	9738	9741	9743	9746	9749	9751	9754	0	1	1	2	2
71	$\bar{1}$·9757	9759	9762	9764	9767	9770	9772	9775	9777	9780	0	1	1	2	2
72	$\bar{1}$·9782	9785	9787	9789	9792	9794	9797	9799	9801	9804	0	1	1	2	2
73	$\bar{1}$·9806	9808	9811	9813	9815	9817	9820	9822	9824	9826	0	1	1	1	2
74	$\bar{1}$·9828	9831	9833	9835	9837	9839	9841	9843	9845	9847	0	1	1	1	2
75	$\bar{1}$·9849	9851	9853	9855	9857	9859	9861	9863	9865	9867	0	1	1	1	2
76	$\bar{1}$·9869	9871	9873	9875	9876	9878	9880	9882	9884	9885	0	1	1	1	2
77	$\bar{1}$·9887	9889	9891	9892	9894	9896	9897	9899	9901	9902	0	1	1	1	1
78	$\bar{1}$·9904	9906	9907	9909	9910	9912	9913	9915	9916	9918	0	1	1	1	1
79	$\bar{1}$·9919	9921	9922	9924	9925	9927	9928	9929	9931	9932	0	0	1	1	1
80	$\bar{1}$·9934	9935	9936	9937	9939	9940	9941	9943	9944	9945	0	0	1	1	1
81	$\bar{1}$·9946	9947	9949	9950	9951	9952	9953	9954	9955	9956	0	0	1	1	1
82	$\bar{1}$·9958	9959	9960	9961	9962	9963	9964	9965	9966	9967	0	0	0	1	1
83	$\bar{1}$·9968	9968	9969	9970	9971	9972	9973	9974	9975	9975					
84	$\bar{1}$·9976	9977	9978	9978	9979	9980	9981	9981	9982	9983					
85	$\bar{1}$·9983	9984	9985	9985	9986	9987	9987	9988	9988	9989					
86	$\bar{1}$·9989	9990	9990	9991	9991	9992	9992	9993	9993	9994					
87	$\bar{1}$·9994	9994	9995	9995	9996	9996	9996	9996	9997	9997					
88	$\bar{1}$·9997	9998	9998	9998	9998	9999	9999	9999	9999	9999					
89	$\bar{1}$·9999	9999	**0000**	**0000**	**0000**	**0000**	**0000**	**0000**	**0000**	**0000**					

The **bold** typeface indicates that the integral part changes

Logarithms of Cosines

	0′	6′	12′	18′	24′	30′	36′	42′	48′	54′	Subtract 1′	2′	3′	4′	5′
0°	0·0000	0000	0000	0000	0000	0000	0000	0000	0000	9999					
1	$\bar{1}$·9999	9999	9999	9999	9999	9999	9998	9998	9998	9998					
2	$\bar{1}$·9997	9997	9997	9996	9996	9996	9996	9995	9995	9994					
3	$\bar{1}$·9994	9994	9993	9993	9992	9992	9991	9991	9990	9990					
4	$\bar{1}$·9989	9989	9988	9988	9987	9987	9986	9985	9985	9984					
5	$\bar{1}$·9983	9983	9982	9981	9981	9980	9979	9978	9978	9977					
6	$\bar{1}$·9976	9975	9975	9974	9973	9972	9971	9970	9969	9968					
7	$\bar{1}$·9968	9967	9966	9965	9964	9963	9962	9961	9960	9959	0	0	0	1	1
8	$\bar{1}$·9958	9956	9955	9954	9953	9952	9951	9950	9949	9947	0	0	1	1	1
9	$\bar{1}$·9946	9945	9944	9943	9941	9940	9939	9937	9936	9935	0	0	1	1	1
10	$\bar{1}$·9934	9932	9931	9929	9928	9927	9925	9924	9922	9921	0	0	1	1	1
11	$\bar{1}$·9919	9918	9916	9915	9913	9912	9910	9909	9907	9906	0	1	1	1	1
12	$\bar{1}$·9904	9902	9901	9899	9897	9896	9894	9892	9891	9889	0	1	1	1	1
13	$\bar{1}$·9887	9885	9884	9882	9880	9878	9876	9875	9873	9871	0	1	1	1	2
14	$\bar{1}$·9869	9867	9865	9863	9861	9859	9857	9855	9853	9851	0	1	1	1	2
15	$\bar{1}$·9849	9847	9845	9843	9841	9839	9837	9835	9833	9831	0	1	1	1	2
16	$\bar{1}$·9828	9826	9824	9822	9820	9817	9815	9813	9811	9808	0	1	1	1	2
17	$\bar{1}$·9806	9804	9801	9799	9797	9794	9792	9789	9787	9785	0	1	1	2	2
18	$\bar{1}$·9782	9780	9777	9775	9772	9770	9767	9764	9762	9759	0	1	1	2	2
19	$\bar{1}$·9757	9754	9751	9749	9746	9743	9741	9738	9735	9733	0	1	1	2	2
20	$\bar{1}$·9730	9727	9724	9722	9719	9716	9713	9710	9707	9704	0	1	1	2	2
21	$\bar{1}$·9702	9699	9696	9693	9690	9687	9684	9681	9678	9675	0	1	1	2	2

	0′	6′	12′	18′	24′	30′	36′	42′	48′	54′	1′	2′	3′	4′	5′
22	$\bar{1}$·9672	9669	9666	9662	9659	9656	9653	9650	9647	9643	1	1	2	2	3
23	$\bar{1}$·9640	9637	9634	9631	9627	9624	9621	9617	9614	9611	1	1	2	2	3
24	$\bar{1}$·9607	9604	9601	9597	9594	9590	9587	9583	9580	9576	1	1	2	2	3
25	$\bar{1}$·9573	9569	9566	9562	9558	9555	9551	9548	9544	9540	1	1	2	2	3
26	$\bar{1}$·9537	9533	9529	9525	9522	9518	9514	9510	9506	9503	1	1	2	3	3
27	$\bar{1}$·9499	9495	9491	9487	9483	9479	9475	9471	9467	9463	1	1	2	3	3
28	$\bar{1}$·9459	9455	9451	9447	9443	9439	9435	9431	9427	9422	1	1	2	3	3
29	$\bar{1}$·9418	9414	9410	9406	9401	9397	9393	9388	9384	9380	1	1	2	3	4
30	$\bar{1}$·9375	9371	9367	9362	9358	9353	9349	9344	9340	9335	1	1	2	3	4
31	$\bar{1}$·9331	9326	9322	9317	9312	9308	9303	9298	9294	9289	1	2	2	3	4
32	$\bar{1}$·9284	9279	9275	9270	9265	9260	9255	9251	9246	9241	1	2	2	3	4
33	$\bar{1}$·9236	9231	9226	9221	9216	9211	9206	9201	9196	9191	1	2	3	3	4
34	$\bar{1}$·9186	9181	9175	9170	9165	9160	9155	9149	9144	9139	1	2	3	3	4
35	$\bar{1}$·9134	9128	9123	9118	9112	9107	9101	9096	9091	9085	1	2	3	4	5
36	$\bar{1}$·9080	9074	9069	9063	9057	9052	9046	9041	9035	9029	1	2	3	4	5
37	$\bar{1}$·9023	9018	9012	9006	9000	8995	8989	8983	8977	8971	1	2	3	4	5
38	$\bar{1}$·8965	8959	8953	8947	8941	8935	8929	8923	8917	8911	1	2	3	4	5
39	$\bar{1}$·8905	8899	8893	8887	8880	8874	8868	8862	8855	8849	1	2	3	4	5
40	$\bar{1}$·8843	8836	8830	8823	8817	8810	8804	8797	8791	8784	1	2	3	4	5
41	$\bar{1}$·8778	8771	8765	8758	8751	8745	8738	8731	8724	8718	1	2	3	4	6
42	$\bar{1}$·8711	8704	8697	8690	8683	8676	8669	8662	8655	8648	1	2	3	5	6
43	$\bar{1}$·8641	8634	8627	8620	8613	8606	8598	8591	8584	8577	1	2	4	5	6
44	$\bar{1}$·8569	8562	8555	8547	8540	8532	8525	8517	8510	8502	1	2	4	5	6

The **bold** typeface indicates that the integral part changes

Logarithms of Cosines (*continued*)

	0′	6′	12′	18′	24′	30′	36′	42′	48′	54′	Subtract 1′	2′	3′	4′	5′
45°	$\bar{1}$·8495	8487	8480	8472	8464	8457	8449	8441	8433	8426	1	3	4	5	6
46	$\bar{1}$·8418	8410	8402	8394	8386	8378	8370	8362	8354	8346	1	3	4	5	7
47	$\bar{1}$·8338	8330	8322	8313	8305	8297	8289	8280	8272	8264	1	3	4	6	7
48	$\bar{1}$·8255	8247	8238	8230	8221	8213	8204	8195	8187	8178	1	3	4	6	7
49	$\bar{1}$·8169	8161	8152	8143	8134	8125	8117	8108	8099	8090	1	3	4	6	7
50	$\bar{1}$·8081	8072	8063	8053	8044	8035	8026	8017	8007	7998	2	3	5	6	8
51	$\bar{1}$·7989	7979	7970	7960	7951	7941	7932	7922	7913	7903	2	3	5	6	8
52	$\bar{1}$·7893	7884	7874	7864	7854	7844	7835	7825	7815	7805	2	3	5	7	8
53	$\bar{1}$·7795	7785	7774	7764	7754	7744	7734	7723	7713	7703	2	3	5	7	9
54	$\bar{1}$·7692	7682	7671	7661	7650	7640	7629	7618	7607	7597	2	4	5	7	9
55	$\bar{1}$·7586	7575	7564	7553	7542	7531	7520	7509	7498	7487	2	4	5	7	9
56	$\bar{1}$·7476	7464	7453	7442	7430	7419	7407	7396	7384	7373	2	4	6	8	10
57	$\bar{1}$·7361	7349	7338	7326	7314	7302	7290	7278	7266	7254	2	4	6	8	10
58	$\bar{1}$·7242	7230	7218	7205	7193	7181	7168	7156	7144	7131	2	4	6	8	10
59	$\bar{1}$·7118	7106	7093	7080	7068	7055	7042	7029	7016	7003	2	4	6	9	11
60	$\bar{1}$·6990	6977	6963	6950	6937	6923	6910	6896	6883	6869	2	4	7	9	11
61	$\bar{1}$·6856	6842	6828	6814	6801	6787	6773	6759	6744	6730	2	5	7	9	12
62	$\bar{1}$·6716	6702	6687	6673	6659	6644	6629	6615	6600	6585	2	5	7	10	12
63	$\bar{1}$·6570	6556	6541	6526	6510	6495	6480	6465	6449	6434	3	5	8	10	13
64	$\bar{1}$·6418	6403	6387	6371	6356	6340	6324	6308	6292	6276	3	5	8	11	13
65	$\bar{1}$·6259	6243	6227	6210	6194	6177	6161	6144	6127	6110	3	6	8	11	14
66	$\bar{1}$·6093	6076	6059	6042	6024	6007	5990	5972	5954	5937	3	6	9	12	15

	0′	6′	12′	18′	24′	30′	36′	42′	48′	54′	1′	2′	3′	4′	5′
67	$\bar{1}$·5919	5901	5883	5865	5847	5828	5810	5792	5773	5754	3	6	9	12	15
68	$\bar{1}$·5736	5717	5698	5679	5660	5641	5621	5602	5583	5563	3	6	10	13	16
69	$\bar{1}$·5543	5523	5504	5484	5463	5443	5423	5402	5382	5361	3	7	10	14	17
70	$\bar{1}$·5341	5320	5299	5278	5256	5235	5213	5192	5170	5148	4	7	11	14	18
71	$\bar{1}$·5126	5104	5082	5060	5037	5015	4992	4969	4946	4923	4	8	11	15	19
72	$\bar{1}$·4900	4876	4853	4829	4805	4781	4757	4733	4709	4684	4	8	12	16	20
73	$\bar{1}$·4659	4634	4609	4584	4559	4533	4508	4482	4456	4430	4	9	13	17	21
74	$\bar{1}$·4403	4377	4350	4323	4296	4269	4242	4214	4186	4158	5	9	14	18	23
75	$\bar{1}$·4130	4102	4073	4044	4015	3986	3957	3927	3897	3867	5	10	15	20	24
76	$\bar{1}$·3837	3806	3775	3745	3713	3682	3650	3618	3586	3554	5	11	16	21	26
77	$\bar{1}$·3521	3488	3455	3421	3387	3353	3319	3284	3250	3214	6	11	17	23	28
78	$\bar{1}$·3179	3143	3107	3070	3034	2997	2959	2921	2883	2845	6	12	19	25	31
79	$\bar{1}$·2806	2767	2727	2687	2647	2606	2565	2524	2482	2439	7	14	20	27	34
80	$\bar{1}$·2397	2353	2310	2266	2221	2176	2131	2085	2038	1991	8	15	23	30	38
81	$\bar{1}$·1943	1895	1847	1797	1747	1697	1646	1594	1542	1489	8	17	25	34	42
82	$\bar{1}$·1436	1381	1326	1271	1214	1157	1099	1040	0981	0920	10	19	29	38	48
83	$\bar{1}$·0859	0797	0734	0670	0605	0539	0472	0403	0334	0264	11	22	33	44	55
84	$\bar{1}$·0192	0120	0046	**9970**	**9894**	**9816**	**9736**	**9655**	**9573**	**9489**	13	26	39	53	66
85	$\bar{2}$·9403	9315	9226	9135	9042	8946	8849	8749	8647	8543	16	32	48	64	81
86	$\bar{2}$·8436	8326	8213	8098	7979	7857	7731	7602	7468	7330	21	41	62	83	103
87	$\bar{2}$·7188	7041	6889	6731	6567	6397	6220	6035	5842	5640					
88	$\bar{2}$·5428	5206	4971	4723	4459	4179	3880	3558	3210	2832					
89	$\bar{2}$·242	$\bar{2}$·196	$\bar{2}$·145	$\bar{2}$·087	$\bar{2}$·020	$\bar{3}$·941	$\bar{3}$·844	$\bar{3}$·719	$\bar{3}$·543	$\bar{3}$·242					

The **bold** typeface indicates that the integral part changes

Logarithms of Tangents

	0′	6′	12′	18′	24′	30′	36′	42′	48′	54′	1′	2′	3′	4′	5′
0°	$-\infty$	$\bar{3}$·242	$\bar{3}$·543	$\bar{3}$·719	$\bar{3}$·844	$\bar{3}$·941	$\bar{2}$·020	$\bar{2}$·087	$\bar{2}$·145	$\bar{2}$·196					
1	$\bar{2}$·2419	2833	3211	3559	3881	4181	4461	4725	4973	5208					
2	$\bar{2}$·5431	5643	5845	6038	6223	6401	6571	6736	6894	7046					
3	$\bar{2}$·7194	7337	7475	7609	7739	7865	7988	8107	8223	8336	21	42	63	83	104
4	$\bar{2}$·8446	8554	8659	8762	8862	8960	9056	9150	9241	9331	16	32	48	65	81
5	$\bar{2}$·9420	9506	9591	9674	9756	9836	9915	9992	**0068**	**0143**	13	26	40	53	66
6	$\bar{1}$·0216	0289	0360	0430	0499	0567	0633	0699	0764	0828	11	22	34	45	56
7	$\bar{1}$·0891	0954	1015	1076	1135	1194	1252	1310	1367	1423	10	20	29	39	49
8	$\bar{1}$·1478	1533	1587	1640	1693	1745	1797	1848	1898	1948	9	17	26	35	43
9	$\bar{1}$·1997	2046	2094	2142	2189	2236	2282	2328	2374	2419	8	16	23	31	39
10	$\bar{1}$·2463	2507	2551	2594	2637	2680	2722	2764	2805	2846	7	14	21	28	35
11	$\bar{1}$·2887	2927	2967	3006	3046	3085	3123	3162	3200	3237	6	13	19	26	32
12	$\bar{1}$·3275	3312	3349	3385	3422	3458	3493	3529	3564	3599	6	12	18	24	30
13	$\bar{1}$·3634	3668	3702	3736	3770	3804	3837	3870	3903	3935	6	11	17	22	28
14	$\bar{1}$·3968	4000	4032	4064	4095	4127	4158	4189	4220	4250	5	10	16	21	26
15	$\bar{1}$·4281	4311	4341	4371	4400	4430	4459	4488	4517	4546	5	10	15	20	25
16	$\bar{1}$·4575	4603	4632	4660	4688	4716	4744	4771	4799	4826	5	9	14	19	23
17	$\bar{1}$·4853	4880	4907	4934	4961	4987	5014	5040	5066	5092	4	9	13	18	22
18	$\bar{1}$·5118	5143	5169	5195	5220	5245	5270	5295	5320	5345	4	8	13	17	21
19	$\bar{1}$·5370	5394	5419	5443	5467	5491	5516	5539	5563	5587	4	8	12	16	20
20	$\bar{1}$·5611	5634	5658	5681	5704	5727	5750	5773	5796	5819	4	8	12	15	19
21	$\bar{1}$·5842	5864	5887	5909	5932	5954	5976	5998	6020	6042	4	7	11	15	19

	0′	6′	12′	18′	24′	30′	36′	42′	48′	54′	1′	2′	3′	4′	5′
22	$\bar{1}$·6064	6086	6108	6129	6151	6172	6194	6215	6236	6257	4	7	11	14	18
23	$\bar{1}$·6279	6300	6321	6341	6362	6383	6404	6424	6445	6465	3	7	10	14	17
24	$\bar{1}$·6486	6506	6527	6547	6567	6587	6607	6627	6647	6667	3	7	10	13	17
25	$\bar{1}$·6687	6706	6726	6746	6765	6785	6804	6824	6843	6863	3	7	10	13	16
26	$\bar{1}$·6882	6901	6920	6939	6958	6977	6996	7015	7034	7053	3	6	9	13	16
27	$\bar{1}$·7072	7090	7109	7128	7146	7165	7183	7202	7220	7238	3	6	9	12	15
28	$\bar{1}$·7257	7275	7293	7311	7330	7348	7366	7384	7402	7420	3	6	9	12	15
29	$\bar{1}$·7438	7455	7473	7491	7509	7526	7544	7562	7579	7597	3	6	9	12	15
30	$\bar{1}$·7614	7632	7649	7667	7684	7701	7719	7736	7753	7771	3	6	9	12	14
31	$\bar{1}$·7788	7805	7822	7839	7856	7873	7890	7907	7924	7941	3	6	9	11	14
32	$\bar{1}$·7958	7975	7992	8008	8025	8042	8059	8075	8092	8109	3	6	8	11	14
33	$\bar{1}$·8125	8142	8158	8175	8191	8208	8224	8241	8257	8274	3	5	8	11	14
34	$\bar{1}$·8290	8306	8323	8339	8355	8371	8388	8404	8420	8436	3	5	8	11	14
35	$\bar{1}$·8452	8468	8484	8501	8517	8533	8549	8565	8581	8597	3	5	8	11	13
36	$\bar{1}$·8613	8629	8644	8660	8676	8692	8708	8724	8740	8755	3	5	8	11	13
37	$\bar{1}$·8771	8787	8803	8818	8834	8850	8865	8881	8897	8912	3	5	8	10	13
38	$\bar{1}$·8928	8944	8959	8975	8990	9006	9022	9037	9053	9068	3	5	8	10	13
39	$\bar{1}$·9084	9099	9115	9130	9146	9161	9176	9192	9207	9223	3	5	8	10	13
40	$\bar{1}$·9238	9254	9269	9284	9300	9315	9330	9346	9361	9376	3	5	8	10	13
41	$\bar{1}$·9392	9407	9422	9438	9453	9468	9483	9499	9514	9529	3	5	8	10	13
42	$\bar{1}$·9544	9560	9575	9590	9605	9621	9636	9651	9666	9681	3	5	8	10	13
43	$\bar{1}$·9697	9712	9727	9742	9757	9772	9788	9803	9818	9833	3	5	8	10	13
44	$\bar{1}$·9848	9864	9879	9894	9909	9924	9939	9955	9970	9985	3	5	8	10	13

The **bold** typeface indicates that the integral part changes

Logarithms of Tangents (*continued*)

	0′	6′	12′	18′	24′	30′	36′	42′	48′	54′	1′	2′	3′	4′	5′
45°	0·0000	0015	0030	0045	0061	0076	0091	0106	0121	0136	3	5	8	10	13
46	0·0152	0167	0182	0197	0212	0228	0243	0258	0273	0288	3	5	8	10	13
47	0·0303	0319	0334	0349	0364	0379	0395	0410	0425	0440	3	5	8	10	13
48	0·0456	0471	0486	0501	0517	0532	0547	0562	0578	0593	3	5	8	10	13
49	0·0608	0624	0639	0654	0670	0685	0700	0716	0731	0746	3	5	8	10	13
50	0·0762	0777	0793	0808	0824	0839	0854	0870	0885	0901	3	5	8	10	13
51	0·0916	0932	0947	0963	0978	0994	1010	1025	1041	1056	3	5	8	10	13
52	0·1072	1088	1103	1119	1135	1150	1166	1182	1197	1213	3	5	8	10	13
53	0·1229	1245	1260	1276	1292	1308	1324	1340	1356	1371	3	5	8	11	13
54	0·1387	1403	1419	1435	1451	1467	1483	1499	1516	1532	3	5	8	11	13
55	0·1548	1564	1580	1596	1612	1629	1645	1661	1677	1694	3	5	8	11	14
56	0·1710	1726	1743	1759	1776	1792	1809	1825	1842	1858	3	5	8	11	14
57	0·1875	1891	1908	1925	1941	1958	1975	1992	2008	2025	3	6	8	11	14
58	0·2042	2059	2076	2093	2110	2127	2144	2161	2178	2195	3	6	9	11	14
59	0·2212	2229	2247	2264	2281	2299	2316	2333	2351	2368	3	6	9	12	14
60	0·2386	2403	2421	2438	2456	2474	2491	2509	2527	2545	3	6	9	12	15
61	0·2562	2580	2598	2616	2634	2652	2670	2689	2707	2725	3	6	9	12	15
62	0·2743	2762	2780	2798	2817	2835	2854	2872	2891	2910	3	6	9	12	15
63	0·2928	2947	2966	2985	3004	3023	3042	3061	3080	3099	3	6	9	13	16
64	0·3118	3137	3157	3176	3196	3215	3235	3254	3274	3294	3	7	10	13	16
65	0·3313	3333	3353	3373	3393	3413	3433	3453	3473	3494	3	7	10	13	17
66	0·3514	3535	3555	3576	3596	3617	3638	3659	3679	3700	3	7	10	14	17

	0′	6′	12′	18′	24′	30′	36′	42′	48′	54′	1′	2′	3′	4′	5′
67	0·3721	3743	3764	3785	3806	3828	3849	3871	3892	3914	4	7	11	14	18
68	0·3936	3958	3980	4002	4024	4046	4068	4091	4113	4136	4	7	11	15	19
69	0·4158	4181	4204	4227	4250	4273	4296	4319	4342	4366	4	8	12	15	19
70	0·4389	4413	4437	4461	4484	4509	4533	4557	4581	4606	4	8	12	16	20
71	0·4630	4655	4680	4705	4730	4755	4780	4805	4831	4857	4	8	13	17	21
72	0·4882	4908	4934	4960	4986	5013	5039	5066	5093	5120	4	9	13	18	22
73	0·5147	5174	5201	5229	5256	5284	5312	5340	5368	5397	5	9	14	19	23
74	0·5425	5454	5483	5512	5541	5570	5600	5629	5659	5689	5	10	15	20	25
75	0·5719	5750	5780	5811	5842	5873	5905	5936	5968	6000	5	10	16	21	26
76	0·6032	6065	6097	6130	6163	6196	6230	6264	6298	6332	6	11	17	22	28
77	0·6366	6401	6436	6471	6507	6542	6578	6615	6651	6688	6	12	18	24	30
78	0·6725	6763	6800	6838	6877	6915	6954	6994	7033	7073	6	13	19	26	32
79	0·7113	7154	7195	7236	7278	7320	7363	7406	7449	7493	7	14	21	28	35
80	0·7537	7581	7626	7672	7718	7764	7811	7858	7906	7954	8	16	23	31	39
81	0·8003	8052	8102	8152	8203	8255	8307	8360	8413	8467	9	17	26	35	43
82	0·8522	8577	8633	8690	8748	8806	8865	8924	8985	9046	10	20	29	39	49
83	0·9109	9172	9236	9301	9367	9433	9501	9570	9640	9711	11	22	34	45	56
84	0·9784	9857	9932	**0008**	**0085**	**0164**	**0244**	**0326**	**0409**	**0494**	13	26	40	53	66
85	1·0580	0669	0759	0850	0944	1040	1138	1238	1341	1446	16	32	48	65	81
86	1·1554	1664	1777	1893	2012	2135	2261	2391	2525	2663	21	42	63	83	104
87	1·2806	2954	3106	3264	3429	3599	3777	3962	4155	4357					
88	1·4569	4792	5027	5275	5539	5819	6119	6441	6789	7167					
89	1·758	1·804	1·855	1·913	1·980	2·059	2·156	2·281	2·457	2·758					

The **bold** typeface indicates that the integral part changes

Index